MARINE NAVIGATION AND SAFETY OF SEA TRANSPORTATION

Marine Navigation and Safety of Sea Transportation

Navigational Problems

Editor

Adam Weintrit

Gdynia Maritime University, Gdynia, Poland

CRC Press is an imprint of the
Taylor & Francis Group, an **informa** business

A BALKEMA BOOK

Published by:
CRC Press/Balkema
P.O. Box 447, 2300 AK Leiden, The Netherlands
e-mail: Pub.NL@taylorandfrancis.com
www.crcpress.com – www.taylorandfrancis.com

First issued in paperback 2020

ISBN 13: 978-0-367-57639-4 (pbk)
ISBN 13: 978-1-138-00107-7 (hbk)

Visit the Taylor & Francis Web site at
http://www.taylorandfrancis.com

and the CRC Press Web site at
http://www.crcpress.com

Typeset by V Publishing Solutions Pvt Ltd., Chennai, India

List of reviewers

Prof. Roland **Akselsson**, Lund University, Sweden
Prof. Yasuo **Arai**, Independent Administrative Institution Marine Technical Education Agency,
Prof. Michael **Baldauf**, Word Maritime University, Malmö, Sweden
Prof. Andrzej **Banachowicz**, West Pomeranian University of Technology, Szczecin, Poland
Prof. Marcin **Barlik**, Warsaw University of Technology, Poland
Prof. Michael **Barnett**, Southampton Solent University, United Kingdom
Prof. Eugen **Barsan**, Constanta Maritime University, Romania
Prof. Milan **Batista**, University of Ljubljana, Ljubljana, Slovenia
Prof. Angelica **Baylon**, Maritime Academy of Asia & the Pacific, Philippines
Prof. Christophe **Berenguer**, Grenoble Institute of Technology, Saint Martin d'Hères, France
Prof. Heinz Peter **Berg**, Bundesamt für Strahlenschutz, Salzgitter, Germany
Prof. Tor Einar **Berg**, Norwegian Marine Technology Research Institute, Trondheim, Norway
Prof. Jarosław **Bosy**, Wroclaw University of Environmental and Life Sciences, Wroclaw, Poland
Prof. Zbigniew **Burciu**, Gdynia Maritime University, Poland
Sr. Jesus Carbajosa **Menendez**, President of Spanish Institute of Navigation, Spain
Prof. Andrzej **Chudzikiewicz**, Warsaw University of Technology, Poland
Prof. Frank **Coolen**, Durham University, UK
Prof. Stephen J. **Cross**, Maritime Institute Willem Barentsz, Leeuwarden, The Netherlands
Prof. Jerzy **Czajkowski**, Gdynia Maritime University, Poland
Prof. Krzysztof **Czaplewski**, Polish Naval Academy, Gdynia, Poland
Prof. Daniel **Duda**, Naval University of Gdynia, Polish Nautological Society, Poland
Prof. Alfonso **Farina**, SELEX-Sistemi Integrati, Rome, Italy
Prof. Andrzej **Fellner**, Silesian University of Technology, Katowice, Poland
Prof. Andrzej **Felski**, Polish Naval Academy, Gdynia, Poland
Prof. Włodzimierz **Filipowicz**, Gdynia Maritime University, Poland
Prof. Börje **Forssell**, Norwegian University of Science and Technology, Trondheim, Norway
Prof. Alberto **Francescutto**, University of Trieste, Trieste, Italy
Prof. Jens **Froese**, Jacobs University Bremen, Germany
Prof. Wiesław **Galor**, Maritime University of Szczecin, Poland
Prof. Jerzy **Gaździcki**, President of the Polish Association for Spatial Information; Warsaw, Poland
Prof. Witold **Gierusz**, Gdynia Maritime University, Poland
Prof. Dorota **Grejner-Brzezinska**, The Ohio State University, United States of America
Prof. Marek **Grzegorzewski**, Polish Air Force Academy, Deblin, Poland
Prof. Lucjan **Gucma**, Maritime University of Szczecin, Poland
Prof. Vladimir **Hahanov**, Kharkov National University of Radio Electronics, Kharkov, Ukraine
Prof. Jerzy **Hajduk**, Maritime University of Szczecin, Poland
Prof. Michał **Holec**, Gdynia Maritime University, Poland
Prof. Stojce Dimov **Ilcev**, Durban University of Technology, South Africa
Prof. Toshio **Iseki**, Tokyo University of Marine Science and Technology, Japan,
Prof. Jacek **Januszewski**, Gdynia Maritime University, Poland
Prof. Tae-Gweon **Jeong**, Korean Maritime University, Pusan, Korea
Prof. Mirosław **Jurdziński**, Gdynia Maritime University, Poland
Prof. John **Kemp**, Royal Institute of Navigation, London, UK
Prof. Andrzej **Królikowski**, Maritime Office in Gdynia; Gdynia Maritime University, Poland
Prof. Pentti **Kujala**, Helsinki University of Technology, Helsinki, Finland
Prof. Jan **Kulczyk**, Wroclaw University of Technology, Poland
Prof. Krzysztof **Kulpa**, Warsaw University of Technology, Warsaw, Poland
Prof. Shashi **Kumar**, U.S. Merchant Marine Academy, New York
Prof. Andrzej **Lenart**, Gdynia Maritime University, Poland
Prof. Nadav **Levanon**, Tel Aviv University, Tel Aviv, Israel
Prof. Andrzej **Lewiński**, University of Technology and Humanities in Radom, Poland
Prof. Józef **Lisowski**, Gdynia Maritime University, Poland
Prof. Vladimir **Loginovsky**, Admiral Makarov State Maritime Academy, St. Petersburg, Russia
Prof. Mirosław **Luft**, University of Technology and Humanities in Radom, Poland
Prof. Evgeniy **Lushnikov**, Maritime University of Szczecin, Poland
Prof. Zbigniew **Łukasik**, University of Technology and Humanities in Radom, Poland
Prof. Marek **Malarski**, Warsaw University of Technology, Poland
Prof. Boyan **Mednikarov**, Nikola Y. Vaptsarov Naval Academy,Varna, Bulgaria
Prof. Jerzy **Mikulski**, Silesian University of Technology, Katowice, Poland
Prof. Józef **Modelski**, Warsaw University of Technology, Poland
Prof. Wacław **Morgaś**, Polish Naval Academy, Gdynia, Poland
Prof. Janusz **Narkiewicz**, Warsaw University of Technology, Poland

TABLE OF CONTENTS

Navigational Problems
Introduction

A. Weintrit
Gdynia Maritime University, Gdynia, Poland

The monograph is addressed to scientists and professionals in order to share their expert knowledge, experience and research results concerning all aspects of navigation, safety at sea and marine transportation.

The contents of the book are partitioned into nine separate chapters: Ship control (covering the chapters 1.1 through 1.4), Decision Support Systems (covering the chapters 2.1 through 2.5), Marine Traffic (covering the chapters 3.1 through 3.5), Search and Rescue (covering the chapters 4.1 through 4.5), Meteorological aspect and weather condition (covering the chapters 5.1 through 5.5), Inland, sea-river, personal and car navigation systems (covering the chapters 6.1 through 6.6), Air navigation (covering the chapters 7.1 through 7.3), Maritime communications (covering the chapters 8.1 through 8.9), and Methods and algorithms (covering the chapters 9.1 through 9.3).

In each of them readers can find a few chapters. Chapters collected in the first chapter, titled 'Ship control', concerning the course-keeping adaptive control system for the nonlinear MIMO model of a container vessel, the multi-step matrix game of safe ship control with different amounts admissible strategies, catastrophe theory in intellectual control system of vessel operational strength, and concept of integrated INS/visual system for autonomous mobile robot operation

In the second chapter there are described problems related to decision support systems: functionality of navigation decision supporting system – NAVDEC, a study on the development of navigation visual supporting system and its sea trial test, application of ant colony optimization in ship's navigational decision support system, issue of making decisions with regard to ship traffic safety in different situations at sea, and ship handling in wind and current with neuroevolutionary decision support system.

Third chapter is about marine traffic. The readers can find some information about development and evaluation of traffic routeing measurements, Świnoujście– Szczecin fairway expert safety evaluation, expert indication of dangerous sections in Świnoujście–Szczecin fairway, traffic incidents analysis as a tool for improvement of transport safety, and vessel traffic stream analysis in vicinity of the Great Belt Bridge.

The fourth chapter deals with Search and Rescue (SAR) problems. The contents of the fourth chapter are partitioned into five subchapters: search and rescue of migrants at sea, ergonomics-based design of a life-saving appliance for search and rescue activities, the signals of marine continuous radar for operation with SART, risk analysis on dutch search and rescue capacity on the North Sea, and the operational Black sea delta regional exercise on oil spill preparedness and search and rescue – GEODELTA 2011.

The fifth chapter deals with meteorological aspect and weather conditions. The contents of the fifth chapter are partitioned into five: operational enhancement of numerical weather prediction with data from real-time satellite images, analysis of the prevailing weather conditions criteria to evaluate the adoption of a future ECA in the Mediterranean Sea, monitoring of ice conditions in the Gulf of Riga using micro class unmanned aerial systems, global warming and its impact on Arctic navigation: the Northern Sea Route shipping season 2012, and unloading operations on the fast ice in the region of Yamal Peninsula as the chapter of transportation operations in the Western Arctic.

In the sixth chapter there are described problems related to inland, sea-river, personal and car navigation systems: the method of the navigation data fusion in inland navigation, PER estimation of AIS in inland rivers based on three dimensional ray tracking, analysis of river – sea transport in the direction of the Danube – Black Sea and the Danube - Rhine River - River Main, study of the usage of car navigation system and navigational information to assist coastal navigational safety, remote spatial database access in the navigation system for the blind, and integration of inertial sensors and GPS system data for the personal navigation in urban area.

Seventh chapter concerns air navigation. The readers can find some information about accuracy of GPS receivers in naval aviation, comparative analysis of the two Polish hyperbolic systems AEGIR and Jemioluszka, and the analysis of implementation needs for automatic dependent surveillance in air traffic in Poland.

The eighth chapter deals with maritime communications. The contents of the eighth chapter are partitioned into nine: Multiple access technique applicable for maritime satellite communications, Classification and characteristics of mobile satellite antennas (MSA) for maritime applications, Development of Cospas-Sarsat satellite distress and safety systems (SDSS) for maritime and other mobile applications, The propagation characteristic of DGPS correction data signal at inland sea – propagation characteristic on LF/MF band radio wave, Communication automation in maritime transport, Audio watermarking in the maritime VHF radiotelephony, Enhancement of VHF radiotelephony in the frame of integrated VHF/DSC – ECDIS/AIS system, Modernization of the GMDSS, and VHF satellite broadcast channel as a complement to the emerging VHF Data Exchange (VDE) system.

The ninth chapter deals with methods and algorithms. The contents of the ninth chapter concerns the overview of the mathematical theory of evidence and its application in navigation, a new method for determining the attitude of a moving object, and simulation of Zermelo navigation on Riemannian manifolds for $\dim(R \times M)=3$

Each subchapter was reviewed at least by three independent reviewers. The Editor would like to express his gratitude to distinguished authors and reviewers of chapters for their great contribution for expected success of the publication. He congratulates the authors for their excellent work.

Chapter 1

Ship Control

The Course-keeping Adaptive Control System for the Nonlinear MIMO Model of a Container Vessel

M. Brasel & P. Dworak
West Pomeranian University of Technology, Szczecin, Poland

ABSTRACT: In the paper an adaptive multi-controller control system for a MIMO nonlinear dynamic process is presented. The problems under study are exemplified by synthesis of a surge velocity and yaw angle control system for a 4-DOF nonlinear MIMO mathematical model of a single-screw high-speed container vessel. The paper presents the complexity of the assumed model to be analyzed and the method of synthesis of the course-keeping control system. In the proposed course-keeping control system use is made of a set of (stable) linear modal controllers that create a multi-controller structure from which a controller appropriate to given operation conditions is chosen on the basis of the measured auxiliary signals. The system synthesis is carried out by means of system pole placement method after having linearized the model 4-DOF motions of the vessel in steady states. The final part of the paper includes simulation results of system operation with an adaptive controller of stepwise varying parameters along with conclusions and final remarks.

1 INTRODUCTION

Nonlinear control systems are commonly encountered in many different areas of science and technology. In particular, problems difficult to solve arise in motion and/or position control of various vessels, like drilling platforms and ships, sea ferries, container ships etc. Complex motions and/or complex-shaped bodies moving in the water, and in case of ships also at the boundary between water and air, give rise to resistance forces dependent in a nonlinear way on velocities and positions, thus causing the floating bodies to become strongly nonlinear dynamic plants.

In general, there are two basic approaches to solve the control problem for nonlinear plants. The first one called "nonlinear" consists in synthesizing a nonlinear controller that would meet certain requirements over the entire range of control signals variability (Fabri & Kadrikamanathan 2001; Huba et al. 2011; Khalil 2001; Tzirkel-Hancock & Fallside 1992; Witkowska et al. 2007). Substantial difficulties encountered in employing this approach are due to the fact that control plants are multivariable (MIMO). The second approach called "linear" consists in designing an adaptive linear controller with varying parameters to be systematically tuned up in keeping with changing plant operating conditions determined by system nominal "operating points". Here, linearization of nonlinear MIMO plants is a prerequisite for the methods to be employed. After linearization local linear models are obtained valid for small deviations from "operating points" of the plant.

Since properties exhibited by linear models at different (distant) "operating points" of the plant may substantially vary, therefore the controllers used should be either robust (Ioannou & Sun 1996) (usually of a very high order as has been observed by (Gierusz 2005)) or adaptive with parameters being tuned in the process of operation (Äström & Wittenmark 1995).

If the description of the nonlinear plant is known, then it is possible to make use of systems with linear controllers prepared earlier for possibly all "operating points" of the plant. Such controllers can create either a set of controllers with switchable outputs from among which one controller designed for the given system "operating point" (Bańka et al. 2010a; Bańka et al. 2010b; Dworak & Pietrusewicz 2010) is chosen, or multi-controller structures the control signal components of which are formed, for example, as weighted means of outputs of a selected controller group according to Takagi-Sugeno-Kang (TSK) rules, i.e. with weights being proportional to the degree of their membership of appropriately fuzzyfied areas of plant outputs or other auxiliary signals (Tanaka & Sugeno 1992; Tatjewski 2007; Dworak et al. 2012a; Dworak et al. 2012b).

What all the above-mentioned multi-controller structures, where not all controllers at the moment are utilized in a closed-loop system, have in common is that all controllers employed in these structures must be stable by themselves, in distinction to a single adaptive controller with varying (tuned) parameters. This means that system strong stability conditions should be fulfilled (Vidyasagar 1985).

In the presented paper an adaptive modal MIMO controller with (stepwise) varying parameters in the process of operation is studied. The controller can be physically realized as a multi-controller structure of modal controllers with switchable outputs. The considered adaptive control system will be designed for all possible "operating points" of the plant. In the simulation studies a 4-DoF nonlinear model of a single-screw high-speed container vessel has been used as a nonlinear MIMO plant. The main goal of the paper is a synthesis of the course-keeping adaptive control system for a container vessel assuming two controlled variables: yaw angle and forward speed of the ship relative to water.

2 NONLINEAR MODEL OF A CONTAINER SHIP

The considered course-keeping control system structure is studied by means of a 4-DOF nonlinear mathematical model of container vessel (Son & Nomoto 1981, Fossen 1994), having L=175m in length, B=25.4m in beam, with an average draught of H=8.5m. The yaw angle and the ship's position are defined in an Earth-based fixed reference system. In contrast, force and speed components with respect to water are determined in a moving system related with the ship's body and the axes directed to the front and the starboard of the ship with the origin placed in its gravity center (G). These are shown in Fig. 1.

Designations for the linear and angular speed of the ship, in the considered degrees of freedom ship motion are as follows: u (surge velocity), v (sway velocity), p (roll rate) and r (yaw rate).

Corresponding designations of the position coordinates of the ship are as follows: x_o (ship position in N-S), y_o (ship position in W-E), ϕ (roll angle), ψ (yaw angle).

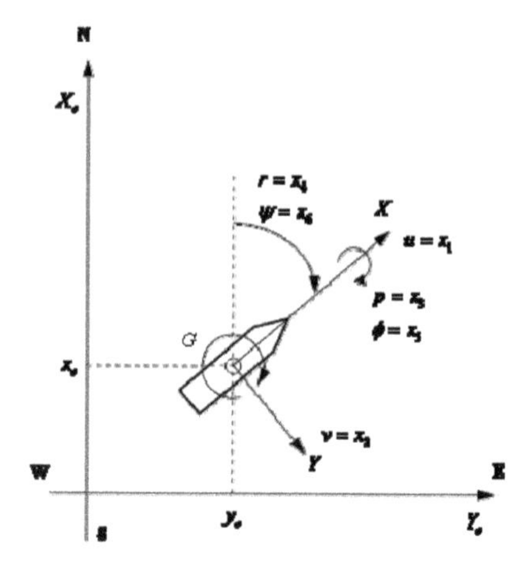

Figure 1. Ship's co-ordinate systems.

General nonlinear equations of motion in surge, sway, roll and yaw (Son & Nomoto 1981, Fossen 1994) are as follows:

$$\begin{aligned}
&(m+m_x)\dot{u}-(m+m_y)vr=X\\
&(m+m_y)\dot{v}+(m+m_x)ur+m_y\alpha_y\dot{r}-m_yl_y\dot{p}=Y\\
&(I_x+J_x)\dot{p}-m_yl_y\dot{v}-m_xl_xur+W\overline{GM}\phi=K\\
&(I_z+J_z)\dot{r}+m_y\alpha_y\dot{v}=N-Yx_G.
\end{aligned} \tag{1}$$

Here m denotes the ship mass; m_x, m_y, J_x, J_z denote the added mass and added moment of inertia in the x and y directions and about the x-axes and z- axes, respectively. I_x and I_z denote moment of inertia about the x-axes and z- axes, respectively. Furthermore, α_y denotes the x-coordinates of the center of m_y, while l_x and l_y denote the z-coordinates of the centers of m_x and m_y, respectively. x_G is the location of the center of gravity in the x-axes, $\overline{GM}$ is the metacentric height and W is the ship displacement.

The hydrodynamic forces X, Y and moments K, N in above equations are given as:

$$\begin{aligned}
&X=X_{uu}|u|u+(1-t)T+X_{vr}vr+X_{vv}v^2+X_{rr}r^2\\
&+X_{\phi\phi}\phi^2+c_{RX}F_N\sin(\delta),
\end{aligned} \tag{2}$$

$$\begin{aligned}
&Y=Y_vv+Y_rr+Y_pp+Y_\phi\phi+Y_{vvv}v^3+Y_{rrr}r^3\\
&+Y_{vvr}v^2r+Y_{vrr}vr^2+Y_{vv\phi}v^2\phi+Y_{v\phi\phi}v\phi^2\\
&+Y_{rr\phi}r^2\phi+Y_{r\phi\phi}r\phi^2+(1+a_H)F_N\cos(\delta),
\end{aligned} \tag{3}$$

$$N = N_v v + N_r r + N_p p + N_\phi \phi + N_{vvv} v^3 + N_{rrr} r^3 + N_{vvr} v^2 r + N_{vrr} v r^2 + N_{vv\phi} v^2 \phi + N_{v\phi\phi} v \phi^2 + N_{rr\phi} r^2 \phi + N_{r\phi\phi} r \phi^2 + (x_R + a_H x_H) F_N \cos(\delta), \quad (4)$$

$$K = K_v v + K_r r + K_p p + K_\phi \phi + K_{vvv} v^3 + K_{rrr} r^3 + K_{vvr} v^2 r + K_{vrr} v r^2 + K_{vv\phi} v^2 \phi + K_{v\phi\phi} v \phi^2 + K_{rr\phi} r^2 \phi + K_{r\phi\phi} r \phi^2 - (1 + a_H) z_R F_N \cos(\delta). \quad (5)$$

Here, the rudder force F_N can be resolved as:

$$F_N = \frac{-6.13\Delta}{\Delta + 2.25} \cdot \frac{A_R}{L^2} \left(u_R^2 + v_R^2\right) \sin(\alpha_R), \quad (6)$$

where:

$$\alpha_R = \delta + \tan^{-1}(v_R / u_R), \quad (7)$$

$$u_R = u_P \varepsilon \sqrt{1 + 8kK_T / (\pi J^2)}, \quad (8)$$

$$v_R = \gamma v + c_{Rr} r + c_{Rrrr} r^3 + c_{Rrrv} r^2 v, \quad (9)$$

where:

$$J = u_P V / (nD), \quad (10)$$

$$K_T = 0.527 - 0.455J, \quad (11)$$

$$u_P = \cos(v)\left[(1 - w_p) + \tau\left\{(v + x_p r)^2 + c_{pv} v + c_{pr} r\right\}\right]. \quad (12)$$

The remaining coefficients and model parameters used in the equations (1) are given by (Fossen 1994). The actual speed of the vessel is designated as $V = \sqrt{u^2 + v^2}$. Control signals of the nonlinear MIMO model of the ship (1) are: δ (rudder angle) and n (propeller shaft speed). In the simulations we assume the following limitations of control signals: the maximum speed of the screw $n_{\max} = 160rpm$, the maximum rudder angle $\delta_{\max} = 15\text{deg}$ and maximum rudder angular velocity $\dot{\delta}_{\max} = 5\text{deg}/s$.

In addition, this model takes into account the dynamics of the actuators described in section 3.

3 COURSE-KEEPING ADAPTIVE CONTROL SYSTEM

The dynamic model of the container ship (1) can be described in the state-space nonlinear form:

$$\begin{aligned} \dot{x}_2(t) &= f(x_2, u) \\ y(t) &= g(x_2, u), \end{aligned} \quad (13)$$

with the semi-state vector $x_2(t)$ defined as shown in Fig 1.

$$x_2(t) = [u \;\; v \;\; p \;\; r \;\; \phi \;\; \psi]^T \quad (14)$$

and output and control signals defined as:

$$\begin{aligned} y(t) &= [u(t) \;\; \psi(t)]^T \\ u(t) &= [\delta(t) \;\; n(t)]^T. \end{aligned} \quad (15)$$

For the synthesis of the control system, the resulting model is linearized in the nominal operating points of the ship, defined as $x_2(t) = x_{2n}(t)$. The nominal state vector of the model (1) in the nominal operating regimes is defined as:

$$x_{2n}(t) = [u_n \;\; v_n \;\; 0 \;\; r_n \;\; \phi_n \;\; \text{var}]^T. \quad (16)$$

The values of state variables: u_n, v_n, r_n, ϕ_n are defined in the turning circle simulation tests carried out in MATLAB/Simulink for various control signals: δ_o and n_o. The range of changes of these signals is as follows: $\delta_o = \langle -15 \div 15 \rangle \text{deg}$ in steps of 1deg and $n_o = \langle 5 \div 160 \rangle rpm$ in steps of $5rpm$, which gives a set of 992 operating points. Any combination of the control signals and their corresponding parameters of ship motions: u_n, v_n, r_n and ϕ_n determines the nominal operating point of the ship. For example, the obtained functions $u_n(\delta, n)$ and $v_n(\delta, n)$ are shown in Figures 2 and 3, respectively.

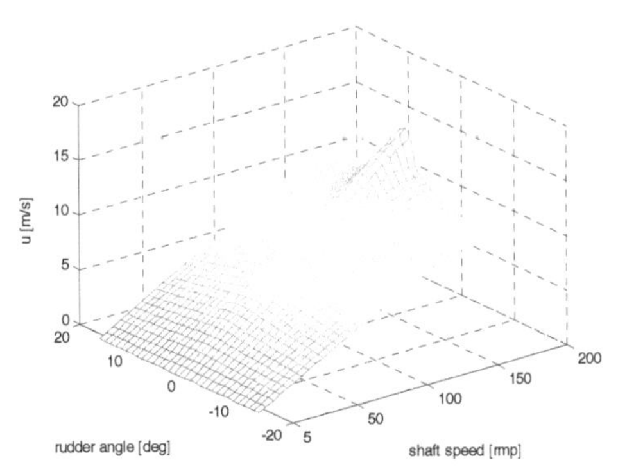

Figure 2. The surge velocity in the nominal operating points.

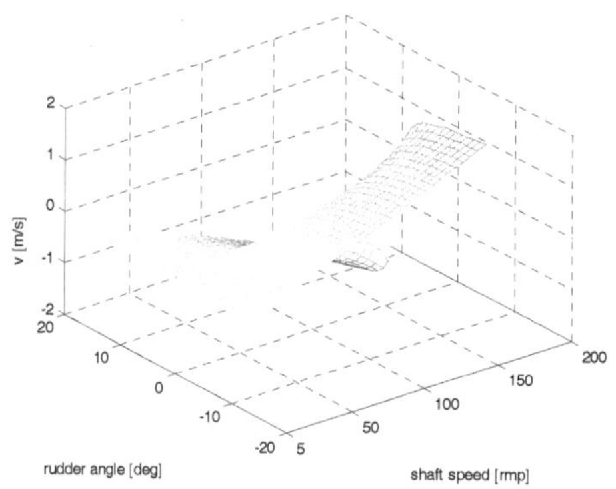

Figure 3. The sway velocity in the nominal operating points.

As a result of the linearization performed in the whole range of the nominal control signals one obtains linear state-space models of the container ship:

$$\begin{aligned}\dot{x}_2(t) &= A_2[x_2(t)-x_{2n}]+B_2[u(t)-u_n]\\ y(t)-y_n &= C_2[x_2(t)-x_{2n}],\end{aligned} \tag{17}$$

where:

$$A_2=\left[\frac{\partial}{\partial x}f^{\mathrm{T}}(x,u)\right]^{\mathrm{T}}_{\substack{x_2=x_{2n}\\u=u_n}}=\begin{bmatrix}a_{11}&a_{12}&0&a_{14}&a_{15}&0\\a_{21}&a_{22}&a_{23}&a_{24}&a_{25}&0\\a_{31}&a_{32}&a_{33}&a_{34}&a_{35}&0\\a_{41}&a_{42}&a_{43}&a_{44}&a_{45}&0\\0&0&1&0&0&0\\0&0&0&a_{64}&a_{65}&0\end{bmatrix},$$

$$B_2=\left[\frac{\partial}{\partial u}f^{\mathrm{T}}(x,u)\right]^{\mathrm{T}}_{\substack{x_2=x_{2n}\\u=u_n}}=\begin{bmatrix}b_{11}&b_{21}&b_{31}&b_{41}&0&0\\b_{12}&b_{22}&b_{32}&b_{42}&0&0\end{bmatrix}^T,$$

$$C_2=\left[\frac{\partial}{\partial x}g^{\mathrm{T}}(x,u)\right]^{\mathrm{T}}_{\substack{x_2=x_{2n}\\u=u_n}}=\begin{bmatrix}1&0&0&0&0&0\\0&0&0&0&0&1\end{bmatrix},$$

with the entries a_{ij} and b_{ij} depending on values of surge velocity u_n, sway velocity v_n, yaw angular velocity r_n, roll angle ϕ_n and control signals $u_n=[\delta_o\ \ n_o]^T$ in the nominal operating points of the container vessel.

For the synthesis of the control system, the steering machine model based on (Fossen 1994) is represented by a first-order dynamic system with time constant $T_\delta=1.8s$ and gain $K_\delta=1$, while the shaft model is represented by a linear model with average time constant $T_m=10.48s$ and gain $K_m=1$. Thus, actuators block shown in Fig. 4 can be described in state-space form as:

$$\begin{aligned}\dot{x}_1(t) &= A_1x_1(t)+B_1u_c(t)\\ y_1(t) &= x_1(t),\end{aligned} \tag{18}$$

where:

$$A_1=\begin{bmatrix}-0.556&0\\0&-0.095\end{bmatrix}, B_1=\begin{bmatrix}0.556&0\\0&0.095\end{bmatrix}.$$

Here $u_c(t)=[\delta_c(t)\ \ n_c(t)]^T$ is a vector of commanded control signals and $x_1(t)=u(t)=[\delta(t)\ \ n(t)]^T$ is a vector of control signals. Now, the full state vector $x(t)$ of the vessel can be written as: $[x_1(t)\ \ x_2(t)]^T$. Thus, the state vector of the ship is as follows:

$$x(t)=[\delta\ \ n\ \ u\ \ v\ \ p\ \ r\ \ \phi\ \ \psi]^T. \tag{19}$$

Finally, the full linearized model of the container vessel is described by the matrices:

$$A=\begin{bmatrix}A_1&0\\B_2&A_2\end{bmatrix}, B=\begin{bmatrix}B_1\\0\end{bmatrix}, C=[0\ \ C_2]. \tag{20}$$

The obtained linear models (21) with known parameters are the starting point for applying many known methods for linear multivariable control system design. When the linear MIMO systems are considered usually multivariable modal (or possibly optimal LQG/LQR) controllers are designed.

In the case of non-measurable state variables, modal controllers used in the proposed control system structure are multivariable dynamic systems with parameters defined in time domain by:

$$\begin{aligned}\dot{x}_r(t) &= A_rx_r(t)+B_re(t)\\ u(t) &= C_rx_r(t)+D_re(t),\end{aligned} \tag{21}$$

where:

$$A_r=A-BF-LC,\ B_r=L,\ C_r=-F,\ D_r=0. \tag{22}$$

Here, F is the matrix of proportional feedback related to state vector components (reconstructed by the observer) of the plant models, and L is the gain matrix of full-order Luenberger observers that reconstruct the state vector of the plant linear models (20). Synthesis of modal controllers is based on using any of the known techniques of pole placement in stable regions of the s-plane (Bańka et al. 2013). If we decide on (strictly causal) modal controllers based on full-order Luenberger observers the design performed directly in time domain (and also in s-domain without solving polynomial matrix equations) boils down to separate determining the feedback matrix F, which forces the closed-loop eigenvalues to the pole locations specified by the adopted (stable) pole values *pole_sys*, and the weight matrix $\boldsymbol{L}$ of the full-order Luenberger observer for appropriately chosen observer poles *pole_obs*. The real parts of the latter should be more negative than those selected for the *pole_sys* set.

In the case of measurable state variables the main step on the road to synthesizing a modal control system in time domain is to determine the state feedback gain matrix $\boldsymbol{F}$. Assuming the modal control plant is given by the linear MIMO system described by matrices (20), the vector of commanded control signals is as follows:

$$\boldsymbol{u}_c(t) = -\boldsymbol{F}\boldsymbol{x}(t) + \boldsymbol{u}_n, \tag{23}$$

which shifts the poles of a linear plant model to desired locations specified by the preassigned a priori values of *pole_sys,* here chosen as: -0.11, -0.12, -0.13, -0.14, -0.15, -0.16, -0.17, -0.18. Such choice of the poles *pole_sys* has been performed experimentally to obtain control processes without excessive overshoots on controlled signals with "reasonable" times needed to achieve reference control conditions and possibly without exceeding the limitations on the control signals.

The block diagram of the proposed course-keeping adaptive control system is depicted in Fig. 4. It consists of an adaptively changed state feedback matrix $\boldsymbol{F}$ with stepwise switchable parameter values, chosen according to the current operating point of the ship. The resulting set of 992 modal controllers has been used to create an adaptive controller with stepwise varying parameters, tuned on the basis of two auxiliary signals measured that are: surge and sway speed components of the ship with respect to water shown in Figures 2 and 3.

If the state vector of the ship model (1) is not measurable the state feedback matrix should be replaced by an adaptive modal controller (21) based on the Luenberger observer or the Kalman filter (Bańka ct al. 2013).

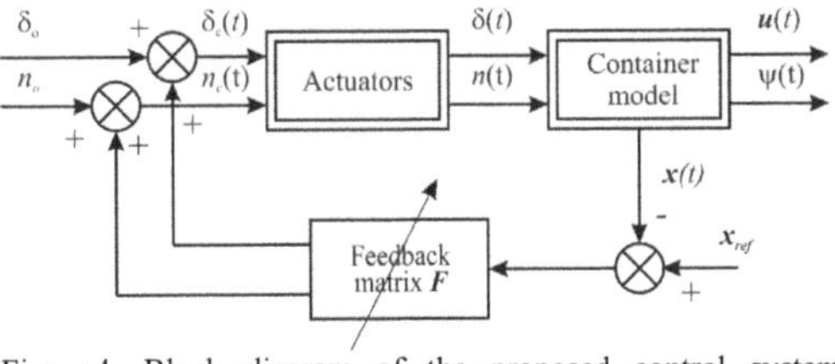

Figure 4. Block diagram of the proposed control system structure.

4 RESULTS OF SIMULATION TESTS

The usefulness of the above presented control structure is proved by an example of a course-keeping adaptive control system for the nonlinear MIMO model of a container vessel (1). The goal of regulation was a simultaneous control of the ship's course and her forward speed. Results of simulations carried out in Matlab/Simulink environment are presented in Fig. 5 and 6. The initial state of the ship has been taken as:

$$\boldsymbol{x}(0) = \begin{bmatrix} 0 & 50 & 10.18 & 0 & 0 & 0 & 0 & 0 \end{bmatrix}^T,$$

which means that the ship goes forward with the speed of 10.18 [knots]. The first maneuver at t=100s was the change of the course angle to 20deg with keeping the ship forward speed at u=10.18knots. Then after 200s the ship was speeded up to u=15.27knots. Both changes have been done according to the assumed ship dynamics with negligible cross coupling of her outputs. The proposed control structure provides the required control quality. All maneuvers have been done with acceptable values of the control signals: rudder angle and shaft speed, presented in Fig. 6.

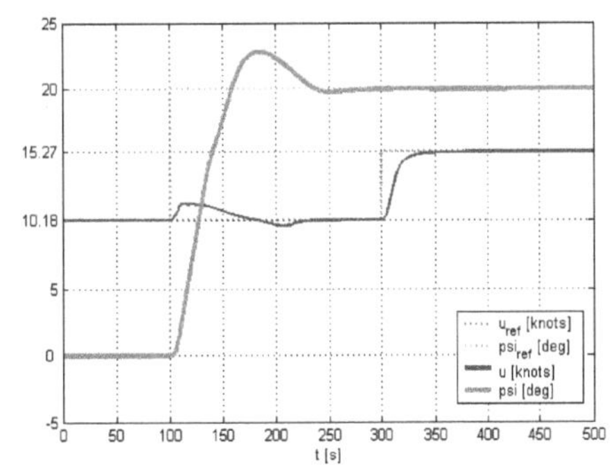

Figure 5. Ship's course angle and forward speed.

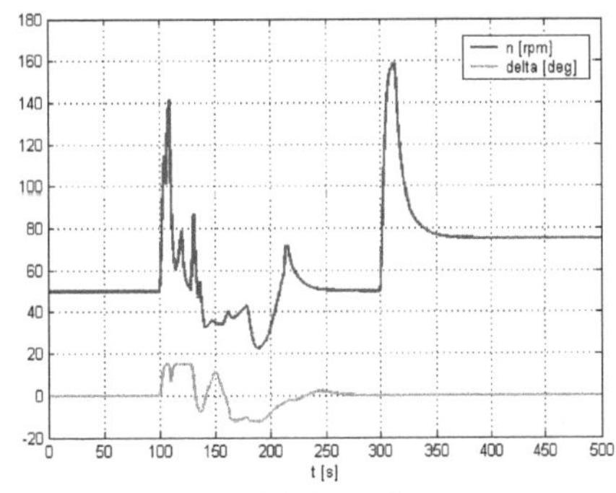

Figure 6. Rudder angle and shaft speed.

Figure 7 presents values of indices i and j which denote the current operating point. Change of their values define moments of switching of the feedback matrix $\boldsymbol{F}$.

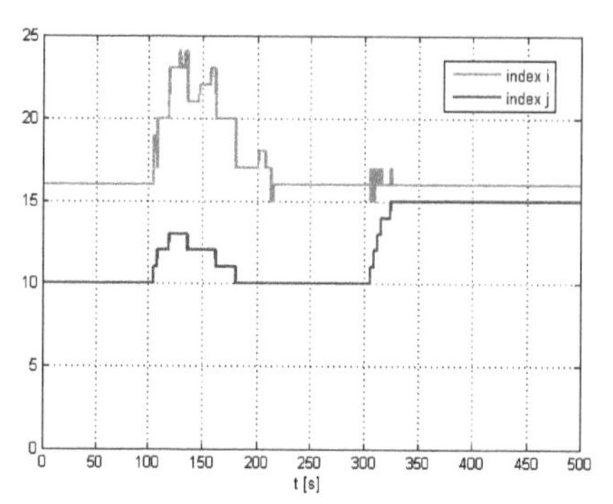

Figure 7. Moments of switching of the feedback matrix ***F***.

5 CONCLUSION

In the paper an adaptive control system for the nonlinear MIMO plant was proposed and tested. The utilized adaptive gain scheduling modal controller allows one to control a strongly nonlinear process, here the model of a container vessel. The synthesis of the controller is based on the linearization of a nonlinear ship model in operating points corresponding to the set of 992 typical operating regimes. The adaptive controller stepwise varies its parameters on the basis of auxiliary signals measured during ship operation. The presented example of course-keeping control of the ship, shows efficiency of this method and the appropriateness of its use to the direct control or as a part of more complex control systems, e.g. a model loop in the MFC control structure (Dworak et al. 2012b).

REFERENCES

Äström, K. & Wittenmark, B. (1995). Adaptive control Addison Wesely.

Bańka, S., Brasel, M., Dworak, P., & Latawiec, J. K. (2010a). Switched-structure of linear MIMO controllers for positioning of a drillship on a sea surface, Międzyzdroje: Methods and Models in Automation and Robitics 2010.

Bańka, S., Dworak, P., & Brasel, M. (2010b). On control of nonlinear dynamic MIMO plants using a switchable structure of linear modal controllers (in Polish). Pomiary, Automatyka, Kontrol, 5, 385-391.

Bańka, S., Dworak, P., & Jaroszewski K. (2013). Linear adaptive structure for control of a nonlinear MIMO dynamic plant. International Journal of Applied Mathematics and Computer Science 23(1), (in printing)

Dworak, P. & Pietrusewicz, K. (2010). A variable structure controller for the MIMO Thermal Plant (in Polish). Przeglad Elektrotechniczny 6, 116-119.

Dworak, P. & Bańka, S. (2012a). Adaptive multi-controller TSK Fuzzy Structure for Control of Nonlinear MIMO Dynamic Plant. 9th IFAC Conference on Manoeuvring and Control of Marine Craft.

Dworak, P., Jaroszewski K. & Brasel. M. (2012b). A fuzzy TSK controller for the MIMO Thermal Plant (in Polish). Przeglad Elektrotechniczny 10a, 83-86.

Fabri, S. & Kadrikamanathan, V. (2001). Functional adaptive control. An intelligent systems approach. Springer Verlag. Berlin.

Fossen T. I. (1994). Guidance and Control of Ocean Vehicles. John Wiley and Sons,1994.

Gierusz, W. (2005). Synthesis of multivariable control systems for precise steering of ship's motion using selected robust systems design methods (in Polish). Gdynia Maritime Academy Press. Gdynia.

Huba, M., Skogestad, S., Fikar, M., Hovd, M., Johansen, T.A., & Rohal'-Ilkiv, B. (2011). Selected topics on constrained and nonlinear control. Slovakia, ROSA. Dolný Kubín.

Ioannou P. and Sun J., 1996, Robust adaptive control: Prentice Hall, 1996.

Khalil, H.K. (2001). Nonlinear systems. Prentice Hall.

Son, K. H., Nomoto K., 1981. On the Coupled Motion of Steering and Rolling of a High Speed Container, J.S.N.A., Japan, Vol. 150, 232-244.

Tanaka, K. & Sugeno, M. (1992). Stability analysis and design of fuzzy control systems. Fuzzy Sets and System 45, 135-156.

Tatjewski, P. (2007). Advanced Control of Industrial Processes. Springer Verlag. London.

Tzirkel-Hancock, E. & Fallside, F. (1992). Stable control of nonlinear systems using neural networks. International Journal of Robust and Nonlinear Control 2(1), 63-86.

Van Amerongen, J., 1982. Adaptive Steering of Ships – A Model Reference Aproach to Improved Maneuvering and Economical Course Keeping, PhD thesis, Delf University of Technology, The Netherlands, 1982.

Vidyasagar, M. (1985). Control system synthesis: A factorization approach. The Massachusetts Institute of Technology Press. Massachusetts.

Witkowska, A., Tomera, M., & Śmierzchalski R. (2007). A backstepping approach to ship course control. International Journal of Applied Mathematics and Computer Science, 17(1), 73-85.

The Multi-step Matrix Game of Safe Ship Control with Different Amounts Admissible Strategies

J. Lisowski
Gdynia Maritime University, Poland

ABSTRACT: This paper describes the process of the safe ship control in a collision situation using a differential game model with j participants. The basic model of the process includes non-linear state equations and non-linear, time varying constraints of the state variables as well as the quality game control index in the forms of the game integral payment and the final payment. As an approximated model of the manoeuvring process, model of multi-step matrix game in the form of dual linear programming problem has been adopted here. The Risk Game Manoeuvring (RGM) computer program has been designed in the Matlab/Simulink software in order to determine the own ship's safe trajectory. These considerations have been illustrated with examples of a computer simulation using an RGM program for determining the safe ship's trajectory in real navigational situation during passing ten objects. Simulation research were passed for five sets of admissible strategies of the own ship and met objects.

1 INTRODUCTION

The process of a ship passing other objects at sea very often occurs in conditions of uncertainty and conflict accompanied by an inadequate co-operation of the ships with regard to the International Regulations for Preventing Collisions at Sea (COLREG). It is, therefore, reasonable to investigate, develop and represent the methods of a ship's safe handling using the rules of theory based on dynamic games and methods of computational intelligence.

In practice, the process of handling a ship as a control object depends both on the accuracy of the details concerning the current navigational situation obtained from the ARPA (Automatic Radar Plotting Aids) anti-collision system and on the form of the process model used for determining the rules of the handling synthesis. The ARPA system ensures automatic monitoring of at least 20 j-th encountered objects, determining their movement parameters (speed V_j, course ψ_j) and elements of approaching to own ship ($D^j_{\min} = DCPA_j$ – Distance of the Closest Point of Approach, $T^j_{\min} = TCPA_j$ – Time to the Closest Point of Approach) and also assess the risk r_j of collision (Bist 2000, Bole et al. 2006, Cahill 2002, Gluver & Olsen 1998).

However, the range of functions of a standard ARPA system ends up with a simulation of a manoeuvre selected by navigator. The problem of selecting such a manoeuvre is very difficult as the process of control is very complex since it is dynamic, non-linear, multi-dimensional and game making in its nature (Figures 1, 2 and 3) (Clark 2003, Fang & Luo 2005, Fossen 2011, Lisowski 2007, Perez 2005).

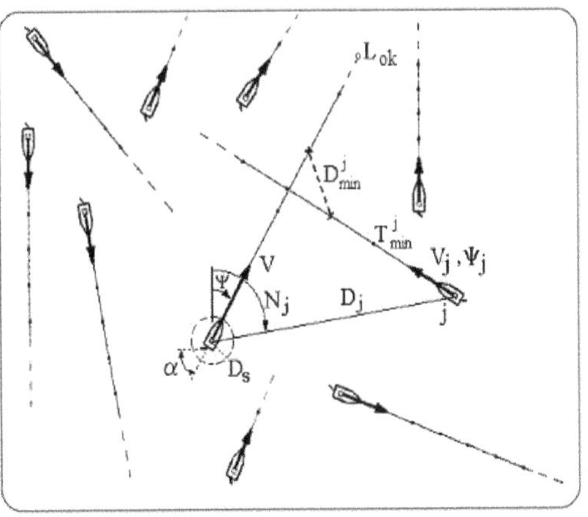

Figure 1. Parameters describing the process of the own ship passing j-th encountered object.

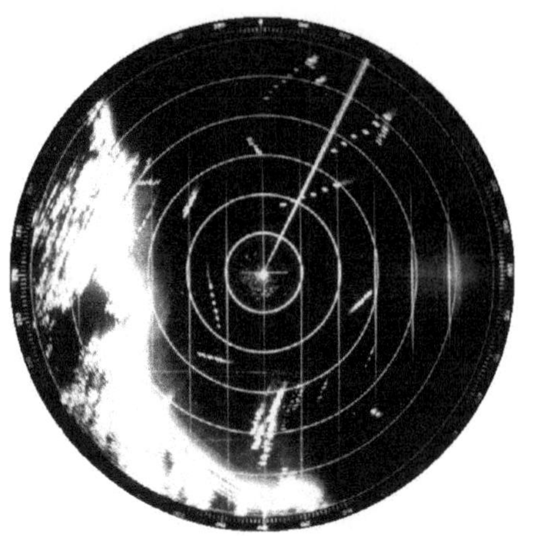

Figure 2. The photo of a radar screen in situation j=12 encountered objects at the Gdansk Bay.

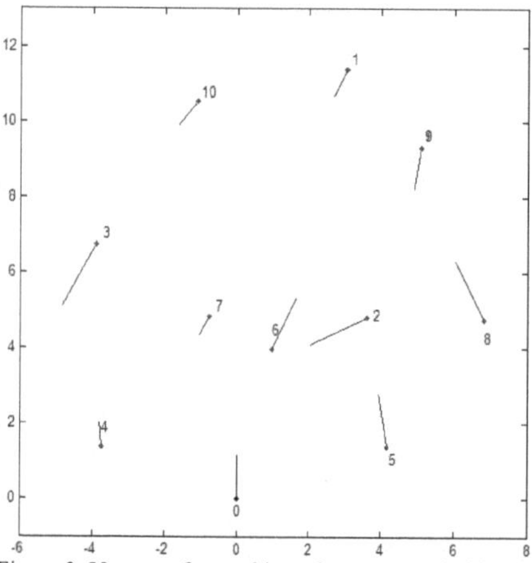

Figure 3. Vectors of own ship and encountered objects.

While formulating the model of the process it is essential to take into consideration both the kinematics and the dynamics of the ship's movement, the disturbances, the strategy of the encountered objects and the formula assumed as the goal of control. The diversity of selection of possible models directly affects the synthesis of the ship's handling algorithms which are afterwards affected by the ship's handling device, directly linked to the ARPA system and, consequently, determines the effects of safe and optimal control.

2 DIFFERENTIAL GAME MODEL OF THE SAFE SHIP CONTROL PROCESS

The most general description of the own ship's passing the j number of other encountered ships is the model of a differential game of a j number of objects (Figure 4).

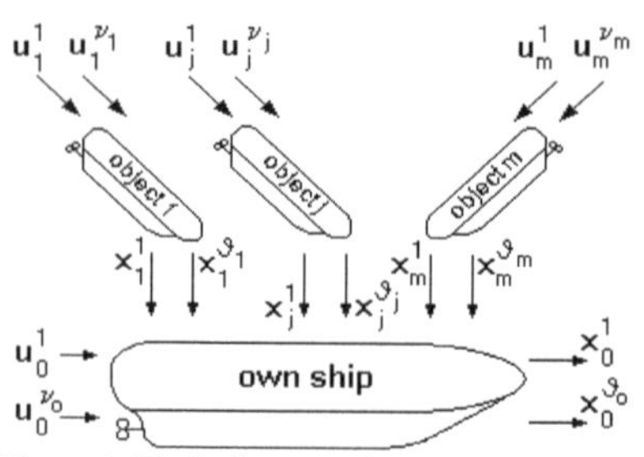

Figure 4. Block diagram of a model ship's differential game including j participants.

General dynamic features of the process are described by a set of state equations in the following form:

$$\dot{x}_i = f_i(x_0^{\vartheta_0},\ x_j^{\vartheta_j},\ u_0^{\nu_0},\ u_j^{\nu_j}, t) \tag{1}$$

where $\vec{x}_0^{\vartheta_0}(t)$ is ϑ_0 dimensional vector of the process state of the own ship determined in a time span $t \in [t_0, t_k]$, $\vec{x}_j^{\vartheta_j}(t)$ is ϑ_j dimensional vector of the process state for the j-th object, $\vec{u}_o^{\nu_0}(t)$ is ν_0 dimensional control vector of the own ship and $\vec{u}_j^{\nu_j}(t)$ is ν_j dimensional control vector of the j-th object (Isaacs 1965, Keesman 2011).

The state variable $x_0^{\vartheta_0}$ is represented by the values: course, angular turning speed, speed, drift angle, rotational speed of the screw propeller and controllable pitch propeller - of the own ship and $x_j^{\vartheta_j}$ by the values: distance, bearing, course and speed - of the j-th object. While the control value $u_0^{\nu_0}$ is represented by: reference rudder angle, reference rotational speed screw propeller and reference controllable pitch propeller - of the own ship and $u_j^{\nu_j}$ by the values: course and speed - of the j-th object (Isil & Koditschek 2001).

The constraints of the control and the state of the process are connected with the basic condition for the safe passing of the ships at a safe distance D_s in compliance with COLREG Rules, generally in the following form (Mesterton-Gibbons 2001):

$$g_j[x_j^{\vartheta_j}(t), u_j^{\nu_j}(t)] \le 0 \qquad j = 1,2,\ldots,m \tag{2}$$

The constraints (2) as „ship's domains" take a form of a circle, ellipse, hexagon or parable and may be generated, for example, by the neural network (Figure 5) (Baba & Jain 2001, Cockcroft & Lameijer 2006, Landau et al. 2011, Lisowski 2008, Millington & Funge 2009, Zio 2009).

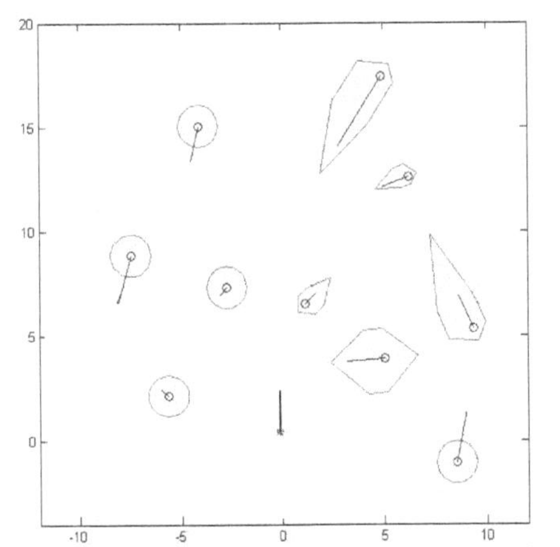

Figure 5. The shapes of the neural ship's domains in the situation of 10 encountered objects.

The synthesis of the decision making pattern of the ship's handling leads to the determination of the optimal strategies of the players who determine the most favourable, under given conditions, conduct of the process. For the class of non-coalition games, often used in the control techniques, the most beneficial conduct of the own ship as a player with j-th object is the minimization of her goal function in the form of the payments – the integral payment and the final one:

$$I_0^j = \int_{t_0}^{t_k} [x_0^{\vartheta_0}(t)]^2 dt + r_j(t_k) + d(t_k) \; \rightarrow \min \qquad (3)$$

The integral payment determines the loss of way of the own ship to reach a safe passing of the encountered objects and the final one determines the risk of collision and final game trajectory deflection from reference trajectory (Straffin 2001).

Generally two types of the steering goals are taken into consideration - programmed steering $u_0(t)$ and positional steering $u_0[x_0(t),t]$. The basis for the decision making steering are the decision making patterns of the positional steering processes, the patterns with the feedback arrangement representing the differential games (Luus 2000).

The application of reductions in the description of the own ship's dynamics and the dynamic of the j-th encountered object and their movement kinematics lead to the approximated matrix game model (Engwerda 2005, Lisowski 2009).

3 THE MULTI-STEP MATRIX GAME MODEL OF SAFE CONTROL PROCESS

3.1 *State and control variables*

The differential game is reduced to a matrix game of a j number of participants who do not co-operate among them (Figure 6) (Lisowski 2010a).

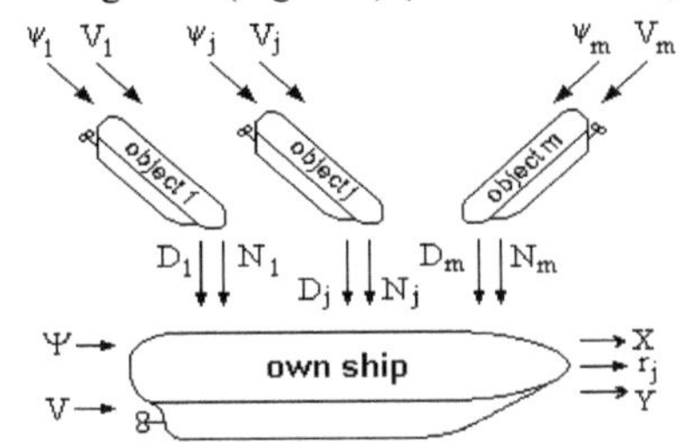

Figure 6. Block diagram of a model ship's approximated game j participants.

The state and control variables are represented by the following values:

$$\begin{aligned} & x_0^1 = X,\; x_0^2 = Y,\; x_j^1 = D_j,\; x_j^2 = N_j, \\ & u_0^1 = \psi,\; u_0^2 = V, \\ & u_j^1 = \psi_j,\; u_j^2 = V_j \\ & j = 1, 2, ..., m \end{aligned} \qquad (4)$$

3.2 *Risk of collision*

The matrix game includes the values determined previously on the basis of data taken from an anticollision system ARPA the value a collision risk r_j with regard to the determined strategies of the own ship and those of the j-th encountered objects (Lisowski 2010b, Osborne 2004).

The form of such a game is represented by the risk matrix $\boldsymbol{R}=[r_j(v_0, v_j)]$ containing the same number of columns as the number of participant I (own ship) strategies. She has; e.g. a constant course and speed, alteration of the course 20° to starboard, to 20° port etc., and contains a number of lines which correspond to a joint number of participant II (j-th object) strategies:

$$R = [r_j(\nu_0, \nu_j)] = \begin{vmatrix} r_{11} & r_{12} & .... & r_{1,\nu_0-1} & r_{1\nu_0} \\ r_{21} & r_{22} & .... & r_{2,\nu_0-1} & r_{2\nu_0} \\ .... & .... & .... & .... & .... \\ r_{\nu_1 1} & r_{\nu_1 2} & .... & r_{\nu_1,\nu_0-1} & r_{\nu_1\nu_0} \\ .... & .... & .... & .... & .... \\ r_{\nu_j 1} & r_{\nu_j 2} & .... & r_{\nu_j,\nu_0-1} & r_{\nu_j\nu_0} \\ .... & .... & .... & .... & .... \\ r_{\nu_m 1} & r_{\nu_m 2} & .... & r_{\nu_m,\nu_0-1} & r_{\nu_m\nu_0} \end{vmatrix} \qquad (5)$$

The value of the risk of the collision r_j is defined as the reference of the current situation of the approach described by the parameters $D^j_{\min}$ and T^j_{min}, to the assumed assessment of the situation defined as safe and determined by the safe distance of approach D_s and the safe time T_s – which are necessary to execute a manoeuvre avoiding a collision with consideration actual distance D_j between own ship and encountered j-th ship:

$$r_j = \left[\varepsilon_1 \left(\frac{D^j_{\min}}{D_s} \right)^2 + \varepsilon_2 \left(\frac{T^j_{\min}}{T_s} \right)^2 + \varepsilon_3 \left(\frac{D_j}{D_s} \right)^2 \right]^{-\frac{1}{2}} \tag{6}$$

where the weight coefficients ε_1, ε_2 and ε_3 are depended on the state visibility at sea (good or restricted), kind of water region (open or restricted), speed V of the ship, static L and dynamic L_d length of ship, static B and dynamic B_d beam of ship, and in practice are equal (Figures 7 and 8):

$$1 \le (\varepsilon_1, \varepsilon_2, \varepsilon_3) \le 20 \tag{7}$$

$$L_d = 1.1 (L + 0.345\, V^{1.6}) \tag{8}$$

$$B_d = 1.1 (B + 0.767\, LV^{0.4}) \tag{9}$$

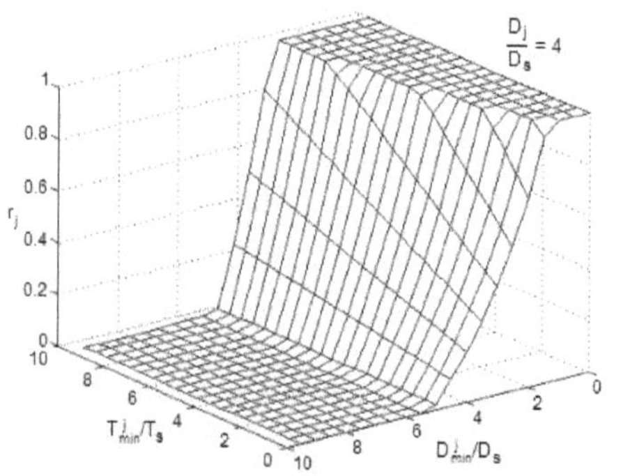

Figure 7. The surface of the collision risk value r_j in dependence on relative values distance and time of j-th object approach.

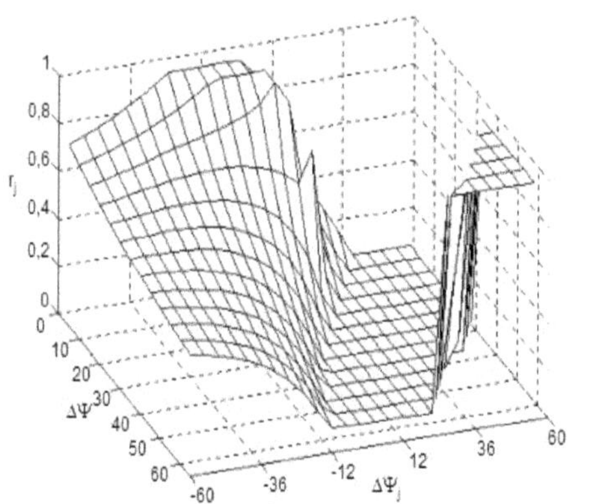

Figure 8a. Dependence of the collision risk on the strategy the own ship and that of the j-th encountered object to approaching from the LB.

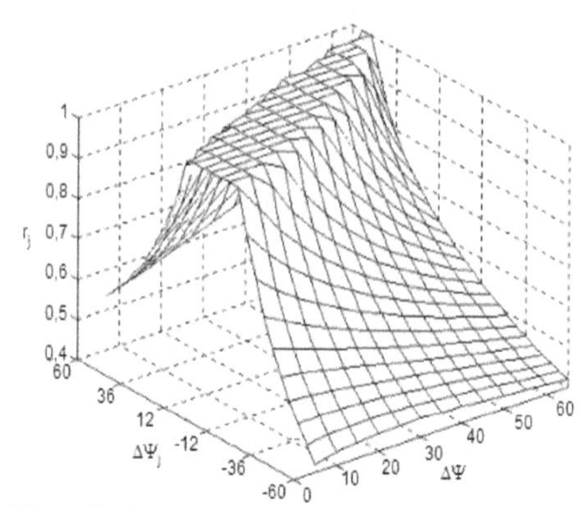

Figure 8b. Dependence of the collision risk on the strategy the own ship and that of the j-th encountered object to approaching from the SB.

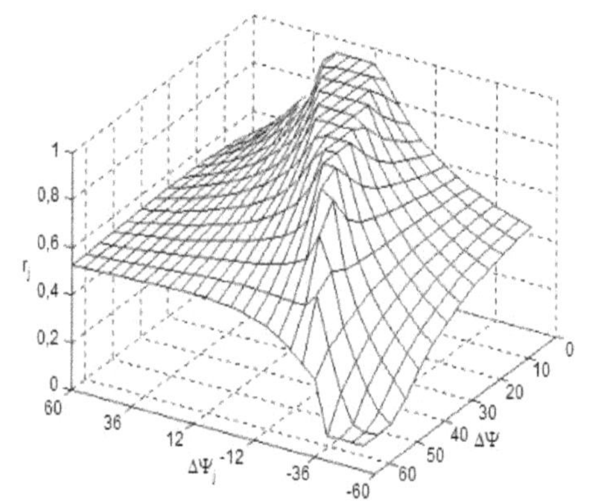

Figure 8c. Dependence of the collision risk on the strategy the own ship and that of the j-th encountered object to approaching from the stern.

The constraints affecting the choice of strategies are a result of the recommendations of the way priority at sea. Player I (own ship) may use ν_0 of various pure strategies in a matrix game and player II (encountered object) has ν_j of various pure strategies (Pietrzykowski 2011).

3.3 *Control algorithm*

As the game, most frequently, does not have saddle point the state of balance is not guaranteed, there is a lack of pure strategies for both players in the game. In order to solve this problem dual linear programming may be used (Pantoja 1988).

In a dual problem player I having ν_0 various strategies to be chosen tries to minimize the risk of collision (Modares 2006):

$$I_0 = \min_{\nu_0} r_j \tag{10}$$

while player II having ν_j strategies to be chosen try to maximize the risk of collision (Mehrotra 1992):

$$I^j = \max_{\nu_j} r_j \tag{11}$$

The problem of determining an optimal strategy may be reduced to the task of solving dual linear programming problem (Basar & Olsder 1982):

$$\left(I_0^j\right)^* = \min_{\nu_0} \max_{\nu_j} r_j \tag{12}$$

Mixed strategy components express the probability distribution $\boldsymbol{P}=[p_j(\nu_0, \nu_j)]$ of using pure strategies by the players (Lisowski 2012a):

$$P = [p_j(\nu_0, \nu_j)] = \begin{vmatrix} p_{11} & p_{12} & \cdots & p_{1,\nu_0-1} & p_{1\nu_0} \\ p_{21} & p_{22} & \cdots & p_{2,\nu_0-1} & p_{2\nu_0} \\ \cdots & \cdots & \cdots & \cdots & \cdots \\ p_{\nu_1 1} & p_{\nu_1 2} & \cdots & p_{\nu_1,\nu_0-1} & p_{\nu_1\nu_0} \\ \cdots & \cdots & \cdots & \cdots & \cdots \\ p_{\nu_j 1} & p_{\nu_j 2} & \cdots & p_{\nu_j,\nu_0-1} & p_{\nu_j\nu_0} \\ \cdots & \cdots & \cdots & \cdots & \cdots \\ p_{\nu_m 1} & p_{\nu_m 2} & \cdots & p_{\nu_m,\nu_0-1} & p_{\nu_m\nu_0} \end{vmatrix} \tag{13}$$

The solution for the steering goal is the strategy of the highest probability and will also be the optimal value approximated to the pure strategy:

$$\left(u_0^{\nu_0}\right)^* = u_o^{\nu_0}\left\{[p_j(\nu_0, \nu_j)]_{\max}\right\} \tag{14}$$

The safe trajectory of the own ship has been treated here as a sequence of changes course and speed (Lisowski 2012b).

The values established are as follows: safe passing distances among the ships under given visibility conditions at sea D_s, time delay of manoeuvring and the duration of one stage of the trajectory as one calculation step. At each step the most dangerous object is determined with regard to the value of the collision risk r_j. Consequently, on the basis of the semantic interpretation of the COLREG Regulations the direction of a turn of the own ship is selected to the most dangerous encountered object (Flechter 1987, Lisowski 2012c).

The collision matrix risk $\boldsymbol{R}$ is determined for the admissible strategies of the own ship ν_0 and those ν_j for j-th object encountered. By applying dual linear programming in order to solve the matrix game you obtain the optimal values of the own course and that of the j-th object at the smallest deviation from their initial values.

If, at a given step, no solution can be found at a speed of the own ship V, the calculations are repeated at the speed reduced by 25% until the game has been solved. The calculations are repeated step by step until the moment when all elements of the matrix $\boldsymbol{R}$ become equal to zero and the own ship, after having passed the encountered objects, returns to her initial course and speed.

In this manner optimal safe trajectory of the ship is obtained in a collision situation (Fadali & Visioli 2009, Gałuszka & Świerniak 2005).

Using the function of *lp – linear programming* from the Optimization Toolbox contained in the Matlab software, the RGM program has been designed for the determination of the safe ship's trajectory in a collision situation (Lisowski 2012d).

4 COMPUTER SIMULATION

4.1 *RGM-1 program*

Simulation tests in Matlab/Simulink of the RGM program have been carried out with reference to real situation at Gdańsk Bay of passing $j=10$ encountered objects, introduced in Figures 2 and 3.

For the first base version RGM-1 of the program, the following values for the strategies have been adopted (Figure 9) (Lisowski & Lazarowska 2013, Nisan et al. 2007):

$\nu_0 = 13 \rightarrow \left|0^o \div 60^o\right|$ for each of the 5°,

$\nu_j = 25 \rightarrow \left(-60^o \div +60^o\right)$ for each of the 5°.

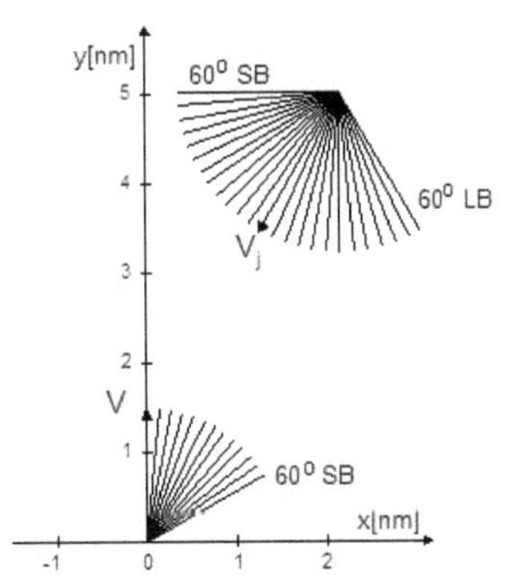

Figure 9. Possible mutual strategies of the own ship and those of the j-th encountered object in program RGM-1.

The computer simulation, performed on version of the RGM-1 program is presented on Figure 10.

Good visibility: $D_s=0.5$ *nm*

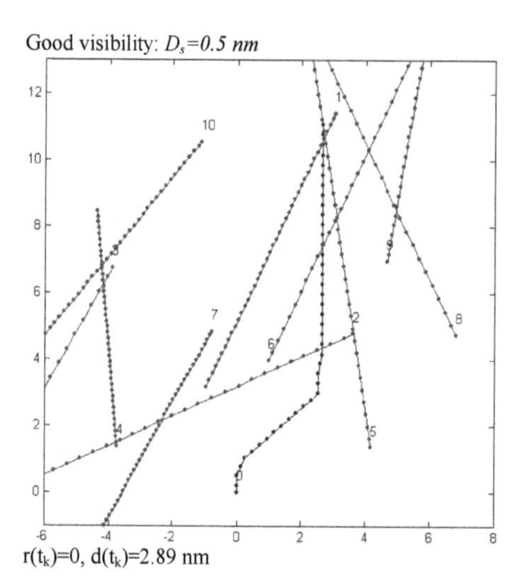

$r(t_k)=0$, $d(t_k)=2.89$ nm

Restricted visibility: $D_s=2.5$ nm

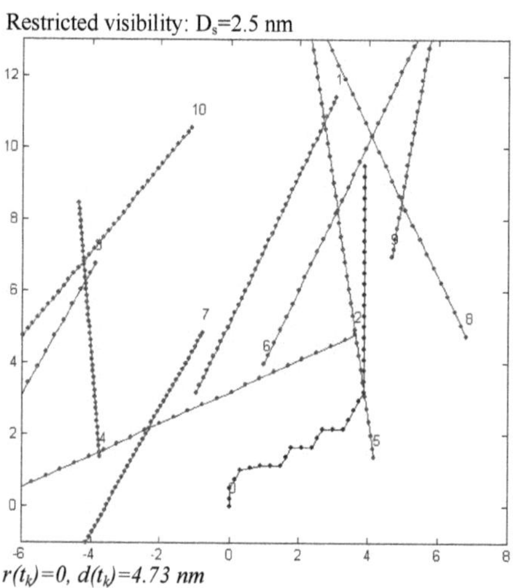

$r(t_k)=0$, $d(t_k)=4.73$ *nm*

Figure 10. The ship's game trajectories for the RGM-1 algorithm.

4.2 *RGM-2 program*

For the second version RGM-2 of the program, the number of own ship strategies has been reduced to (Figure 11):

$$\nu_0 = 13 \rightarrow \left|0^o \div 60^o\right| \text{ for each of the } 5^o,$$

$$\nu_j = 3 \rightarrow \left(-30^o, 0^o, +30^o\right).$$

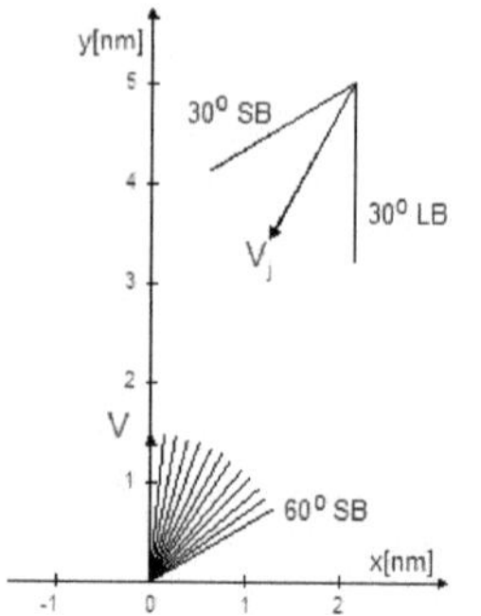

Figure 11. Possible mutual strategies of the own ship and those of the *j*-th encountered object in program RGM-2.

The computer simulation, performed on version of the RGM-2 program is presented on Figure 12.

Good visibility: $D_s=0.5$ *nm*

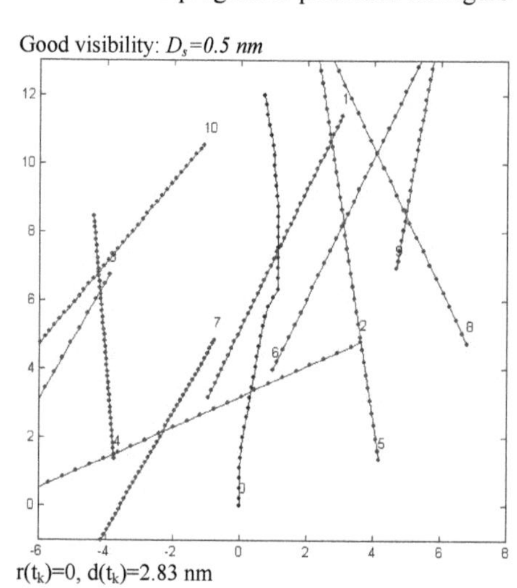

$r(t_k)=0$, $d(t_k)=2.83$ nm

Restricted visibility: $D_s=2.5$ *nm*

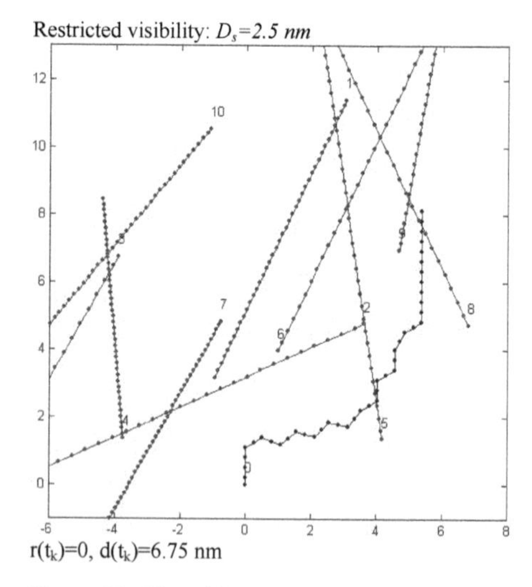

$r(t_k)=0$, $d(t_k)=6.75$ nm

Figure 12. The ship's game trajectories for the RGM-2 algorithm.

4.3 *RGM-3 program*

For the version RGM-3 of the program, the number of own ship strategies has been reduced to (Figure 13) (Szłapczyński & Śmierzchalski 2009):

$$v_0 = 4 \rightarrow \left|0^o, 20^o, 40^o, 60^o\right|,$$

$$v_j = 3 \rightarrow \left(-30^o, 0^o, +30^o\right),$$

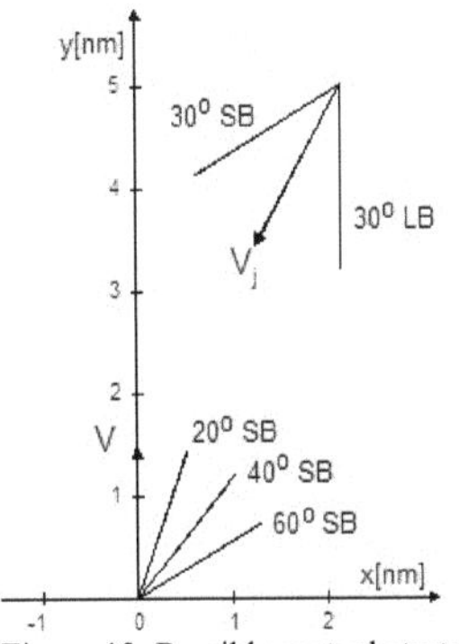

Figure 13. Possible mutual strategies of the own ship and those of the *j*-th encountered object in program RGM-3.

The computer simulation, performed on version of the RGM-3 program is presented on Figure 14.

Good visibility: D_s=*0.5 nm*

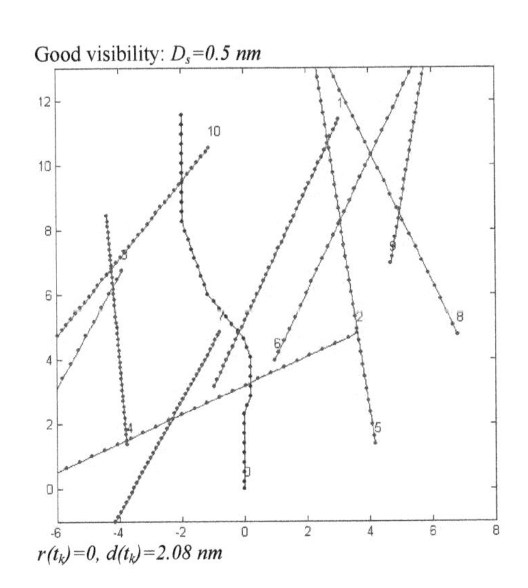

$r(t_k)$=0, $d(t_k)$=2.08 nm

Restricted visibility: D_s=*2.5 nm*

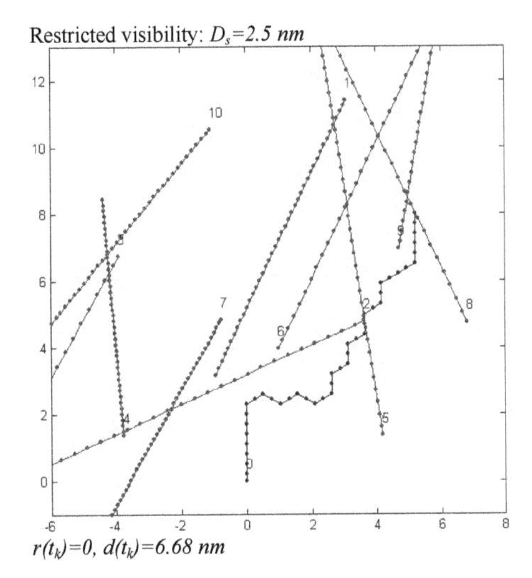

$r(t_k)$=0, $d(t_k)$=6.68 nm

Figure 14. The ship's game trajectories for the RGM-3 algorithm.

4.4 *RGM-4 program*

For the version RGM-4 of the program, the number of own ship strategies has been reduced to (Figure 15):

$$v_0 = 3 \rightarrow \left|0^o, 30^o, 60^o\right|,$$

$$v_j = 3 \rightarrow \left(-30^o, 0^o, +30^o\right),$$

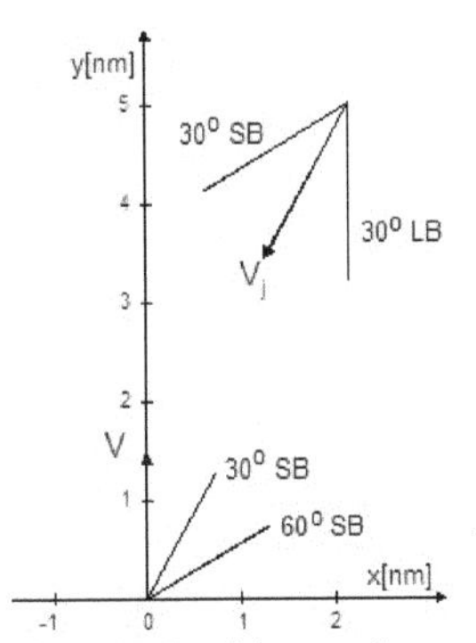

Figure 15. Possible mutual strategies of the own ship and those of the *j*-th encountered object in program RGM-4.

The computer simulation, performed on version of the RGM-4 program is presented on Figure 16.

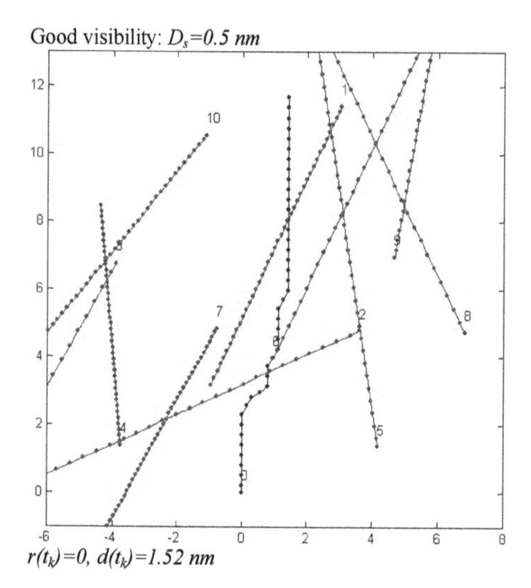

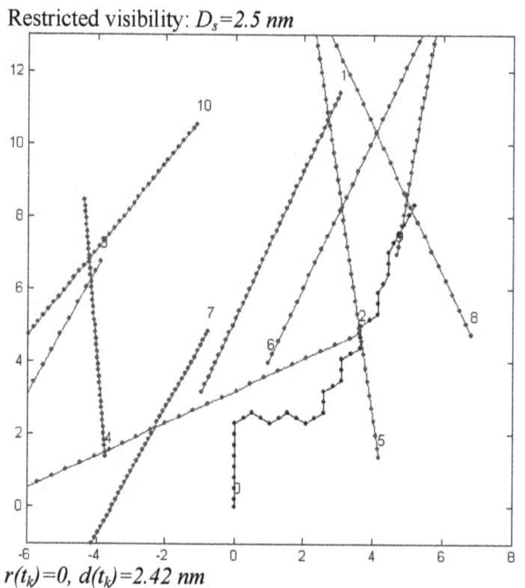

Figure 16. The ship's game trajectories for the RGM-4 algorithm.

4.5 *RGM-5 program*

For the version RGM-5, the number of the own ship strategies has been reduced to (Figure 17):

$$\mathrm{v}_0 = 2 \rightarrow \left|0^o,\ 60^o\right|,$$

$$\mathrm{v}_j = 3 \rightarrow \left(-30^o,\ 0^o,\ +30^o\right).$$

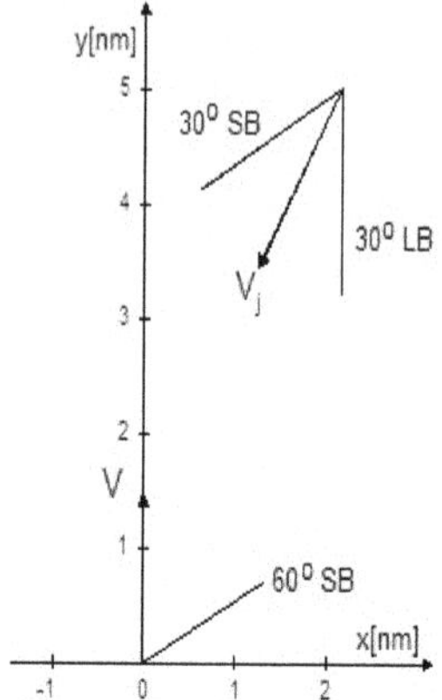

Figure 17. Possible mutual strategies of the own ship and those of the *j*-th encountered object in program RGM-5.

The computer simulation, performed on version of the RGM-5 program is presented on Figure 18.

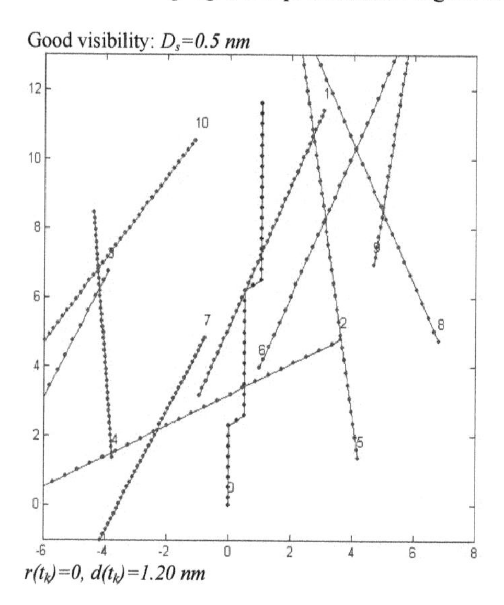

Restricted visibility: D_s=2.5 nm

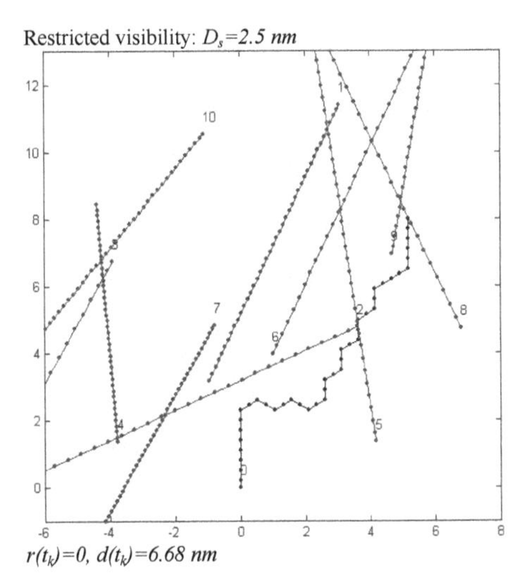

$r(t_k)=0$, $d(t_k)=6.68$ nm

Figure 18. The ship's game trajectories for the RGM-5 algorithm.

5 CONCLUSIONS

Analysis of the computer simulation studies of RGM program for different amounts of possible strategies of own ship and met objects allows to draw the following conclusions:

- The synthesis of an optimal on-line control on the base of model of a multi-step matrix game makes it possible to determine the safe game trajectory of the own ship in situations when she passes a greater *j* number of the encountered objects;
- The trajectory has been described as a certain sequence of manoeuvres with the course and speed;
- The RGM computer program designed in the Matlab also takes into consideration the following: regulations of the Convention on the International Regulations for Preventing Collisions at Sea, advance time for a manoeuvre calculated with regard to the ship's dynamic features and the assessment of the final deflection between the real trajectory and its assumed values;
- The essential influence to form of safe and optimal trajectory and value of deflection between game and reference trajectories has the number of admissible strategies of own ship and encountered objects;
- It results from the performed simulation testing this algorithm is able to determine the correct game trajectory when the ship is not in a situation when she approaches too large number of the observed objects or the said objects are found at long distances among them;
- In the case of the high traffic congestion the program is not able to determine the safe game manoeuvre. This sometimes results in the backing of the own object which is continued until the time when a hazardous situation improves.

REFERENCES

Baba, N. & Jain, L.C. 2001. *Computational intelligence in games*. New York: Physica-Verlag.

Basar, T. & Olsder, G.J. 1982. *Dynamic non-cooperative game theory*. New York: Academic Press.

Bist, D.S. 2000. *Safety and security at sea*. Oxford-New Delhi: Butter Heinemann.

Bole, A., Dineley, B. & Wall, A. 2006. *Radar and ARPA manual*. Amsterdam-Tokyo: Elsevier.

Cahill, R.A. 2002. *Collisions and thair causes*. London: The Nautical Institute.

Clarke, D. 2003. The foundations of steering and manoeuvering, *Proc. of the IFAC Conference on Manoeuvering and Control Marine Crafts, Girona*: 10-25.

Cockcroft, A.N. & Lameijer, J.N.F. 2006. *The collision avoidance rules*. Amsterdam-Tokyo: Elsevier.

Engwerda, J.C. 2005. *LQ dynamic optimization and differential games*. West Sussex: John Wiley & Sons.

Fadali, M.S. & Visioli, A. 2009. *Digital control engineering*. Amsterdam-Tokyo: Elsevier.

Fang, M.C. & Luo, J.H. 2005. The nonlinear hydrodynamic model for simulating a ship steering in waves with autopilot system. *Ocean Engineering* 11-12(32):1486-1502.

Fletcher, R. 1987. *Practical methods of optimization*. New York: John Wiley and Sons.

Fossen, T.I. 2011. *Marine craft hydrodynamics and motion control*. Trondheim: Wiley.

Gałuszka, A. & Świerniak, A. 2005. Non-cooperative game approach to multi-robot planning. *International Journal of Applied Mathematics and Comuter Science* 15(3):359-367.

Gluver, H. & Olsen, D. 1998. *Ship collision analysis*. Rotterdam-Brookfield. A.A. Balkema.

Isaacs, R. 1965. *Differential games*. New York: John Wiley & Sons.

Isil Bozma, H. & Koditschek, D.E. 2001. Assembly as a non-cooperative game of its pieces: Analysis of ID sphere assemblies. *Robotica* 19:93-108.

Keesman, K.J. 2011. *System identification*. London-NewYork: Springer.

Landau, I.D., Lozano, R., M'Saad, M. & Karimi, A. 2011. *Adapive control*. London-New York: Springer.

Lisowski, J. 2007. The dynamic game models of safe navigation. In A. Weintrit (ed), *Marine navigation and safety of sea transportation*, Gdynia Maritime University and The Nautical Institute in London: 23-30.

Lisowski, J. 2008. Computer support of navigator manoeuvring decision in congested water. *Polish Journal of Environmental Studies* 5A(17): 1-9.

Lisowski, J. 2009. Sensitivity of safe game ship control on base information from ARPA radar. In G. Kouemou (ed), *Radar Technology*: 61-86. Vukovar: In-Teh.

Lisowski, J. 2010a: Optimization of safe ship control using Matlab/Simulink. *Polish Journal of Environmental Studies* 4A(19): 73-76.

Lisowski, J. 2010b. Optimization decision support system for safe ship control. In C. A. Brebbia (ed), *Risk Analysis*: 259-272. Southampton-Boston: WIT Press.
Lisowski, J. 2012a: The multistage positional game of marine objects with different degree of cooperation. *Solid State Phenomena* 180: 56-63.
Lisowski, J. 2012b: The optimal and safe trajectories for different forms of neural state constraints. *Solid State Phenomena*, 180:64-69.
Lisowski, J. 2012c: Game control methods in avoidance of ships collisions. *Polish Maritime Research* 74(19): 3-10.
Lisowski, J. 2012d: The sensitivity of safe ship control in restricted visibility at sea. *TransNav - International Journal on Marine Navigation and Safety of Sea Transportation* 1(6):35-45.
Lisowski, J. & Lazarowska, A. 2013: The radar data transmission to computer support system of ship safety. *Solid State Phenomena* (in printing).
Luus, R. (2000). *Iterative dynamic programming*, CRC Press, Boca Raton.
Mehrotra, S. 1992. On the implementation of a primal-dual interior point method. *SIAM Journal on Optimization* 4(2):575-601.
Mesterton-Gibbons, M. 2001. *An introduction to game theoretic modeling*. Providence: American Mathematical Society.
Millington, I. & Funge, J. 2009. *Artificial intelligence for games*. Amsterdam-Tokyo: Elsevier.
Modarres, M. 2006. *Risk analysis in engineering*. Boca Raton: Taylor & Francis Group.
Nisan, N., Roughgarden, T., Tardos, E. & Vazirani, V.V. 2007. *Algorithmic game theory*. New York: Cambridge University Press.
Osborne, M.J. 2004. *An introduction to game theory*. New York: Oxford University Press.
Pantoja, J.F.A. 1988. Differential dynamic programming and Newton's method. *International Journal of Control* 5(47):1539-1553.
Perez, T. 2005. *Ship motion control*. London: Springer.
Pietrzykowski, Z. 2011. *The navigational decision support system on a sea-going vessel*. Szczecin: Maritime University.
Straffin, P.D. 2001. *Game theory and strategy*. Warszawa: Scholar (in polish).
Szłapczyński, R. & Śmierzchalski, R. 2009. Supporting navigators decisions by visualizing ship collision risk. *Polish Maritime Research* 1(59): 83-88.
Zio, E. 2009. Computational methods for reliability and risk analysis. *Series on Quality, Reliability and Engineering Statistics* 14: 295-334.

Catastrophe Theory in Intelligent Control System of Vessel Operational Strength

E.P. Burakovskiy, Yu.I. Nechaev, P.E. Burakovskiy & V.P. Prokhnich
Kaliningrad State Technical University, Kaliningrad, Russia

ABSTRACT: The calculation paradigm at extreme situation modeling onboard intelligent control systems of marine vessels strength are discussed. Special attention is paid for solving complexity problems and adequacy of mathematical models in uncertainty situations and insufficiency of initial information.

1 COMPUTER MATHEMATICS AT REALIZING MODERN CATASTROPHE THEORY

At realizing catastrophe theory methods on the basis of highly computation general principles and structure of information model are taken into account, which secure analysis and forecast of situations being investigated. Models of situations control are developed within the framework of fuzzy logical basis and formalized analysis methods and forecast of interaction dynamics in various operational conditions [1]- [9].

This article discusses application of the developed concept of interpreting current situations in complex dynamic environment by method of catastrophe theory at vessel strength control onboard intelligent system (IS). Situations arising in operating marine dynamic object (DO) are typical examples of non-standard situations being characterized by uncertainty and insufficiency of initial information.

Non-linear dynamics of investigated objects is generated by complex hydrodynamics interaction of vessel with the ambient environment in the conditions of continual changing of object and environment conditions [1].

A modified catastrophe model depicting geometrical interpretation of current situations on the basis of paradigm for processing information in multiprocessor computer, is a universal construction of dynamic catastrophe image, containing typical elements of complex system behaviour [3].

Modeling and interpreting current situations in onboard IS of new generations is performed in complex dynamic environment which makes it necessary to use all accessible arsenal of analysis methods on the basis of modern high performance computing. The analyzed situations are often distinguished by prominent non-linearity, non-stationary and uncertainty characteristic for a broad range of self-organizing systems. In these conditions construction of interpreting models is performed on the basis of assumptions, hypotheses and simplifying suppositions [3].

Let us formulate demands which are necessary when constructing and using programming complex within the IS concept of controlling complex DO [1]. These demands present 3 key provisions which determine calculations paradigms in complex dynamic environment:

1 Situation control principle determining strategy: each class of possible environment conditions and DO corresponds to a certain class of acceptable solutions, proceeding from analysis of analytical and geometrical interpreting of current situations.
2 Principle of hierarchy IS organization, including strategic planning level of behaviour, tactic level of actions planning, performance level (decision making) and a complex of information-measurement devices providing optimum DO control.
3 Principle of founded choice of intelligence technologies used in solving tasks on the basis of modern catastrophe theory methods for hierarchy levels of decision making for controlling DO in complex dynamic environment.

Practical applications of catastrophe theory at interpreting current situations involve solving tasks difficult to be formalized. Complexity problem at developing IS on the basis of catastrophe theory methods is of paramount importance. It is closely connected with the information compression problem and singling out that part of it which determines situation analysis and developing practical recommendations [1], [7].

One of the effective trends of solving these problems is associated with using method of minimum description length (MDL) formulated by A.N. Kolmogorov within framework of algorithm information theory. This method proves rather fruitful at constructing and analyzing mathematical models of dynamics for IS functioning on a real time basis. In contrast to Shannon theory assuming extraction of optimum codes from knowledge of messages source model [3], Kolmogorov theory, on the contrary, discusses solution of construction task model of events source on the basis optimum codes search and optimum data presentation. Among models multitude a precise model is chosen which describes investigated DO without information loss. An approach based on MDL in IS is broadly used at constructing particular mathematical models of vessel dynamics on the basis of general model.

By way of data processing model of mathematics and physics modeling an information model [1] at figure 1 may be discussed.

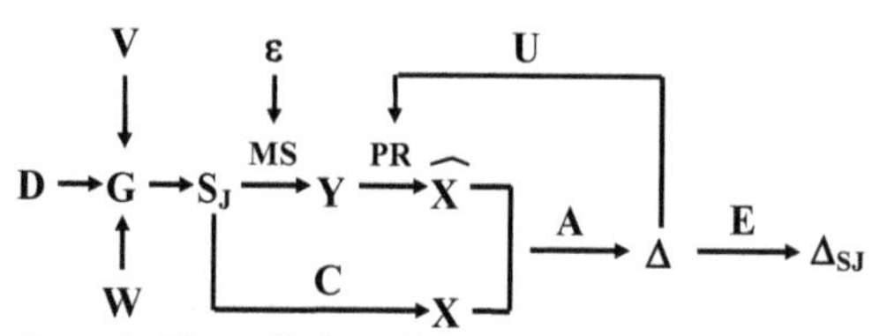

Figure 1. Chart of information model of interaction between vessel and environment

Here: D– dynamic object (DO); V, W– environment (wind, waves); G– situations generation model; S_j– particular situation; MS– measurement system with instruments for monitoring and measurement properties of an investigated object (cinematic and dynamic characteristics) in a J situation; ε– measurement error; Y– monitoring results; PR– processor, performing information transforming by means of mathematics and linguistics modeling; $\hat{X}$– imitation modeling results (new knowledge about dynamics interaction); C– interaction model (target operator) forming reliable assessment X (physics modeling results); A– adequator, comparing $\hat{X}$ and X and producing assessment adequacy Δ for obtained values X; U– control, correcting linguistic model and adjusting mathematics models coefficients, and when necessary– choosing a more fitting mathematical description; E– operator, producing a maximum possible adequacy assessment Δ_{SJ}.

Realization of algorithm for information processing is performed on the basis of high performance computing.

2 CALCULATIONS PARADIGM AND COMPLEXITY THEORY IN IS FOR SHIP STRENGTH CONTROL

One of the main problems in the decision making systems is the necessity of producing a great deal of calculations. It is especially characteristic of interpreting complex situations by means of catastrophe theory which are to be dealt with when formalizing knowledge in IS of ship strength control. When the number of attributes of investigated situation is large the use of conventional calculations methods brings about a sharp rise of calculations volume ("the curse of dimensions").

Speeding up of information processing is facilitated by transition from principle of sequence calculations (locality principle) to principle of parallel and combined processing, when intercoordinated information processing in a set of algorithms or elements of calculation process is performed (non-local information processing).

Contradiction between increasing complexity of the models being developed and necessity of using traditional methods of their using determines one of the most important tasks of interpreting dynamic situations– development of mathematics modeling methods for controlled movement of marine DO with taking into account demands not only their adequacy but also complexity of the model itself. Solution of this task involves developing methods and algorithms, realizing in conditions of uncertainty and lack of correctness of information provision, directed search of optimum models.

The formation of mathematical models multitude is based on involving such mathematical ship's behaviour, which correspond to the set modeling purpose. The models chosen as a result of the analysis are united into initial multitude [3]:

$$M(W,S)=\{m(w,s)\}, \tag{1}$$

which may make possible comparison of mathematical models' elements among themselves for analysis and choice of a preferable variant. The initial set M(W,S) is a functional space, constituting parameters relations of environment W and DO S. Each element of this set $m(w,s)\in M(W,S)$ corresponds to the aim of the modeling aim[m(w,s)].

Relation

$$R[m_1(w,s)],\ R[m_2(w,s)] \tag{2}$$

is a relation of binary equivalency, if

$$aim[m_1(w,s)]=aim[m_2(w,s)];$$

$$m_1(w,s),m_2(w,s)\in M(W,S). \tag{3}$$

All elements of the set M(W,S) according to the modeling aims of the ship's dynamics must be equivalent. Other equivalency relations R on the set M(W,S) are possible. But for some set elements M(W,S)

$$m_1(w,s), m_2(w,s) \in M(W,S) \quad (4)$$

may be given relation of a partial order $[m_1(w,s)]P[m_2(w,s)]$.

A set M(W,S) of all mathematic models, having a common modeling aim in a special task with the predetermined equivalency relations in this set may be called an aim models space (AEM) of the ship's dynamics in rough waters. Such space may be presented as a procession

$$AEM = < M(W,S), \{R\}, \{P\} >, \quad (5)$$

under condition, that

$$R(A) = aim\ [m_1(w,s)],\ R(A) \in \{R\} \quad (6)$$

and the set {R} is encircled.

Space (5) may unite ship's behaviour models in rough waters for various extreme situations. It provides solution of constructing algorithms for solving specific tasks of assessing ship's safety in a predetermined operation area.

Complexity principle is oriented at fulfilling ever increasing needs of complex systems theory, but at mathematical modeling of controlled objects. A general theory of complex systems is based on using complexity principle, especially in a non-stationary dynamic environment. The use of complexity principle at mathematical modeling of DO behaviour in uncertain and not correct conditions of information provision need defining a target model together with its complexity assessment: "model realization- model complexity" as a whole unit. Such approach corresponds Zadeh concept about transition to taking into account non-distinct sets theory and neural-non-distinct systems at mathematical modeling.

Interaction dynamics of complex object with environment may be generally described by a mathematical model [1]:

$$dx/dt = f(X,Y,t),\ x(t_0)=X_0,\ F(X,t) \le 0,\ t \in [t_0,T], \quad (7)$$

where X – n-dimensional vector of phase coordinate; Y – m-dimensional vector of occasional stirs; F(X,t) – is an area of changing phase coordinate vector, determining safe operational conditions; $x(t_0)=X_0$– occasional initial conditions, t- time.

The task may be solved by limited values of output parameters which are the criteria basis

$$\begin{aligned} & R_j \in Q_j,\ j=1,\ldots,J; \\ & Q_1 \in \Omega_1\ (z_1{*} < Z_{G1} < Z_j{**}); \\ & \quad\ldots \\ & Q_1 \in \Omega_J\ (z_1{*} < Z_{G1} < Z_j{**}); \\ & Z_{CR} = (Z_G)_{min}. \end{aligned} \quad (8)$$

Here $\Omega_1,\ldots,\Omega_J$ – area of changing criteria relations with taking into account uncertainty of initial information; Z_{CR} – critical value of investigated characteristics.

Thus, the task (7), (8) is to synthesize an algorithm of situation analysis as well as assess the correctness measure of criteria relations in uncertainty conditions of initial information. The suggested methodology of interpreting current situations based on catastrophe theory methods assumes an all-round analysis of vessel and environment dynamic interaction on the basis of a priory information. This realizes the chain of information transformation: "physical model"- "analytical model"- "geometrical model". The final stage of interpretation is a "cognitive model", presented as a simple image easy to interpret [3].

3 DYNAMIC ENVIRONMENTS DETERMINING CALCULATION TECHNOLOGY

A concept basis of the supplement under consideration is based on using the paradigm of processing information in multiprocessor computer environment [1], and achievements in the field of intelligence technologies of XXI century [3]. The basic principles of information transforming in media difficult to formalize are formulated in [1]. A formalized nucleus of an intelligence support system for processes of construction and use of knowledge models at analyzing current situations on the basis of analytical and geometrical components of current catastrophe theory is realized within the framework of non-distinct logical basis. The fundamental basis of such interpretation is a concept of non-distinct aims and limitations [3].

Effectiveness rise of functioning a procedure component is achieved by using a principle of competition and formalizing procession of non-distinct information in a highly productive computation media [1]. The other principles of new generation IS effectiveness rise are the principle of openness, a principle of complexity and that of non-linear self-organization. Realization of these principles are executed within the framework of soft computing concept, integrating fuzzy logic, neural networks and genetic algorithm [5].

A general approach is discussed media classification in relation to onboard IS supporting calculation technology of modern catastrophe theory [3]. Intelligence modeling medium and visualizing complex dynamic situations is a key basis for composing analytical and geometrical components by means of logical knowledge system, providing IS functioning. Bellow a classification of dynamic environments in IS for supporting process of modeling and visualizing current situations.

1. A partially formalized environment, constituting non-distinct logical basis oriented at presenting outer stirring by climatic spectrum. Uncertainty of environment lies in complexity of formalizing interaction dynamics with taking into account all operating factors, especially wind gusts, approximated in the form of standard calculation charts, accepted at forming criterial assessment basis of ships safety, and floating technical devices by various classification societies and international standards [3].
2. Considerable uncertainty in complex conditions of interaction between an object and environment. The interaction model in these conditions constitutes fuzzy logical basis oriented at presenting an outward stirring by a sequence of non-regular waves packets of different form and intensiveness. Environment uncertainty lies in complexity of formalizing interaction model with taking into account a real pattern of ship's behavior as a non-linear non-standard system.
3. Full uncertainty, determined by lack of interaction model and constituting a unique case of investigating current situation on the basis of hypothesis and simplified suppositions. Environment uncertainty lies in complexity of constructing a formal interaction model with accounting for real pattern of ship's behavior at different level of outer stirs.

Demarcations of environments mentioned above is associated with solving the problem of choosing the boarder of uncertainty area "where begins and finishes inadmissibility". Solution of this task is possible only depending on peculiar features of interaction of ship and environment. An algorithm of transforming information is realized on the basis of modern catastrophe theory within the framework of fractal geometry. This algorithm accounts for catastrophe dynamic structure peculiarities. For rising effectiveness of reflecting the current situation in complex dynamic environments the geometrical images of fractals are complemented by structures, realized on the cognitive paradigm basis. By way of illustration figure 2 presents two scenarios of developing the current situation on the basis of fractal geometry and a corresponding dynamic model of catastrophe.

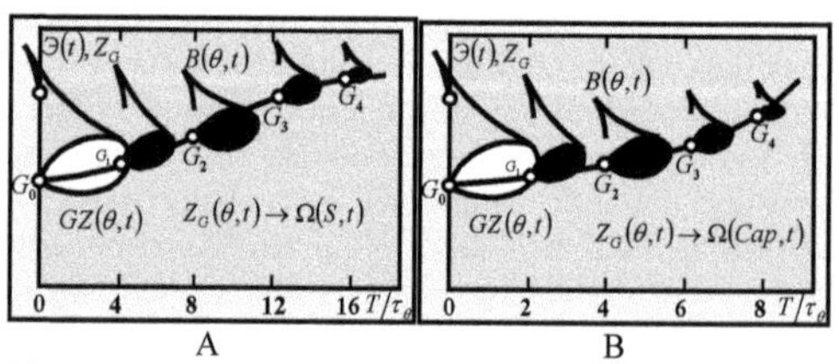

Figure 2. Evolution of dynamic system in conditions of situation stabilization (A) and at a loss of movement stability (B).

The first scenario corresponds to the case of situation stabilization in the process of DO movement to the aim attractor (stable system condition), the second one– to the loss of stability (catastrophe emergence). The designations on the figure are: t – time; Э(t) – process entropy; Z_G – applicata of mass centre of DO; G_0, G_1,..,G_4 – the mass centre position; GZ(θ,t) – an area integrating dynamic environment by means of fractal geometry; Ω(St) and Ω(Cap) – areas, reflecting stabilization situation and a loss of stability (capsizing).

The control solutions performed by logical system of knowledge brought about transformation of geometrical scene in direction of movement to an aim attractor, which is formed by means of sequential transforming of information on the basis of dynamic basis of IS knowledge. At investigating dynamic system evolution on the basis of fractal geometry a theoretical and practical interest presents a problem of falling the system outside the admitted limits, determined by peculiarities of DO behavior at interaction with environment.

A formal apparatus of transition of dynamic system conditions is based on presenting the process within the framework of non-distinct logical basis. An algorithm of DO control is realized by means of possessive functions, determined by a non-distinct logic system with the property of universal functions approximator [1], [5].

4 THE MODEL OF FUNCTIONAL RELATIONS AND SYNTHESIS OF CONCEPTUAL MODEL FOR CALCULATIONS ORGANIZATION

Let us discuss from positions of system analysis the principles of construction and synthesis of conceptual model of the DO strength control. Main attention will be paid to singling out functional dependences and model of functional relations, determining ship- environment interaction [5].

In relation to the task of presenting and investigating basic components of ship's strength at realizing catastrophe theory methods a network of dependencies allows to single out combination of factors and to construct functional dependencies corresponding to the level of task being solved at different stages of analysis and situation interpretation. In simple cases the solution is achieved on the basis of statistical methods in criterial relations, in a more complex ones – non-traditional procedures in the framework of soft calculations concept are used.

Let us discuss the use of functional relations method [4] at constructing mathematical models getting more complex in the tasks of ship's strength control. As an algorithm of transforming a structure depicted at figure 3 will be discussed. On the basis of this structure a typical tasks of realizing solutions with the use Data Mining procedures is presented. Here X_1,X_2,X_3 present vectors of initial information, describing dynamics of environment F(V,W) (wind V, waving W) and interaction parameters F(D). Dark circles characterize procedures A_1 – A_5 providing information procession on the basis of statistic analysis.

A_1 procedure realizes disperse analysis of factors of X_1 and X_2 vectors, and A_2 procedure– a correction analysis of factors of vectors X_2 and X_3. Further procedures A_3, A_4, A_5 realize construction regression models getting more complex. Procedure A_3 here provides linear regression analysis, procedure A_4– non-linear regression analysis, and finally procedure A_5– an expanded regression model.

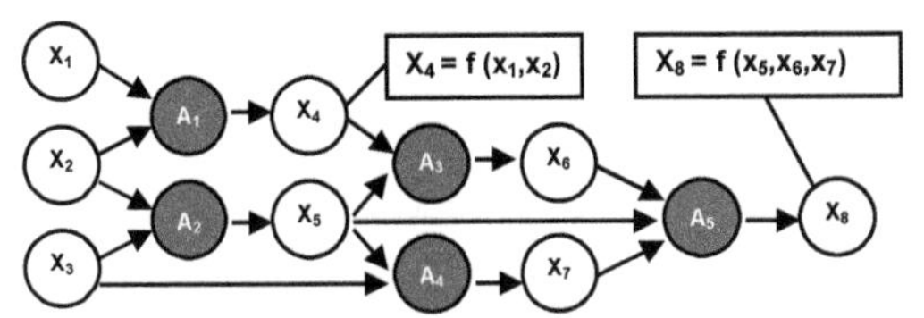

Figure 3. The model of functional relations, realizing construction of models getting more complex in IS of ship's strength control

Thus, information structure at figure 3 formalizes procedures of organizing calculations on the basis of functional relations method with the use of sequential statistics analysis. The advantage of such analysis is in a greatest formalization of a phenomenon at solving practical tasks.

5 ASSESSMENT OF ADEQUACY IN THE FRAMEWORK OF DYNAMIC ENVIRONMENTS FORMALIZATION PARADIGM

Construction of mathematic model and assessment of its adequacy are based on using standard procedures realization of which in complex dynamic environments it is necessary to take into account peculiarities of interaction dynamics of an investigated object and environment. The problem of adequacy of methods and models realizing information procession in IS of strength control acquires new meaning and content taking into account real data flow and peculiarities of highly productive calculations.

Conceptual basis of mathematic models adequacy assessment functioning in conditions of uncertainty and incompleteness of initial information is determined according to the following statements.

Statement 1. Adequacy of mathematical models at availability of physical modeling data is assessed in accordance with the traditional calculation patterns accepted in classical mathematics.

Statement 2. Adequacy of mathematical models constructed on the basis of assumptions may be checked with an approach suggested in paper [3] and allowing to single out a "pattern" model taking into account most fully peculiarities of an investigated physical process.

Statement 3. Adequacy of mathematical models constructed on the basis of hypotheses about physical regulation of an investigated phenomenon or a process may be checked by constructing alternatives area and using method of choosing solutions in a non-distinct environment on the basis of competing calculations technologies.

Realization of above statements is performed taking into account demands to mathematical model – non-contradiction and submission to all laws of mathematical logic. Model validity is determined by the ability to describe adequately an investigated situation and to forecast new results and phenomenon properties. These forecasts may refer to events which experimental investigation is difficult to carry out or altogether impossible. The solution of the set task depends also on physical regularities of an investigated situation and criterial basis of its interpretation.

The task of adequacy assessment, especially mathematical models, describing complex evolution of an investigated system in a non-stationary environment, constitutes multistage iteration process of obtaining evidence of conclusions correctness as to the system's behaviour. One of the popular patterns of models' validation is O. Balci pattern [6], which is modernized taking into account specific supplements with the purpose of taking into account data of physical and neural-non-distinct modeling. Difficulties of using O. Balci pattern at assessing adequacy in conditions of full uncertainty brought about an all-round analysis of similar situations and a search for relevant models of reflecting interaction dynamics. As one of the approaches to assessing adequacy a method may be used, based on parameters identification of non-distinct model with the use of expert knowledge [1]. But a more preferable approach in this situation is development of non-distinct basis for a procedure of assessing adequacy in conditions of full uncertainty of interaction environment. The developed method assumes calculation of deviation of a response of an investigated model and pattern responses, obtained at realizing a non-distinct conclusion by precedent [1],[5]. Parameters adjustment of the model is performed in such a way, as to make a minimum model response.

A mechanism of non-distinct conclusion by precedent is based on transforming a priori data within framework of information procession paradigm in a multiprocessor computation medium (figure 4). Here NNA– neural network ensembles; KBP –precedent knowledge basis; MCP– modeling and comparative analysis block; MS– measuring system; CT– competitive technologies; AA– alternatives analysis; $\Phi_1(\cdot)$,..., $\Phi_N(\cdot)$ – initial data applied on standard (SA) and neural network (ANN) algorithms; $\alpha_1\beta_1$,...,α_N β_N – output data for SA and ANN; $F_1(\cdot)$,...,$F_N(\cdot)$ – situation models determined as a result of alternatives analysis.

Thus, a model obtained as a result of non-distinct conclusion by precedent M(S*) is considered as adequate to investigated situation M(S), of condition of adequacy criterion is observed:

$$A(M(S^*), M(S)) \geq t, \qquad (9)$$

where t – threshold of non-distinct situations equation S* and S, which depends on demands to model accuracy and may be accepted in the range $t \in [0.7;0.9]$.

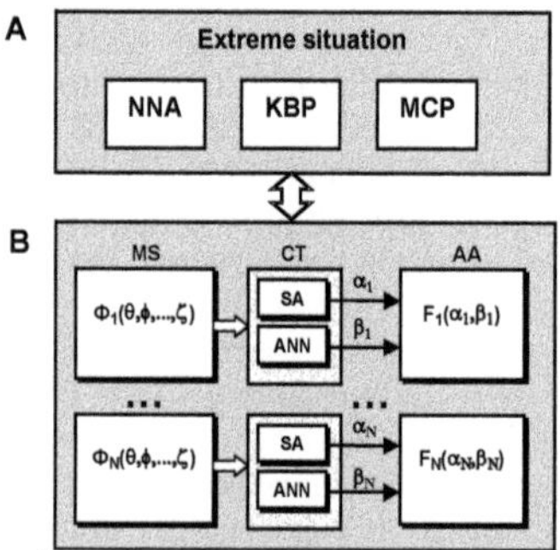

Figure 4. Information flow at forming non-distinct conclusion model by precedent (A) in a multiprocessor computation medium (B)

6 CONCLUSIONS

Thus, a new paradigm of calculation technology for dynamics of complex objects, realized in IS of ship strength control brings about the following advantages:

1 Expanding traditional approaches to information procession on the basis of new methods, models and algorithms of taking decisions support in complex dynamic environments.
2 Accounting for indefiniteness and insufficiency of initial information at interpreting complex decisions in multimode dynamic systems.
3 Development of inner potential of taking decision theory on the basis of competition principle and alternatives analysis at choosing a preferable calculation technology.

REFERENCES

1. Onboard intelligence systems. Part 2. Ships systems. – Moskow: Radiotechnik, 2006.
2. Kolmogorov A.N. About representation of continuous functions several variable as a superposition of continuous functions one variable and addition // the Reports AH USSR. 1957. т.114. Vol.5, p.p. 953-956.
3. Nechaev Yu.I. Catastrophe theory: modern approach to decision-making. – St.-Petersburg: Art-Express, 2011.
4. Silich M.P., Khabibulina N.Yu. Search of the decisions on model of the functional attitudes(relations) // Information technologies. №9. 2004, p.p.27-33.
5. System of artificial intelligence in intellectual technology of XXI century. – St.-Petersburg: Art-Express, 2011.
6. Balci O. Verification, validation and accreditation // Proceedings of the 1998 Winter Simulation Conference. – 1998, p.p.41-48.
7. Zadeh L. Fuzzy logic, neural networks and soft computing // Commutation on the ASM-1994. Vol.37. №3, p.p.77-84.
8. A. Lebkowski, R. Smierzchalski, W. Gierusz, K. Dziedzicki. Intelligent Ship Control System. TransNav – International Journal on Marine Navigation and Safety of Sea Transportation, 2(1), 2008, 63-68.
9. Z. Pietrzykowski, J. Uriasz. Knowledge Representation in a Ship's Navigational Decision Support System. TransNav – International Journal on Marine Navigation and Safety of Sea Transportation, 4(3), 2010, 265-270.

Concept of Integrated INS/Visual System for Autonomous Mobile Robot Operation

P. Kicman & J. Narkiewicz
Warsaw University of Technology, Warsaw, Poland

ABSTRACT: In the paper we are presenting method for integration of feature based visual odometry algorithm with low-cost IMU. The algorithm is developed for operation on small mobile robot investigating crime scenes. Detailed literature review of navigation systems based on visual odometry is provided along with out-line of the implemented algorithms. System architecture and current development state are described. Plans for further work are summarized.

1 INTRODUCTION

1.1 *General*

The navigation system described in this paper is developed for mobile robot Gryf (Figure 1) that will be used for investigation of criminal scenes. The robot will be operated remotely, but after the communication failure it should return to the operator autonomously.

The robot will be used to explore the areas to which the human access is not possible, either due to confined space or due to CBRN (chemical, biological, radiological and nuclear) threats.

To properly execute the autonomous return the robot requires accurate navigation system that will robustly operate in a previously unknown indoor environment. The navigation should be very precise, to not destroy the crime evidences and to find the return way. There is no guarantee for GNSS availability during operations. So it was decided that visual and inertial sensors will be combined to perform the task.

Returning to the operator the robot follows the path selected during the way to the operation area. It means that the navigation system has to provide accurate log of the driven path and then should be capable of following this path during its way back.

The cameras, which will be used during operation are usually low-cost sensors; the efficiency of the system depends on effective software to process the images. We are implementing visual odometry (VO) approach that is based on a dead reckoning principle. Inherent to the dead-reckoning method are errors that accumulate with time.

To diminish error accumulation integration of visual odometry with low-cost inertial navigation system (INS) will be applied. It is not perfect solution to integrate two dead reckoning systems, but there is no other navigation data sensor which might be used in the areas without well known landmarks. Fussing signals from these two sensors should improve the overall navigation, as their errors are independent of each other. INS measurements can be also used to provide current information used for scale recovery procedure which has to be performed in visual odometry, when monocular camera is used.

Figure 1. Gryf - mobile platform.

1.2 *Literature review*

Visual odometry sometimes also called ego-motion estimation is an incremental method that estimates the vehicle motion parameters using differences of displacement of selected items on consecutive video frames. Using this method both relative position as well as orientation of vehicle can be estimated. It is possible to use monocular camera for visual odometry, however stereo-camera provides more stable features, as the information about the third dimension (i.e.the depth) can be extracted from single frame using triangulation.

The review of the methods and current state-of-the-art is summarized in recently published survey papers by Scaramuzza and Fraundorfer [Scaramuzza & Fraundorfer, 2011, Fraundorfer & Scaramuzza, 2012]. Previously the visual odometry was successfully reported in a series of classic papers by Nister who examined scenarios for monocular and stereo-vision [Nister et al., 2004, Nister et al., 2006]. These papers initiated the rapid expansion of the method and the term visual odometry has gained common acceptance.

In many cases the visual odometry is superior to the traditional odometry based on wheel encoders, as visual system does not suffer from slippage problems and provides significantly better estimate of direction (for instance heading [Nourani-Vatani et al., 2009]). This feature is especially important in outdoor vehicle operation, when encoder-based odometry may be unreliable. The visual odometry methodology drawbacks include high computational cost and sensitivity to poor texture and to changes in lightning, etc. [Johnson et al., 2008].

Three main methodologies can be used to calculate visual odometry.

The *first* and the most popular of them is *feature tracking*. This technique is based on use of point features detected and tracked along the images sequence. There are usually three main steps of feature tracking in odometry implementation: feature extraction, feature matching and motion estimation. In the first step the selected frame features are detected. If stereo camera is used these features are matched with the corresponding points in the second stereo frame providing 3D position of the points in space. Then points are matched with features from the previous frame. Finally the motion of the camera is estimated based on the features displacement. This scheme is very similar to the Structure from Motion (SfM) type solutions [Koenderink & van Doorn, 1991]. Relative poses of cameras and features can be estimated for instance from 5 matching features as it was derived and demonstrated in [Nister, 2004]. Algorithms using 6, 7 and 8 feature pairs are also available [Stewenius et al., 2006].

Feature tracking approach was developed and revised by many researchers. Significant improvement to this approach was utilization of landmark matching techniques [Zhu et al., 2007]. In this approach a robot builds global landmarks of group of points in places that have been visited. When the location is revisited the re-observed features are used to correct the position. Improvement can be also made with Sparse Bundle Adjustment (SBA) performed on a couple of recent frames [Sunderhauf et al., 2006]. In this approach several recent frames are stored in the memory and local optimization of vehicle trajectory is performed for them. This allows to reduce the drift error significantly. In [Konolige et al., 2010] authors presented very accurate visual odometry system (with less than 0.1% error) on 10 km long track. This solution is improved by use of SBA which reduced the error by the factor of 2 to 5. The final navigation information is then fused with data from inertial measurement unit (IMU) using EKF in a loose coupling paradigm. The IMU was used as an inclinometer (information on roll and pitch) and yaw rate sensor. It was shown that the fusion of visual odometry with IMU improved the positioning by additional factor of 10. In [Tardif et al., 2008] authors provided solution with use of omnidirectional camera. They also decoupled estimation of rotational and translational motion making use of epipolar constraint [Hartley & Zisserman, 2004]. This approach enabled accurate motion estimation without use of computationally intensive iterative optimization. In [Scaramuzza & Siegwart, 2008] authors also use omnidirectional camera and track SIFT points to estimate motion of the vehicle. They also use the concept of appearance-based visual compass to improve estimation of the rotation. They assume pure rotational movement which is good approximation for small displacements and extract the rotation using various similarity measures. Visual Odometry based on feature tracking has been also successfully used on the surface of Mars as a secondary navigation system of Mars Exploration Rovers [Maimone et al., 2007] as well as during the recent mission of the Mars Science Laboratory [Johnson et al., 2008].

The second methodology for calculating visual odometry is based on the *optical flow*. In this approach change of brightness of image pixels over the consecutive frames is tracked. The calculated optical flow reflects the motion of the image from which the motion of the camera can be extracted. This method is computationally cheaper than feature tracking, however it is less accurate over time. To improve the robustness to the image noise, the algorithm called sparse optical flow has been developed. It is used to calculate the flow only for the chosen features in the images [Nourani-Vatani & Borges, 2011]. The optical flow visual odometry was demonstrated with downward looking camera in

[Dille et al., 2010]. In [Campbell et al., 2005] authors used optical flow measurements from monocular camera to estimate motion of the vehicle and to detect obstacles The system was tested on various surfaces. In [Corke et al., 2004] authors compared two methods for visual odometry for planetary rover using omnidirectional camera. First one was based on optical flow and second one was a full structure-from-motion solution. As expected the structure-from-motion solution provided higher precision estimates but at larger computational cost.

The third methodology is based on *template matching*. The estimation of motion is based on the template that is extracted from the image and searched for in the next frame. The displacement of the template is used to calculate the displacement of the vehicle. The method is superior over the previous two methods as it works reliably with almost no texture surfaces when feature tracking and optical flow methods do not work well [Nourani-Vatani & Borges, 2011]. However, the appearance of shadows and obstructions of view pose significant problem in applications of this method, which is not an issue for the previous two techniques. This drawback makes that approach impractical in most real-life scenarios. The solution with downward looking camera has also been presented [Nourani-Vatani et al., 2009, Nourani-Vatani & Borges, 2011].

Use of cameras for navigation have been also investigated in marine navigation. For example author of [Bobkiewicz, 2008] is considering use of digital camera for tracking celestial bodies.

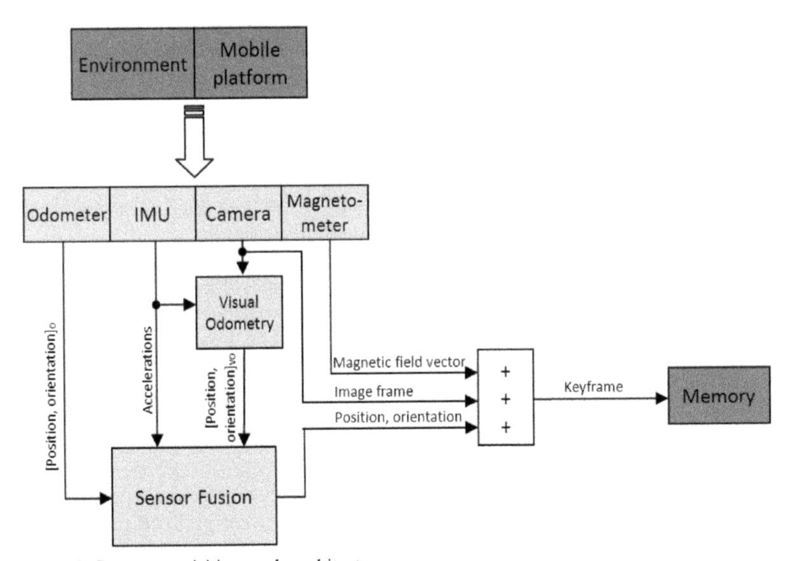

Figure 2. System acquisition mode architecture.

1.3 *In this paper*

The paper is structured as follows. In chapter 2 general overview of the developed navigation system is presented. Chapter 3 describes details of the developed visual odometry algorithm and chapter 4 contains information about integration of visual odometry with INS. Chapter 5 contains description of current development state of the system and finally, conclusions and plans for further work are described in chapter 6.

2 NAVIGATION SYSTEM CONCEPT

2.1 *General concept*

The navigation system developed operates in two modes. The first mode is a passive acquisition mode (Figure 2), when the navigation system only gathers information from surrounding areas, calculates the robot path and saves it to the memory. In this mode the robot is teleoperated and the navigation system does not provide any information externally. The second mode is vehicle autonomous operation (Figure 3) during which the robot returns to a starting position. The currently calculated position (calculated by the same algorithm as in the acquisition mode) is compared with path information stored in the memory of the robot. Based on the difference between observation and expectance, the commands are elaborated in the vehicle control system, which ensure that the robot follows the previous path.

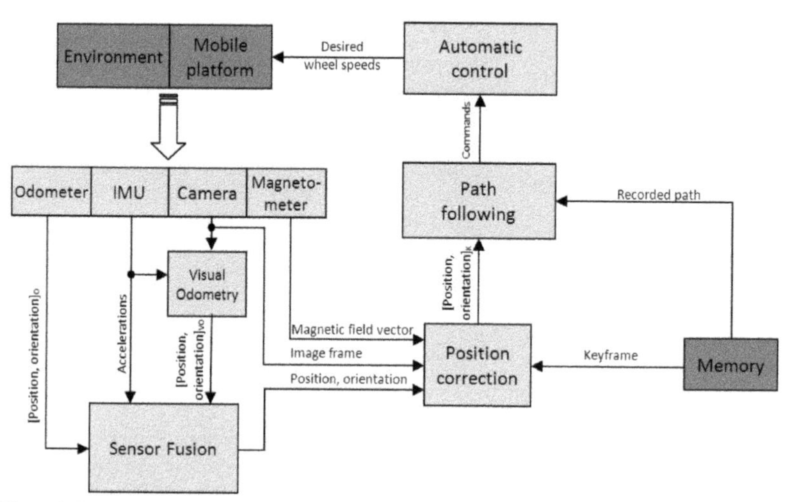

Figure 3. System autonomous mode architecture.

2.2 *System architecture*

The system consists of four sensors: odometer, IMU, camera and magnetometer. Images from the camera are fed into the visual odometry algorithm which forms the core of the navigation system. Calculated position is augmented with data from odometers placed on robot wheels and from low cost IMU sensor. The magnetometer is used only as additional reference sensor for corrections that are calculated during autonomous operation phase. All the data is processed by the on board computer. Position of the robot is being determined during entire operation of the robot.

Additionally, during autonomous mode the computer calculates control commands, so the robot stays on the right path towards the starting position. The main function for control algorithm is to minimize the difference between the real and desired path. The commands are calculated based on the difference between current position of the robot and desired path stored in the memory. To ensure that the information about current position is accurate it is corrected based on the comparison of images stored in the memory and currently observed by the camera. Differences between the two are used to update the current robot position. The reference images were saved during the acquisition phase.

3 VISUAL ODOMETRY

3.1 *Introduction*

The core of navigation system is build around visual odometry algorithm that processes visual data from the monocular camera. Our approach is based on the feature tracking methodology. The camera is directed at 45 degrees angle from direction of motion (the centre line of the vehicle). The image frames are processed consecutively providing the translation and the rotation matrices with up-to-scale accuracy. Precise description of the algorithm is provided in the following chapter.

3.2 *Algorithm description*

The algorithm for visual odometry calculation consists of several steps. The flowchart of the algorithm is presented on figure 4. Individual steps of the algorithm are explained below.

First steps consist of the initial pre-processing of the image retrieved from camera. This includes rectification of the image that removes distortions introduced by the camera lens. It is required that the camera is previously calibrated. It is also possible to obtain visual odometry in non-calibrated case, however initial calibration simplifies the motion estimation process [Stewenius et al., 2006]. After the rectification, an image is converted to a gray scale and smoothed to remove a noise.

In the following steps the point features are extracted from the frame. Ideally those features should be invariant to changes in lightning, perspective and scale. The good analysis of point image features regarding application in visual navigation can be found in [Agrawal et al., 2008, Bakambu et al., 2012]. In the presented case FAST features [Rosten & Drummond, 2006] were used for speed and simplicity purposes. Next, those features are matched in pairs with the points extracted from previous frame. Then they are grouped into random sets consisting of five matched points.

The five pairs of matched points is a minimal set that can be used to calculate finite number of solutions and to generate essential matrix E [Stewenius et al., 2006]. This essential matrix is representing relative orientation changes between the two views. It is calculated using implementation of algorithms provided by Nister in [Nister, 2004]. Finally this matrix is used to extract rotation and translation matrices with up to scale accuracy. For this procedure Singular Value Decomposition is used (SVD) [Hartley & Zisserman, 2004].

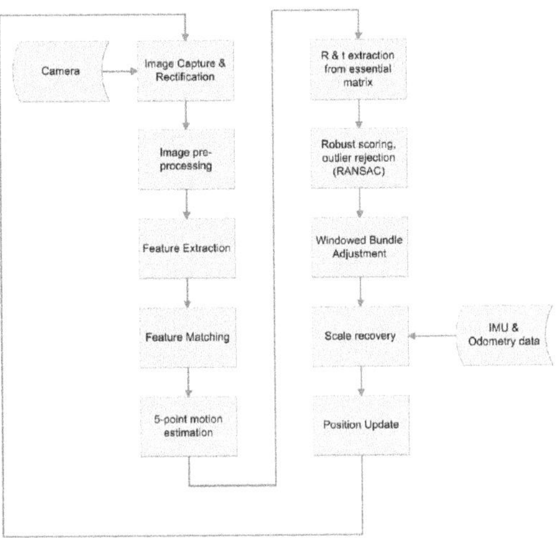

Figure 4. Visual odometry algorithm

Further steps of the algorithm are dedicated to improve the overall results and to make the calculations more robust. First step utilizes robust scoring algorithm RANSAC [Fischler & Bolles, 1981] that enables to reject solutions that do not match overall consensus. This prevents use of the incorrectly matched features as this would significantly deteriorate the solution. Final, polishing, step for the estimation includes Windowed Bundle Adjustmet [Hartley & Zisserman, 2004]. This is process of iterative optimization with goal of reducing reprojection error on several (usually up to 5) previous frames. The Bundle Adjustment algorithm is a standard solution in photogrammetry and for solving structure-from-motion problem. However, in its basic form it performs multiple iterations over entire available set of images. This prevents its use in real-time calculations and cannot be applied in straightforward manner for visual odometry. Hence, the 'Windowed' approach uses only few recent frames. This modification does not provide optimal solution for entire path, but enables online calculation while still improving the results by minimization of errors.

The final step of the algorithm requires to calculate the scale. As it was mentioned before, visual odometry with use of monocular camera provides constraints for only 5 degrees of freedom, therefore information for scale calculation must come from another source. In our case the other sensors are providing this data. Translation of the robot between two views for which VO was calculated is estimated based on data coming from odometry and robot movement model. When this information is available, the position is updated through concatenation and the algorithm repeats itself for new image.

4 INTEGRATION WITH INS

Integration with INS includes use of the IMU measurements for scale recovery in visual odometry. For that purpose the distance travelled by the vehicle obtained from IMU will be used to approximate scale of the motion estimation calculated by visual odometry.

At the sensor fusion level, the navigation parameters calculated by both VO and IMU will be fused using Kalman filtering methodology [Kalman, 1960]. The series of tests will be performed to determine the best version of the filter. Several variations of original filter such as EKF, IEKF, UKF or SPKF are planned to be tested. This comparative study will help to adjust the statistical model for the processes representing the navigation system.

5 CURRENT DEVELOPMENT

The preliminary tests of the visual odometry algorithm have been performed. However, the code is still going through debugging process and no conclusive results have been achieved so far. The implementation includes only the basic steps of visual odometry algorithm. RANSAC scoring and bundle adjustment optimization have not been programmed yet. As there are no other sensors available at the moment - the calculations are being made with up-to-scale accuracy.

6 CONCLUSIONS AND FUTURE WORK

The basic assumptions behind the visual odometry methodology was verified and it was concluded that there are efficient methods to estimate the vehicle path using the monocular camera. The basic version of the algorithm was prepared and is going through tests and debugging process. The next steps will focus on the development of the more advanced parts of the algorithm such as RANSAC scoring and local optimization with use of Bundle Adjustment. These methods are expected to significantly improve the positioning and to make the solution of the visual odometry system more robust. Implementation and testing of the advanced Kalman filters will also be done to integrate the visual and INS sensors.

ACKNOWLEDGMENT

The research is funded within the project "Development of mobile platform to support forensic investigation of crime scenes where CBRN threat may occur", grant NCBR No. 0015/R/I01/2011/01.

REFERENCES

Agrawal, Motilal, Konolige, Kurt, & Blas, Morten. 2008. CenSurE: Center Surround Extremas for Realtime Feature Detection and Matching. *Pages 102–115 of:* Forsyth, David, Torr, Philip, & Zisserman, Andrew (eds), *Computer Vision ECCV 2008*. Lecture Notes in Computer Science, vol. 5305. Springer Berlin / Heidelberg. 10.1007/978-3-540-88693-8_8.

Bakambu, Joseph Nsasi, Langley, Chris, Pushpanathan, Giri, MacLean, W. James, Mukherji, Raja, & Dupuis, Eric. 2012. Field Trial Results of Planetary Rover Visual Motion Estimation in Mars Analogue Terrain. *Journal of Field Robotics*, **29**(3), 413–425.

Bobkiewicz, P. 2008. Estimation of Altitude Accuracy of Punctual Celestial Bodies Measured with Help of Digital Still Camera. *Pages 279-284 vol. 2 of: TransNav - International Journal on Marine Navigation and Safety of Sea Transportation.*

Campbell, J., Sukthankar, R., Nourbakhsh, I., & Pahwa, A. 2005 (april). A Robust Visual Odometry and Precipice Detection System Using Consumer-grade Monocular Vision. *Pages 3421 – 3427 of: Robotics and Automation, 2005. ICRA 2005. Proceedings of the 2005 IEEE International Conference on.*

Corke, P., Strelow, D., & Singh, S. 2004 (sept.-2 oct.). Omnidirectional Visual Odometry for a Planetary Rover. *Pages 4007 – 4012 vol.4 of: Intelligent Robots and Systems, 2004. (IROS 2004). Proceedings. 2004 IEEE/RSJ International Conference on*, vol. 4.

Dille, Michael, Grocholsky, Ben, & Singh, Sanjiv. 2010. Outdoor Downward-Facing Optical Flow Odometry with Commodity Sensors. *Pages 183–193 of:* Howard, Andrew, Iagnemma, Karl, & Kelly, Alonzo (eds), *Field and Service Robotics*. Springer Tracts in Advanced Robotics, vol. 62. Springer Berlin / Heidelberg. 10.1007/978-3-642-13408-1_17.

Fischler, Martin A., & Bolles, Robert C. 1981. Random sample consensus: a paradigm for model fitting with applications to image analysis and automated cartography. *Commun. ACM*, **24**(6), 381–395.

Fraundorfer, F., & Scaramuzza, D. 2012. Visual Odometry : Part II: Matching, Robustness, Optimization, and Applications. *Robotics Automation Magazine, IEEE*, **19**(2), 78 –90.

Hartley, R. I., & Zisserman, A. 2004. *Multiple View Geometry in Computer Vision*. Second edn. Cambridge University Press, ISBN: 0521540518.

Johnson, Andrew E., Goldberg, Steven B., Cheng, Yang, & Matthies, Larry H. 2008 (May). Robust and Efficient Stereo Feature Tracking for Visual Odometry. *Pages 39–46 of: IEEE International Conference on Robotics and Automation.*

Kalman, Rudolph Emil. 1960. A New Approach to Linear Filtering and Prediction Problems. *Transactions of the ASME–Journal of Basic Engineering*, **82**(Series D), 35–45.

Koenderink, Jan J., & van Doorn, Andrea J. 1991. Affine structure from motion. *J. Opt. Soc. Am. A*, **8**(2), 377–385.

Konolige, Kurt, Agrawal, Motilal, & Sola, Joan. 2010. Large-Scale Visual Odometry for Rough Terrain. *Pages 201–212 of:* Kaneko, Makoto, & Nakamura, Yoshihiko (eds), *Robotics Research*. Springer Tracts in Advanced Robotics, vol. 66. Springer Berlin / Heidelberg. 10.1007/978-3-642-14743-2_18.

Maimone, Mark, Cheng, Yang, & Matthies, Larry. 2007. Two Years of Visual Odometry on the Mars Exploration Rovers. *Journal of Field Robotics*, **24**(3), 169–186.

Nister, D., Naroditsky, O., & Bergen, J. 2004 (june-2 july). Visual odometry. *Pages I–652 – I–659 Vol.1 of: Computer Vision and Pattern Recognition, 2004. CVPR 2004. Proceedings of the 2004 IEEE Computer Society Conference on*, vol. 1.

Nister, David. 2004 (June). An Efficient Solution to the Five-Point Relative Pose Problem. *In: IEEE Transactions on Pattern Analysis and Machine Intelligence*, vol. 26.

Nister, David, Naroditsky, Oleg, & Bergen, James. 2006. Visual odometry for ground vehicle applications. *Journal of Field Robotics*, **23**(1), 3–20.

Nourani-Vatani, Navid, & Borges, Paulo Vinicius Koerich. 2011. Correlation-Based Visual Odometry for Ground Vehicles. *Journal of Field Robotics*, **28**(5), 742–768.

Nourani-Vatani, Navid, Roberts, Jonathan, & Srinivasan, Mandiam V. 2009 (May). Practical Visual Odometry for Car-like Vehicles. *In: IEEE International Conference on Robotics and Automation.*

Rosten, Edward, & Drummond, Tom. 2006. Machine Learning for High-Speed Corner Detection. *Pages 430–443 of:* Leonardis, AleĹˇ, Bischof, Horst, & Pinz, Axel (eds), *Computer Vision â€" ECCV 2006*. Lecture Notes in Computer Science, vol. 3951. Springer Berlin / Heidelberg. 10.1007/11744023_34.

Scaramuzza, D., & Fraundorfer, F. 2011. Visual Odometry [Tutorial]. *Robotics Automation Magazine, IEEE*, **18**(4), 80 –92.

Scaramuzza, D., & Siegwart, R. 2008. Appearance-Guided Monocular Omnidirectional Visual Odometry for Outdoor Ground Vehicles. *Robotics, IEEE Transactions on*, **24**(5), 1015 –1026.

Stewenius, Henrik, Engels, Christopher, & Nister, David. 2006. Recent developments on direct relative orientation. *ISPRS Journal of Photogrammetry and Remote Sensing*, **60**(4), 284 – 294.

Sunderhauf, Niko, Konolige, Kurt, Lacroix, Simon, & Protzel, Peter. 2006. Visual Odometry Using Sparse Bundle Adjustment on an Autonomous Outdoor Vehicle. *Pages 157–163 of:* Levi, Paul, Schanz, Michael, Lafrenz, Reinhard, & Avrutin, Viktor (eds), *Autonome Mobile Systeme 2005*. Informatik aktuell. Springer Berlin Heidelberg. 10.1007/3-540-30292-1_20.

Tardif, Jean-Philippe, Pavlidis, Yanis, & Daniilidis, Kostas. 2008 (September). Monocular Visual Odometry in Urban Environments Using an Omnidirectional Camera. *Pages 2531–2538 of: IEEE/RSJ International Conference on Intelligent Robots and Systems.*

Zhu, Zhiwei, Oskiper, T., Samarasekera, S., Kumar, R., & Sawhney, H.S. 2007 (oct.). Ten-fold Improvement in Visual Odometry Using Landmark Matching. *Pages 1 –8 of: Computer Vision, 2007. ICCV 2007. IEEE 11th International Conference on.*

Chapter 2

Decision Support Systems

Functionality of Navigation Decision Supporting System – NAVDEC

P. Wołejsza
Maritime University of Szczecin, Poland

ABSTRACT: The known navigational systems in use and methods of navigational decision support perform information functions and as such are helpful in the process of safe conduct of a vessel. However, none of these known systems provides a navigator with ready solutions of collision situations taking account of all the vessels in the proximity of own ship, where the Collision Regulations apply. Another shortcoming of these systems is that they do not explain the assessment of a navigational situation and proposed manoeuvre parameters. This paper presents functionality of NAVDEC – new Navigational Decision Supporting System both for ocean going ships and pleasure crafts.

1 INTRODUCTION

At present there are no requirements obliging sea-going vessels to be equipped with a decision support system that would assist navigators in collision situations. Consequently, vessels do not carry such systems. Besides, manufacturers of navigational equipment and specialized software are not interested in developing and implementing decision support systems as shipowners show no demand for them. Unfortunately, most shipping companies share an opinion that if a device or software is not required by law, it will not be purchased.

In this connection, it seems purposeful to launch a widespread campaign aimed at decision makers promoting mandatory installation of a navigational decision support system. The navigator able to use a system that correctly qualifies a situation in compliance with the COLREGs and submits possible solutions would not make errors as was in cases presented in papers (Magaj & Wołejsza 2007, Banachowicz & Wołejsza 2007, Magaj & Wołejsza 2010). It goes without saying that the implementation of such systems would enhance the safety of navigation.

2 STATISTICS

According to the reports from the States in Baltic region there were 124 ship accidents in the HELCOM area in 2010 (Figure 1), which is 19 more than the year before (increase of 18%) and 11 less than in 2008 (decrease of 8%).

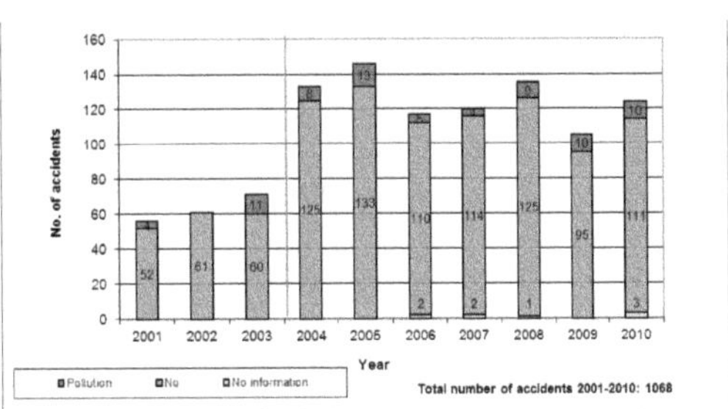

Figure 1. Accidents in Baltic region in the period 2001-2010[helcom]

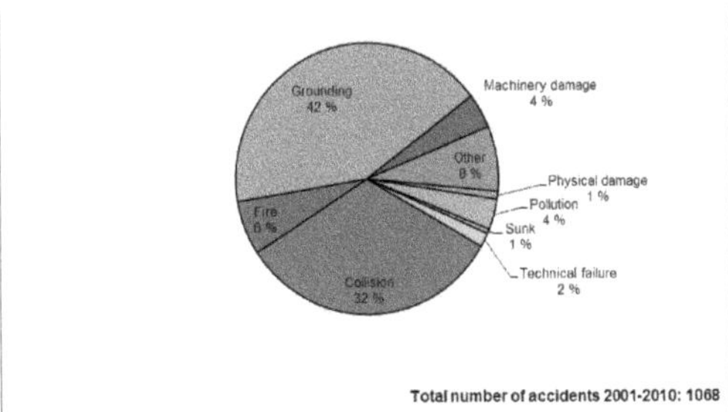

Figure 2. Types of accidents in Baltic region in the period 2001-2010[helcom]

The share of collision accidents (32%) in 2010 equals their share in the total number of accidents during 2001-2010 (Figure 2). The share of groundings (29%) was significantly lower in 2010 compared to their share of the past 10-year period (42%).

Amounting to 40 cases (32%) of all accidents; collisions were the most frequent type of shipping accidents in the Baltic in 2010. This was the first time since 2006 that collisions were more common than groundings in the Baltic Sea. The number of reported collisions has been decreasing since 2005-2006 but increased by 18% in 2010 from the lowest reported number of collisions in 2009 (34 collisions), now equaling the number of collisions in 2007 (Figure 3).

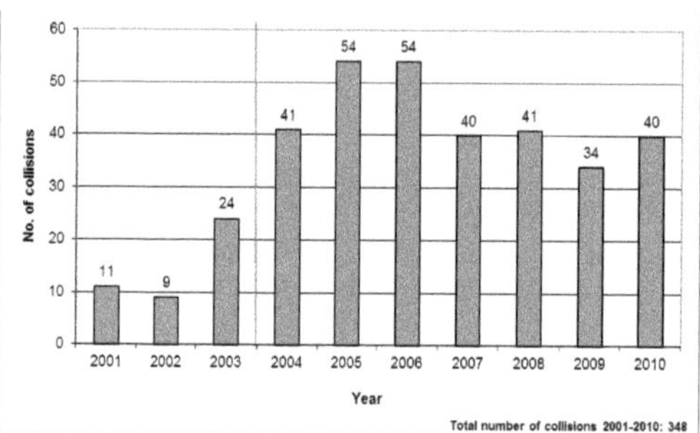

Figure 3. Collisions in Baltic region in the period 2001-2010[helcom]

Ship to ship collisions accounted for 50% of all collision cases in 2010 and the rest of the cases were collisions with fixed and/or floating structures, e.g. peers, navigation signs etc. The number of ship to ship collisions in 2010 was higher than in the last three years but still 30% less than in 2005-2006. The number of collisions with objects has remained largely unchanged in previous years but decreased by roughly 20% in 2010 compared to 2005-2009 (Figure 4).

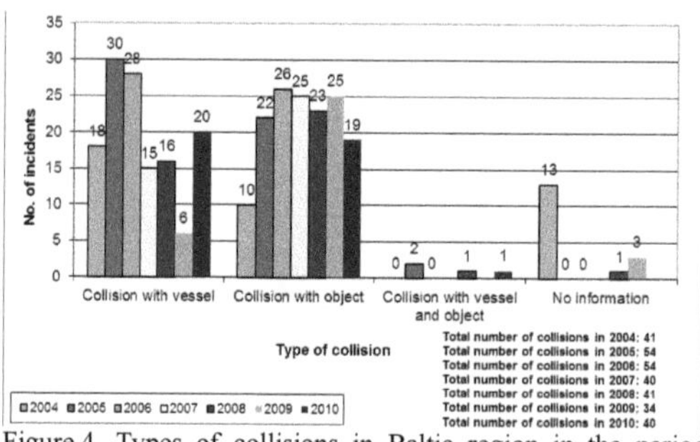

Figure 4. Types of collisions in Baltic region in the period 2004-2010[helcom]

On the picture below there are statistical information from insurance company The Swedish Club.

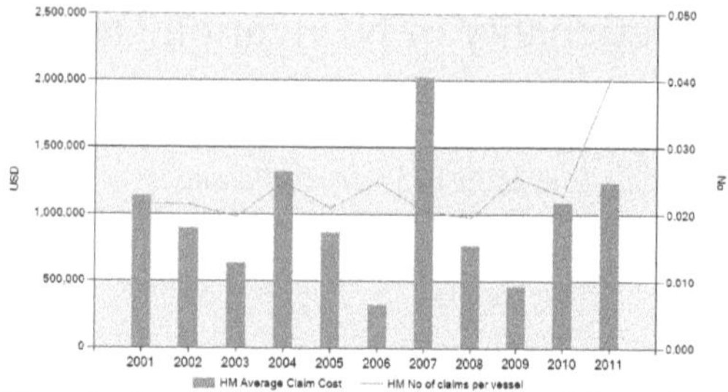

Figure 5. Average claim cost & frequency 2001 — 2011, limit >= USD 10 000 [Swedish]

According data presented above, average cost of collision is more than 1,000,000 USD. The Swedish Club shares 13.6% (2010) of hull and machinery insurance global market. According to [Equasis] there were 77,768 merchant vessels in 2010. According Figure 5, around 2.5% of vessels are in collision every year i.e. over 1,900. In this situation the total cost of collisions is around 2,000,000,000 USD per year.

3 FUNCTIONALITY OF NAVDEC

Presently systems used on board vessels are information systems only. NAVDEC is not only information system. It's also decision support system.

Figure 6 presents a reconstructed situation of collision situation between Gotland Carolina (own vessel) and Conti Harmony (target) at 0900 hrs. The range of courses that assure safe passing at the preset CPA or larger is marked yellow on the circle. The recommended manoeuvre is indicated as 'NEW COURSE' and enables the ships to pass each other at the assumed CPA. The speed range satisfying the assumed criteria is marked green, and proceeding at 'NEW SPEED' will result in the ships' distance during passing being equal the assumed CPA. At operator's request, the system can display the recommended trajectory based on the generated solutions and the next waypoint (Figure 7). Moreover NAVDEC presents status of both ships according to COLREG. Such information would be very helpful for inexperienced duty officer from *Conti Harmony* and for sure, enables to solve collision situation.

The third officer of *Conti Harmony* had signed on one day before and it was his first **independent** bridge watch in his life. The third officer on the *Gotland Carolina* had held this position in the company for 3 years and despite sea experience was not able to qualify the encounter situation correctly.

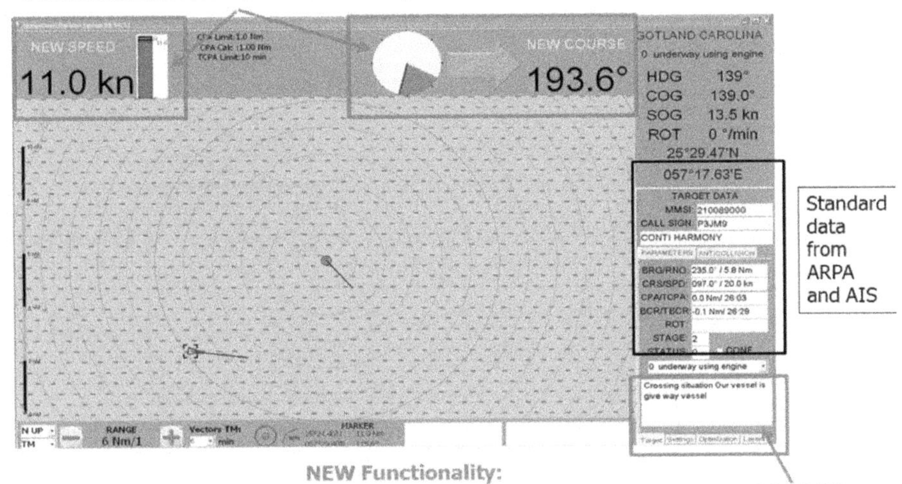

Figure 6. New functionality of NAVDEC

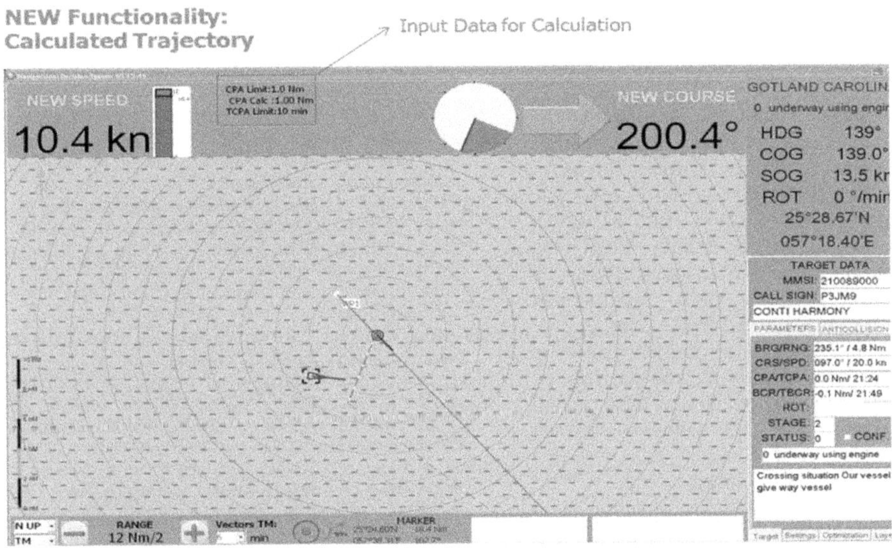

Figure 7. New functionality of NAVDEC

He presumably made a mistake other navigators happen to make. It is authors' opinion that the navigator on board the *Gotland Carolina* had probably come to a conclusion that if a faster ship than his was located below his beam, then it was a case of overtaking (Rule 13).

Consequently, he took no preventive action as prescribed by Rule 17 of COLREGs. The *Conti Harmony* was in fact faster and was approaching the *Gotland Carolina* from behind her beam (right course angle 097^0), but the regulations specify that the limit between overtaking and crossing courses is set up on course angle 112.5^0, a fact navigators neglect only too often.

The watch officer on the *Gotland Carolina* correctly qualified the situation and in the first stage of the encounter followed Rule 17. However, he did not take advantage of the possibility provided by paragraph a) ii), and the most importantly, he did not take action as prescribed by paragraph b) of the mentioned rule. What is most shocking in the event: neither of the vessel took any preventive action till the very moment of collision!!!

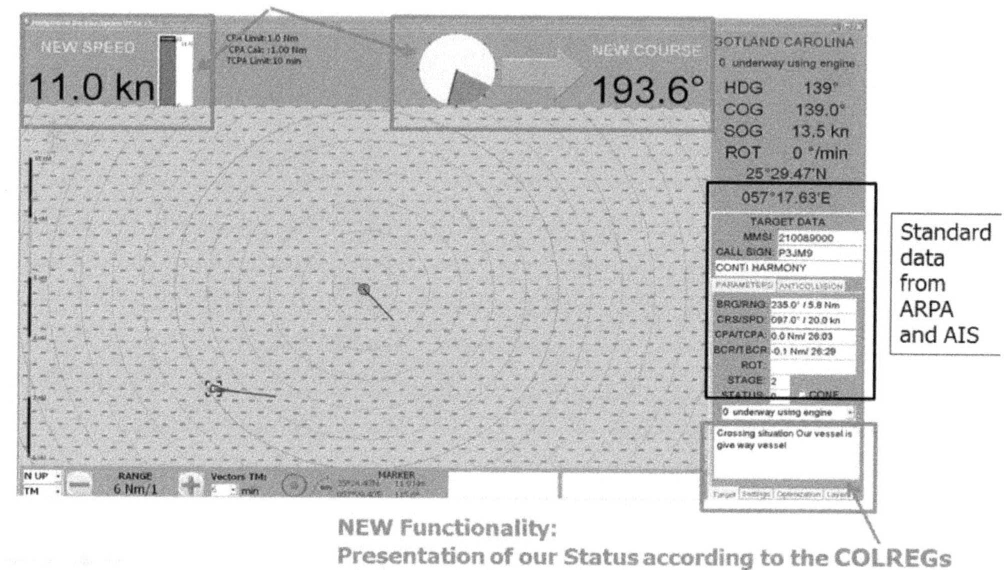

Figure 8. New functionality of NAVDEC

4 CONCLUSIONS

NAVDEC gradually increases functionality of existing navigational systems. First of all it qualifies encounter situations according COLREG. Navigator receives notification if she is stand-on or give way vessel and from which rule it comes from. Moreover system gives ready solution of collision situation i.e. save courses and speeds which enable to pass with other targets on assumed CPA. Additionally suggested trajectory is presented on the chart.
If mandatory installation of a navigational decision support system reduces number of collision only by 1%, total savings, only for insurers, will be around 20,000,000 USD per year. The collision between Gotland Carolina and Conti Harmony is a proof that this percentage will be much higher.

NAVDEC also solves collision situation with more than one target, which is presented on figure 8. Rosette sette shown on top of the screen presents save (yellow) and dangerous (red) courses in relation to all targets.

REFERENCES

Banachowicz & Wołejsza "The analysis of possibilities how the collision between m/v Gdynia and m/v Fu Shan Hai could have been avoided", Advances in marine navigation and safety of sea transportation, 2007.

Pietrzykowski Z., Uriasz J.: Knowledge Representation in a Ship's Navigational Decision Support System. TransNav - International Journal on Marine Navigation and Safety of Sea Transportation, Vol. 4, No. 3, pp. 265-270, 2010

Equasis Statistics, The world merchant fleet 2010, www.equasis.org

Helsinki Commission, "Report on shipping accidents in the Baltic Sea area during 2010, www.helcom.fi

Magaj & Wołejsza "Using AIS information for anticollision manoeuvre examplified by the collision between m/v Ziemia Łódzka and m/v Vertigo", Polish Journal of Environmental Studies Vol.16, No. 6B, 2007.

Magaj & Wołejsza, "Analysis of possible avoidance of the collision between m/v Gotland Carolina and m/v Conti Harmony", Annual of Navigation No 16, pp 165-172, Gdynia 2010.

The Swedish Club, "Collisions and groundings 2011", www.swedishclub.com

The Swedish Club, "Basic facts 2012", www.swedishclub.com

Pietrzykowski & Borkowski & Wołejsza, Maritime Intelligent Transport Systems, Communication in Computer and Information Sciences, Telematics in the Transport Environment, Springer Verlag Berlin Heidelberg, pp. 284-292, 2012.

Pietrzykowski & Borkowski & Wołejsza & Uriasz, Marine navigational decision support system, International symposium Information on Ships ISIS, Hamburg, 2012, CD, ISSN 2191-8392

A Study on the Development of Navigation Visual Supporting System and its Sea Trial Test

N. Im, E.K. Kim, S.H. Han & J.S. Jeong
Mokpo National Maritime University, Mokpo, South Korea

ABSTRACT: This paper shows the developed navigation visual supporting system that will help watch officers recognize small floating objects such as sea marks on the sea more easily. Sea trial tests were carried out to verify its performance on the sea. The system consists of a composite video sensor for video signal, a laser distance measuring unit, a Pan/Tilt unit, and a central control unit. The video signal and the laser distance measuring unit were installed inside of the Pan/Tilt unit. The developed system was mounted in a training ship for a sea trail test to evaluate the recognition ability for buoyage and lighthouses. The image from the system was compared with that of a binocular telescope. It was found that the image from the system was more clear and recognizable than that of a binocular telescope within a 5 km radius. The developed system could be used to recognize small buoys or dangerous floating objects in a harbor area more easily and quickly.

1 INTRODUCTION

Marine traffic near the Korean sea coast has been increased by economic growth. The increase of marine traffic has caused congestion and has been one of the causes of marine accident [1]. According to the marine accident statistical data from the Korea Coast Guard, the number of marine accidents has increased every year since 2004 showing that 4,172 ships encountered marine accidents between 2004 and 2008. The annual average number of marine accidents in the past five years was 834 [2]. The most frequent marine accident was ship collisions caused by the failure of radar detecting. Human error was a large part of the reason for collision accidents. Therefore, we can easily find related research in order to prevent ship collision [3].

Recently, sea pirate attacks and armed robbery against ships have increased considerably, according to the International Maritime Bureau (IMB) [4]. According to a report from The United Nations International Maritime Organization (IMO) and the IMB, 293 sea pirate attacks were reported and of those 49 ships were kidnapped [5]. The IMO recommends the installation of marine surveillance equipment (Day/Night Vision) to secure navigation safety from pirate attacks and ship collisions [6]. In addition, the United States enacted the ISPS code after the 911 terror attacks and required that ships be equipped with the security and surveillance system [7][8]. Consequently, shipping companies started to install the marine surveillance equipment and the related industrial market is expected to expand gradually.

Nowadays, many studies are carried out on navigation visual supporting systems. A study used the data of RADAR ARPA for development of navigation observation system [9]. According to the E-Navigaton strategic plan of IMO, the importance of maritime situation awareness for safe navigation is emphasized [10]. However the reality is that the major mean of maritime situation awareness is binocular telescopes with the help of ARPA and AIS.

This study describes the development of navigation visual supporting systems that can improve recognition ability for sea marks. Sea trial tests were carried out to verify the performance. It was found that the images from the system were more clear and recognizable than that of a binocular telescope within a 5 km radius.

2 CONFUGURATION AND FUNCTION

Radar has been used as one of the essential tools of navigation equipment. However it has trouble recognizing small floating objects on the sea. In case of a binocular telescope, it has limitations of detecting and recognizing floating objects several nautical miles away.

In this study, the navigation visual supporting system was developed to make up for the weakness of existing navigation equipment. Since visual information such as a video image is provided on a monitor, navigators can easily recognize small dangerous objects such as marine marks. The purpose of the system is to improve the recognition ability for maritime situation awareness, one of the important factors for navigation safety on the sea.

The system provides visual information about dangerous floating objects around a ship using a laser distance measuring technique and video signals from composite video sensors. The system consists of a composite video sensor for video signals, a laser distance measuring unit, a Pan/Tilt unit, and a central control unit. The configuration is shown in figure. 1. The operations of the joy stick in the central control unit give commands to each unit through the controller in the Pan/Tilt unit. Each unit carries out the order according to the signals from the controller of the Pan/Tilt unit.

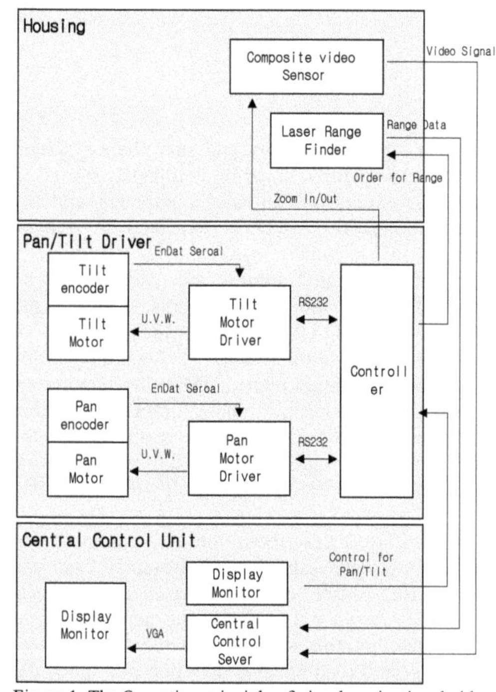

Figure 1. The Operating principle of visual navigational aid

At this time, the collected data is transferred to the central control unit and the related information such as a target image and the distance between the ship and the target are displayed on the monitor. The particulars of navigation visual supporting system are shown in table 1.

Table 1. Figures of Real time front observation navigation System

Parameter	Unit	Value
Driving range	deg	360
Accuracy of angle	mrad	1mrad
Detection range of rager	m	10,000
Accuracy of distance	m	±5
Power of rager	mJ,ns	4, 30
wavelength of rager	㎛	1.54
type of rager	-	Erglass
weight	kg	1.5
Power loss	W	25
loss information	pixel	1.3M
Optical zoom	-	12
Distance of detection	km	< 5.0

2.1 *Composite video Sensor*

The Image sensor module for navigation equipment basically has the same observation function compared with a general camera CCD. However, it requires durability and waterproofed structure and design since it is operated in the harsh conditions of sea environment. In addition, visual information connected to ship operations should be provided in real time.

In this study, the composite image sensor is used for the front monitoring and gets the image from camera CCD and LLCCD. With a telephoto lens, magnification can be adjusted appropriately. As shown in figure 2, the image sensor for the front monitoring consists of five main parts. The specifications of each part are explained in Table 2.

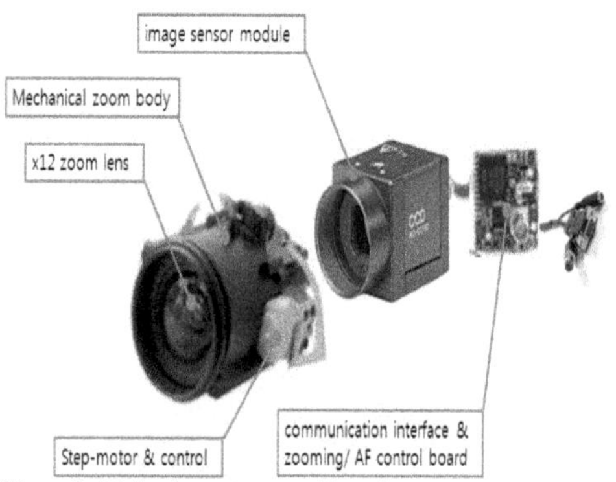

Figure 2. Main element parts of imaging sensor

Table 2. Features of complex imaging sensor

Parameter	Value	Unit
Image Format	½	inch
Focal Length	8~96	mm
Iris	F1.6~1000	Video
Mount Type	C	-
Temp. Range	-10 ~ +50	℃
Size	130×77×87	mm
Weight	1.2	kg
Power	12	VDC
External communication	UART, 12C, SPI	

2.2 *Laser distance measuring unit*

The laser distance measuring unitprovides the distance information between targets and a ship. The bearing of the targets can be easily obtained bythe Pan/Tilt unit. However, it is necessary to get the distance information from the laser distance measuring unit in order to fix the target's position on a sea chart.

The laser distance measuring unit calculates the distance between a target and a ship by measuring the time of flight for a radiated laser. It uses a high output and small pulse laser because it is safer for human eyes. The distance measured by the laser distance measuring unit is more accurate compared to that of existing Radar systems. It has a distance error of less than several meters. Besides, this unit has the advantage of being compact and light.

In this study, the laser distance measuring unit detected and recognized small floating objects and ships within 10km. The specification of the laser distance measuring unit is explained in table 3.

Table 3. Features of laser range finder

Parameter	Value	Unit
Wavelength	1.54	μm
Supply power	5	mJ
Pulse width	30	nsec
Beam divergence angle	0.5	mrad
Repetition ratio	2 (Max.)	Hz
Size	155×110×225	mm
Weight	3.5	kg

2.3 *Pan/Tilt unit*

The navigation visual supporting system is required to detect dangerous objects from every direction.

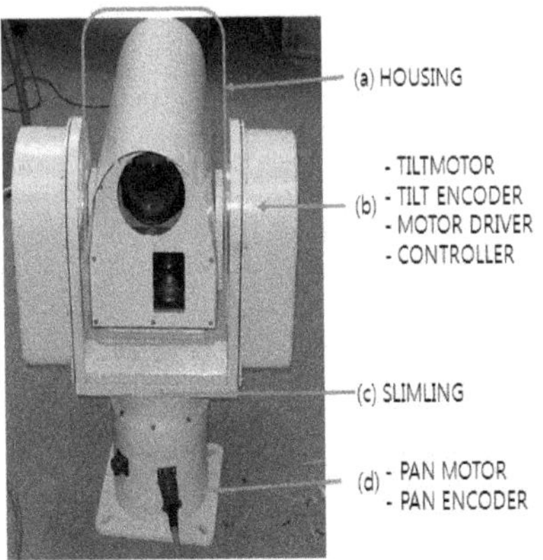

Figure 3. Body of Pan/Tilt

The Pan/Tilt unit enables the system to move vertically and to rotate a full 360-degree. The unit consists of an observation window, controlling algorithms, a drive, and a special housing that protects the system in the harsh conditions of a sea environment. Figure 3 shows the main body of the developed Pan/Tilt unit. The image signal sensor and the laser distance measuring unit are mounted inside the housing. A type of slip ring was adopted in the rotary Pan to secure rotary motions. The specification of the Pan/Tilt unit is shown in Table 4.

Table 4. Features of Pan/Tilt

Parameter	Value	Unit
Acuator	DD(Direct drive) SERVO SYSTEM-	
Supply power	85V AC ~ 265V AC	V AC
Action speed	PAN : 0.1~90 TILT : 0.1~40	°/sec
Driving range	PAN : 360 Endless / TILT : -30~+30	°
Temp. Range	-40 ~ +50	°c
Size	530(L) X 280(W) X 390(H)	mm
Weight	60	Kg
Texture	SUS316, AL6061	-
Optical device	f:10~500mm vidual zoom camera	-
Type of Control Signaling	RS-422	-
Signaling speed	9600bps	-
Type of protocol	PELCO-D	-
Software	GUI Software	-

2.4 *Central control unit*

The central control unit receives distance information and bearing data from the laser distance measuring unit and the composite image sensor unit respectively. The information is displayed on the monitor of the central control unit. The monitor also displays the visual image detected by the composite image sensor. Figure 4 shows the central control unit that consists of a monitor, a joystick controller, a control server, and a rack. Each part is accommodated in the rack.

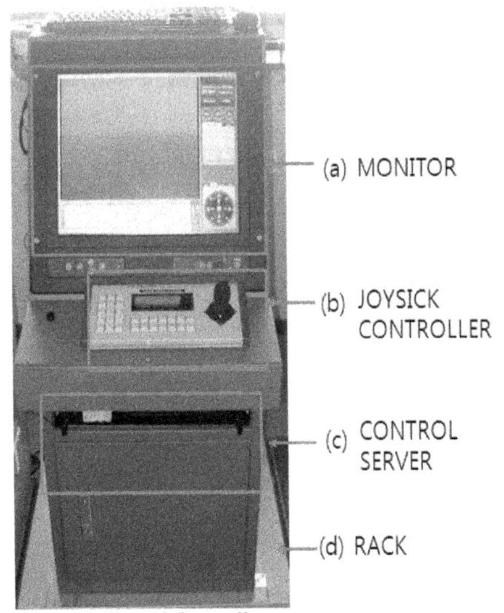

Figure 4. Central Controller

2.4.1 *Monitor*

A marine type monitor is adopted in order to identify the screen data more easily in navigation environments. The monitor display consists of a main screen and menu tools that enable users to zoom in and out of the main screen. Figure 5 shows the picture of the monitor and its specification is shown in table. 5.

Figure 5. The monitor of Central Controller

Table 5. Features of the monitor of central controller

Parameter	Value	Unit
Size	20.1	inch
Active Area	408*306	mm
Resolution	1600 * 1200	dpi
Contrast	300:1	-
Light	250	Cd/m
Field of View	85	o

2.4.2 *Keyboard*

The keyboard of this system has multi-function controls since it is specifically designed to control a CCTV system. A receiver can control the Pan/Tilt motions, its power systems and the laser distance measuring unit. The status of the system is displayed on the LCD monitor. Figure 6 shows the details of the keyboard and its specification is shown in table 6.

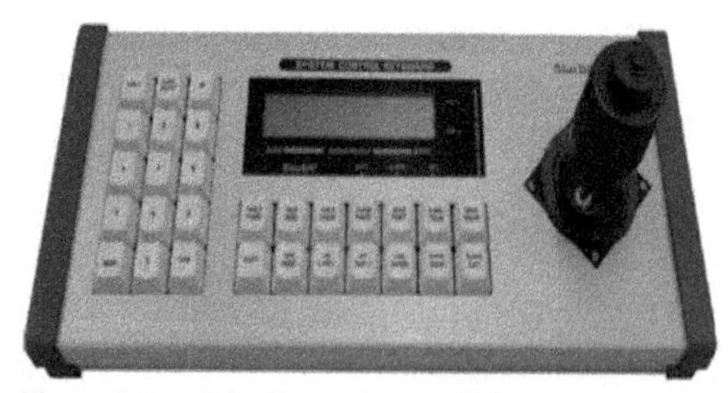

Figure 6. Joystick of central controller

Table 6. Features of the Joystick of central controller

Parameter	Value
Type of Control Signaling	RS232 , RS485 , RS422
Preset	256
Step of Control Speed	10
Supply power	12

2.4.3 *Control sever*

The control server receives an analog image signal from the composite image sensor. The analog image is saved in a built-in HDD using the compression algorithm of the H. 262. The control server displays the saved image in the monitor in real time. The saved image can be replayed and searched by the built-in software. The control software displays split screens and provides several functions like the control of Pan/Tilt through the GUI, the logging of status data, the adjusting of a search range, and the display of a laser distance. The specification is shown in table 7.

Table 7. Features of the control sever

Parameter	Values
Type of Processor	Intel CoreTM 2 duo
HDD	2.5" 160
Embedded Cache	256KB ′ 4, L2
processor FSB	1333
Main Memory	2 x 240-pin DDR2 667 to 2.0GB.
Onbord Lan Features	Two Realtek RTL8111C PCI Express Gigabit controllers
Audio Features	Realtek ALC662 High Definition audio CODEC

3 SEA TRIAL TEST

Sea trial test were carried out in order to verify the performance of the system. One of the training ships of Mokpo maritime university was used for the sea test in the area between Mokpo and Kwang-Yang.

3.1 *The installation of the system on board*

The Pan/Tilt was mounted on the campus deck, the highest floor on the training ship, in order to easily detect targets from a high position. The central control unit was installed on the bridge deck where navigation officers watch the image on the system's monitor. Figures 7 and 8 show the Pan/Tilt unit on the campus deck and the central control unit on the bridge deck.

Figure 7. The Pan-tilt on compass deck

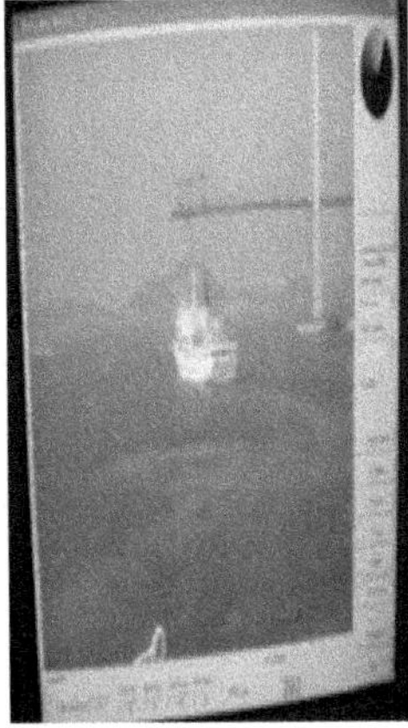

Figure 8. The central controller in the bridge

3.2 *The result of sea trial tests*

The sea trial tests were carried out on the condition of the weather and the sea state as shown in Table. 8.

First, lighthouses and sea marks were designated as targets of which distances were measured by radar and a sea chart. Then, these targets were detected by the system and their images were displayed on the monitor in order to verify the system's recognition ability. Finally, the images were compared with that of a binocular telescope and naked eyes. The details of targets are shown in table. 9. The detected images from the system are shown in figures 9 -12.

Target A in table 9 was a lighthouse 5200m away from the sea trial ship. The target was dimly visible but not clearly recognizable by a binocular telescope and naked eyes at that time. However it was found that the system provided the image as shown in figure 9 where the target was more recognizable. Target B in table 9 was a small lighthouse which was not visible with naked eyes at the distance of 3700 meters. A vague figure was just seen even by a binocular telescope. However, the system showed more a clear and recognizable image as shown in figure 10. The image provided by the system made it possible to recognize the target as a lighthouse.

Target C and D in table 9 were small floating sea marks. It was relatively difficult to recognize these targets with naked eyes or a binocular telescope. Figure 10 shows the image of the target C obtained by the system. It was possible to recognize the target at the distance of 3300 meters. The image of Target D is shown in figure 11. It was easily recognizable by the system at the distance of 5370 meters. The results of the sea trial tests indicated that the images from the system were more clear and recognizable than that of a binocular telescope within a 5 km radius. The system detected and recognized the targets more effectively and easily compared to a binocular telescope or naked eyes.

These tests indicated that the developed system could be used to recognize small buoys or dangerous floating objects in a harbor area more easily and quickly

Table 8. The state of sea

Items	Dimensions
Sea Condition	Beaufort Scale 3
Wind Seed	5 knot
Visibility Scale	4
Sea Wave	0.6 m

Table 9. The features of targets

	Target-A	Target-B	Target-C	Target-D
Type	Light house		Lighted Buoy	
Distance(m)	5200	3700	3300	5370
Hight(m)	11	9.3	5.6	5.6
Breath	4-5	2.3	2.0	2.0

Figure 9. The snapshot of target A

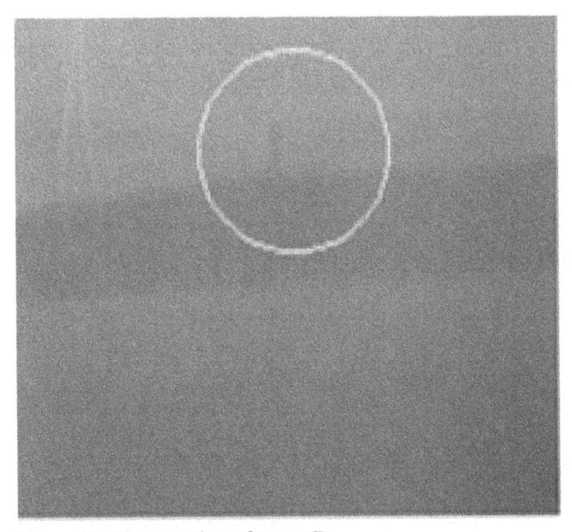

Figure 10. The snapshot of target B

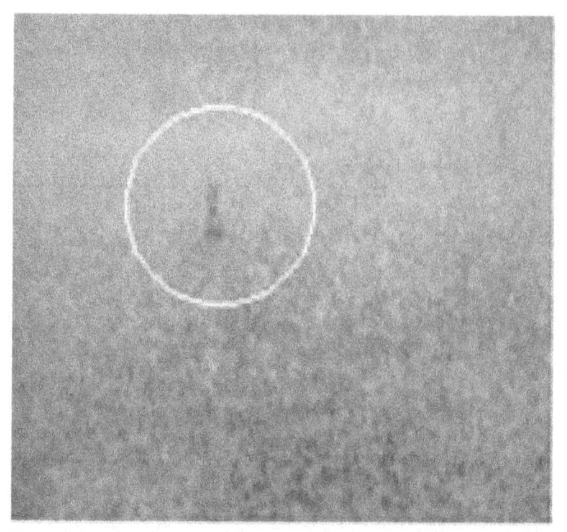

Figure 11. The snapshot of target C

Figure 12. The snapshot of target D

4 CONCLUSIONS AND FUTURE WORKS

This study described the developed navigation visual supporting system which will improve maritime situation awareness. The system has the merit of showing targets in a monitor visually and recognizing small targets in real time. It could be used to make up for the weakness of existing Radar, since Radar can't provide any information about the recognition of targets.

The developed system was mounted in a training ship for sea trail tests to evaluate the recognition ability for buoyage and lighthouses. The images from the system were compared with that of a binocular telescope. It was found that the images from the system were more clear and recognizable than that of a binocular telescope within a 5 km radius. The developed system could be used to recognize small buoys or dangerous floating objects in a harbor area more easily and quickly.

ACKNOWLEDGEMENT

This research was a part of the project titled "Development of the Intelligent Technology for Marine Accident Protection & Salvage" funded by the Ministry of Land, Transport and Maritime Affairs, Korea.

REFERENCES

[1] Woon-Jae Jang, Jong-SooKeum(2004), An Analysis on the Models of Occurrence Probability of Marine Casualties, Vol.10, No. 2, pp.29-34.

[2] The Korea Coust Guard (2010), Marine accident annul data.

[3] Ji-Hyun Yoon, Seung-Keon Lee, Namkyun Im(2005), A Study on Automatic Control for Collision Avoidance of a Ship under Appearance of Multi-vessels, Journal of Korean Navigation and Port Research, Vol.29, No.1, pp.29-34.

[4] IMB(2003), Piracy and Armed Robbery Against Ships, Annual Report

[5] IMO Annual report(2009), REPORT ON ACTS OF PIRACY AND ARMED ROBBERY AGAINST SHIPS, MSC.4/Circ.152, 29 March 2010, pp.1-40.

[6] IMO Resolution MSC. 94(72), Performance standards for night vision equipment for high speed craft(HSC).

[7] Suk-Yoon Choi(2006), Countermeasures against Maritime Terrorism and Criminal Law, Journal of Korea Institute of Maritime Law, Vol.18, No.2, p.29.

[8] Han-Yong Kwen (2010), A study on International Law research to prevent sea error attacks and sea pirates, Journal of Dong-A Law, Vol.48, pp.895-926.

[9] KenjiroHikida, Nobuo Mitomo, JunjiFukuto, Kenji Yoshimura(2009), Development of a support system for recognition and operation of RADAR/ARPA information with Head-up display, Japan Institute of Navigation, Vol 121, pp.7-12.

[10] IMO Report of the Working Group(2010), DEVELOPMENT OF AN E-NAVIGATION STRATEGY IMPLEMENTATION PLAN, NAV 56/WP.5/Rev.1, 28 July 2010, pp. 1-65.

Application of Ant Colony Optimization in Ship's Navigational Decision Support System

A. Lazarowska
Gdynia Maritime University, Gdynia, Poland

ABSTRACT: The aim of navigational decision support system is to aid in the process of conducting navigation. Determination of global optimal route between harbours, monitoring of current navigational situation and determination of safe manoeuvre or trajectory of own ship in collision situation constitute the tasks of the system. The paper introduces an innovative approach of ant colony optimization algorithm implementation to determine safe optimal route of passage between harbours. Static and dynamic constraints such as land, channels, shallows as well as encountered ships are taken into account. Assumptions of the system, description of the algorithm and recent results are presented.

1 INTRODUCTION

The function of anti-collision system onboard a ship is currently accomplished by Automatic Radar Plotting Aid (ARPA). According to International Maritime Organization (IMO), all ships of 10 000 gross tonnage and over constructed on and after 1 July 2002 have to be equipped with Automatic Radar Plotting Aid. These requirements are described in Safety of Life at Sea Convention (SOLAS), enacted by IMO. ARPA system enables automatic tracking of at least 20 targets. It also generates dangerous target alarm, which indicates that the computed values of Time to the Closest Point of Approach (TCPA) and Distance of the Closest Point of Approach (DCPA) exceed the specified safe limits. A collision avoidance support function performed by the ARPA system is called the trial manoeuvre. This function enables the deck officer to check the effects of planned manoeuvre. It is performed by the simulation of course change or speed change of own vessel.

Despite the use of many electronic devices in the process of maritime navigation, ship accidents caused by human error still occur. Rothblum (2000) in her report states that between 75 to 96% of marine casualties are caused at least partly by a human error. Martinez de Oses et al. (2003) point out that 74 % of sea accidents between years 1981 and 1992 were caused by human element, 45 % of which were classified as misjudgement and 23% as lack of attention. This implies the need for decision support system application. Applied system should not only analyze data and calculate the collision risk, but should also propose solutions.

The paper presents the system developed to aid the deck officer in the process of conducting navigation. Maritime navigation includes tasks (Jurdziński, 2003) such as:

- planning of safe, optimal route of passage between harbours;
- monitoring of current navigational situation;
- detection of collision risk;
- correction of the route of passage in order to achieve safe situation.

Presented system fulfills the functions mentioned above.

Application of the system would contribute to increase maritime safety and also reduce operating costs of the ship.

2 ASSUMPTIONS AND SYSTEM DESIGN

Proposal of navigational decision support system application does not constitute a new approach. Harris et al. proposed a collision avoidance system called Maritime Avoidance Navigation, Totally Integrated System (MANTIS), which is composed of the Localization of Vessel States and its Environment (LVSE), Automatic Collision Avoidance Advisory Service (ACAAS), an Integrated Display System (IDS), Path Planning and Scheduling Service (PPSS) and Automated Ship Guidance and Control

(ASGC). Śmierzchalski et al. (2008) presented Intelligent Ship Control System (ISCS) , where evolutionary algorithms are used to determine safe, optimal path of the ship and fuzzy controller to steer the ship along the determined path.

Pietrzykowski et al. (2008, 2010, 2011) introduced model of navigational decision support system with the use of Unified Modeling Language (UML). Designed system provides following functions:

- monitoring, analysis and assessment of navigational situation;
- activating alarm in dangerous situations;
- presenting current level of navigational safety;
- solving collision situations;
- clear presentation of current navigational situation
- interaction with the navigator via the graphical user interface.

Proposed system has been tested on board the research/training ship "Nawigator XXI" owned by Maritime University of Szczecin. The tests were performed during 4 days of voyage on the Baltic Sea from 31st of August 2009 to the 4th of September 2009 (Magaj, 2011). The system integrates signals from the following navigational equipment: Automatic Identification System (AIS), ARPA, Global Positioning System (GPS), log and gyrocompass. The system presents current navigational situation in a graphical form. It also displays recommended course alteration.

Researchers interested in the determination of anti-collision manoeuvres introduced various methods. Lisowski (Lisowski, 2007, 2009, 2012) defined safe ship control process with the use of game theory. He proposed a differential game model to be applied for the process of safe steering of a ship in collision situations. Due to high complexity of the presented model, resulting from non-linear and multidimensional state equations and non-linear and changing in time state and steering constraints, he proposed simplified models with respect to the basic model. In the positional game model the differential game is reduced to the multi-stage game of j non-cooperating players. The vital assumption of positional game is that the strategies of own ship depend on the current position p(tk) of the target ships at current step k. The optimal steering of own ship at each stage is calculated by applying simplex method to solve the linear programming problem. The second model introduced by Lisowski is based on the reduction of differential game into multi-step matrix game. In this approach a collision risk matrix is created including values of collision risk rj determined for acceptable strategies of own ship and acceptable strategies of the target ships. The problem of determining an optimal strategy is also reduced to solving dual linear programming problem. Lisowski introduced also application of dynamic programming to determine ship's safe optimal trajectory. He defined the Bellman's functional equation for the process of ship's control by alteration of the rudder angle and the rotational speed of the propeller. In this approach he proposed also the use of neural networks for the determination of the ship domain, which constitutes an area around the target ship that cannot be transgressed by own ship. Domains in the form of circle, ellipse, parabola and hexagon were proposed by various researchers. Goodwin proposed a circle domain with the use of safe distance parameter Ds, while Śmierzchalski defined a domain in the form of hexagon. Pierzykowski et al. introduced ship fuzzy domain (Pietrzykowski, 2006). Safe distance is the smallest acceptable distance from own ship to the navigational obstacle. Śmierzchalski developed a trajectory planning system based on evolutionary algorithms (Śmierzchalski, 1998). In this method it is assumed that own ship moves from the starting point (xo, yo) to the end point (xe, ye) along a safe path S with a uniform speed. The first stage of the algorithm constitutes random generation of an initial population of paths. After that each path is evaluated. In this procedure a value of fitness function is calculated for each trajectory. This is a measure of the path's adaptation to the navigational environment. Next, selected paths are modified with the use of specialized set of following operators: soft mutation, mutation, adding a gene, swapping gene location, crossing, smoothing, deleting a gene and individual repair. Genetic operators are used to create new offsprings (paths), which are likely to have better qualities in the next generation. The process of multiple creating, evaluating and transforming paths is carried out until a stop test in the form of a defined number of generations is completed. To sum up, this approach takes into account static constraints representing lands, channels, shallows, water lines, as well as dynamic constraints representing target ships. Further development of this method was presented by Szłapczyński (Szłapczyński, 2010) and Śmierzchalski et al. (Śmierzchalski, 2011). Perera et al. introduced fuzzy logic based decision making system for the collision avoidance at sea. The fuzzy rules were developed with respect to the COLREGs. The output of the system consist of collision risk assessment and fuzzy decisions, which are then defuzzified to obtain control actions of vessel course and/or speed change. The results of the designed system were presented for two vessels crossing situation, overtake and heading situation (Perera, 2011). To conclude, various methods have been developed to solve the problem of safe manoeuvres determination. In the designed system ant colony optimization is used for the determination of safe trajectory of the ship with consideration of static and dynamic constraints, which constitutes a new approach to this issue.

Vital assumptions of decision support system at sea presented in this paper were based on approaches introduced by Lisowski (Lisowski, 2012) and Śmierzchalski (Śmierzchalski, 2005).

Assumptions included in the system consist of:
- static constraints representing land modelled as concave and convex polygons;
- dynamic constrains representing met ships in the form of hexagon;
- rectilinear uniform motion of met objects;
- defined start point and end point of anti-collision trajectory;
- optimality criterion in the form of the smallest lost of way on overtaking other ships;
- International Regulations for Preventing Collisions at Sea – COLREGs (Cockroft, 2012) taken into account;
- lead time and manoeuvre time taken into account;
- large alteration of ship course from 10° to 90°;
- information about current navigational situation obtained from ARPA system (Lisowski, in press);
- real time system.

Figure 1 presents the system design (Lazarowska, 2012).

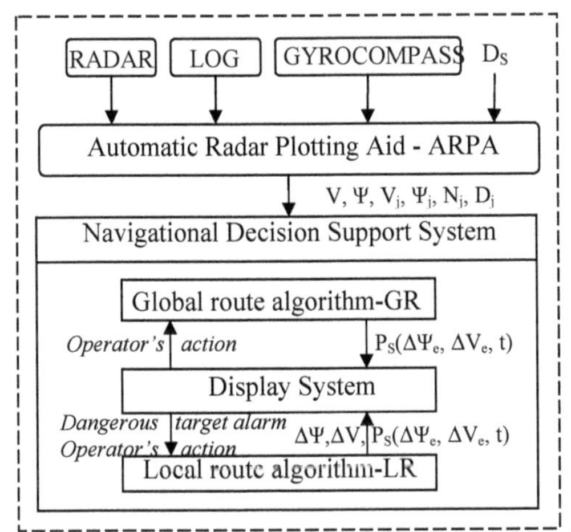

Figure 1. Block diagram of decision support system (Ds -safe distance, Ψ – own ship course, V – own ship speed, Ψj – tracked target course, Vj – tracked target speed, Nj – tracked target bearing from own ship, Dj – tracked target distance from own ship, ΔΨ- manoeuvre of course change, ΔV- manoeuvre of speed change, Ps(ΔΨe, ΔVe, t) - safe ship trajectory)

3 ANT COLONY OPTIMIZATION PRINCIPLE

Ants, while moving between food source and their nest, deposit some chemicals on the ground. These chemicals are called pheromones. Other ants choose the path marked by stronger pheromone concentrations. This trail-laying and trail-following behaviour of ants, which constitutes a form of indirect communication is called stigmergy. It has also been developed that ants behaving in this way, find the shortest path between food source and their nest. Ant colonies foraging behaviour inspired Dorigo, Colorni and Maniezzo (Dorigo, 2011). In the early 1990's they introduced first Ant Colony Optimization algorithm based on indirect communication of a colony of simple agents called artificial ants to solve the travelling salesman problem. The goal of this path optimization problem it to find a minimal length closed route connecting j given cities, with the assumption that each city can be visited only once.

Problem representation with the use of ACO.

A minimization problem (S, f, Ω) is considered, where:
S is the set of candidate solutions,
f is the objective,
Ω is the set of constraints.

An objective function cost value $f(s, t)$ is assigned by the objective function to each candidate solution $s \in S$.

The aim is to find a globally optimal solution $s_{opt} \in S$, it means a minimum cost solution that satisfies the constraints Ω.

Problem description exploited by ants include a finite set $C=\{c_1, c_2, \dots, c_N\}$ of components.

The problem states constitute sequences $x=\langle c_i, c_j, \dots, c_k, \dots \rangle$ over the elements of C.

X is a set of all possible sequences.

The length of a sequence x, that is the number of components in the sequence x, is expressed by $|x|$.

The finite set of constraints Ω defines the set of feasible states $\widetilde{X}$, with $\widetilde{X} \subseteq X$.

The set of feasible solutions is denoted by S^* (with $S^* \subseteq \widetilde{X}$ and $S^* \subseteq S$.

It spite of above-mentioned definition of the problem, it has to be considered, whether to implement constraints Ω in a hard way allowing ants to build only feasible solutions or in a soft way, where ants can also build infeasible solutions, which will be penalized then, in dependence of their degree of infeasibility.

The problem definition can be described as a graph (N,E), where the nodes N are defined by the cities and the connections between the cities are presented as the edges E of the graph.

During every iteration of the algorithm, each ant k (k=1,...,m) moves on the problem graph from one city to another, until it completes the tour. Ant's next move is based on a probabilistic transition rule expressed by:

$$p_{ij}^{k}(t) = \frac{[\tau_{ij}(t)]^{\alpha} \cdot [\eta_{ij}]^{\beta}}{\sum\limits_{l \in N_k^i} [\tau_{il}(t)]^{\alpha} \cdot [\eta_{il}]^{\beta}} \quad (1)$$

where $\tau_{ij}(t)$ is the amount of virtual pheromone trail on the edge connecting city i to city j, which reflects the experience acquired by ants during problem solving, η_{ij} is a heuristic information called visibility. Parameters α and β determine the relative influence of pheromone trail and heuristic information and N_i^k is the set of cities which ant k has not yet visited, when it is in city i. When all the ants construct their solutions at present iteration, the pheromone trails are updated according to the expression given by:

$$\tau_{ij}(t+1) = (1-\rho) \cdot \tau_{ij}(t) + \sum_{k=1}^{m} \Delta \tau_{ij}^{k}(t) \quad (2)$$

where ρ is the pheromone trail evaporation rate, which enables the algorithm to "forget" bad decisions previously made by the ants, $\Delta\tau_{ij}^{k}(t)$ is the amount of pheromone ant k deposits on arcs belonging to its tour (Dorigo, 2010).

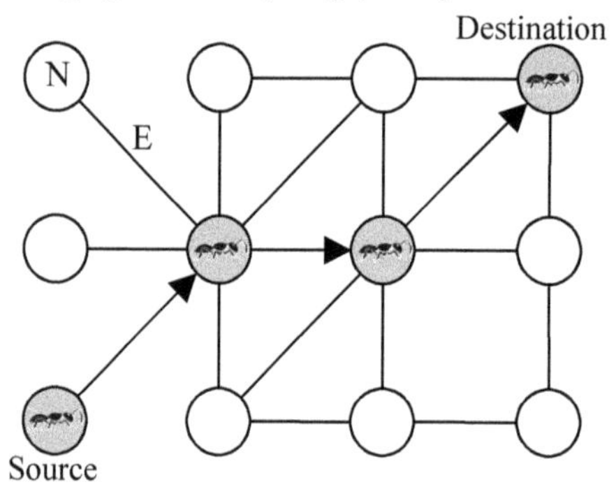

Figure 2. Problem representation in the form of a graph with nodes N and edges E.

4 GLOBAL ROUTE

Global path planning takes into account information about static environment. All of the constraints that does not change in time are considered as static constraints, which include land, channels, shallows, fairways. The lands are modelled as a set of points which form convex and concave polygons. The function of this subsystem is the determination of ship's safe, optimal route of passage between harbours, which means safe trajectory from the departure harbour defined by coordinates (xo, yo) to the destination harbour defined by coordinates (xe, ye). Global route determination is based on modified Ant Colony Optimization algorithm in order to adjust it to considered issue. It means that the algorithm uses Ant Colony Optimization operation principle, but none of developed types of ant algorithms is directly applied. Developed algorithm constitutes a modified version of ACO that uses some elements of ant-cycle type Ant System and part of Ant Colony System concept. Similarly to approach applied in ant-cycle type Ant System (AS) pheromone update is done after the ant had constructed its tour and the amount of pheromone deposited is set to be a function of the tour quality. In ant-density and ant-quality type the ants update a pheromone directly after a move. Sequential implementation of tour construction is applied, that means that an ant has built a complete tour before the next one starts to build another one. Update of pheromone trails consists of two stages. At first pheromone evaporation takes place. It is accomplished by lowering the pheromone value of all arcs by a constant factor. Then deposit of pheromone on the arcs that the ant has crossed in its tour occurs. The amount of deposited pheromone depends on the length of the tour. Ant moves from one node to another according to pseudorandom proportional rule as it is presented in Ant Colony System concept. Constraints are implemented in the way enabling the ants to build only feasible solutions. Ship's path planning problem is represented by a graph (N,E), where nodes N are placed in acceptable waypoints and edges E connect these nodes, for which passing of a ship is possible without transgressing constraints. Solution contains coordinates of the waypoints and the distance of each line segment of the path. Course at each line segment of the path is also calculated. Global route algorithm in the form of pseudocode and an example of obtained results are placed below.

Simulation results of global path planning algorithm are presented in figures 3 to 8. Computation time for presented example situation results ranges from 6 to 7 minutes, what is acceptable for considered problem. Calculations were performed on computer with Intel Core 2 Duo Processor E7500 (2.93 GHz), Windows system. For each example situation a table with results including course of a ship at every line segment connecting two consecutive waypoints and distance of every line segment creating the path is presented - tables 1 to 6.

Input: start point and destination point, static constraints
Compute possible nodes
Initialize pheromone amount
For iteration =1,2,...iteration_number
  For ant=1,2,....ant_numer
    For step=1,2, ... max_step
    Compute ant's next move probability
    Select ant's next move based on computed probability
    Save selected node in the history of ants past positions
      if *the current point is equal to the destination point*
      Update of the pheromone amount
      Break the step loop
      End if
    End For
  End For
Smooth the determined path
Save the path determined in the iteration
End For
Select the shortest path
Output: coordinates of waypoints and distance of each line segment belonging to the path

Table 1. Results for example 1.

line segment	CSE [°]	distance [nm]
1-2	39	12,9
2-3	322	62,4
3-4	270	176,8
4-5	0	50,4
5-6	322	181,7

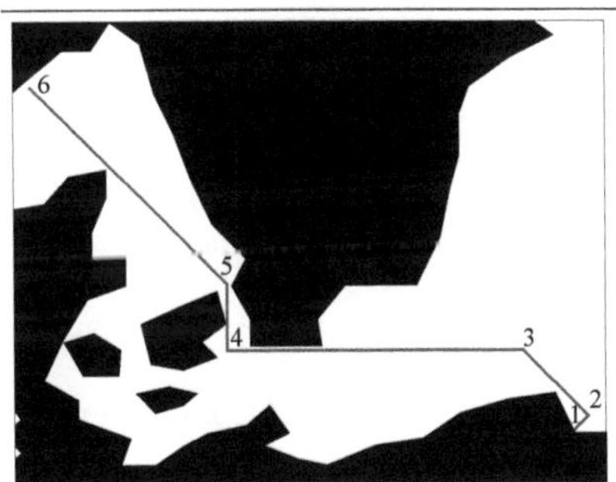

Figure 3. Example 1 of a solution obtained by application of the global route algorithm, distance of path - 484,2 nm.

Table 2. Results for example 2.

line segment	CSE [°]	distance [nm]
1-2	39	12,9
2-3	26	216,8
3-4	37	89,3
4-5	0	23,7
5-6	71	145,6
6-7	90	52,9

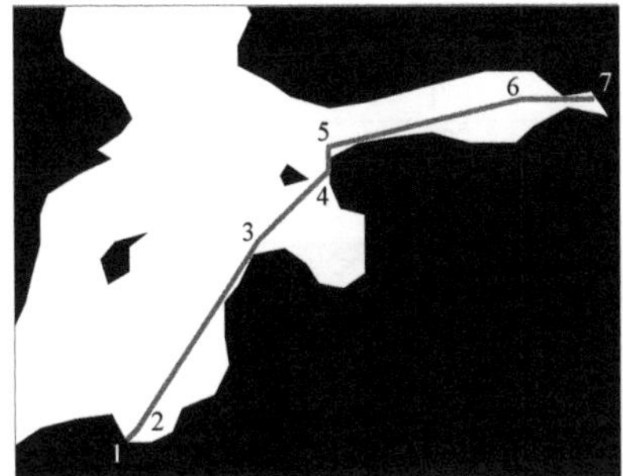

Figure 4. Example 2 of a solution obtained by application of the global route algorithm, distance of path - 541,2 nm.

Table 3. Results for example 3.

line segment	CSE [°]	distance [nm]
1-2	39	12,9
2-3	8	518,5
3-4	30	175,8

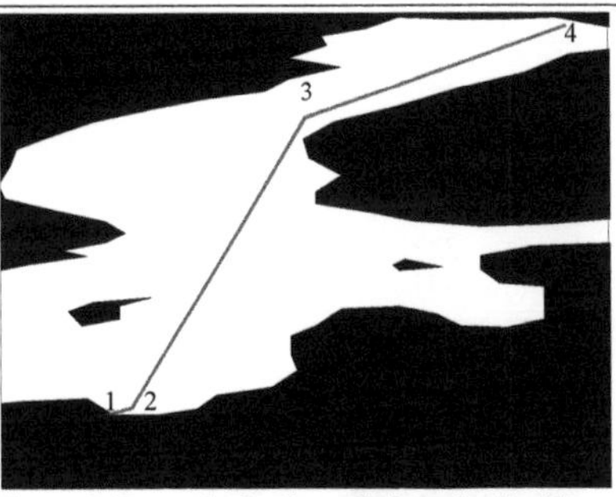

Figure 5. Example 3 of a solution obtained by application of the global route algorithm, distance of path - 707,2 nm.

Table 4. Results for example 4.

line segment	CSE [°]	distance [nm]
1-2	34	176,6
2-3	21	203,1

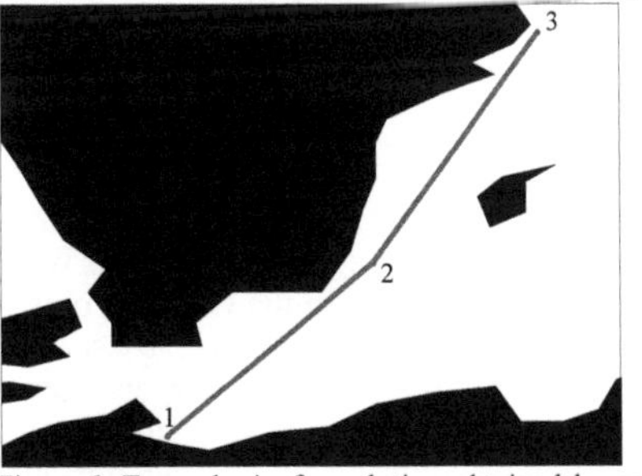

Figure 6. Example 4 of a solution obtained by application of the global route algorithm, distance of path - 379,7 nm.

Table 5. Results for example 5.

line segment	CSE [°]	distance [nm]
1-2	90	118,6
2-3	63	108,9
3-4	26	212,8

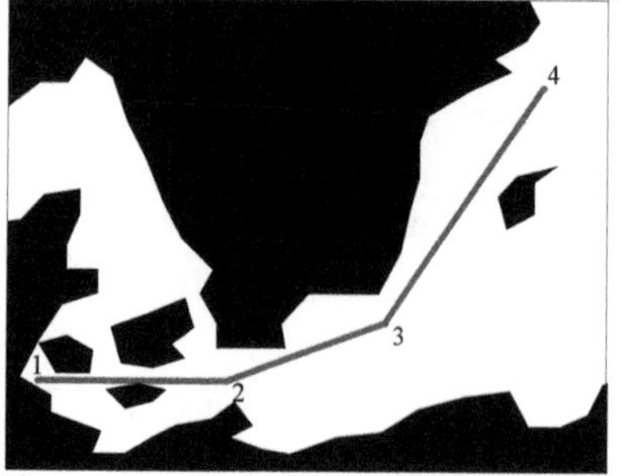

Figure 7. Example 5 of a solution obtained by application of the global route algorithm, distance of path - 440,3 nm.

Table 6. Results for example 6.

line segment	CSE [°]	distance [nm]
1-2	203	235,4
2-3	270	117,4
3-4	0	49,2
4-5	322	121,8
5-6	293	59,5

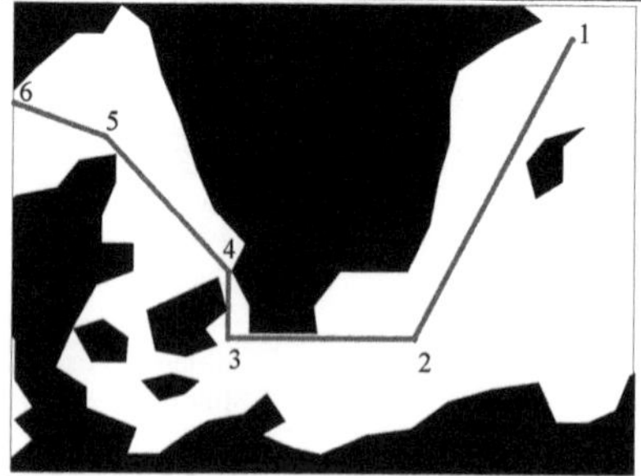

Figure 8. Example 6 of a solution obtained by application of the global route algorithm - distance of path - 583,3 nm.

5 LOCAL ROUTE

Local route is determined when dangerous target alarm is activated in ARPA system. It happens when the computed values of Time to the Closest Point of Approach (TCPA) and Distance of the Closest Point of Approach (DCPA) for any of the targets exceed the specified safe limits. Local route determination can also be initialized manually by the operator. Local route algorithm calculates anti-collision trajectory from current position of own ship to the next waypoint determined by global path planner. This approach is applicable in situations in restricted waters, where safe trajectory should be determined from actual position of own ship (xo, yo) to the defined ending point (xe, ye). Most of collision situations occur in narrow passages for example in Danish straits. For situations in open waters the problem can be reduced to finding solution including defined ending course. Moreover, the solution should also fulfill requirements of COLREGS.

According to rule 8b of International Regulations for Preventing Collisions at Sea (COLREGS) "any alteration of course and/or speed to avoid collision shall, if the circumstances of the case admit, be large enough to be readily apparent to another vessel observing visually or by radar; a succession of small alterations of course and/or speed should be avoided". However COLREGS define rules only for two ships in good visibility. In restricted visibility it defines only general recommendations (Cockcroft, 2012).

Local route determination is also based on ant colony optimization algorithm modified to fit the considered problem. Algorithm applied for global route determination was developed by implementation of procedures taking into account the dynamic constraints in the form of encountered vessels. Target ships are modelled in the form of hexagons. Dimensions of the hexagon forces the application of COLREGs. Solution contains ship course at every line segment of the trajectory, the distance of the line segment and the time of passage. Local route algorithm in the form of pseudocode and an example of obtained results are placed below.

Input: current navigational situation data
Compute own ship and targets position after lead time
Compute possible nodes
Initialize pheromone amount
For iteration =1,2,...iteration_number
  For ant=1,2,....ant_numer
    For step=1,2, ... max_step
    Compute ant's next move probability
    Select ant's next move based on computed probability
    Save selected node in the history of ants past positions
      if *the current point is equal to the destination point*
      Update of the pheromone amount
      Break the step loop
      End if
    End For
  End For
Smooth the determined path
Save the path determined in the iteration
End For
Remove unfeasible solutions
Select the shortest path fulfilling constraints

Output: own ship course at every line segment of path, the distance of the trajectory and the time of passage(t*)

Table 7. Situation 1 motion parameters of all ships.

Setting	CSE [°]	SPD [kn]	BRG [°]	RNG [nm]
Own ship	0	10	—	—
Target 1	270	10	45	7

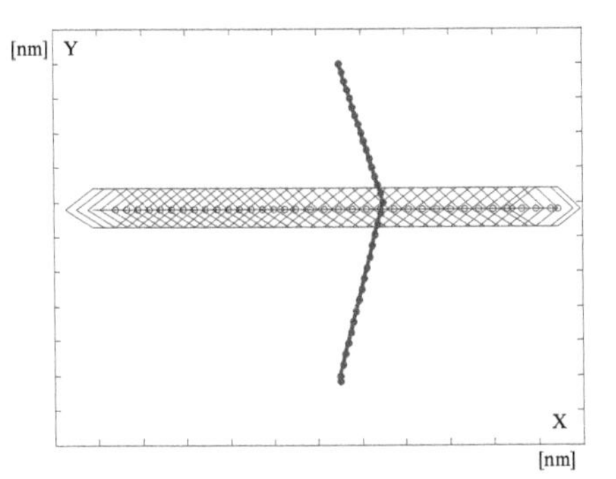

Figure 9. Determined trajectory for situation 1: t* = 55.3328 minutes, distance = 9.2221nm, own ship course = 11°, 346°.

Table 8. Situation 2 motion parameters of all ships.

Setting	CSE [°]	SPD [kn]	BRG [°]	RNG [nm]
Own ship	0	14	—	—
Target 1	180	10	0	4

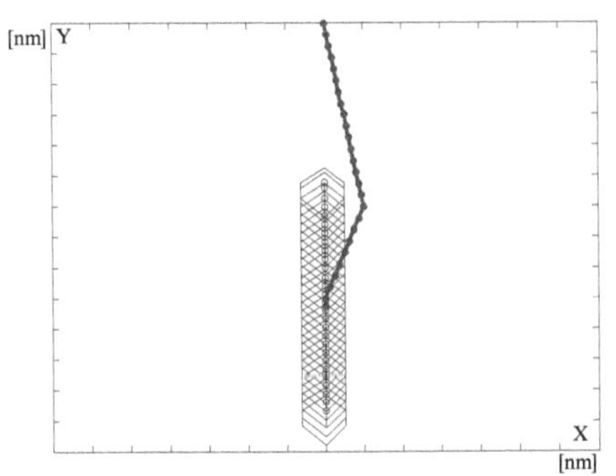

Figure 10. Determined trajectory for situation 2: t* = 39.6216 minutes, distance = 9.245 nm, own ship course = 18°, 351°.

Table 9. Situation 3 motion parameters of all ships.

Setting	CSE [°]	SPD [kn]	BRG [°]	RNG [nm]
Own ship	0	12	—	—
Target 1	90	10	315	5

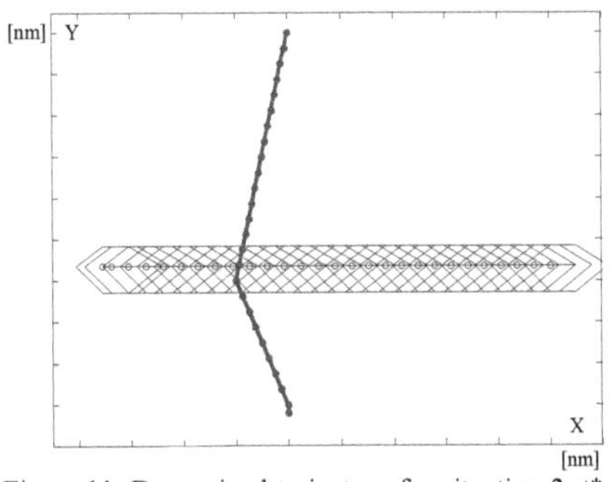

Figure 11. Determined trajectory for situation 3: t* = 46.2252 minutes, distance = 9.245 nm, own ship course = 342°, 9°.

Table 10. Situation 4 motion parameters of all ships.

Setting	CSE [°]	SPD [kn]	BRG [°]	RNG [nm]
Own ship	0	10	—	—
Target 1	270	10	45	5
Target 2	90	12	315	7

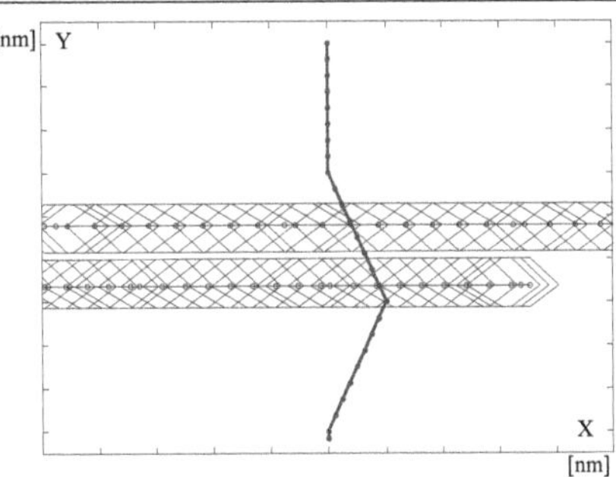

Figure 12. Determined trajectory for situation 4: t* = 55.9473 minutes, distance = 9.3246 nm, own ship course = 18°, 342°, 0°.

Table 11. Situation 5 motion parameters of all ships.

Setting	CSE [°]	SPD [kn]	BRG [°]	RNG [nm]
Own ship	0	10	—	—
Target 1	246	14	41	5
Target 2	287	10	79	7
Target 3	191	12	2	4.6

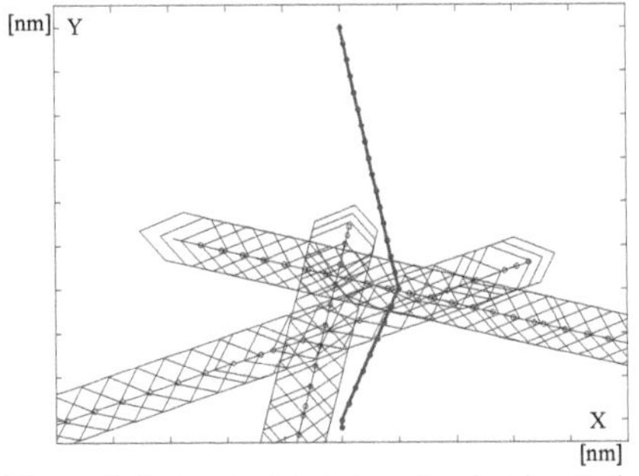

Figure 13. Determined trajectory for situation 5: t* = 55.4702 minutes, distance = 9.245 nm, own ship course = 18°, 351°.

Table 12. Situation 6 motion parameters of all ships.

Setting	CSE [°]	SPD [kn]	BRG [°]	RNG [nm]
Own ship	0	10	–	–
Target 1	220	16	22.5	7
Target 2	180	14	0	8
Target 3	90	10	320	6
Target 4	345	12	180	4.5

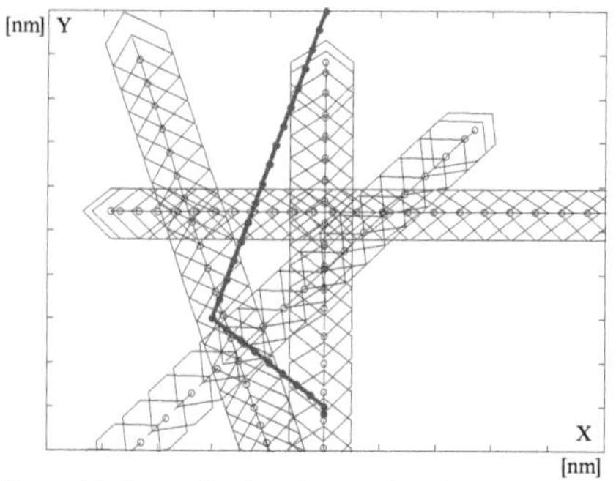

Figure 14. Determined trajectory for situation 6: t* = 60.6512 minutes, distance = 10.1085 nm, own ship course = 315°, 16°.

Table 13. Situation 7 motion parameters of all ships.

Setting	CSE [°]	SPD [kn]	BRG [°]	RNG [nm]
Own ship	0	10	–	–
Target 1	270	10	45	5
Target 2	275	9	75	7
Target 3	272	10	57.5	9
Target 4	90	8	327	7
Target 5	95	7	315	9

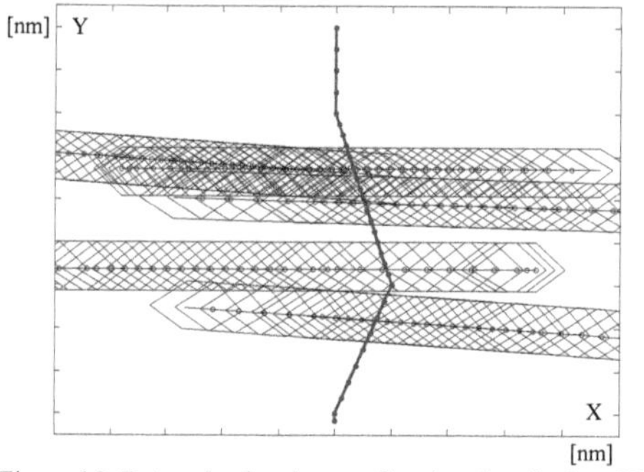

Figure 15. Determined trajectory for situation 7: t* = 55.7123 minutes, distance = 9.2854 nm, own ship course = 18°, 346°, 0°.

Table 14. Situation 8 motion parameters of all ships.

Setting	CSE [°]	SPD [kn]	BRG [°]	RNG [nm]
Own ship	0	10	–	–
Target 1	45	14	5	45
Target 2	90	10	7	315
Target 3	135	12	4.6	1
Target 4	180	5	3.9	25
Target 5	225	14.5	4	78
Target 6	270	11	3.5	12
Target 7	315	9	6.5	3

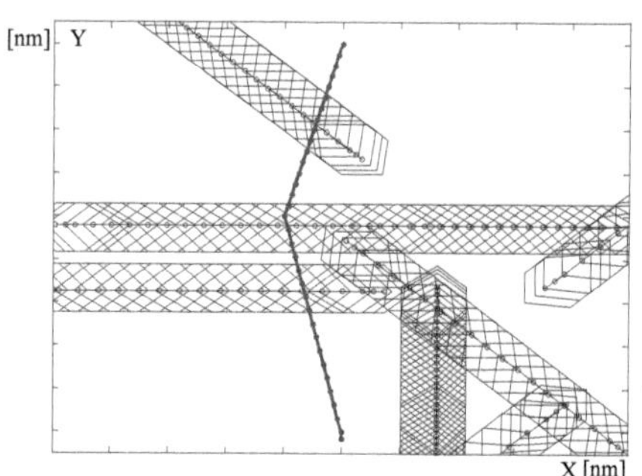

Figure 16. Determined trajectory for situation 8: t* = 55.3328 minutes, distance = 9.2221 nm, own ship course = 349°, 14°.

Table 15. Situation 9 motion parameters of all ships.

Setting	CSE [°]	SPD [kn]	BRG [°]	RNG [nm]
Own ship	0	14	–	–
Target 1	90	8.5	320	5
Target 2	110	11	331	7.5
Target 3	175	12	1	5
Target 4	250	14.5	34	10
Target 5	270	12.5	63	6.5
Target 6	310	9.8	122	5.5
Target 7	30	16	189	5
Target 8	25	15	234	6

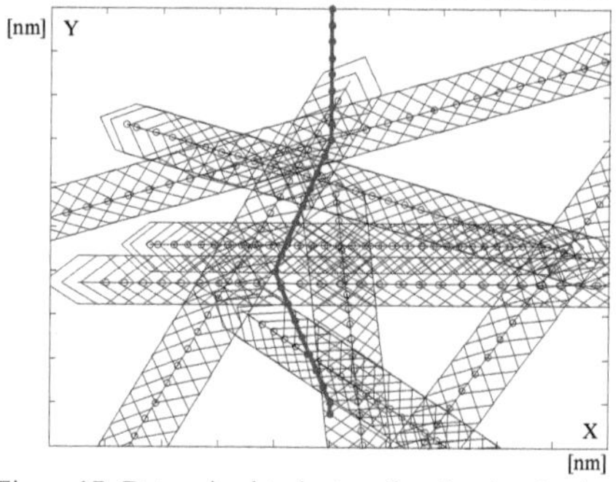

Figure 17. Determined trajectory for situation 9: t* = 39.9624 minutes, distance = 9.3246 nm, own ship course = 342°, 18°, 0°.

Table 16. Situation 10 motion parameters of all ships.

Setting	CSE [°]	SPD [kn]	BRG [°]	RNG [nm]
Own ship	0	10	–	–
Target 1	100	6	300	6
Target 2	280	11	130	12
Target 3	350	19	140	4
Target 4	20	20	220	5
Target 5	160	15	270	3
Target 6	50	6	330	12
Target 7	190	8	40	6
Target 8	330	20	120	9
Target 9	80	13	350	8
Target 10	270	16	60	15

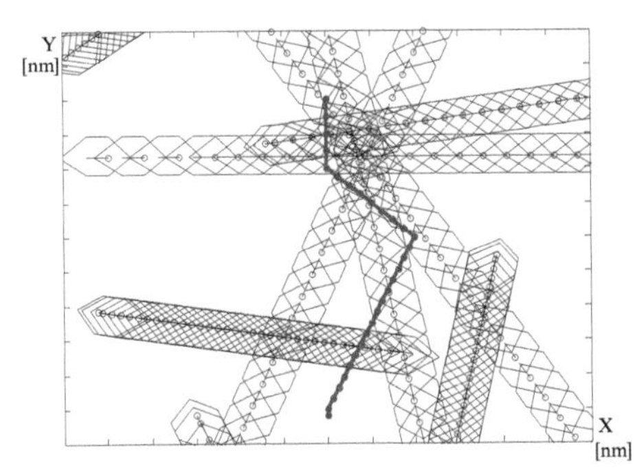

Figure 18. Determined trajectory for situation 10: t* = 61.2816 minutes, distance = 10.2136 nm, own ship course = 22°, 315°, 0°.

Table 17. Situation 11 motion parameters of all ships.

Setting	CSE [°]	SPD [kn]	BRG [°]	RNG [nm]
Own ship	350	12	–	–
Target 1	141	11.4	7	8.6
Target 2	171	7.7	340	11.5
Target 3	112	9.7	34	10.5
Target 4	173	10.2	46	12.3
Target 5	186	12.1	350	9.4
Target 6	240	10.3	45	15.1
Target 7	242	14.5	355	6.4
Target 8	246	16	37	5.3
Target 9	243	9.2	60	16
Target 10	190	8.5	86	12.9
Target 11	337	10.5	152	9
Target 12	270	9.4	47	10.8
Target 13	0	12.5	180	7.6
Target 14	340	12.3	182	5.4
Target 15	20	20.7	230	5.9
Target 16	50	9.3	233	3.7
Target 17	24	5.5	234	6.3
Target 18	177	8.8	278	4.4
Target 19	200	11.5	330	6.8

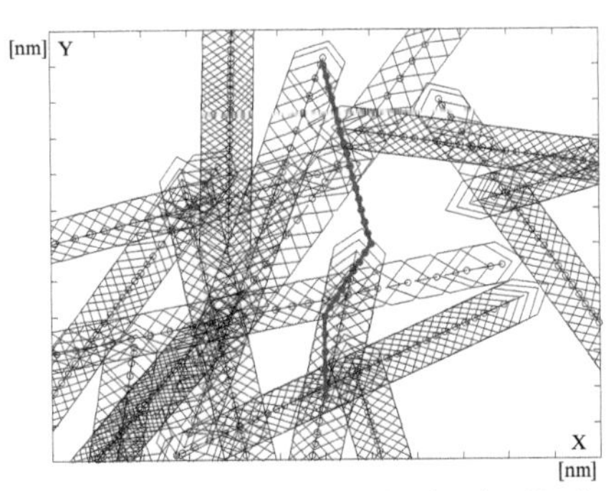

Figure 19. Determined trajectory for situation 11: t* = 46.6754 minutes, distance = 9.3351 nm, own ship course = 350°, 17°, 339°.

Table 18. Situation 12 motion parameters of all ships.

Setting	CSE [°]	SPD [kn]	BRG [°]	RNG [nm]
Own ship	20	14	–	–
Target 1	0	15	220	3.9
Target 2	27	4	287	6.8
Target 3	180	5	315	4.2
Target 4	150	10	330	7
Target 5	297	11	335	3.3
Target 6	90	10	339	5.3
Target 7	95	12.5	341	7.4
Target 8	162	16	352	13.1
Target 9	180	2	0	10
Target 10	117	10	1	7
Target 11	225	14.4	7	8.1
Target 12	223	14	20	8.8
Target 13	135	10.5	21	7
Target 14	137	7	37	5
Target 15	45	14	26	2.2
Target 16	285	14.5	45	5.9
Target 17	0	15	142	6.4
Target 18	27	10.5	161	3.1
Target 19	153	11	309	3.2
Target 20	210	2	50	3.9

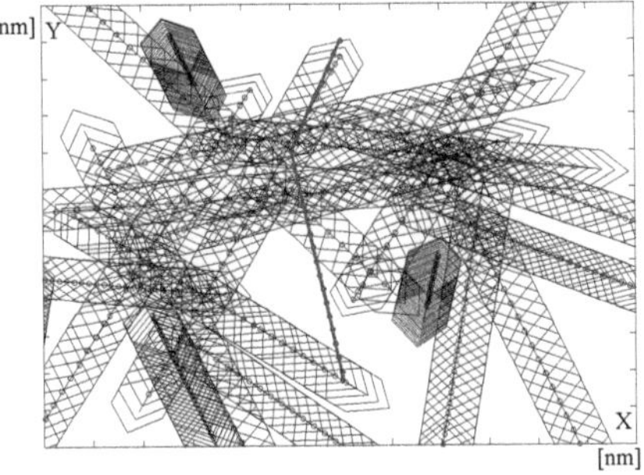

Figure 20. Determined trajectory for situation 12: t* = 39.6216 minutes, distance = 9.245 nm, own ship course = 11°, 38°.

Computation time for presented example situation results ranges from 12 to 60 seconds. All of received solutions fulfill rule 8b of COLREGs. Course alternation are in range from 10° to 90° and are readily apparent to another vessel. Situation 1 presented in figure 9 fulfills rule 15 of COLREGs, considering two vessels crossing situation. According to this rule " the vessel which has the other on her starboard side shall keep out of the way and shell, if the circumstances of the case admit, avoid crossing ahead of the other vessel". Compliance of rule 14 is presented in figure 10 for situation 2. This rule concerns head-on situation. According to this rule " when two power-driven vessels are meeting on reciprocal or nearly reciprocal courses so as to involve risk of collision each shall alter her course to starboard so that each shall pass on the port side of the other". In situation 3 according to COLREGs target ship should keep out of the way, because she has own ship on her starboard, but if the target vessel does not take action to avoid collision, own ship has to alter the course as is shown in figure 11. In situation presented in figure 12 own vessel first alters course according to rule 15 of COLREGs in order to allow passage of target 1 and then returns to its initial course avoiding also target 2. As it is presented in figures 13 to 20, the algorithm deals also with more complex situations of meeting from 3 up to 20 target ships.

6 SUMMARY AND CONCLUSIONS

The paper presents an innovative approach of ant colony optimization algorithm implementation in ship's navigational decision support system. Presented results constitute confirmation of a successful implementation of the method in considered issue. All of the vital assumptions for the considered problem were taken into account, such as obeying COLREGs, minimization of way loss on overtaking other ships and great course changes of anti-collision manouvres. Introduced approach to determination of safe optimal route of passage between harbours can be applied in real electro-navigational system onboard a ship, because of its low computational time, up to 1 minute for local route and a few minutes for the global path. Application of the system will cause the number of ship accidents to decrease, providing protection of human life and health, transported cargo and natural environment. Moreover the system will also assure more effective transport by taking into account optimality criterion.

REFERENCES

Blum C. 2005: Ant colony optimization: Introduction and recent trends, *Physics of life Reviews*, No 2(4), p.353-373.

Blum C., Dorigo M. 2005: Ant colony optimization theory: A survey, *Theoretical Computer Science* , No 344, p.243-278.

Cockcroft A.N., Lameijer J.N.F. 2012: A Guide to the Collision Avoidance Rules, Butterworth-Heinemann Ltd.

Dorigo M., Montes de Oca M., Oliveira S., Stützle T. 2011: Ant Colony Optimization.,In J. J. Cochran, editor, Wiley Encyclopedia of Operations Research and Management Science, p. 114-125, John Wiley & Sons.

Dorigo M., Stützle T. 2010: Ant Colony Optimization: Overview and Recent Advances, In M. Gendreau and J.-Y. Potvin, editors, Handbook of Metaheuristics, volume 146 of International Series in Operations Research & Management Science, chapter 8, p. 227-263. Springer, New York.

Jurdziński M. 2003: Podstawy nawigacji morskiej, Wydawnictwo Akademii Morskiej w Gdyni, Gdynia.

Lazarowska A. 2012: Decision support system for collision avoidance at sea, *Polish Maritime Research*, Vol. 19, No 74, p. 19-24.

Lisowski J. 2012: The Sensitivity of Safe Ship control in Restricted Visibility at Sea, *TransNav - International Journal of Marine Navigation and Safety of Sea Transportation*, Vol.6, No.1, p. 35-45.

Lisowski J. 2012:Game control methods in avoidance of ships collisions,*Polish Maritime Research*, Vol.19, No74, p.3-10.

Lisowski J. 2009: The Comparison of Safe Control Methods in Marine Navigation in Congested Waters, *TransNav - International Journal of Marine Navigation and Safety of Sea Transportation*, Vol.3, No.2, p. 163-172.

Lisowski J. 2007: The Dynamic Game Models of Safe Navigation, *TransNav - International Journal of Marine Navigation and Safety of Sea Transportation*, Vol.1, No.1, p. 11-18.

Lisowski J. 2001: Computational intelligence methods in the safe ship control process, *Polish Maritime Research*, No. 1,Vol. 8, p. 18-24.

Lisowski J., Lazarowska A. (in press): The radar data transmission to computer support system of ship safety. *Solid State Phenomena*, Vol. 196,Trans Tech Publications, Switzerland.

Magaj J., Uriasz J. 2011: Performance of the decision support system in real conditions, *Scientific Journals Maritime University of Szczecin*, 27(99), pp. 28-35.

Martínez de Osés F.X., Ventikos N. P. 2003: A Critical Assessment of Human Element Regarding Maritime Safety: Issues of Planning, Policy and Practice.

Perera L.P., Carvalho J.P., Guedes Soares C. 2011: Fuzzy logic based decision making system for collision avoidance of ocean navigation under critical collision conditions, *Journal of Marine Science and Technology*, Vol. 16, No. 1, p.84-99.

Pietrzykowski Z. 2011: Navigational decision support system as an element of intelligent transport systems, *Scientific Journals Maritime University of Szczecin*, 25(97), pp.41-47.

Pietrzykowski Z., Uriasz J. 2010: Knowledge Representation in a Ship's Navigational Decision Support System. *TransNav - International Journal of Marine Navigation and Safety of Sea Transportation*, Vol.4, No.3, p. 265-270.

Pietrzykowski Z., Magaj J., Chomski J. 2008: Model of navigational decision support system on a seagoing vessel, *Scientific Journals Maritime University of Szczecin*, 13(85), pp. 65-73.

Pietrzykowski Z., Uriasz J. 2006: Ship domain in navigational situation assessment in an open sea area. Proceedings of COMPIT'06 5th International Conference on Computer Applications and Information Technology in the Maritime Industries, 8-10 May 2006, Oegstgeest, The Netherlands, p. 237-244.

Rothblum, A.M. 2000: Human Error and Marine Safety, Proceedings of the Maritime Human Factors Conference, Maryland, USA: pp. 1–10.

Śmierzchalski R., Kolendo P., Jaworski B. 2011: Experimental Research on Evolutionary Path Planning Algorithm with Fitness Function Scaling for Collision Scenarios, *TransNav - International Journal of Marine Navigation and Safety of Sea Transportation*, Vol.5, No.4, p. 489-495.

Śmierzchalski R., Łebkowski A., Gierusz W., Dziedzicki K. 2008: Intelligent Ship Control System, *TransNav - International Journal of Marine Navigation and Safety of Sea Transportation*, Vol.2, No.1, p. 63-68.

Śmierzchalski R. , Michalewicz Z. 2005: Path Planning in Dynamic Environments, Innovations in Robot Mobility and Control , p. 135-153.

Śmierzchalski R. 1998: Synteza metod i algorytmów wspomagania decyzji nawigatora w sytuacji kolizyjnej na morzu, Prace Naukowe Wyższej Szkoły Morskiej w Gdyni.

Szłapczyński R. 2010: Evolutionary sets of Cooperating Trajectories in Multi-Ship Encounter Situations - Use Cases, *TransNav - International Journal of Marine Navigation and Safety of Sea Transportation*, Vol.4, No.2, p. 191-196.

Tran T., Harris C.J., Wilson P.A. 1999: Maritime Avoidance Navigation, Totally Integrated system (MANTIS), 12th SCSS - 12th Ship Control Systems Symposium, The Hague, The Netherlands, Royal Dutch Navy.

Issue of Making Decisions with Regard to Ship Traffic Safety in Different Situations at Sea

J. Girtler
Technical University of Gdańsk, Poland

ABSTRACT: The paper refers to possibilities of making operational decisions that would enable to ensure safety to a ship in case of application of the statistical decision theory with consideration of an expected value of consequences as a criterion for making such decisions. General description includes conditions for carrying out transportation tasks by ships and it has been shown that following this description it is possible to develop a five-state set of situations at sea, which a ship can face. It has been proved that the situations (elements of the set) can be considered as states of the process of occurring the particular situations, consecutively one by one. The paper provides definitions for probabilities of occurring the particular states (situations) contained in this set, in which a ship can carry out its cruise. The probabilities have been defined for a long operating time for a ship ($t \to \infty$), which means that they are a limiting distribution of the process of state transitions. The theory of semi-Markov processes has been applied to determine the distribution. Based on the distribution and decision-making rules, a set of decisions that possibly may be taken in accordance with the assumed criterion of optimization, has been developed. Also the paper presents a simple example of calculation in order to demonstrate that determination of the consequences of making a decision is more important than estimation of the probability of occurrence of particular situations during a ship cruise.

1 INTRODUCTION

Seagoing ships perform transportation tasks often in very difficult sea conditions. The conditions are influenced by various factors, usually with random properties, such as, for example, [2, 4, 5, 20]: high water waves at strong winds, tides and tidal currents or topside icing. Tidal fluctuations in waters and associated tidal currents are important because they can vary within wide limits in both: the waterbody depth and speed of ocean (water) currents. In addition to the tidal currents, the hydro-meteorological factors generate a horizontal flow of water, which is called an ocean current [4, 5, 20].

A ship going a course in the opposite direction of the wind, is exposed to being drifted which consists in that the winds push the ship off the course line. Sometimes the winds induce large aerodynamic forces and moments on the topside surface. Therefore, the need occurs to compensate the wind action, which requires an adequate rudder angle. During stormy weather, however, it may turn out that even a maximum rudder angle is not sufficient to keep the ship on course. The ship then loses its navigability (course stability) and thus is not able to move in the set direction [21, 22]. It is particularly difficult to maintain the course stability when the adequate high pushing power is impossible to generate due to wear of the ship's main engines and the rudder angle is not possible to reach fast enough because of wear of the steering system. This is severely experienced during stormy weather. Catastrophic consequences may be brought by seismic waves, called "tsunami" [4]. Dangerous for a ship is also flooding of the deck and, in particular, the rapid water penetration into the deck, which can cause a damage to the ship's superstructure and getting the water to the interior [21, 22]. Icebergs are particularly hazardous for the ship transport as they can move long distances, being drifted by ocean currents, respectively north and south, in the direction of the equator, creating a serious threat even to large and modern ships, especially when it is difficult to avoid them. Situations of ship collisions often happen, as well. [19]

Conclusions from these considerations arise that each ship may be caught in different situations classified to the following types of situations at sea (as in the case of each shipborne system), like: normal, complicated, dangerous, emergency and catastrophic [6, 9, 12, 19, 17]. Obviously, the situations are affected by the technical condition of the systems which have a major influence on the ship safety. Such systems, for each ship, include undoubtedly a main engine, steering system and radar equipment (radar). A rational operational decision can be made when knowing the probability of occurrence of a specific sea situation during a ship's cruise and probability of a failure in at least one of the mentioned systems which influence the ship safety. In practice, decisions that can be made are as follows: 1) start performing the task, 2) perform a renovation of the systems which particularly influence the ship safety and then, start performing the task, and 3) not to leave the harbor until the reasons that threaten the ship safety, are eliminated. The issue of making a rational decision can be characterized in an analytical aspect by using the theory of the semi-Markov processes to determine the probabilities of occurring the sea situations and failures of the said systems, as well as the statistical decision theory to make an operational decision for the specific transportation task to be taken by the ship. This requires a development of a model of occurring changes in the technical state of these systems which are fundamental for the ship safety, a model of changes of situations at sea, in which the ship performs transportation tasks and a model of making operational decisions.

2 STATE TRANSITION MODEL FOR THE TECHNICAL SYSTEMS OF FUNDAMENTAL SIGNIFICANCE FOR SHIP SAFETY

The model assumes that the systems of fundamental significance for ship safety includes a main engine, steering system and radar. All of the systems may fail during their lifetime. As a rule the systems are usually renovated (of course, if it is profitable) by providing a proper maintenance.

For these systems, as well as for other shipborne systems, the model of the process of changing their reliability states can be considered as a semi-Markov process $\{W(t): t \geq 0\}$ with a set of states $S = s_i$; $i = 0, 1, 2, 3$. The interpretation of the states $s_i \in S(i = 0, 1, 2, 3)$ is as follows: s_0 – state of full ability for all systems under consideration (main engine, steering system and radar), s_1 – disabled state of the main engine, s_2 – disabled state of the steering system, s_3 – disabled state of radar. Changes of the said states s_i $(i = 0, 1, 2, 3)$ proceed in successive times t_n $(n \in N)$, wherein at time $t_0 = 0$ all the systems are in state s_0. The s_0 state lasts until any of these systems gets a failure. While the states $s_i(i = 1, 2, 3)$ last as long as one of the systems becomes renovated or replaced with another one, if renovation is unprofitable. State transitions from s_i to s_j $(i, j = 0, 1, 2, 3;\ i \neq j)$ occur after the time T_{ij} which is a random variable.

Consideration of the situation in the ship operation phase requires a probabilistic description of the operation process including the probabilities of possible occurrence of the states $s_i(i = 0, 1, 2, 3)$ at particular times $t_0, t_1, ..., t_{n-1}, t_n$ in the operating time of the ship [1, 2, 3, 16, 18, 21].

An assumption can be made that the state of the considered systems at time t_{n+1} and time interval of duration of the state at time t_n do not depend on the states at times $t_0, t_1, ..., t_{n-1}$ or the time intervals of their duration. Thus, it can be assumed that the process $\{W(t): t \geq 0\}$ is a semi-Markov process [5, 7, 8, 10, 11]. A graph of state transitions for this process is shown in Fig.1.

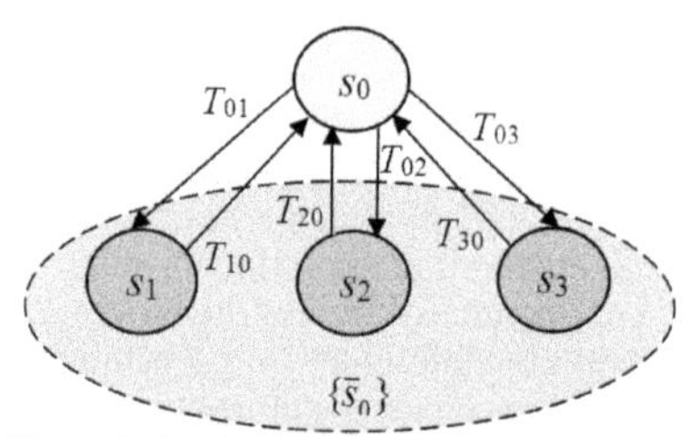

Figure 1. Graph of state transitions for the process $\{W(t): t \in T\}$: s_0 – state of ability of systems, $\{\overline{s}_0\}$ – set of disabled states of systems (engine, steering system, radar): $\{\overline{s}_0\} = \{s_1, s_2, s_3\}$, $s_i \in S(i = 1, 2, 3)$ – states with the following interpretation: s_0 – state of full ability of all considered systems (main engine, steering system and radar), s_1 – disabled state of engine, s_2 – disabled state of steering system, s_3 – disabled state of radar, T_{ij} – time of duration of the state s_i, provided that the next state is s_j $(i, j = 0, 1, 2, 3;\ i \neq j)$

The initial distribution of the process is as follows:

$$P\{W(0) = s_i\} = \begin{cases} 1 & dla \;\; i = 0 \\ 0 & dla \;\; i = 1,\ 2,\ 3 \end{cases} \tag{1}$$

while the matrix function takes the form:

$$\mathbf{Q(t)} = \begin{bmatrix} 0 & Q_{01}(t) & Q_{02}(t) & Q_{03}(t) \\ Q_{10}(t) & 0 & 0 & 0 \\ Q_{20}(t) & 0 & 0 & 0 \\ Q_{20}(t) & 0 & 0 & 0 \end{bmatrix} \tag{2}$$

The matrix function $\mathbf{Q(t)}$ is a model of transition of reliability states for the considered (mentioned earlier) systems. Non-zero elements $Q_{ij}(t)$ of the matrix $Q(t)$ depend on the distributions of random variables which are the time intervals of staying the processes $\{W(t): t \geq 0\}$ in the states s_i $S(i = 0, 1, 2, 3)$. Elements of the matrix function $\mathbf{Q(t)}$ are the probabilities of the process transition from state s_i to s_j $(s_i, s_j \in S)$ at the less time than t, defined as follows:

$$Q_{ij}(t) = P\{W(\tau_{n+1}) = s_j, \\ \tau_{n+1} - \tau_n < t | W(\tau_n) = s_i\} = p_{ij}F_{ij}(t) \tag{3}$$

where:

p_{ij} – one-step transition probability of the homogeneous Markov chain;

$p_{ij} = P\{W(\tau_{n+1}) = s_j | W(\tau_n) = s_i = \lim_{t\to\infty} Q_{ij}(t)$;

$F_{ij}(t)$ - distribution of the random variable T_{ij} representing the duration of the state s_i of the process $\{W(t): t \geq 0\}$, provided that the next state of the process is s_j.

The matrix **P** of transition probabilities for the embedded Markov chain to the process, as it results from the matrix function **Q(t)** (2), is as follows [4, 5, 7]:

$$\mathbf{P} = \begin{bmatrix} 0 & p_{01} & p_{02} & p_{03} \\ 1 & 0 & 0 & 0 \\ 1 & 0 & 0 & 0 \\ 1 & 0 & 0 & 0 \end{bmatrix} \tag{4}$$

In the process $\{W(t): t \geq 0\}$ the random variables T_{ij} have finite, positive expected values [4, 5, 7, 15]. Therefore, its limiting distribution

$$P_j = \lim_{t\to\infty} P_{ij}(t) = \lim_{t\to\infty} P\{W(t) = s_j\},$$
$$s_j \in S(j = 0, 1, 2, 3) \tag{5}$$

gets the following formula:

$$P_j = \frac{\pi_j \mathrm{E}(T_j)}{\sum_{k=0}^{3} \pi_k \mathrm{E}(T_k)} \tag{6}$$

The probabilities $\pi_j(j = 0, 1, 2, 3)$ in the formula (6) are limiting probabilities of the embedded Markov chain to the process$\{W(t): t \geq 0\}$. However, $\mathrm{E}(T_j)$ and $\mathrm{E}(T_k)$ are expected values of the random variables for T_j and T_k respectively, which are the times of being the systems in the state s_j and s_k respectively, regardless of which state will be the next.

Determination of the limiting distribution (6) requires solution of the system of equations that include the said limiting probabilities $\pi_j(j = 0, 1,..., 10)$ of the embedded Markov chain and the matrix **P** of the probabilities of transition from the state s_i to s_j, defined by the formula (4). Such a system is a system with the following form:

$$\left.\begin{aligned} [\pi_0, \pi_1, \pi_2, \pi_3] &= [\pi_0, \pi_1, \pi_2, \pi_3] \cdot \mathbf{P} \\ \sum_{k=1}^{4} \pi_k &= 1 \end{aligned}\right\} \tag{7}$$

As a result of solving the system of equations (7), by using the formula (6), the following relationships can be obtained:

$$P_0 = \frac{E(T_0)}{E(T_0) + \sum_{k=0}^{3} p_{0k} E(T_k)}, \quad P_1 = \frac{p_{01} E(T_1)}{E(T_0) + \sum_{k=0}^{3} p_{0k} E(T_k)},$$
$$P_2 = \frac{p_{02} E(T_2)}{E(T_0) + \sum_{k=0}^{3} p_{0k} E(T_k)}, \quad P_3 = \frac{p_{03} E(T_3)}{E(T_0) + \sum_{k=0}^{3} p_{0k} E(T_k)} \tag{8}$$

P_0 is a limiting probability that in a longer operating time (in theory at $t \to \infty$) the considered systems (main engine, steering system, radar) are in the state s_0. Thus, this probability defines the factor of the systems' technical readiness. However, the probabilities $P_j(j = 1, 2, 3)$ are limiting probabilities for the states $s_j \in S(j = 1, 2, 3)$ of the systems at $t \to \infty$, thus the probabilities that they are in disabled states.

An exemplary realization of the process $\{W(t): t \geq 0\}$ illustrating the occurrence of the disabled states of the systems in the operating time, is shown in Fig. 2. To obtain (obviously in approximation) the values of the probabilities $P_j(j = 0, 1, 2, 3,)$ estimation of p_{ij} and $\mathrm{E}(T_j)$ is required.

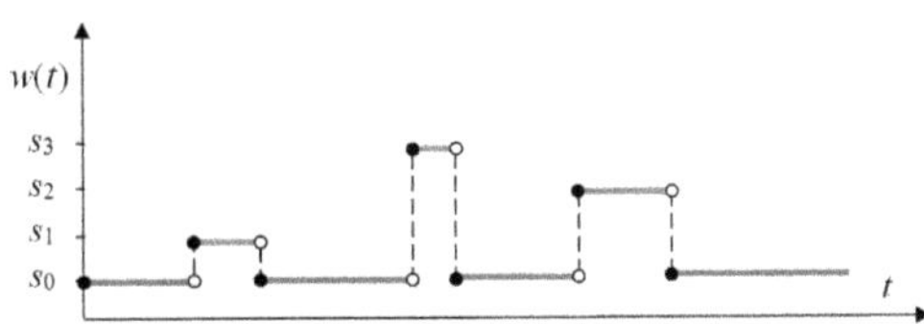

Figure 2. An exemplary realization of the process $\{W(t): t \geq 0\}$ for the considered systems: s_0 – state of ability of systems, s_1 – disabled state of main engine, s_2 – disabled state of clutch, s_3 – disabled state of steering system, s_4 – disabled state of radar

Estimation of the probabilities p_{ij} and expected values $\mathrm{E}(T_j)$ is possible after having the realization $w(t)$ of the process $\{W(t): t \in 0\}$ in sufficiently long time period of investigation, thus for $t \in [0, t_b]$, where the time of investigation for the process: $t_b >> 0$. Then, it is possible to determine the numbers of $n_{ij}(i,j = 0, 1, 2, 3; i \neq j)$, transitions of the process$\{W(t): t \in 0\}$ from state s_i to s_j in sufficiently long time and define the values of the estimator $\hat{P}_{ij}$ for the unknown probability p_{ij}. The estimator with the highest reliability of the transition probability p_{ij} is statistics [7, 15]:

$$\hat{P}_{ij} = \frac{N_{ij}}{\sum_j N_{ij}}, \quad i \neq j; \; i, j = 0, 1, 2, 3, \tag{9}$$

of which the value $\hat{p}_{ij} = \frac{n_{ij}}{\sum_j n_{ij}}$ is an estimation for the unknown transition probability p_{ij}.

From the $w(t)$ course of the process $W(t)$ there can also be obtained the realizations $t_j^{(m)}$, $m = 1, 2, ..., n_{ij}$ of the random variables T_j. Application of a point estimation allows to calculate easily $E(T_j)$ as an arithmetic mean value for the realization $t_j^{(m)}$. Acquisition of the necessary information in order to calculate the probabilities requires employing adequate diagnosing tools (*SDG*) for the mentioned systems which in this case become diagnosed systems (*SDN*) [7, 11].

3 MODEL OF SITUATIONS EMERGING TO A SHIP AT SEA

In the presented situation, making a rational decision is possible in case of application of the statistical decision theory and thus - the expected value of consequences, as a criterion for making this decision [1, 10, 14]. However, specification of a set of decisions that may be taken in accordance with the adopted optimization criterion, requires identification of the current decision-making situation in the ship operation phase.

When ships are under operation, there may appear different situations resulting from their use as intended. The situations can be divided with regard to the degree of risk to the ship safety, into [4, 9, 12, 17]: complicated, dangerous, emergency and catastrophic. Users of the ships, either direct (e.g. a crew) or indirect (e.g. an owner), tend to ensure a normal situation, i.e. such one in which safe motion of a ship is possible, which means no threat to the ship, health (especially life) of any individual crew member, etc.

Changes in particular situations, in which the ship motion is possible to continue, can be considered as a discrete-state continuous-time random process $\{Y(t): t \geq 0\}$ with a four-state set of states $S^* = \{s_j^*; i = 0, 1, 2, 3, 4\}$, where t – time of the process realization, which in practice means duration of a cruise of the given ship [4, 10, 11].

A general interpretation to the states $s_j^* \in S^*$ (j = 0, 1, 2, 3, 4) is, in this case, as follows: s_0^* – normal situation, s_1^* – complicated situation, s_2^* – dangerous situation, s_3^* – emergency situation, s_4^* – catastrophic situation.

These states occur at random times and extend within the time intervals $[\tau_0, \tau_1), [\tau_1, \tau_2),..., [\tau_n, \tau_{n+1})$, which are random variables.

In a normal situation (s_0*) ship crews carry out their duties in conditions which they are accustomed to. Such conditions do not cause any excessive stress to the crews and do not force them to any excessive physical or intellectual effort.

A complicated situation (s_1^*) occurs when some events appear that hinder performance of the task being carried out by the ship. Such events may include failures of: main engine, steering system, radar. In such cases the crews are forced to some physical and intellectual emergency effort in order to satisfy the demands arising from the need to eliminate the causes of the emerged situation.

A dangerous situation (s_2^*) occurs when some events appear that make performance of the task impossible. Such events may include, for instance, failures of the mentioned systems in deteriorating weather conditions (approaching storm). In such cases it is necessary to increase the crew's efforts in order to restore the ship's ability to move safely.

An emergency situation (s_3^*) occurs when the ship's crew is not able to restore the technical state of the damaged systems in deteriorating weather conditions and rising storm.

A catastrophic situation (s_4^*) is a situation where there is no chance to prevent sinking the ship with a part of or the whole crew, or without the crew which sometimes is successful to be all rescued.

A transition from state s_0^* to s_1^* proceeds with the probability p_{01}, after the time interval being a realization of the random variable T_{01}, representing the time of duration of the situation (state) s_0^*, provided that the next situation is s_1^*. A proper action taken by the crew may, of course, regain the situation s_0^*, which can proceed with the probability p_{10} and after the time interval being realization of the random variable T_{10}. In the event of further deterioration in factors it can turn out that reaching the situation (state) s_0^* is impossible, which inevitably leads to a dangerous situation (state) s_2^*. This means a transition of the process $\{Y(t): t \geq 0\}$ from state s_1^* to s_2^*. Such a change in the situation proceeds with the probability p_{12}, and after the time interval being realization of the random variable T_{12}. Actions of the crew, adequate for this situation, can lead back to s_0^*, but only after getting (prior regaining) the situation s_1^*. Otherwise, due to deterioration in the shipborne systems' technical state and conditions for ship motion, a situation s_3 (thus emergency) will emerge with the probability p_{23} and after the period of time being realization of the random variable T_{23} representing duration of the situation (state) s_2^*, provided that situation s_3^* will be the next. A return from this situation to the previously mentioned ones is possible in some cases, however, also the situation s_4^* (thus catastrophic) may emerge when preventing the loss of human life and / or extensive damage to a ship is practically impossible [4, 11, 12]. A transition from s_3^* to s_4^* proceeds with the probability p_{34}, after the time interval being realization of the random variable T_{34} representing duration of the state s_3^*, provided that s_4^* will be the next situation. A return from this state to the preceding ones is also still possible in some cases when the object moves.

Therefore, we can assume that there is a sense in considering the process $\{Y(t): t \geq 0\}$ with the graph of transitions of its states $s_j^* \in S$ (j = 0, 1, 2, 3, 4), as shown in Fig. 3.

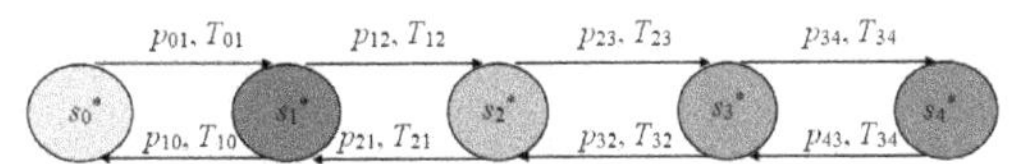

Figure 3. Graph of state transitions for the process $\{Y(t): t \geq 0\}$

The presented graph of state transitions for the process $\{Y(t):\ t \geq 0\}$ (Fig. 3) is of principal importance for the proposed solution for the problem of determining probabilities of staying the process in the specified states. It enables determination of the matrix function which is necessary to establish the formulas needed to estimate these probabilities.

Changes of states enclosed in the set $S = \{s_i^*;\ i = 0, 1, 2, 3, 4\}$ during motion of each ship, can be considered as a process $\{Y(t):\ t \geq 0\}$ with constant (equal) realizations in particular right-side continuous time intervals [1, 6]. The lengths of the time intervals for the particular states of the process $\{Y(t):\ t \geq 0\}$ are random variables T_{ij} representing the time of duration of the state $s_i^* \in S$ of the process, provided that the next state will be $s_j^* \in S$, where $i,j = 0, 1, 2, 3, 4$ and $i \neq j$. These variables are independent random variables with finite expected values $E(T_{ij})$. Additionally, the process has a property consisting in that duration of the current state s_i^*, which existed at the time τ_n and the current state at the time τ_{n+1} do not depend stochastically on the preceding states or the time interval of their duration. Thus, it can be assumed that future states (situations) depend only on the current situation. This means that the process $\{Y(t):\ t \geq 0\}$ can be regarded as a semi-Markov process with the graph of state transitions illustrated in Fig.3. Determination of the process requires to define its initial distribution P_i and matrix function $Q(t)$.

The initial distribution of the process $\{Y(t):\ t \geq 0\}$is as follows:

$$P_i = P\{Y(0) = s_i^*\} = \begin{cases} 1 & dla \quad i = 0 \\ 0 & dla \quad i = 1,\ 2,\ 3,\ 4 \end{cases} \tag{10}$$

In accordance with the graph of state transitions shown in Fig.1, the matrix function takes the following form:

$$Q(t) = \begin{bmatrix} 0 & Q_{01}(t) & 0 & 0 & 0 \\ Q_{10}(t) & 0 & Q_{12}(t) & 0 & 0 \\ 0 & Q_{21}(t) & 0 & Q_{23}(t) & 0 \\ 0 & 0 & Q_{32}(t) & 0 & Q_{34}(t) \\ 0 & 0 & 0 & Q_{43}(t) & 0 \end{bmatrix} \tag{11}$$

The presented initial distribution (10) and matrix function (11) of the process $\{Y(t): t \geq 0\}$ enable, just like in case of studying the previously described process $\{W(t):\ t \in 0\}$, to define a limiting distribution of the process $\{Y(t):\ t \geq 0\}$, with the following interpretation:

$$P_j = \lim_{t \to \infty} P\{Y(t) = s_j^*\}, \qquad j = \overline{0,4}$$

using in this case the formula [15]:

$$P_j = \frac{\pi_j E(T_j)}{\sum_{k=0}^{4} \pi_k E(T_k)}, \qquad j = 0,\ 1,\ ...,\ 4 \tag{12}$$

where: $\pi_j = \lim_{n \to \infty} \frac{1}{n} \sum_{k=1}^{n} P\{Y(\tau_n) = s_j | Y(0) = s_i\}$, and $[\pi_j; j = 0, 1, 2, 3, 4]$ is a stationary distribution of the embedded Markov chain $\{Y(\tau_n):\ n \in N\}$ to the process$\{Y(t): t \geq 0\}$.

This distribution satisfies the following system of equations (13) and (14) [15]:

$$\sum_{i=0}^{4} \pi_i p_{ij} = \pi_j; \quad i, j = 0,1,...,4 \tag{13}$$

$$\sum_{i=0}^{4} \pi_i = 1 \tag{14}$$

Solving the system of equations (13) and (14) with regard to the matrix (11) in accordance with the formula (12) the following relationships are obtained:

$$\left.\begin{aligned} P_0 = \frac{p_{10}p_{21}p_{32}E(T_0)}{M};\ P_1 = \frac{p_{21}p_{32}E(T_1)}{M};\ P_2 = \frac{p_{12}p_{32}E(T_2)}{M}; \\ P_3 = \frac{p_{12}p_{23}E(T_3)}{M};\ P_4 = \frac{p_{12}p_{23}p_{34}E(T_4)}{M} \end{aligned}\right\} \tag{15}$$

where:
$M = p_{10}p_{21}p_{32}E(T_0) + p_{21}p_{32}E(T_1) + p_{12}p_{32}E(T_2) + p_{12}p_{23}E(T_3) + p_{12}p_{23}p_{34}E(T_4)$ where: p_{ij} – probability of the transition process $\{Y(t): t \geq 0\}$ from s_i^* to s_j^* ($s_i^*, s_j^* \in S;\ i, j = 0, 1, 2, 3, 4;\ i \neq j$);
$E(T_j)$ – expected value of random variable $T_j (j = 0, 1, 2, 3, 4)$ being the duration time of the state $s_j^* \in S (j = 0, 1, 2, 3, 4)$ of the process $\{Y(t):\ t \geq 0\}$ independently of which state the process transits to.

The expected values $E(T_j)$ depend on the expected values $E(T_{ij})$ and probabilities p_{ij} as follows:

$$E(T_j) = E(T_i) = \sum_j p_{ij} E(T_{ij}), \quad i, j = \overline{0,4};\ i \neq j \tag{16}$$

The particular probabilities $P_j (j = 0, 1, 2, 3, 4)$ are of the following interpretation:

$P_0 = \lim_{t \to \infty} P\{Y(t) = s_0^*\}$, $P_1 = \lim_{t \to \infty} P\{Y(t) = s_1^*\}$, $P_2 = \lim_{t \to \infty} P\{Y(t) = s_3^*\}$, $P_3 = \lim_{t \to \infty} P\{Y(t) = s_3^*\}$, $P_4 = \lim_{t \to \infty} P\{Y(t) = s_4\}$.

The probability P_0 can be considered as a measure of a safe ship motion. Also the probability P_1 can be regarded as a measure of an almost safe motion of an object. But the other probabilities can be assumed as measures of a hazard to the ship which considerably increases from the occurrence of the state s_2^*, i.e. dangerous situation. Thus, the probability $P_B = P_0 + P_1$ can be assumed as a measure of a safe ship motion.

To obtain (obviously in approximation) the values of probabilities $P_j (j = 0, 1, 2, 3, 4)$ an estimation of p_{ij} and $E(T_j)$ is required, as mentioned when considerations referred to the failures of such systems as a main engine, steering system and radar.

4 MODEL OF MAKING OPERATIONAL DECISIONS FOR A SHIP

Having information on the technical state of systems (main engine, steering system, radar) affecting essentially the ship safety, which is in form of probabilities of their effectiveness and their failures, as well as information in form of probabilities of occurrence of situations at sea during a cruise, enables to take, by using the statistical decision theory, a rational decision from among the following three possibilities:

- decision d_1 – start performing the ordered task without prior preventive maintenance to renovate the systems,
- decision d_2 – firstly, make a proper (appropriate to the established diagnosis) preventive maintenance for particular systems in order to provide a full renovation to them and simultaneously regain their properties which are indispensable to perform the task, and then start performing the task at the time fixed by the orderer,
- decision d_3 – delay the date of starting performance of the ordered task until the reasons threatening the ship safety are eliminated.

The last from the presented decisions (d_3) must be taken always when not a complete but only partial renovation of the systems affecting essentially the ship safety, is possible. This is due to the fact that the partial renovation of such systems can ensure a safe ship motion only when the external conditions do not create any significant hazard in the event of failure of these systems.

For the considered situation, there can be defined, by using the statistical decision theory, the following expected values for the consequences of making each of the specified decisions d_k ($k = 1, 2, 3$):

$$E\left(c\left(d_k\right)\right)=\sum_{i=1}^{4} E\left(c\left(d_k, s_i\right) P\left(s_i\right); \quad k=1,\ 2,\ 3 \qquad (17)$$

where:

$E(c|d_1, s_0) = P(s_0^*|d_1, s_0)c(d_1, s_0^*, s_0) + P(s_1^*|d_1, s_0)c(d_1, s_1^*, s_0) + P(s_2^*|d_1, s_0)c(d_1, s_2^*, s_0) + P(s_3^*|d_1, s_0)c(d_1, s_3^*, s_0) + P(s_4^*|d_1, s_0)c(d_1, s_4^*, s_0)$

$E(c|d_1, s_1) = P(s_0^*|d_1, s_1)c(d_1, s_0^*, s_1) + P(s_1^*|d_1, s_1)c(d_1, s_1^*, s_1) + P(s_2^*|d_1, s_1)c(d_1, s_2^*, s_1) + P(s_3^*|d_1, s_1)c(d_1, s_3^*, s_1) + P(s_4^*|d_1, s_1)c(d_1, s_4^*, s_1)$

...

...

$E(c|d_3, s_2) = P(s_0^*|d_3, s_2)c(d_3, s_0^*, s_2) + P(s_1^*|d_3, s_2)c(d_3, s_1^*, s_2) + P(s_2^*|d_3, s_2)c(d_3, s_2^*, s_2) + P(s_3^*|d_3, s_2)c(d_3, s_3^*, s_1) + P(s_4^*|d_3, s_2)c(d_3, s_4^*, s_2)$

$E(c|d_3, s_3) = P(s_0^*|d_3, s_3)c(d_3, s_0^*, s_3) + P(s_1^*|d_3, s_3)c(d_3, s_1^*, s_3) + P(s_2^*|d_3, s_3)c(d_3, s_2^*, s_3) + P(s_3^*|d_3, s_3)c(d_3, s_3^*, s_3) + P(s_4^*|d_3, s_3)c(d_3, s_4^*, s_3)$

(18)

$$E\left(c|d_1\right)=P\left(c|d_1\right)$$

Estimation of the expected values $E(c|d_k, s_i)$, for $k = 1, 2, 3$ and $i = 0, 1, 2, 3$, and then $E(c|d_k)$ enables further application of the following decision-making logic: *from among decisions $d_k (k = 1, 2, 3)$ this one should be selected, which the highest value $E(c|d_k)$ has been assigned to* [1, 12].

The decision-making situations can be presented in form of a decision tree. Because of its complexity such a tree is shown in a simplified version in Fig. 4.

Application of the logic is justified since the Bayesian statistical decision theory shows that the expected value of consequences (benefits or losses) may be a criterion for selection of the best possible decision, according to the previously formulated rule, provided that the values of consequences (results, outcomes) following from making the particular decisions, have been properly defined [1].

Taking either of these decisions $d_k (k = 1, 2, 3)$ which belong to the set [5, 7] $D = \{d_1, d_2, d_3\}$, with regard to a four-element set of states S and five-element set of situations S^*, brings on certain consequences $c(d_k, s_i, s_j^*) \equiv c_m \in C(m = 1, 2, ..., 60)$ which should be assessed before the decision is taken. The consequences are obviously dependent on the specified states $s_i (i = 0, 1, 2, 3)$ and situations $s_j^* (j = 0, 1, 2, 3, 4)$. These can be monetary values (costs, incomes or profits), or any benefits or losses significant for the user of the ship.

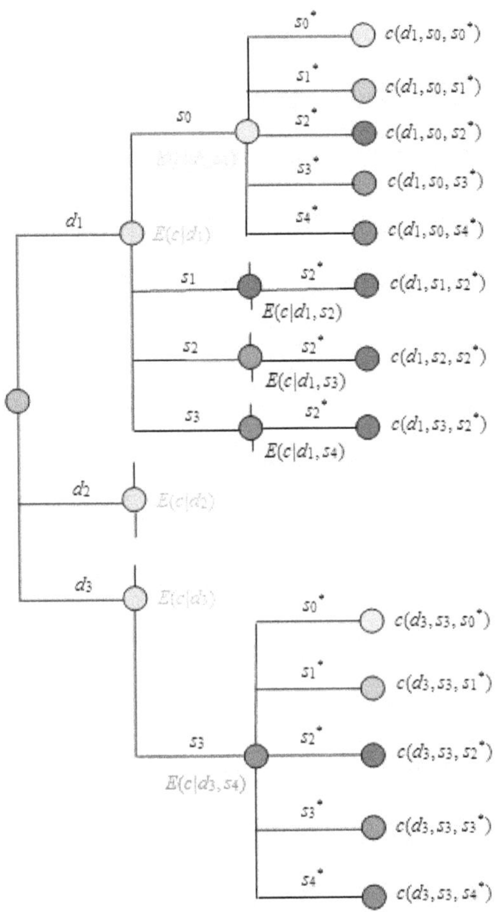

Figure 4. A decision tree for operational decisions d_k ($k = 1, 2, 3$) with regard to states s_i($i = 0, 1, 2, 3$), situations s_j^*($j = 0,1, 2, 3, 4$) and consequences $c(d_k, s_i, s_j^*)$.

Selection of the best from among the three specified decisions, in conditions resulting from the possibility of occurrence (obviously with certain probabilities) of the states s_i($i = 0, 1, 2, 3$) and situations s_j^* ($j = 0, 1, 2, 3, 4$), requires taking into account the decision criteria such as:

- expected values $E(c|d_k, s_i)$ of consequences $c(d_k, s_i, s_j^*)$, i.e. expected values of consequences at assumption that the decision d_k was taken when the ship's systems were in the state s_i, of which defining requires a knowledge of the probabilities $P_j = P(s_j^*)$, i.e. probabilities of occurrence of particular sea situations s_j^* ($i = 0, 1, 2, 3, 4$),
- expected values $E(c|d_k)$ corresponding to each of the decisions d_k as the values representing the product of the expected value $E(c|d_k, s_i)$ and probabilities $P_i = P(s_i)$, i.e. probabilities of occurrence of particular states s_i($i = 0, 1, 2, 3$).

The presented decision-making situation shows that the more difficult may be rather to assess the consequences of the decisions $c(d_k, s_i, s_j^*)$ than the probabilities $P(s_i)$ or $P(s_j^*)$. We can tend to an accurate estimation of the probabilities or accept the less accurate estimations for them. It can also be shown that for the needs of making operational decisions, an accurate assessment of effectiveness of the ship's systems in terms of their reliability, is not necessary. Therefore, the currently known distributions of random variables can be used for assessing the reliability of the ship's systems [1, 2, 3, 21].

5 REMARKS AND CONCLUSIONS

Making rational operational decisions in the phase of ship operation, especially in reference to its safety, is not possible without the knowledge of, among others,:

- probabilities $P_j = P(s_j^*)$ of occurrence of situations, especially these threatening the ship safety;
- probabilities $P_i = P(s_i)$ of occurrence of states, especially these ones that make impossible the use of the systems influencing fundamentally the ship safety, like main engines, steering systems and radars;
- probabilistic decision-making logic, e.g. formulated in the Bayesian decision theory.

Usefulness of the above probabilities for determining a decision-making situation with regard to safety of the systems has been showed in the papers [4, 6] presenting the rule for making an optimal decision by considering the expected values of consequences as a criterion function.

To determine these probabilities and rates there can be used: the theory of semi-Markov processes and statistical decision theory, as well as the theory of probability and mathematical statistics.

Application of the semi-Markov process as a model of changes of the specified situations in which the ship motion may be realized, instead of the Markov process, results from that it should not be expected that the random variables T_{ij} and T_i have any distributions concentrated in the set $R_+ = [0, +\infty)$. Application of the Markov process in this case would be justified if an assumption could be made that the random variables T_{ij} and T_i have exponential distributions.

The presented models can be of considerable practical importance because of ease in determining the estimators of the transition probabilities p_{ij} and ease in estimating the expected values $E(T_j)$. The considerations should include the fact that the point estimation of the expected value $E(T_j)$ does not enable to define the accuracy of its estimation. Such accuracy is possible by the interval estimation, where a confidence interval $[t_{dj}, t_{gj}]$ is determined with random end points, which includes, with a certain probability (confidence level) β, the unknown expected value $E(T_j)$.

The statistical decision theory can be useful for making operational decisions in the phase of ship operation. This theory enables formulation of mathematical models for making decisions under uncertainty. It explains how to make a choice from a set of possible decisions, when (among others) the state of a system indispensable to ensure the ship safety, cannot be precisely defined and also the sea situation that may arise during the ship's cruise, cannot be accurately predicted [1, 4].

REFERENCES

[1] Benjamin J. R., Cornell C. A.: Probability, statistics, and decision for civil engineers. WNT, Warszawa 1977.

[2] Firkowicz S.: Statystyczna ocena jakości i niezawodności lamp elektronowych. WNT, Warszawa 1963.

[3] Gercbach I. B., Kordonski Ch. B.: Reliability models for technical objects. WNT, Warsaw 1968.

[4] Girtler J., Kuszmider S., Plewiński L.: Selected problems of operation of sea going ships in the aspekt of navigation safety. Monograph. WSM, Szczecin 2003.

[5] Girtler J., Kitowski Z., Kuriata A.: Safety of a ship at sea. Systemic approach. WKiŁ, Warsaw 1995.

[6] Girtler J.: Zastosowanie bayesowskiej statystycznej teorii decyzji do sterowania procesem eksploatacji urządzeń. Materiały XXII Zimowej Szkoły Niezawodności nt. Wartościowanie niezawodnościowe w procesach realizacji zadań technologicznych w ujęciu logistycznym. SPE KBM PAN, Szczyrk 1994, s.55□62.

[7] Girtler J.: Sterowanie procesem eksploatacji okrętowych silników spalinowych na podstawie diagnostycznego modelu decyzyjnego. Zeszyty Naukowe AMW, Nr 100A, Gdynia 1989.

[8] Girtler J.: Prawdopodobieństwo jako miara bezpieczeństwa ruchu obiektów technicznych. Materiały VII Konferencji Naukowej „Bezpieczeństwo systemów'98". PTBiN, PTC, SPE KBM PAN, STŚT KT PAN, Wydział Transportu PW, ITWL, Warszawa□Zakopane-Kościelisko 1998, s.163□168.

[9] Girtler J.: Suggestion of interpretation of action and the rule of taking decision with regard to safety of sea ships traffic. Zagadnienia Eksploatacji Maszyn. Z. 2(122), vol. 35, 2000, pp177-191.

[10] Girtler J.: Koncepcja podejmowania decyzji z uwzględnieniem niezawodności i bezpieczeństwa morskich środków transportowych. Materiały Krajowej Konferencji Bezpieczeństwa i Niezawodności, t.1. ITWL, Warszawa-Zakopane-Koscielisko 1999, s.289□296.

[11] Girtler J.: Diagnostics as a condition of shipborne diesel engines operation control. Studies No 28. WSM, Szczecin 1997.

[12] Girtler J.: Availability of sea transport means. Archives of Transport. Polish Academy of Sciences Committee of Transport. Quarterly, vol. 9, iss. 3-4, Warsaw 1997.

[13] Girtler J.: Possibility of valuation of marine diesel engines. Journal of POLISH CIMAC, Vol. 4, No 1, Gdańsk 2009, pp 29-40.

[14] Girtler J.: Conception of selecting reliability models for taking operating decisions related to sea-going ship's systems. Zeszyty Naukowe WSM, Nr 66, Szczecin 2002 s.127-136.

[15] Grabski F.: Theory of semi-Markov processes of operation of technical objects" Research Journal WSMW (AMW), Nr 75A, Gdynia 1982.

[16] Gniedenko B.W., Bielajew J.K., Sołowiew A.D: Mathematical methods in reliability theory. WNT, Warsaw 1968.

[17] Jaźwiński J., Borgoń J.: Operational reliability and aviation safety. WKiŁ, Warsaw 1989.

[18] Krzysztofiak M., Urbanek D.: Metody statystyczne. PWN, Warszawa 1979.

[19] Plewiński L.: Wypadki na morzu. Zderzenia statków. Wyd. WSM w Szczecinie, 2000.

[20] Wiśniewski B.: Wind wave. Treatises and studies t. (CCCIV) 230. Szczecin University, Szczecin 1998.

[21] Reliability Handbook. Collective work edited by J. Migdalski. Machine Industry Publishing House „WEMA", Warsaw 1982.

Ship Handling in Wind and Current with Neuroevolutionary Decision Support System

M. Łącki
Gdynia Maritime University, Poland

ABSTRACT: This paper describes the advanced intelligent ship handling system which simulates a learning process of an autonomous control unit. This unit, called an artificial helmsman, controls model of ship passing through restricted water area. Simulated helmsmen are treated as individuals in population, which through environmental sensing and evolutionary algorithms learn to perform given task efficiently. The task is: safe navigation through restricted water, regarding an influence of winds and currents. Neuroevolutionary methods, which develop artificial neural networks with evolutionary algorithms, have been applied in this system.

1 INTRODUCTION

The safety of people, equipment and the environment is of primary meaning during decision-making processes, especially those occurring in the transport. Such important decisions should be taken with a minimum of uncertainty of the decision maker. This uncertainty may occur due to the existence of a number of factors, such as: the experience of decision-maker, the lack of information regarding the situation of the surrounding area and sometimes an excess of information provided to decision-maker simultaneously from multiple sources.

Computer aid processing systems become more efficient over time regarding constant increase of computing power available to standard users. That allows developing advanced systems which collect and analyze relevant data to support decisions that minimize the risk of taking the wrong ones.

These advanced systems may be used in support decisions on the real ships maneuvering and ship models and simulators used during the training of future officers at training centers.

The neuroevolutionary algorithms are methods that still are intensively tested and developed. Thanks to theirs efficiency, they are becoming widely used in many fields of science and technology, such as:

- automation and robotics systems, e.g. control of a robot arm (Siebel and Sommer 2007);
- designing and diagnostic systems, e.g. mobile hardware acceleration (Larkin, Kinane and O'Connor 2006), search hull damage (Kappatos, Georgoulas, Stylios and Dermatas 2009),
- control systems: for example, a helicopter flight stabilization (De Nardi, Togelius, Holland and Lucas 2006);
- decision support systems, e.g. systems with applied artificial intelligence in computer games (Stanley, Bryant and Risto 2005).

A large number of positive results of the implementation of neuroevolutionary methods obtained in many areas of science encouraged the author of this paper to undertake research and develop his own decision support system intended for use in maritime transport (Łącki 2010b, a).

In this paper, the extension of the functionality of navigational decision support systems is proposed. This solution generates specifications of maneuvering decisions (rudder angle and propeller thrust) that maintain a safe ship trajectory computed in the available water channel. In addition to rudder angle and propeller thrust this system also includes information about time of their execution. It is possible that all maneuvering decisions may be calculated and presented in real time for a given ship dynamics in the presence of certain external disturbances implemented in the system.

2 REINFORCEMENT LEARNING ALGORITHMS

In machine learning it is important to create the advanced computational system that can effectively find a solution of given problem and improve it over time. Reinforcement learning is a kind of machine learning, in which an autonomous unit, called a robot or agent, performs actions in a given environment. Through interaction and the observation of the environment (by input signals) and performing an action he affects this environment and receives an immediate score called reinforcement or reward (Figure 1.).

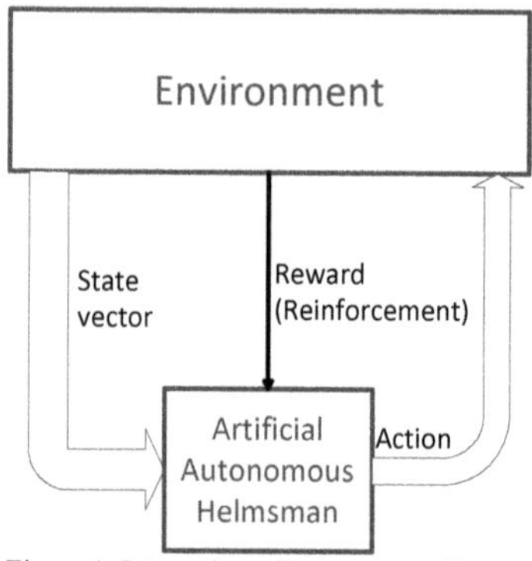

Figure 1. Interaction of helmsman with an environment.

The main task of the agent is to take such actions to adjust the value R which is the sum of the partial reinforcement as much as possible (1).

$$R_T = r_{t+1} + r_{t+2} + \dots\ r_T \qquad (1)$$

Such abilities are very important for simulating helmsman behavior in ship maneuvering on restricted waters.

For simpler layouts learning process can be performed using classic approach, i.e. Temporal Difference Reinforcement Learning (Tesauro 1995; Kaelbling, Littman and Moore 1996) or Artificial Neural Networks with fixed structures. Dealing with high-dimensional spaces is a known challenge in Reinforcement Learning approach (Łącki 2007) which predicts the long-term reward for taking actions in different states (Sutton and Barto 1998).

Evolving neural networks with genetic algorithms has been highly effective in tasks with continuous hidden states (Stanley, Bryant and Risto 2005). Neuroevolution gives an advantage from evolving neural network topologies along with weights which can effectively store action values in machine learning tasks. The main idea of using evolutionary neural networks in ship handling is based on evolving population of helmsmen.

The artificial neural network is the helmsman's brain making him capable of observing actual navigational situation by input signals and choosing an appropriate action. These input signals are calculated and encoded from current situation of the environment.

In every time step the network calculates its output from signals received on the input layer. Output signal is then transformed into one of the available actions influencing helmsman's environment. In this case the vessel on route within the restricted waters is part of the helmsman's environment. Main goal of the helmsmen is to maximize their fitness values. These values are calculated from helmsmen behavior during simulation. The best-fitted individuals, who react properly to wind effect, become parents for next generation.

3 THE FORCES ACTING ON THE SHIP

External disturbances acting on a seagoing ship are mainly wind, wave and current. In this paper the author assumes that the waves in the harbor area have little effect on ship maneuvering characteristics, and this type of interference is omitted in the system.

As the ship moves forward on the straight path (assuming there aren't any significant distorting external forces) there are the two major forces acting on her - the force from the propellers and the force of water resistance. At a constant speed, these forces are equal, but with the opposite direction.

When the ship is turning the additional forces act on the rudder and lateral forces from water pressure appear. During this maneuver, the ship loses a little of her velocity and the pivot point moves back toward amidships.

3.1 *Effect of wind in the ship handling*

Wind is defined as a movement of air relative to the surface of the Earth. Implementation of mathematical model of wind to the motion control system increases its performance and robustness in simulated environment.

Under pressure of wind force, depending of the ships' design (location of the superstructure, the deployment of cargo and on-board equipment, etc.) she tends to deviate from the course, with the wind or into the wind. The smaller the speed and draft of the ship, the greater the influence of wind. Of course, the size of the side surface exposed to wind is essential to the ships movement.

When ship moves forward the center of effort of the wind (wind point, WP) is generally close to amidships, away from pivot point (PP). This difference creates a substantial turning lever between PP and WP thus making the ship to swing of the bow into the wind (with the superstructure deployment at stern) (Figure 2.).

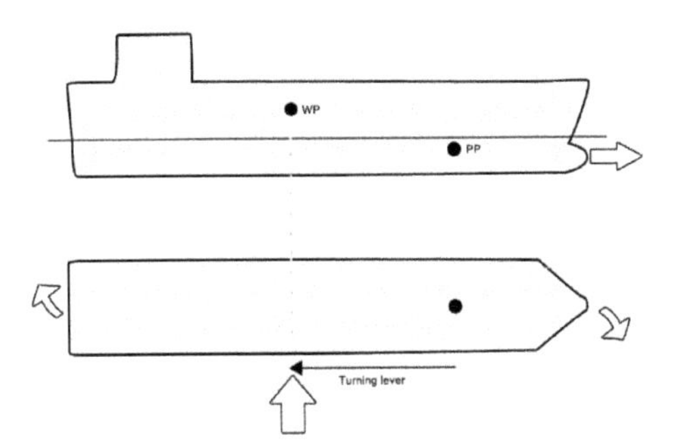

Figure 2. Wind effect and turning lever of ship with stern superstructure, moving forward.

For ship moving forward there are defined terms of relative wind speed V_{rw} and angle of attack γ_{rw} as follows (Isherwood 1973):

$$V_{rw} = \sqrt{u^2_{rw} + v^2_{rw}} \tag{2}$$

$$\gamma_{rw} = -\arctan\left(v_{rw}, u_{rw}\right) \tag{3}$$

where:

$$u_{rw} = u - u_w \tag{4}$$

$$v_{rw} = v - v_w \tag{5}$$

where: u, u_w, v, v_w are longitudinal and lateral velocities of ship and wind, respectively.

Wind forces acting on symmetrical ship are in general calculated as follows:

$$X_{wind} = \frac{1}{2}\rho_{air}V^2_{rw}C_X\left(\gamma_{rw}\right)A_{Fw} \tag{6}$$

$$Y_{wind} = \frac{1}{2}\rho_{air}V^2_{rw}C_Y\left(\gamma_{rw}\right)A_{Lw} \tag{7}$$

$$N_{wind} = \frac{1}{2}\rho_{air}V^2_{rw}C_N\left(\gamma_{rw}\right)A_{Fw}L_{0a} \tag{8}$$

where:

- ρ_{air} – air density,
- A_x – surfaces affected by wind,
- L_{0a} – ship's length,
- C_n – coefficients calculated from available characteristics of VLCC model (Figure 3.), according to OCIMF wind loads (OCIMF 1977).

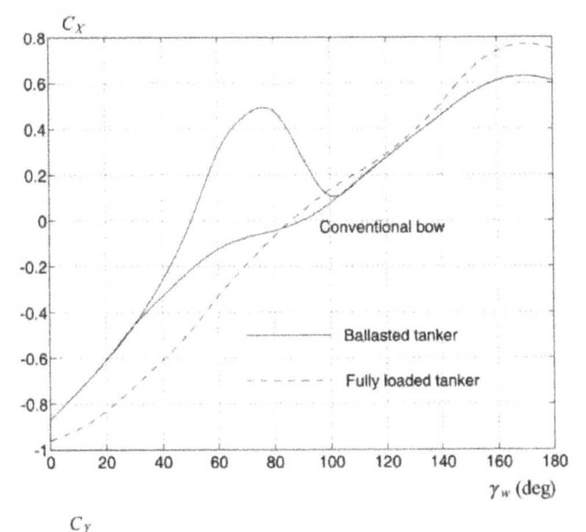

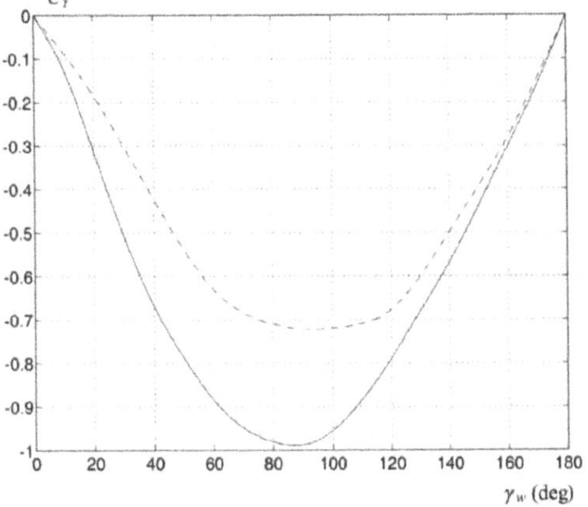

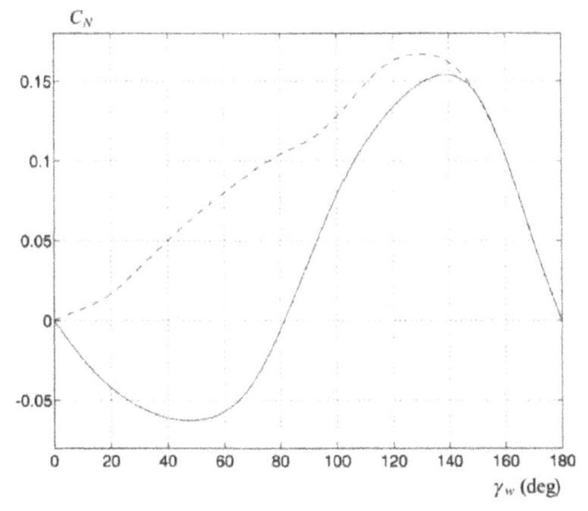

Figure 3. Equations coefficients for relative wind.

3.2 *Effect of water current*

Water current is defined as a flow of a water relative to the bottom of a river, sea or an ocean.

Under pressure of current a ship is drifting together with the water, relative to the ground and any fixed objects. When the ship is moving in current the speed over ground is resultant velocity of ship speed and velocity of the current.

Stronger currents may cause the tendency to swing the ship with the stern or bow towards the side, which may create dangerous situation. This might be particularly difficult to overcome when working in following tide because of the small effectiveness of the rudder at low speed.

4 EVOLUTION OF ARTIFICIAL NEURAL NETWORKS

Neuroevolution of Augmenting Topologies (NEAT) method is one of the Topology and Weight Evolving Artificial Neural Networks (TWEANN's) method (Stanley and Risto 2002b). In this method the whole population begins evolution with minimal networks structures and adds nodes and connections to them over generations, allowing complex problems to be solved gradually starting from simple ones.

The modified NEAT method consist four fundamental procedures which deal with challenges that exist in evolving efficient neural network topology:

1 Begin with a minimal structure and add neurons and connections between them gradually to discover most efficient solutions throughout evolution.
2 Cross-over disparate topologies in a meaningful way by matching up genes with the same historical markings.
3 Separate each innovative individual into a different species to protect it disappearing from the population prematurely.
4 Reduce oversized topologies by removing neurons and connection between them to provide and sustain good overall performance of a whole population of helmsmen.

4.1 *Genetic Encoding*

Evolving structure requires a flexible genetic encoding. In order to allow structures to increase their complexity, their representations must be dynamic and expandable (Braun and Weisbrod 1993). Each genome in NEAT includes a number of inputs, neurons and outputs, as well as a list of connection genes, each of which refers to two nodes being connected (Figure 4.).

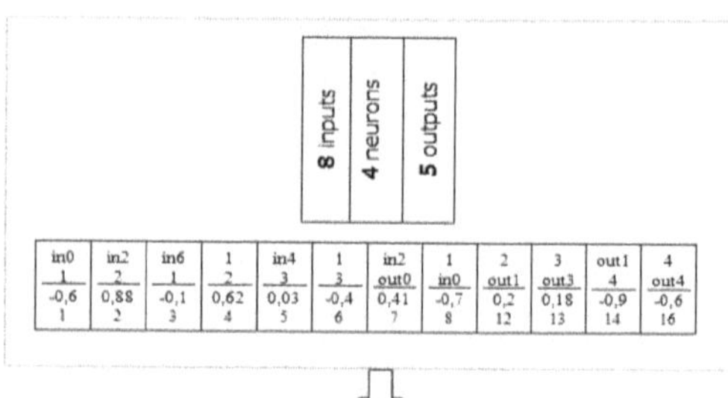

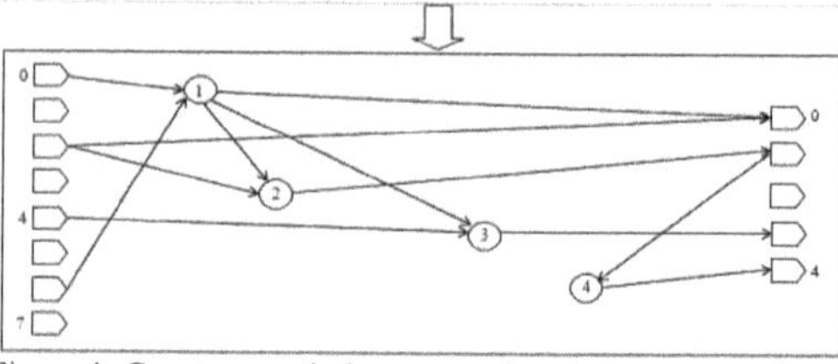

Figure 4. Genotype and phenotype of evolutionary neural network.

In this approach each connection gene specifies the output node, the input node, the weight of the connection, and an innovation number, which allows finding corresponding genes during crossover. Connection loopbacks are also allowed, as shown in figure 4.

4.2 *Genetic operations*

There are two main genetic operations: cross-over and mutation. During cross-over two individuals (parents), exchange their genetic material in purpose of creating new individual (an offspring). The system knows exactly which genes match up with which through innovation numbers. Genes that do not match are either disjoint or excess, depending on whether they occur within or outside the range of the other parent's innovation numbers.

In crossing over operation, the genes with the same innovation numbers are lined up. The offspring is then formed in one of three ways:

- In uniform crossover: matching genes are randomly chosen for the offspring genome, with all disjoints and excesses from both parents.
- In blended crossover: the connection weights of matching genes are averaged, disjoints and excesses are chosen randomly.
- In elite crossover: disjoints and excesses are taken from more fit parent only, all redundant genes from less fit parent are discarded. All matching genes are averaged.

These types of crossover were found to be most effective in evolution of neural networks in extensive testing compared to other methods of crossover (Stanley and Risto 2002a).

Disabled genes have a chance of being re-enabled during mutation, allowing networks to make use of older genes once again.

Evolutionary neural network can keep historic trails of the origin of every gene in the population, allowing matching genes to be found and identified even in different genome structures. Old behaviors encoded in the pre-existing network structure have a chance to not to be destroyed and pass their properties through evolution to the new structures, thus provide an opportunity to elaborate on these original behaviors.

Through mutation, the genomes in modified NEAT will gradually get larger for complex tasks and lower their size in simpler ones. Genomes of varying sizes will result, sometimes with different connections at the same positions. Any crossover operator must be able to recombine networks with differing topologies, which can be difficult. Historical markings represented by innovation numbers allow NEAT to perform crossover without analyzing topologies. Genomes of different organizations and sizes stay compatible throughout evolution, and the variable-length genome problem is essentially solved. This procedure allows NEAT to increase complexity of structure while different networks still remain compatible.

During elitist selection process the system eliminates the lowest performing members from the population. In the next step the offspring replaces eliminated worst individual.

5 SIMULATION RESULTS

The main goal of authors work is to make a system able to simulate a safe passage of ship moving through a restricted coastal area in variable wind and current conditions. This goal may be achieved with Evolutionary Neural Networks.

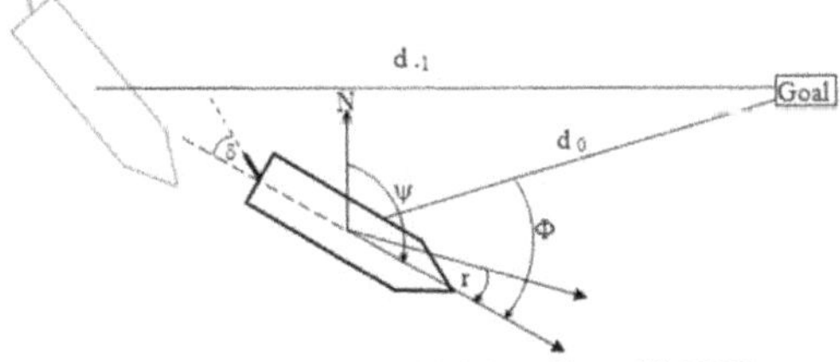

Figure 5. Basic input signals of ship handling with ENN.

Navigational situation of a moving vessel can be described in many ways. Most important is to define proper state vector from abundant range of data signals (Fig. 5.) and arbitrary determine fitness function values received by the helmsman.

The main input signals are gathered from data listed below:

- Ships course over ground,
- Ships angular velocity,
- The ship is on the collision course with an obstacle,
- Distance to collision,
- The ship is approaching destination,
- Ships angle to destination,
- The ship is heading out of the area,
- Distance to canal borders,
- Ship is heading on goal,
- Distance to goal,
- Angle and velocity of current,
- Angle and velocity of wind.

All of the 50 input signals are encoded either binary (0 or 1) or as a real values between 0 and 1. Some of the input signals may be calculated as multi-criteria values (Filipowicz, Łącki and Szłapczyńska 2006). Neural network output values are signals for rudder angle (δ) [deg] and thrust control [rpm].

Fitness calculation defines helmsman ability to avoid obstacles and react to wind forces while sailing toward designated goal. The fitness value of an individual is calculated from arbitrary set action values, i.e.:

- -10 when ship is on the collision course (with an obstacle or shallow waters),
- +10 when she's heading to goal without any obstacles on course,
- -100 when she hits an obstacle or run aground,
- +100 when ship reaches a goal,
- -100 when she departs from the area in any other way, etc.

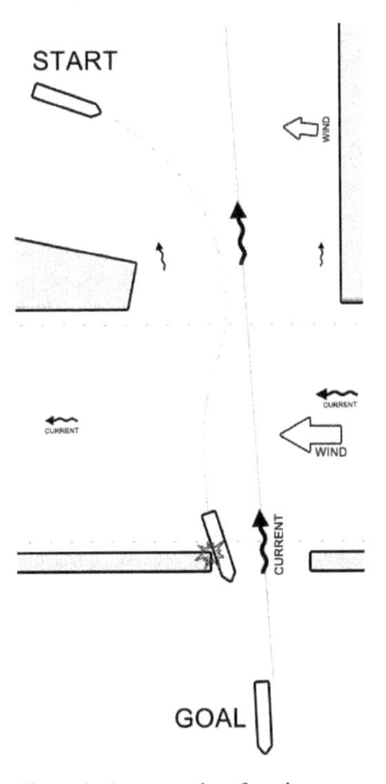

Figure 6. An example of a dangerous situation in narrow passage of simulated coastal environment.

In the simulation of safe passage through restricted waters there are no moving vessels in the area (Fig. 6.). In this situation when ship enters a heavy side wind channel with strong opposite current, there is a high risk to lose maneuverability and hit an obstacle if no preventive action is being made by the helmsman. In designed system an artificial helmsman observes current situation which is encoded as input signals for his neural network and calculates the best (in his opinion and according to his experience) rudder angle (Fig. 7. - 9.).

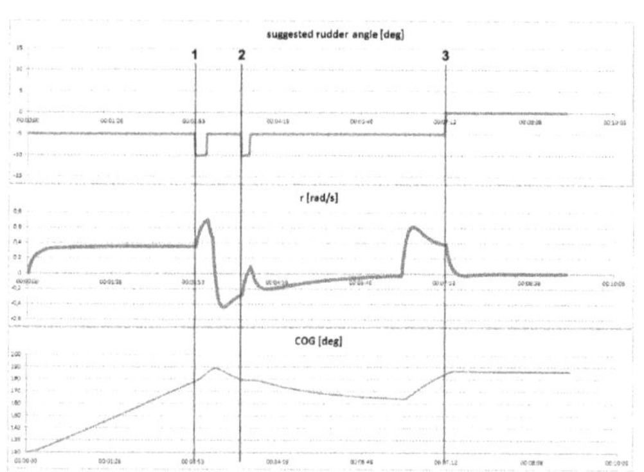

Figure 7. Simulation results of the systems performance in vary wind and current conditions.

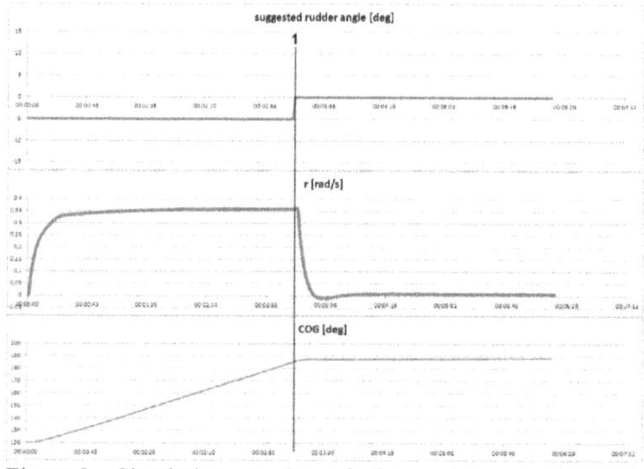

Figure 8. Simulation results of the systems performance without wind and current in the area.

During this simulation additional reinforcement values were introduced to the system, which were: negative reward for high values of suggested rudder's angle (±10° or above) and negative reward for rapid changes of the rudder's angle. This policy resulted in elimination of individuals that took too impetuous actions and favored the gentle ones.

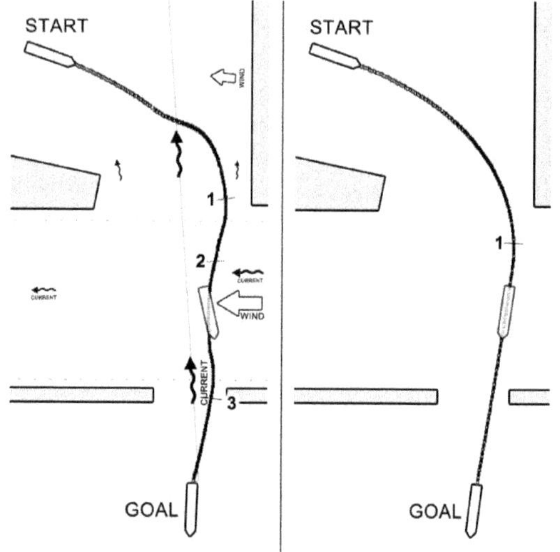

Figure 9. Comparison of simulation results of two situations: passage through restricted area with winds and currents (on the left) and without them (on the right).

6 REMARKS

Neuroevolution approach to intelligent agents training tasks can effectively improve learning process of simulated helmsman behavior in ship handling (Łącki 2008). Artificial neural networks based on NEAT increase complexity and performance of considered model of ship maneuvering in restricted waters.

Implementation of additional disturbances from wind and current in neuroevolutionary system allows to simulate complex behavior of the helmsman in the environment with much larger state space than it was possible in a classic state machine learning algorithms (Łącki 2007). Positive simulation results of maneuvers in variable wind and current conditions encourage to implement this method into navigational decision support systems to increase safety of navigation.

REFERENCES

Braun H. and Weisbrod J. (1993). Evolving Feedforward Neural Networks. International Conference on Artificial Neural Networks and Genetic Algorithms, Innsbruck, Springer-Verlag.

De Nardi R., Togelius J., Holland O. E. and Lucas S. M. (2006). "Evolution of Neural Networks for Helicopter Control: Why Modularity Matters." Evolutionary Computation, 2006. CEC 2006. IEEE Congress on: 1799-1806.

Filipowicz W., Łącki M. and Szłapczyńska J. (2006). "Multicriteria Decision Support for Vessels Routing." Archives of Transport 17: 71-83.

Isherwood J. W. (1973). "Wind resistance of merchant ships." Transactions of Royal Institution on Naval Architects 115: 327-332.

Kaelbling L. P., Littman M. L. and Moore A. W. (1996). "Reinforcement Learning: A Survey." Journal of Artificial Intelligence Research cs.AI/9605: 237-285.

Kappatos V. A., Georgoulas G., Stylios C. D. and Dermatas E. S. (2009). Evolutionary dimensionality reduction for crack localization in ship structures using a hybrid computational intelligent approach. Proceedings of the 2009 international joint conference on Neural Networks. Atlanta, Georgia, USA, IEEE Press: 1907-1914.

Larkin D., Kinane A. and O'Connor N. (2006). Towards hardware acceleration of neuroevolution for multimedia processing applications on mobile devices. Proceedings of the 13th international conference on Neural information processing - Volume Part III. Hong Kong, China, Springer-Verlag: 1178-1188.

Łącki M. (2007). Machine Learning Algorithms in Decision Making Support in Ship Handling. TST, Katowice-Ustroń, WKŁ.

Łącki M. (2008). Neuroevolutionary approach towards ship handling. TST, Katowice-Ustroń, WKŁ.

Łącki M. (2010a). "Model środowiska wieloagentowego w neuroewolucyjnym sterowaniu statkiem." Zeszyty Naukowe Akademii Morskiej w Gdyni 67: 31-37.

Łącki M. (2010b). "Wyznaczanie punktów trasy w neuroewolucyjnym sterowaniu statkiem." Logistyka 6.

OCIMF (1977). Prediction of Wind and Current Loads on VLCCs, Oil Companies International Marine Forum.

Siebel N. T. and Sommer G. (2007). "Evolutionary reinforcement learning of artificial neural networks." International Journal of Hybrid Intelligent Systems - Hybridization of Intelligent Systems 4(3): 171-183.

Stanley K. O., Bryant B. D. and Risto M. (2005). "Real-time neuroevolution in the NERO video game." IEEE Transactions on Evolutionary Computation 9(6): 653-668.

Stanley K. O. and Risto M. (2002a). Efficient evolution of neural network topologies. Proceedings of the Evolutionary Computation on 2002. CEC '02. Proceedings of the 2002 Congress - Volume 02, IEEE Computer Society: 1757-1762.

Stanley K. O. and Risto M. (2002b). Efficient Reinforcement Learning Through Evolving Neural Network Topologies. Proceedings of the Genetic and Evolutionary Computation Conference, Morgan Kaufmann Publishers Inc.: 569-577.

Sutton R. and Barto A. (1998). Reinforcement Learning: An Introduction, The MIT Press.

Tesauro G. (1995). "Temporal difference learning and TD-Gammon." Communications of the ACM 38(3): 58-68.

Chapter 3

Marine Traffic

Development and Evaluation of Traffic Routing Measurements

R. Müller
Hochschule Wismar, University of Technology, Business and Design, Germany

M. Demuth
Schiffahrtsinstitut Warnemünde e.V., Institute at Hochschule Wismar, Germany

ABSTRACT: Traffic Routing Measurements (TRM) like Traffic Separation Schemas, Deep Water routes, recommended waterways with crossing sections or recommended directions of traffic flow are approved methods for enhancing the safety and lightness in transit shipping. This may be reached by demerging of traffic flow directions, by bundling of waterways or by allocating of no-go zones. Mostly the regulations instruments are marked in sea charts as magenta colored lines or objects; sometime they are manifested on water's surface by buoys.
While passing sea areas with TRM they have to be taken into account by the ship master as mandatory recommendations. However, the process of measurement establishing is a long-term international procedure under the leadership of the IMO. On its very beginning the characteristic and outline of the TRM has to be developed conforming to the arrangement's needs and the risk assessment.
This paper will introduce a tool for TRM layout evaluation by traffic flow simulation. The evaluation will be carried out by the identification of a conflict potential caused by the traffic flow regarding regulation measurements.

1 BACKGROUND

Under the leadership of HELCOM Expert Working Group on Transit Routing [1], [3], [7] the German Ministry of Transport Building and Urban Affairs initiated a study to cast lights on the use of traffic route recommendations for enhancement of the navigational safety in Baltic Sea areas. The objectives of the investigations were:

- to suggest a general procedure for evaluation the present real traffic in area between the Kadetrenden and the Bornholms Gatt using Vessels AIS data,
- development of different traffic measurements proposals
- to calculate the different traffic scenario proposals benefits by statistical investigation
- to prognoses the validity of favored route proposals for expected traffic profiles or density in future
- to illustrate the efforts of traffic route measurements by comparison of various route versions

The goal is to prevent and protect the natural environment against a catastrophe caused by growing up traffic density, dangerous of cargos or the installation of offshore wind farms in the Baltic Sea area. Traffic flow conditions enforced by passive Traffic Routeing Measurements (TRM) have to be reached a very high standard and compliance of safety and lightness in shipping for each of all traffic participants.

In analysing historical AIS– data, passive traffic standards have proven to be problematic because they do not necessarily reflect the intricacies or individual aspects of real-time traffic situations. Not enough substantiated, detailed information concerning trafficked routes, navigational spaces of these routes, nor about the frequency of usage of the routes has been gathered in the past which could have been applicable to real-time traffic situations; although information concerning traffic congestion, the regulation of traffic, and the ships which traffic these waters has been gathered by renowned institutes [8]. Primarily and most importantly, gathering information and researching ship traffic should be applied to real-time traffic situations in order to ascertain potential hazards; formally, data has not accomplished this task [2].

This paper will describe how safe, efficient traffic measures can be applied to an a priori traffic situation by using AIS information without having to undergo a cost and time intensive trial and error period.

In the initial phase of research, AIS – data was provided by the Water and Shipping Direction (WSD) North which had been gathered in the Baltic Sea over a long period of time. Also information was made available by the Federal Agency of Shipping and Hydrograph (BSH) including detailed information from Electronical Chart Display and Information System (ECDIS) data banks. For the first time, based on the available data, such extensive information on transit traffic could be laid out in a real-time format which would also facilitate further theoretical investigations.

The following description and conclusions are in reference to the consequences of the applied AIS – data on traffic measures implemented in the southern Baltic Sea region.

2 DATA ACQUISITION

The main advantage of standard and automated data transfer and data logging – in comparison to analogue or manual data entry (for example counting radar targets) – is that all recorded data is delivered in a consistent, high-quality manner, and the information can be provided in an extensive and statistically secure bulk. The plausibility and consistency of gathered AIS raw data for a long period is checked by undergoing a preliminary filter process. The filtering process aims at pinpointing:

- mistakes
- loopholes
- uncorrelated time stamps

The National Marine Electronics Association (NMEA) data flow is revised after assessing:

- status and administration
- own ship data
- faulty speed and course values (data)
- faulty position data

The final analysis consists of different aspects of pertinent ship data – this means the summary contains unequal Maritime Mobile Service Identity (MMSI) in which numbers are simultaneously examined.

In Fig. 1 the position data on all traffic, received in time intervals, is embedded in an ECDIS environment. This image is already historically, because the traffic flows in its urban manner before establishing new traffic route measurements in 2006.

The used transit routes are represented by a grey-sliding scale depending on frequency of position reports. The route will be darker if the number of reports with the same geographical position is growing.

The black lines stand for transit routes with an increased frequency of use.

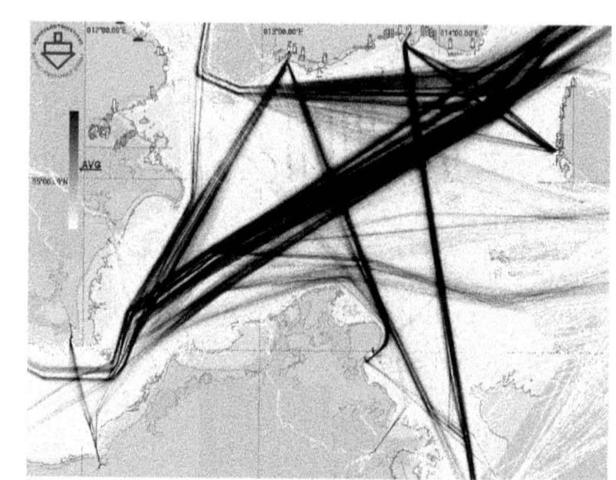

Figure 1. Visualization of real traffic flow scaled by traffic density

An analysis of post-processed information and the first statistical analysis of locally passed transit areas have resulted in descriptive characteristics concerning lateral and chronological distribution, frequency and density of traffic which can be described individually (see Fig 2).

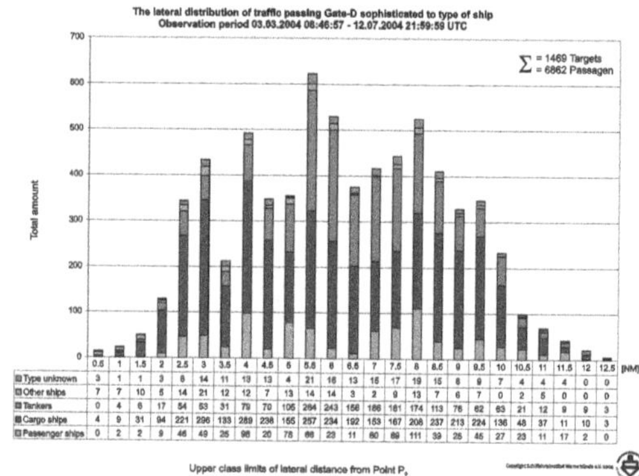

Figure 2. Ship type profile of the transit traffic

3 THE TRAFFIC FLOW SIMULATION MODEL

The tool for traffic flow simulation consists of two main blocks: creating the traffic route conditions and assigning the traffic flow profile. This may be based on artificial or real data bases or settings.

The layout of any traffic scenario may be designed by so called gates which mark the traffic lanes. The width, location, number and distance between the gates are arbitrary. At any gate targets may be transmitted or received solely or both transmitted and received. So, the route conditions are created individually corresponding to the investigation needs like one-way lanes, separated lanes, crossing lanes and junctions.

The traffic flow of each lane is stored in the corresponding gate of the lane as a profile which includes the amount, density, distribution of traffic and the types, dimensions, draughts and velocities of ships.

An overall time controlled transmitting function manages the traffic flow. That means each gate gets its operation order by the transmitting function. So, depending on the profile stored at the gate one or more targets will send out or not in each timeslot (see Fig. 3). The target moves to its receiving gate on an automatically generated path described by a list of waypoints [1], [4].

On the end of the simulation process the traffic is evenly and randomized distributed according to the given traffic flow characteristic. The simulation process can be repeated as often as necessary for reducing the statistic error.

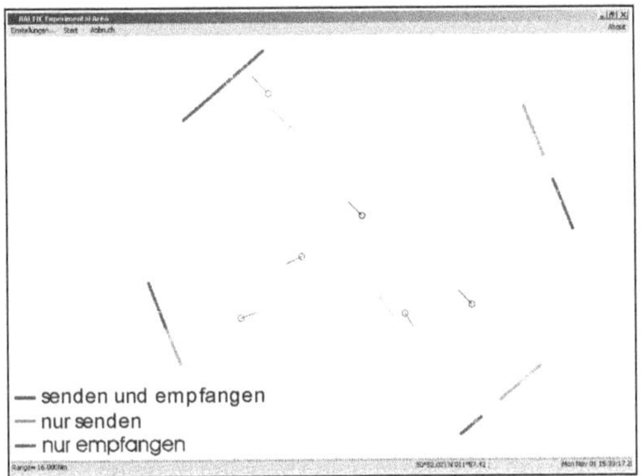

Figure 3. Snapshot of a traffic flow situation

4 TRAFFIC FLOW EVALUATION

The evaluation of the traffic flow is based on the investigation of all time- corresponding waypoint lists of all traffic participants.

If **G** is the set of all passages tgt_i which reflect the whole traffic flow within the simulation:

$$\mathbf{G} = \bigcup_{i=0}^{n} tgt_i ,$$

n = total number of passages

Each passage i marks the path of a transmitted target and consists of a time discrete waypoint list $WP_i = \{wp_i(t_l);\ i, l \in N_0\}$. Each element of the waypoint list includes dynamically information of time, position, course and speed. All synchronous waypoints at time stamp l of different waypoint lists WP_i are linked to a process list

$$\mathbf{P}_l = \{wp_i(t), \text{with } t \equiv l, i = 0,..,n\}.$$

Within the list P_l the algorithm for conflict detection will be executed.

5 CONFLICT DETECTION

A conflict is considered as the violation of a virtual safety zone. The safety zone is defined as a curved surface around a target or reference point with a predefined radius function. The parameter of the radius function are the dimension of target's length and width, the velocity of the object and a defined weight reflecting the special hazardous potential of the target like dangerous cargo or restricted maneuverability. Thus, the curved surface is fitted individually and dynamically by the circumstances of the object to be protected.

Two moving targets come into an encountering conflict if the distance and bearing to each other is less than the targets largest safety zone.

In Fig. 4 the simplest layout of a safety zone is shown. The accessibility of the ship is evenly spread around its reference position [5]. This circle is considered in cases of slow or zero movement of a target.

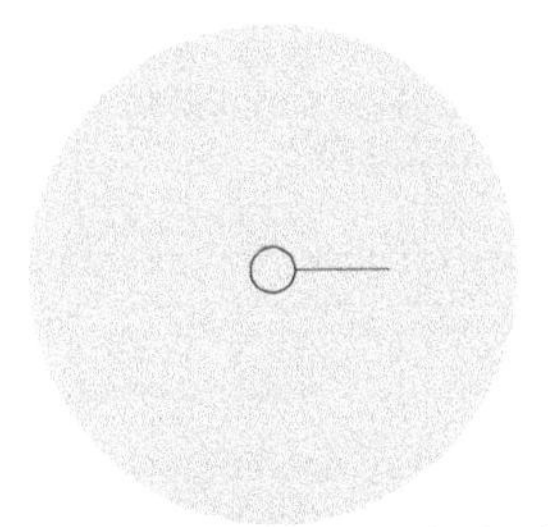
Figure 4. A safety zone of a dead slow moving target

In general, the potentiality of an encountering conflict situation increases to the given transit traffic density proportionally. First of all the conflict duration of an ongoing encountering situation depends on the mutual angle of the encountering objects. The period of a ship passing another with the same route direction is subjected to whose velocity gradient and will be a multiple higher then in oncoming route direction situations.

Accordingly, there is a distinction between different conflict situations:

- passing (similar ship courses ± 10 degrees)
- oncoming (reverse ship courses at ± 10 degrees)
- crossing (remaining ship courses difference)
- travelling past (immobile obstacles)

The first result of the conflict detection algorithm is the set of safety violations generated by the traffic flow simulation. Each conflict may characterized and classified by

- Type of encountering situation
- Duration of conflict
- Number of involved targets
- Risk potential caused by shipping goods or anything else depending on the focus of investigation.

6 AN EXAMPLE

Both the eastbound and westbound transit traffic between the end of deep-water route "DW 17" and the entrance to Bornholmsgat were analyzed. The underlying traffic data were collected at the Stubbenkammer AIS station and made available for the study by the Waterways and Shipping Directorate North in Kiel. The relevant data pool reflects the traffic flow over a period of 131 consecutive days in 2004. As a routine matter, the data provided were checked for consistency and plausibility, duly filtered, and then re-formatted and archived for further investigation.

In a first step, the data material was used as a basis for making a statistical description of the current traffic flows, including a description of their distribution over time and of the numbers of vessels involved in different potential conflict scenarios. The data were processed using the traffic flow simulation model in the defined environment reflecting the origin routing situation at the first stage and a proposed routing measure at the second stage. The scenarios of traffic flow simulation used in the analysis were defined as follows:

Before: This term stands for a traffic condition currently existent in reality (the "present state", see Fig 1).

After: This term means a traffic condition as may be expected to exist upon the implementation of the routing measures proposed by Denmark, Estonia, Finland, Latvia, Lithuania, Poland, Sweden and Germany [6].

After simulation runs the characteristic of detected conflicts were analyzed and compared by the powerful SPSS software package for stochastically analyzing.

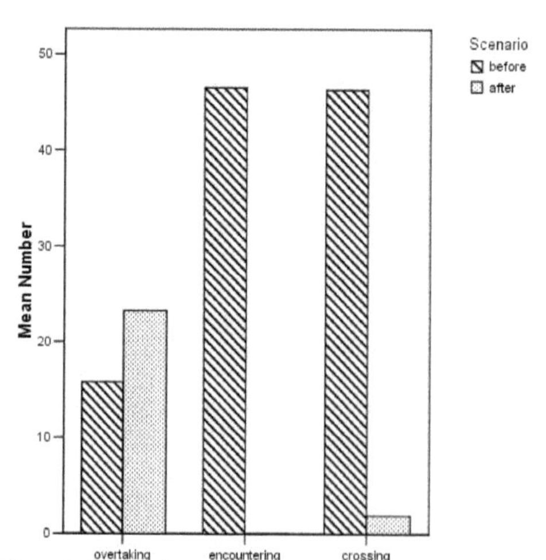

Figure 5. Bars of detected conflict types by scenario

The problem to be considered is as follows: The proposed routing measures will separate the traffic flows according to the traffic direction while, at the same time, drastically reducing the space currently occupied by the traffic routes. That means that shipping traffic, separated by the direction of traffic flow, offers a minimized potential for conflict situations while, at the other hand, the reduced space available for navigation may increase the conflict potential, and the number of targets involved in any one conflict may also increase.

Test Statistics (a, b)

	number
Mann-Whitney U	7664.500
Wilcoxon W	18989.500
Z	-4.776
Asymp. sig. (two-tailed)	.000

a Grouping variable = scenario
b Situation = overtaking

Test Statistics (a, b)

	number
Mann-Whitney U	.000
Wilcoxon W	11325.000
Z	-16.010
Asymp. sig. (two-tailed)	.000

a Grouping Variable = Scenario
b Situation = encountering

Test Statistics (a, b)

	number
Mann-Whitney U	5687.000
Wilcoxon W	17012.000
Z	-7.638
Asymp. sig. (two-tailed)	.000

a Grouping Variable = Scenario
b Situation = crossing

The striking figure in the above tables is the significant difference between the two scenarios, *Before* and *After*, in every single situation. The significant level has been set at 5 per cent ($p < 0.05$). While the number of detected conflicts in overtaking situations is higher in scenario *After* than in scenario *Before*, there is a nil value in the pool of scenario *After* for encountering situations, i.e., no encountering conflict was detected. In crossing situations the number of detected conflicts is extremely reduced in scenario *After*.

As a result of this basic study the following is quite evident:

The implementation of the proposed routing measures will provide a demonstrable decrease in the traffic characteristics that are of relevance to potential conflicts.

The current conflict potential may be enormously reduced by the separation of traffic through the implementation of a traffic separation scheme with separate traffic lanes according to the direction of traffic flow. Such conflict potential reduction is particularly evident as regards encountering situations. The statistical data outlined in the present document go to prove this point most impressively.

The proposed routing measure was adopted by the IMO and was established as the Traffic Separation Scheme "North of Ruegen" on 1 July

2006. At this time an AIS based study was made to diagnose ship master's behavior when passing this transit area.

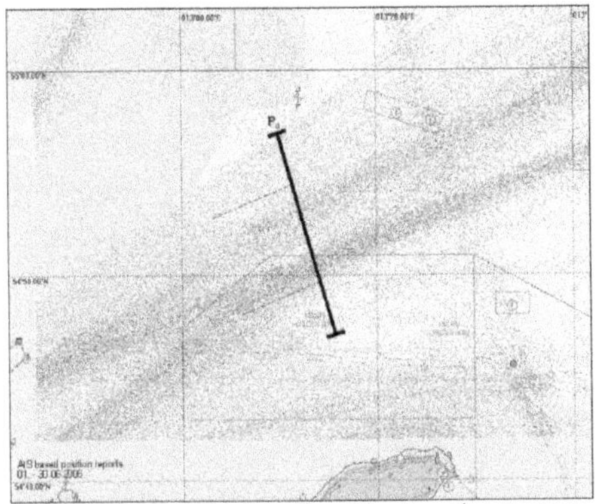
Figure 6. Spread transit traffic without routing measure

The traffic space was reduced within two-third by the installation of the T.S.S.

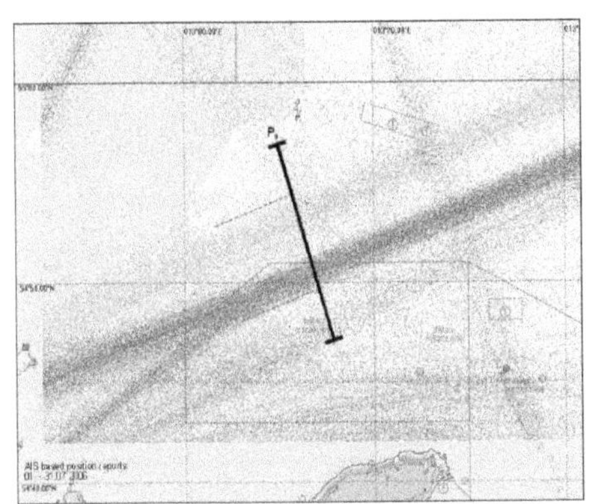
Figure 7. Transit traffic bundled by routing measure

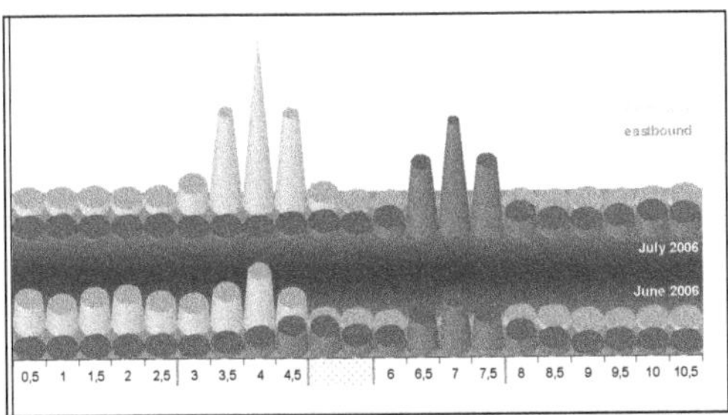

Figure 8. Distribution of transit traffic before and after establishment of transit measure

First of all the study showed that the ship masters are disciplined on a high level. Over 95% of all passages were navigated in respect to the traffic routing measure from the very first time of its recommendation; although this navigational area is not monitored or supervised by Vessel Traffic Services.

The number of encountering situations with reverse ship courses was reduced close to zero by the routing measure. The number of overtaking situations of ships with similar route directions was decreased on 1.5 times. Comparing this result with the outcomes of the evaluation based on the traffic flow simulation it is shown that the prognostic issues came realized.

7 CONCLUSION

In the paper a model of a scientific approach for the assessment and evaluation of traffic routing measures was introduced and an applied example and its benefits were illustrated.

The model permits the simulation of a multitude of traffic situations. This facilitates the realization of forward-looking assessments concerning the implications of possible time and kinetic coordination of ship traffic in concentrated spaces, harbor entrances and channel routes.

In the scientific analysis targeted new route guidance can determine the effects of a modified traffic flow situation. This trend is derived from the foundation of a statistical evaluation of produced simulated data pertaining to the cumulatively examined sea traffic area. The results deliver direct evidence concerning the efficiency of current and planned traffic measures for enhancing the safety and lightness in shipping. Significant indicators for describing and evaluation of traffic flow situations can be deduced from the scientific enquiries.

The outcome of the research yields a scientifically founded and authorized safety and risk analysis and allows the projected modifications in traffic guidance to be objectively evaluated and made comparable.

LITERATURE

[1] R. Müller, A. Zölder, F. Hartmann: Historical AIS Data Use for Navigational Aids. HANSA, 01/2005.
[2] http://www.merikotka.fi/mimic/
[3] R. Müller, M. Demuth, F. Hartmann: Multisensor-gestützte Bewegungssteuerung zur Navigation schneller Schiffe (MUBES). Abschlussbericht, BMBF, 2003.
[4] R. Müller, M. Demuth: Knowledge-Based Advisory System for Manoeuvring Large Ships in Restricted Waters. COMPIT '05, Hamburg, 08.-11.05.2005.
[5] B. Beutel: Simulationstechnische Bewertung von Navigationsübungen. Diplomarbeit, 2003.
[6] IMO, NAV 51/INF.2: Routeing of ships, ship reporting and related matters. Submitted by Germany, 4 March 2005.
[7] HELCOM EWG "Transit Route": HELCOM future task discussion. St. Petersburg (Russia), 20.-21.09.2004
[8] Seeverkehrsmarkt Baltic Sea. Deutsche Ostseehäfen. THB Deutsche Schiffahrts-Zeitung (54) 2001, H. BE 26.01., S. 1-12.

Świnoujście – Szczecin Fairway Expert Safety Evaluation

P. Górtowski & A. Bąk
Maritime University of Szczecin, Poland

ABSTRACT: Paper presents the results of research on safety of navigation in the Świnoujście - Szczecin fairway as the part of the project "Marine Transport Safety Management System in Poland". The research conducted with participation of pilots who are working in that fairway. Aims of this research was to assess the current state of safety of navigation in the fairway Świnoujście-Szczecin and the waters of the both ports as well as the effect of some external conditions on the overall assessment of the safety maneuver. The task was to determine whether kind of maneuver was depending on where you maneuver and identify hazardous sections.

1 ASSESSMENT OF THE CURRENT SITUATION

In the first part of the research attempted to systematize the factors that may be relevant to the safety of navigation. The aim of this section which consists of 24 questions included in the five subjects groups was to identify and analyze areas which, in the pilot judge, may be a potential source of danger.

The analysis included:

- The fairway
- Partnered / Cooperate with the VTS system,
- State aids to navigation,
- Partnered / Cooperate with tugs,
- Partnered / Cooperate with the administrators of the quays.

In any case, the pilot gave the weight (importance) to each of the factors and carried out its assessment, considering the size of the ships which are currently entering into the ports of Świnoujście, Szczecin and Police and ship traffic on the fairway. In addition, pilots could indicate additional factors in each subject group, not included in the list of factors. Such factors will be presented in the results, if it were showed by at least 10% of experts. To determine the weight of the tested factor authors were used five-factor ordinal scale while to determine the assessment was used the bipolar scale with neutral position zero. In this part of the questionnaire were placed the question of particularly dangerous part of the waterway and waterfront and reasons for the choice. Such dangerous part of the waterway and waterfront were placed in the results if it were showed by at least 10% of the experts.

Each questionnaire also included questions for a period of employment in the pilot station (three intervals) and the port in which pilot was working.

2 RESULTS

On the figures from 1 to 6 are showed answers of the pilots arranged in subject groups. In the second group (Cooperation with the VTS system) and in the fifth (Cooperation with the administrators of the quays) more than 10% of the pilots indicated additional factors that have been placed on the charts (with the number of responses). Results obtained in the second group were also presented in two parts depending on professional experience.

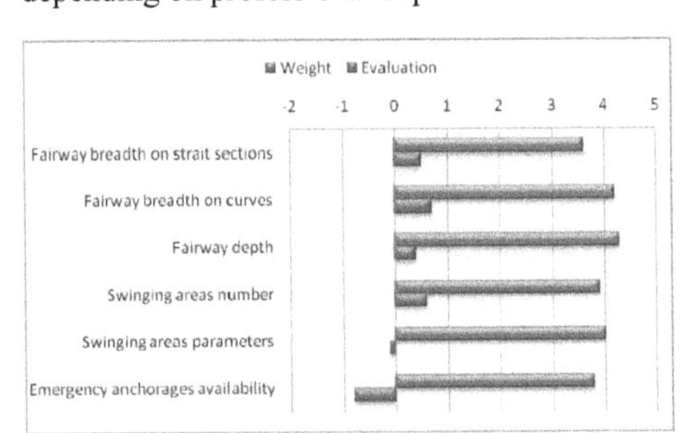

Figure 1. Weights and experts' evaluations in the group of "fairway parameters".

In the first group of factors related to the parameters of the fairway, the lowest rated were two factors - the availability of emergency anchorages and parameters of current swinging areas (Figure 1).

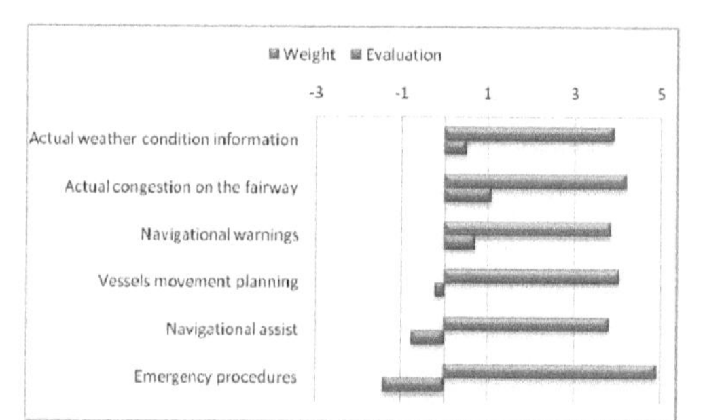

Figure 2. Weights and experts' evaluation in the group of "VTS cooperation".

In the second group was positively evaluated only that part of the activities VTS related to the transfer of information to traffic participants about the seaway and the situation on the fairway. Planning and organization of traffic was rated below then expectations. Much worse was assessed navigational assistance, the transfer of information to the vessels about the current position of the axis of the fairway. This service is not currently provided by the VTS system, but the weight attributed to it by the experts may indicate the need for action to support the position of a ship (Figure 2).

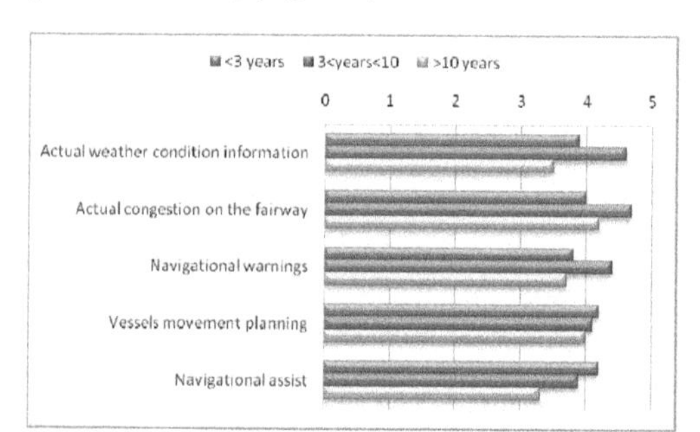

Figure 3. Weights in the group of "VTS cooperation depending on the length of experience.

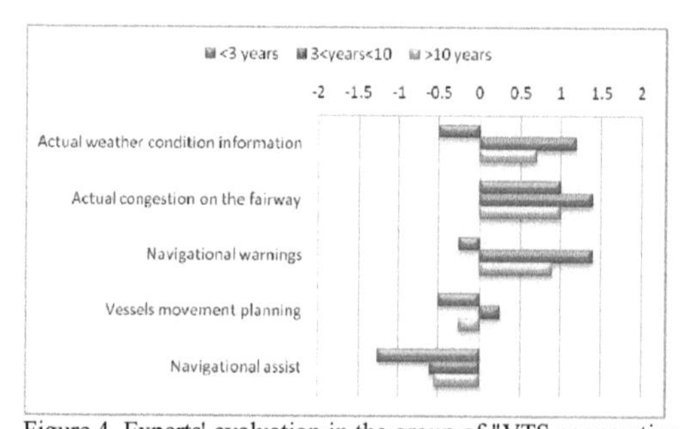

Figure 4. Experts' evaluation in the group of "VTS cooperation depending on the length of experience

Figure 3 and 4 show the opinions of the experts in two groups, depending on the length of experience. In a group of pilots with the lowest experience weight of assistance is the highest, while the lowest rating of all groups. Cooperate with the VTS system was rated lowest by pilots less experienced (under 3 years). One factor that was further indicated by the experts in this subject group are the procedures in emergency situations (known as the powers and competence in emergency situations). Weight given to them and the evaluation showed that it is a very high factor for the safety of navigation, which in the opinion of the experts is not sufficiently resolved.

In the third group was evaluated the status of aids to navigation. The only one factor was negatively assessed. It was marking of navigational hazards, but the weight attributed to this factor is the lowest in the group (Figure 5)

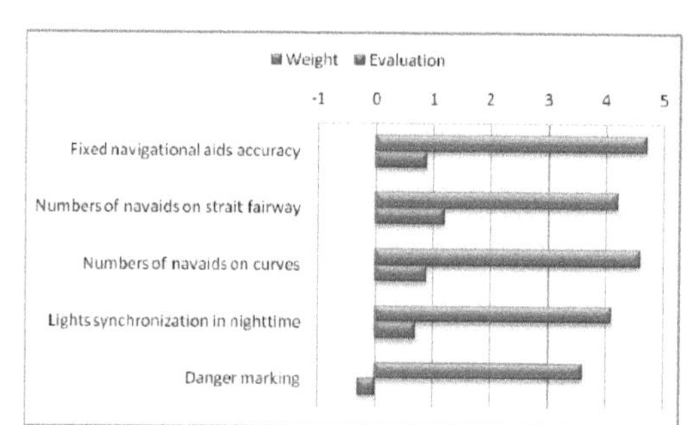

Figure 5. Weights and experts' evaluation in the group of "Navigational aids condition".

In the next group cooperating with tugs was evaluated. Two of the five factors were positively evaluated It were trained tugs crew and number of tugs required by the port's regulations. The lowest rates received the maneuverability of available tugs in port (Figure 6).

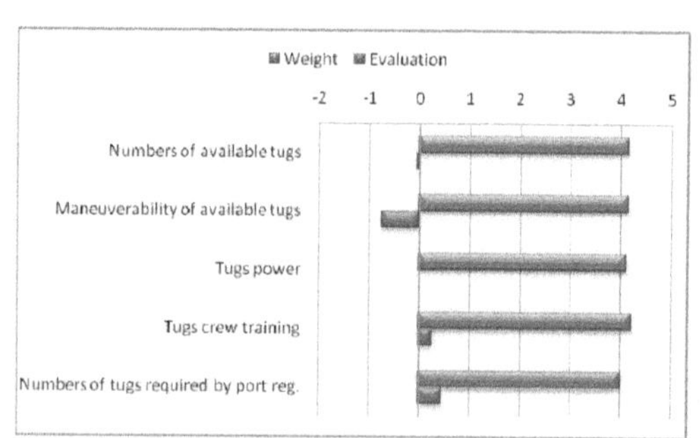

Figure 6. Weights and experts' evaluation in the group of "Cooperation with tugs".

The last group includes estimated working with administrators of the quays. The only positive factor assessed is the number of linesmen. Other factors evaluated were negative. Additional factor identified by participants is faulty or inadequate radio communication with administrators of the waterfront (Figure 7).

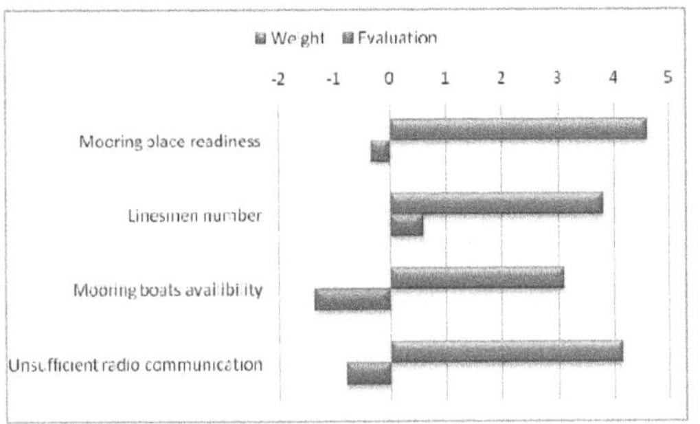

Figure 7. Weights and experts' evaluation in the group of "Cooperation with quays administrators".

Indicated in questionnaires particularly dangerous sections of the fairway or especially dangerous waterfront ware not repeated in more than 10% of the experts.

3 ASSESSMENT OF HYDROMETEOROLOGICAL CONDITION INFLUENCE ON SAFETY MANOEUVRE

Chose scientific method was based on the following assumptions:

- maneuver safety assessment is influenced not only by the dimensions of the vessel but the current hydrometeorological conditions on the fairway as well,
- the assessment of the safety of the maneuver is also affected by the place where the maneuver is executed,
- the impact of the current hydrometeorological conditions can be determined by experts based on their knowledge and experience. The impact of each factor is determined by the weight which is the numerical estimate,
- expert knowledge can be acquired with use of expert questionnaire. The Expert Questionnaire ensure the independence and anonymity expressed opinion by separation each expert during the test. Properly conducted questionnaire provides reliable information. This method is easy to conduct and not too time-consuming, so can be questioned. On the other hand, it is difficult to expect a higher accuracy than the one that has the interviewed person. If interviewed person is highly qualified, the questionnaire gives accurate results.

Questionnaire used in the research was based on the following assumptions:

- expert opinion is formal, subjective judgment of an expert quantitative estimation, which is the real for expert. This opinion is based on an assessment of the facts, and overall on its own associative scheme;
- the opinion expressed by the expert assessment - the number, contained in the set of ratings, ranging from 1 (negligible impact) to 5 (significant influence) - is a quantitative estimate (measurement) of impact of examined factor on safety maneuver which is tested,
- assessment is made on the basis of their knowledge and known expert general rules and methods that have been acquired during the practice.

Dividing into sections of the fairway allows to establish if influence of the examined factors are significantly different for different sections of the fairway. Such sections should be as homogeneous as possible in nature, acting as dredged fairway of similar depth beyond the edge of the track, or artificially constructed channel or dredged part of the river. In addition, in each section should be similar possibility to estimate positions in both a high and low visibility.

Maintained depth of the fairway in sea part and in the port of Świnoujście, is 14.3. From kilometer 3.7 to the port of Szczecin (km 67.7) maintained depth of the fairway is 10, 5 m. Width of the fairway, (with some exceptions) is 90m for a depth of 10.5 m, but the sections are:

- dredged fairways with different natural depth beyond the edge of the fairway (Szczecin Lagoon and Roztoka Odrzańska),
- artificially constructed channels (Kanał Piastowski, Przekop Mieleński),
- dredged sections of the river Oder and Świna.

The distances to the edges and fixed and floating aids to navigation, which can be used to determine the position are significantly different in sections. As part of the fairway that runs through the Szczecin Lagoon, in the winter, when summer's navigational system is removed, the distance between the Trackside Gates, which are the only navigations signs are about 4 Mm. As part of the fairway that runs channels or river beds the distance between the edges are up to 200 m Therefore, the impact of hydro-meteorological factors such as visibility on specific sections can be very different. For the research, the fairway is divided into 11 sections.

The research attempted to determine the effects of four variables on the overall safety level of the performing maneuver, based on the assessment made by the pilots:

- limited visibility (2 Mm),
- fog (visibility 0.1 Mm),
- the strength of the side wind 7° B,
- time of day (day or night).

The impact of this variable was defined to maneuver of ship crossing the fairway, and for passing and overtaking maneuvers vessels. In addition, for maneuver of ship crossing the fairway, will effect of the variable being evaluated by depend on size of the ship. The pilots task was to ratings in the group of small ships (80 - 100m), medium (about 140m) and large (approx. 170 m).

Five-step rating scale from 1 (minor impact) to 5 (significant impact) was adopted.

4 SHIP NAVIGATING THE FAIRWAY - RESULTS

In Figures 8 - 11 shows the average weights (impact) of factors on individual sections of the fairway.

The sections are:

- A - Świnoujście entrance,
- B - Harbour office tower to Mielin island,
- C - Mielin island to Karsibórz corner,
- D - Karsibórz corner to Fairway Gate no 1,
- E - Fairway Gate no 1 to buoys no 7-8,
- F - Buoys no 7-8 to Mańków corner,
- G - Mańków corner to Krępa Dolna beacon,
- H - Krępa Dolna beacon to Raduń Dolna,
- I - Raduń Dolna to Inoujście,
- J - Inoujście to Orli Przesmyk,
- K - Orli Przesmyk to Basen Górniczy.

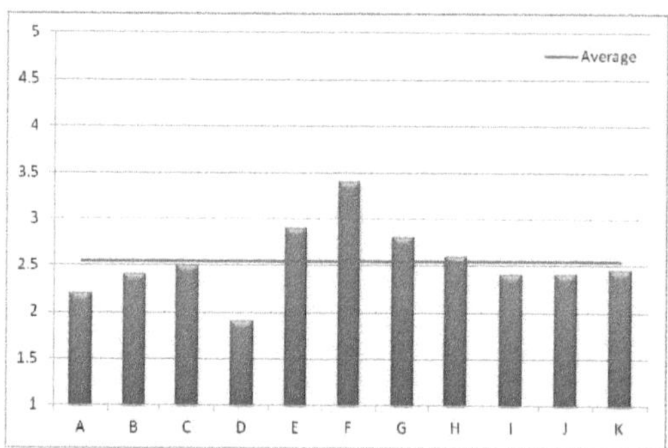

Figure 8. Vessel passage through fairway. Average evaluations of restricted visibility (2Nm) influence.

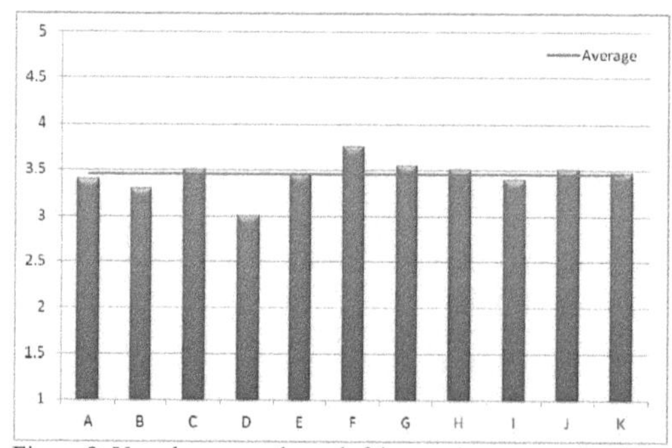

Figure 9. Vessel passage through fairway. Average evaluations of highly restricted visibility (0,1Nm) influence

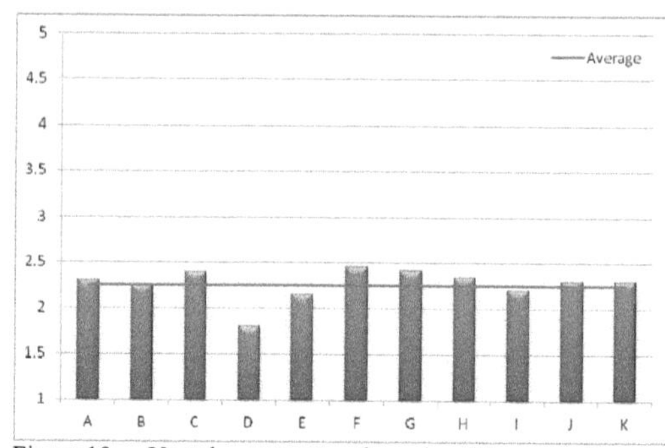

Figure 10. Vessel passage through fairway. Average evaluations of night time influence.

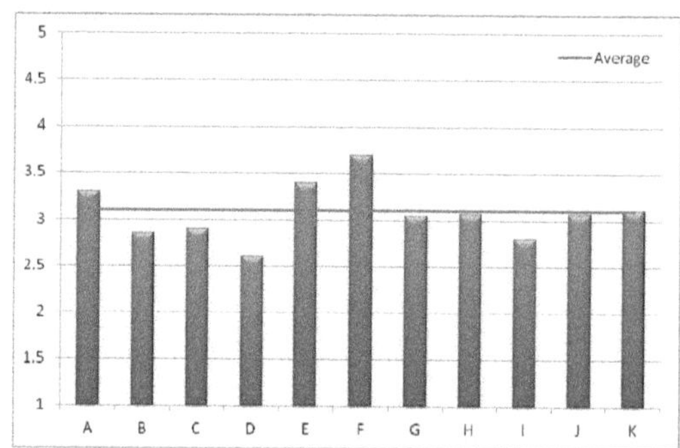

Figure 11. Vessel passage through fairway. Average evaluations of strong side wind 7°B influence.

Comparison charts provided takes some preliminary conclusions:

- the greatest diversity of ratings can be seen when visibility is reduced to 2 Mm,
- the highest weight (the largest impact on the safe vessel passing the fairway) experts attributed to two factors - visibility reduced to 0.1 Mm and side wind,
- the lowest weight was assigned to the night time.

To resolve, with the help of statistical methods, whether there are differences between the average weights assigned factors considered by the experts on a particular parts of the fairway method of analysis of variance was used. This method allows for the sharing of variation observed among pilots ratings on separate parts, each of which can be attributed to any other source. For the analysis of changes induced by only one factor which is the section of the fairway method for the analysis of variance with univariate classification was used. In order to determine whether significant impact on the variability of the ratings has a size of the vessel two-way analysis of variance was used.

Hypothesis tested in the analysis of variance are as follows:

$\mathbf{H_0}$: $\mu_1=\mu_2=...=\mu_i$,

$\mathbf{H_1}$: not all μ_i are equal.

where: μ - average weight (impact assessment study factor) in each section.

If the average weights are different, it is likely that the deviation of expert assessment from the corresponding average weight will be small in comparison with the deviations of the mean ratings of the total average value. The total sum of squared deviations can be determined according to the formula:

$$Q = Q_A + Q_w$$

where:
Q - the total sum of squared deviations,
Q_A - the sum of squared deviations between the groups (sections),
Q_W - the sum of the squares within the group.

Mean square error between the groups:

$$s_A^2 = \frac{Q_A}{r-1} \qquad (1)$$

Mean square error inside the groups:

$$s_W^2 = \frac{Q_W}{n-r} \qquad (2)$$

If the hypothesis zero is true, then the quotient s_A^2/s_W^2 has distribution F with degrees of freedom even r-1 in the numerator and n-r in the denominator:

$$F_{(r-1,n-r)} = \frac{s_A^2}{s_W^2} \qquad (3)$$

Table 1 shows analysis of variance scheme for the influence of visibility.

Table 1. analysis of variance scheme for the restricted visibility up to 2 Nm.

Variance	Sum of squared deviations	Degree of freedom	Mean square error
Between groups	$Q_A = 51,148$	$r\ 1 = 10$	$s_A^2 = 5,415$
Inside groups	$Q_W = 437,031$	$n\text{-}r = 341$	$s_W^2 = 1,282$
$F = 4,225;\ p = 0,000015$			

The obtained results allows to reject the hypothesis of equal effects of limited visibility to 2 Mm in all parts of the fairway.

The method of analysis of variance cannot allow identify what is the nature of inequality and the average of which or from which segments are significantly different from one another. For further analysis it is possible to compare the average pairs using Tukey's test. This test can determine between which average are differences. Hypotheses of the equality of the individual pairs of medium-sized sections to be subjected to control.

The results obtained by the Tukey test indicate the section 6 (from the buoys 7-8 to Mańków corner), where visibility limited to 2 Mm is significantly different from the impact on the majority of the remaining segments of the fairway.

There is a possibility to perform simultaneous testing of two variables - the section of the fairway and the size of the vessel (two base classification). Using the method of two-way analysis of variance can determine the impact of visibility depends on the section of fairway (first base classification) and the size of the vessel (the second base classification). It is also possible to determine whether there is an interaction between the two variables, i.e. whether the impact of the resulting visibility on a particular section depends on the size of the ship. The results of the analysis for the maneuver of ship crossing the fairway when visibility is reduced to 2 nm are shown in Table 2. Decomposition of the overall effect of the variable variation weight and conditional confidence intervals are shown in Picture 11

Table 2. F and p values for two-way variance analysis for vessel passing in restricted visibility (2 Nm).

Variance	Sum of squared deviations	Degree of freedom	Mean square error	Quotient F	p
Section	51,571	10	5,157	4,463	0,000007
Vessel size	49,455	2	24,727	21,400	0,000000
Section/ Vessel size	18,977	20	0,949	0,821	0,687607

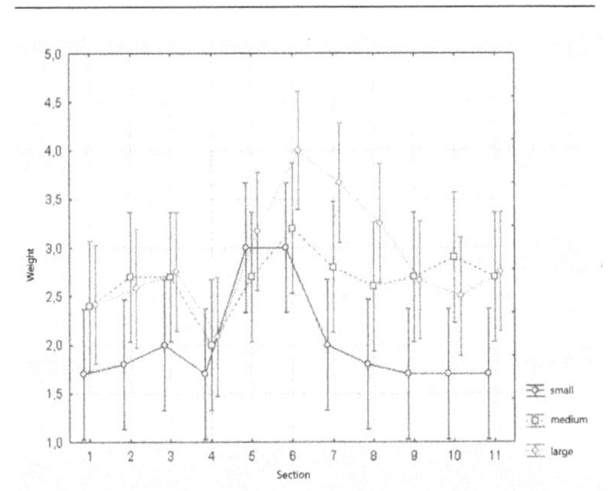

Figure 12. Vessel passage through fairway. Diversity of weights for all 11 sections of fairway and for different size of the vessel (small, medium, large).

The obtained results allow to reject the zero hypothesis of no effect the size of the ship and fairway section, and assume that both the size of the vessel and the section of the fairway is an important factor in classification affects restricted visibility. But do not allow to reject the hypothesis of lack of relationship between the two factors.

The results of the double analysis of variance to determine the effect of a strong reduction of visibility, influence of night time and strong side wind 7° B are shown in Tables 3-5.

Table 3. F and p values for two-way variance analysis for vessel passing in restricted visibility (0,1 Nm).

Variance	Sum of squared deviations	Degree of freedom	Mean square error	Quotient F	p
Section	10,817	10	1,082	0,743	0,683501
Vessel size	35,255	2	17,628	12,111	0,000009
Section/ Vessel size	16,757	20	0,838	0,576	0,928329

Table 4. F and p values for two-way variance analysis for vessel passing in night time.

Variance	Sum of squared deviations	Degree of freedom	Mean square error	Quotient F	p
Section	9,522	10	0,952	0,967	0,472334
Vessel size	29,906	2	14,953	15,183	0,000001
Section/ Vessel size	10,271	20	0,514	0,521	0,957088

Table 5. F and p values for two-way variance analysis for vessel passing during string side wind 7°B.

Variance	Sum of squared deviations	Degree of freedom	Mean square error	Quotient F	p
Section	24,226	10	2,423	1,932	0,040395
Vessel size	29,365	2	14,683	11,708	0,000012
Section/ Vessel size	8,258	20	0,413	0,329	0,997558

The obtained results allow to conclude that:

- possibility of reject the zero hypothesis of no effect the size of the ship and fairway section, and assume that both the size of the vessel and the section of the fairway is an important factor in classification only for effect of the strong side wind,
- for the strong reduction of visibility and night time the only important factor is the size of the ship. Section of the fairway is not a significant classification factor, i.e. the effect of night-time and strong visibility reduction does not differ significantly for different sections of the fairway;
- cannot reject the hypothesis that there is lack of relationship between the size of the ship and the section of the fairway for any of the factors studied.

5 PASSING AND OVERTAKING OF VESSELS ON THE FAIRWAY

In Figures 13-16 shows the average weights (impact) of the studied variables in each section of the fairway.

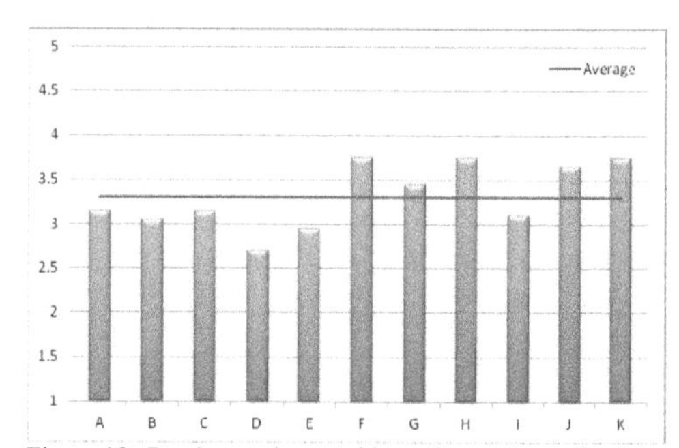

Figure 13. Passing and overtaking of vessels on the fairway. Average evaluations of restricted visibility (2Nm) influence.

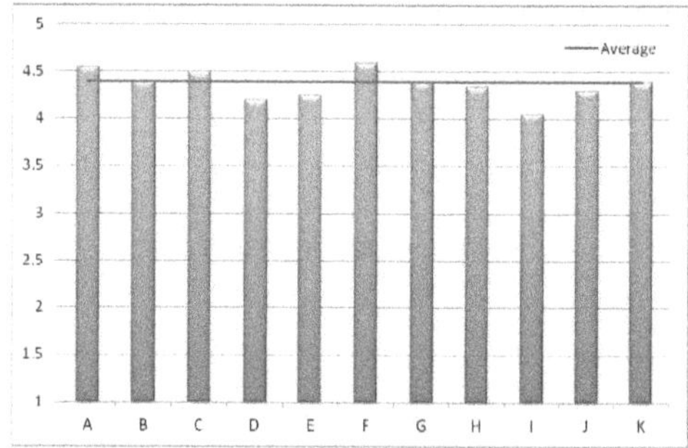

Figure 14. Passing and overtaking of vessels on the fairway. Average evaluations of restricted visibility (0.1Nm) influence.

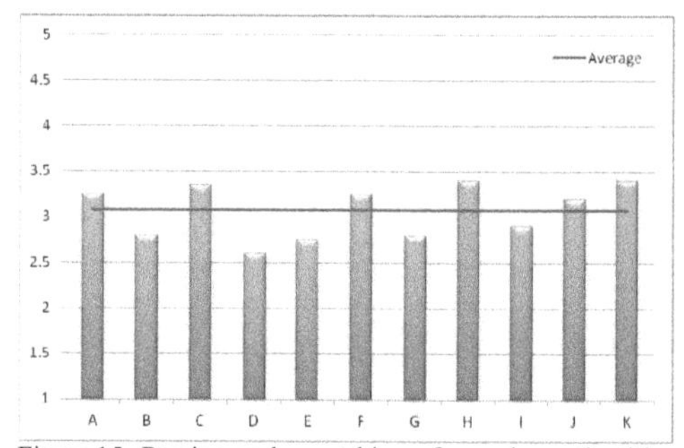

Figure 15. Passing and overtaking of vessels on the fairway. Average evaluations of night time influence.

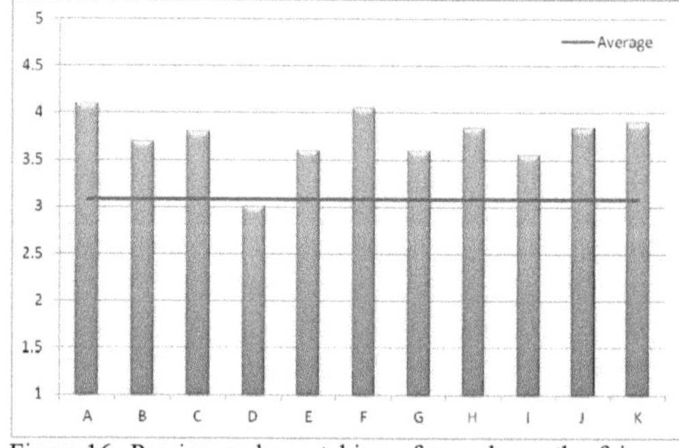

Figure 16. Passing and overtaking of vessels on the fairway. Average evaluations of strong side wind 7°B influence.

Analysis of provided results allows accept some preliminary conclusions:
- the smallest variation ratings can be observed with strong visibility reduced to 0.1 Mm,
- the highest weight (the largest impact on the safe passage of the vessel fairway) participants attributed to two factors - visibility reduced to 0.1 Mm and strong side wind,
- the lowest weight is assigned to the navigation during night time.

6 IMPLEMENTATION

All above presented results of research, but not only those were implemented in Transport Safety Management System. The system itself were divided into two subsystems:
- safety management on open waters in the Polish economic zone,
- safety management in restricted areas.

The Safety Management subsystem in restricted waters covers the following areas:
- Zatoka Pomorska with ports of Szczecin, Świnoujście and Police,
- Zatoka Gdańska with ports of Gdańsk, Gdańsk Port Północny and Gdynia.

The research was very vast included all above mentioned areas. Here in this paper only Świnoujście - Szczecin fairway is presented as the sample. Within the second subsystem the navigational risk assessment model was built. That model serves as a support tool concerns the navigational safety in restricted areas. The model allows to assess the navigational risk periodically and compare the results with acceptable values. Additionally model can predict the risk after making any change in the particular area including port regulations, fixed navigational aids parameters or fairway parameters. In effect it allows to take the optimal decision regarding navigational safety in restricted areas.

That evaluation is taken in two stages:
- assessment and evaluation of actual safety level on the basis of accident statistics and periodic expert questionnaire,
- assessment and evaluation of actual safety level on the basis of marine traffic engineering methods.

The presented research partly fulfill the first stage (Figure 17).

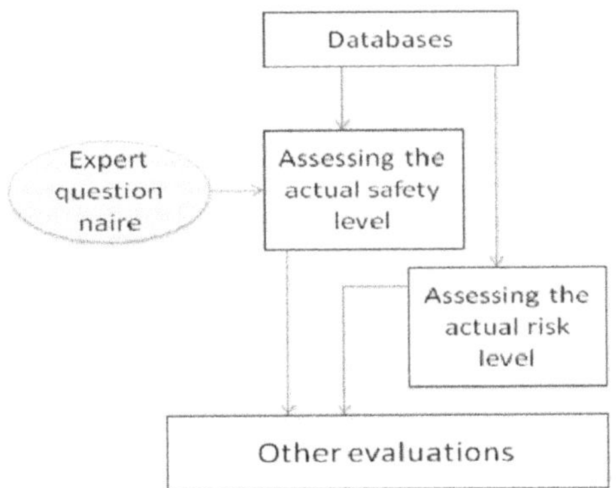

Figure 17. Expert questionnaire as a part of Transport Safety Management System.

In order to keep the system always in actual state, the expert research are performed among pilots every two years. For each port there are special questionnaire form is prepared. The results as it was presented above in the paper are prepared with exploitation of specialist statistical method. It allows to keep the overall system up to date regarding actual condition.

7 CONCLUSIONS

Based from the results of the survey, expert questionnaires play an important role in safety evaluations, which are based on experts' knowledge and experience. The results were treated as starting point in designing Transport Safety Management System as the proper a priori estimation of safety is the crucial factor in later precise monitoring.

REFERENCES

PIANC PTC II (1997): Approach Channels A Guide for Design. Report of Working Group II—30. Supplement to Bulletin No.95, Brussels, Belgium.

W.Feller (2007): An Introduction To Probability Theory and Its Applications, PWN, Warszawa.

S.Gucma (2001): Inżynieria ruchu morskiego, Okrętownictwo i Żegluga Sp. z o. o., Gdańsk.

Expert Indication of Dangerous Sections in Świnoujście - Szczecin Fairway

P. Górtowski & A. Bąk
Maritime University Szczecin, Poland

ABSTRACT: Paper presents the results of research on dangerous sections in the Świnoujście - Szczecin fairway as the part of the project "Marine Transport Safety Management System in Poland". The research conducted with participation of pilots who are working in that fairway. Aims of this research was to find out the most vulnerable areas in the fairway Świnoujście-Szczecin and the waters of the both ports. The results are subjected to affect the fairway division into homogenous section in the light of navigational safety.

1 INTRODUCTION

The experts task was to assessed the indicated maneuver (for example, the vessel crossing the fairway in described hydrometeorological conditions). Assessment was made by experts based on assess factors and own individual associational scheme, their knowledge and well-known expert general specialist rules and methods, gained during practice. Permissible was one of two responses: safe or unsafe.

Indication of dangerous sections is based on a comparison of the observed number with a set of the theoretical numbers The theoretical number (expected) is the number of all evaluated maneuvers. The research assumes that the maneuver is safe as will being evaluated by all the experts. The observed numbers is the number obtained in research (indications made by the pilots). The differences between the theoretical number and the observed numbers were assessed by using the statistical test The zero hypothesis assumes that between theoretical number and observed numbers there is no significant differences.

The test was performed using a four – panel board 2x2 (Table 1).

Table 1. Calculation scheme.

	Safe maneuver	Unsafe maneuver	
Theoretical number	A	B	A+B
Observed number	C	D	C+D
	A+C	B+D	N

2 SHIP NAVIGATING THE FAIRWAY - RESULTS

The pilots task was to evaluate among the group of small ships (80 - 100m, ten assessors), medium (about 140m, ten assessors) and large (approximately 170 m, twelve assessors). Tables 2-6 shows the number of observed indications of safe maneuvers and test results for the passing by maneuver at fairway. Changes of number were evaluated in different groups. Values of test χ^2 are presented only for group of large vessels. In the group of small and medium-sized vessel changes are not significant.

The obtained results allow to conclude that:

- In the tested range (the highest rated ship: length 172m and draft of 8.8 m) in a ship crossing in condition zero and at night time is rated as a safe maneuver.
- In restricted visibility (2mm), the only section in which passing of large ship was not considered as safe is section from the buoys 7-8 to Mańków. Passing in groups of small and medium sized vessels have been assessed as safe. The size of the vessel that can safely move in a fairway in reduce visibility (on indicated section) should be limited.
- In very reduced restricted visibility (0,1 Nm) and strong side wind 7 ° B passing a large ships was not considered as safe. Passing in groups of small and medium sized vessels have been assessed as safe. The size of the vessel that can safely pass on the fairway under these conditions should be limited.

Table 2. Observed number value, test χ^2 and p value in the large vessels group for passing in zero condition.

Section	sum	sum small	sum medium	sum large	χ2	p
Świnoujście entrance	32	10	10	12	0,000000	1,000000
Harbour office tower to Mielin	32	10	10	12	0,000000	1,000000
Mielin to Karsibórz	32	10	10	12	0,000000	1,000000
Karsibórz to Fairway Gate no 1	32	10	10	12	0,000000	1,000000
Fairway Gate no 1 to buoys 7-8	31	10	10	11	1,043478	0,307014
Buoys 7-8 to Mańków	29	10	10	9	3,428571	0,064078
Mańków to Krępa Dolna	32	10	10	12	0,000000	1,000000
Krępa Dolna to Raduń Dolna	30	10	10	10	2,181818	0,139649
Raduń Dolna to Inoujście	30	10	10	10	2,181818	0,139649
Inoujście to Orli Przesmyk	30	10	10	10	2,181818	0,139649
Orli Przesmyk to Basen Górniczy	30	10	10	10	2,181818	0,139649

Table 3. Observed number value, test χ^2 and p value in the large vessels group for passing in restricted visibility (2 Nm) condition.

Section	sum	sum small	sum medium	sum large	χ2	p
Świnoujście entrance	31	10	10	11	1,043478	0,307014
Harbour office tower to Mielin	31	10	10	11	1,043478	0,307014
Mielin to Karsibórz	31	10	10	11	1,043478	0,307014
Karsibórz to Fairway Gate no 1	31	10	10	11	1,043478	0,307014
Fairway Gate no 1 to buoys 7-8	30	10	10	10	2,181818	0,139649
Buoys 7-8 to Mańków	26	10	10	6	8,000000	0,004678
Mańków to Krępa Dolna	31	10	10	11	1,043478	0,307014
Krępa Dolna to Raduń Dolna	29	10	10	9	3,428571	0,064078
Raduń Dolna to Inoujście	29	10	10	9	3,428571	0,064078
Inoujście to Orli Przesmyk	30	10	10	10	2,181818	0,139649
Orli Przesmyk to Basen Górniczy	30	10	10	10	2,181818	0,139649

Table 4. Observed number value, test χ^2 and p value in the large vessels group for passing in restricted visibility (0,1 Nm) condition

Section	sum	sum small	sum medium	sum large	χ2	p
Świnoujście entrance	21	10	10	1	20,307692	0,000007
Harbour office tower to Mielin	19	10	8	1	20,307692	0,000007
Mielin to Karsibórz	18	8	10	0	24,000000	0,000001
Karsibórz to Fairway Gate no 1	23	10	10	3	14,400000	0,000148
Fairway Gate no 1 to buoys 7-8	21	10	8	3	14,400000	0,000148
Buoys 7-8 to Mańków	18	10	8	0	24,000000	0,000001
Mańków to Krępa Dolna	17	8	9	0	24,000000	0,000001
Krępa Dolna to Raduń Dolna	17	8	9	0	24,000000	0,000001
Raduń Dolna to Inoujście	17	8	9	0	24,000000	0,000001
Inoujście to Orli Przesmyk	17	9	8	0	24,000000	0,000001
Orli Przesmyk to Basen Górniczy	18	9	9	0	24,000000	0,000001

Table 5. Observed number value, test χ^2 and p value in the large vessels group for passing in night time condition.

Section	sum	sum small	sum medium	sum large	χ2	p
Świnoujście entrance	30	10	10	10	2,181818	0,139649
Harbour office tower to Mielin	32	10	10	12	0,000000	1,000000
Mielin to Karsibórz	29	10	10	9	3,428571	0,064078
Karsibórz to Fairway Gate no 1	32	10	10	12	0,000000	1,000000
Fairway Gate no 1 to buoys 7-8	32	10	10	12	1,043478	0,307014
Buoys 7-8 to Mańków	29	10	10	9	0,000000	1,000000
Mańków to Krępa Dolna	31	10	10	11	1,043478	0,307014
Krępa Dolna to Raduń Dolna	29	10	10	9	0,252632	0,615228
Raduń Dolna to Inoujście	30	10	10	10	0,000000	1,000000
Inoujście to Orli Przesmyk	30	10	10	10	0,000000	1,000000
Orli Przesmyk to Basen Górniczy	31	10	10	11	0,380952	0,537094

Table 6. Observed number value, test χ^2 and p value in the large vessels group for passing in strong side wind 7°B.

Section	sum	sum small	sum medium	sum large	χ2	p
Świnoujście entrance	25	10	10	5	9,882353	0,001669
Harbour office tower to Mielin	27	10	10	7	6,315789	0,011967
Mielin to Karsibórz	25	10	8	7	6,315789	0,011967
Karsibórz to Fairway Gate no 1	27	10	10	7	6,315789	0,011967
Fairway Gate no 1 to buoys 7-8	24	10	8	6	8,000000	0,004678
Buoys 7-8 to Mańków	22	10	8	4	12,000000	0,000532
Mańków to Krępa Dolna	27	10	10	7	6,315789	0,011967
Krępa Dolna to Raduń Dolna	23	10	8	5	9,882353	0,001669
Raduń Dolna to Inoujście	25	10	10	5	9,882353	0,001669
Inoujście to Orli Przesmyk	23	10	8	5	9,882353	0,001669
Orli Przesmyk to Basen Górniczy	24	10	9	5	9,882353	0,001669

Table 7. Observed number value for passing by in zero condition

Section	sum	sum small	sum medium	sum large
Świnoujście entrance	14	7	4	3
Harbour office tower to Mielin	31	11	11	9
Mielin to Karsibórz	14	8	4	2
Karsibórz to Fairway Gate no 1	31	11	11	9
Fairway Gate no 1 to buoys 7-8	30	11	11	8
Buoys 7-8 to Mańków	11	5	5	1
Mańków to Krępa Dolna	29	11	9	9
Krępa Dolna to Raduń Dolna	22	9	8	5
Raduń Dolna to Inoujście	31	11	11	9
Inoujście to Orli Przesmyk	20	10	8	2
Orli Przesmyk to Basen Górniczy	19	10	7	2

Table 8. Test χ^2 and p values for passing by in zero condition.

Section	χ2 small	p small	χ2 medium	p medium	χ2 large	p large
Świnoujście entrance	5,333333	0,020921	11,200000	0,000818	12,923077	0,000325
Harbour office tower to Mielin	0,000000	1,000000	0,000000	1,000000	1,263158	0,261054
Mielin to Karsibórz	3,789474	0,051576	11,200000	0,000818	16,000000	0,000063
Karsibórz to Fairway Gate no 1	0,000000	1,000000	0,000000	1,000000	1,263158	0,261054
Fairway Gate no 1 to buoys 7-8	0,000000	1,000000	0,000000	1,000000	2,666667	0,102470
Buoys 7-8 to Mańków	3,789474	0,051576	9,000000	0,002700	19,636364	0,000009
Mańków to Krępa Dolna	0,000000	1,000000	2,400000	0,121335	1,263158	0,261054
Krępa Dolna to Raduń Dolna	2,400000	0,121335	3,789474	0,051576	8,000000	0,004678
Raduń Dolna to Inoujście	0,000000	1,000000	0,000000	1,000000	1,263158	0,261054
Inoujście to Orli Przesmyk	1,142857	0,285050	3,789474	0,051576	16,000000	0,000063
Orli Przesmyk to Basen Górniczy	1,142857	0,285050	5,333333	0,020921	16,000000	0,000063

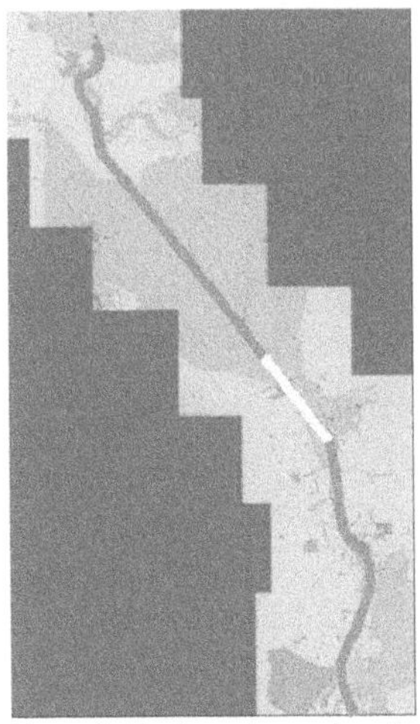

Figure 1. Section where when the visibility is restricted to 2 Nm the size of the vessels that can safely access the fairway should be limited

Figure 1 shows the section where when the visibility is restricted to 2 Nm the size of the vessels that can safely access the fairway should be limited.

3 PASSING OF THE VESSEL ON THE FAIRWAY - RESULTS

The pilots task was to assess the safety of vessels passing by maneuvers in the group of small ships (eleven judges), medium (eleven judges) and large (ten judges). Tables 7 - 8 shows the number of observed indications of safe maneuvers and test results for passing by maneuver of two ships on the fairway in a zero condition. Changes of number were evaluated in different groups (small, medium and large ships).

In Figures 2-6 there are presented the sections of the fairway, in which:

- passing by maneuver has been assessed as safe in all three groups (green),

- passing by maneuver was assessed as safe in the group of small and medium-sized vessels (dark yellow color)
- passing by maneuver is was assessed as safe in the group of small ships (light yellow),
- passing by maneuver is not been assessed as safe in any group (red).

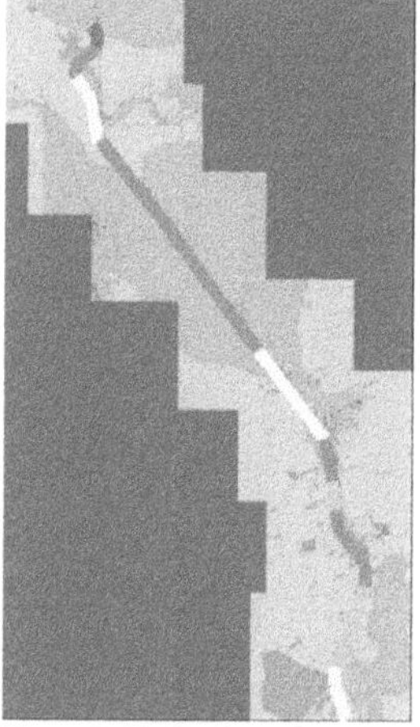

Figure 2. Evaluation of passing by manoeuvre on the fairway in zero condition.

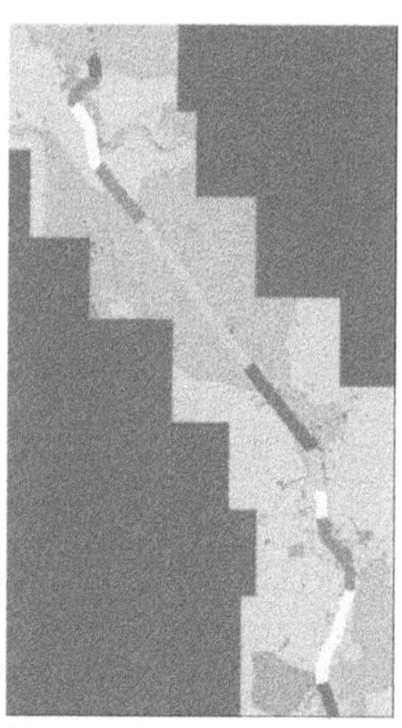

Figure 3. Evaluation of passing by manoeuvre on the fairway in restricted visibility (2 Nm) condition.

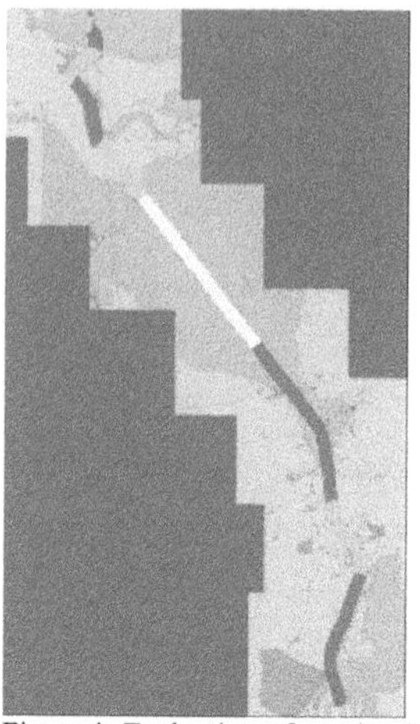

Figure 4. Evaluation of passing by manoeuvre on the fairway in restricted visibility (2 Nm) condition.

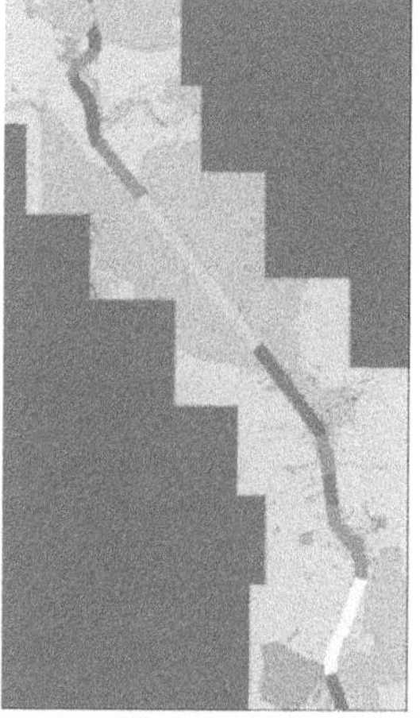

Figure 5. Evaluation of passing by manoeuvre on the fairway during night time condition.

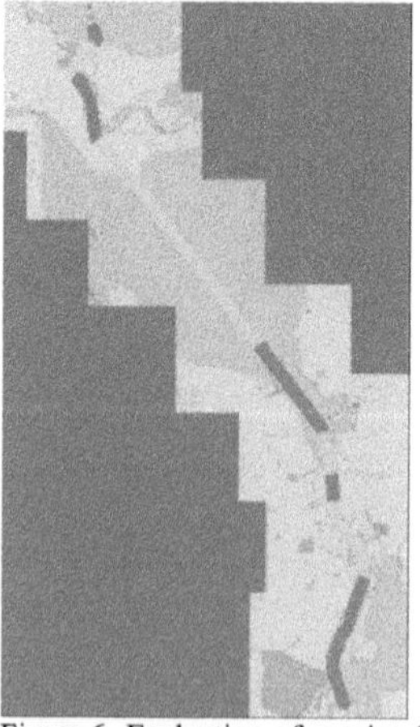

Figure 6. Evaluation of passing by manoeuvre on the fairway in strong side wind 7°B condition.

The obtained results allow to conclude that:

- Taking into consideration the researched scope (the largest ships, which passing by maneuver were evaluated: length of 172m and draft of 8.8 m and a length of 147 and draft 6.8 m) passing by maneuver is assessed as safe maneuver only in certain areas and only under certain conditions.
- With strong reducing visibility (0.1 nm) and a strong side wind is passing by maneuver large ships is not rated as safe in any section of the fairway.

4 CONCLUSIONS

The research allows to indicate the homogenous areas in navigational safety context. As an effect that sections were implemented in "Neptun" application which was created as a part of Marine Transport Safety Management System (Figure 7). Sections are used for calculating and presentation the results of estimated navigational risk. In addition the allowable breadth of fairway for particular vessel is showed.

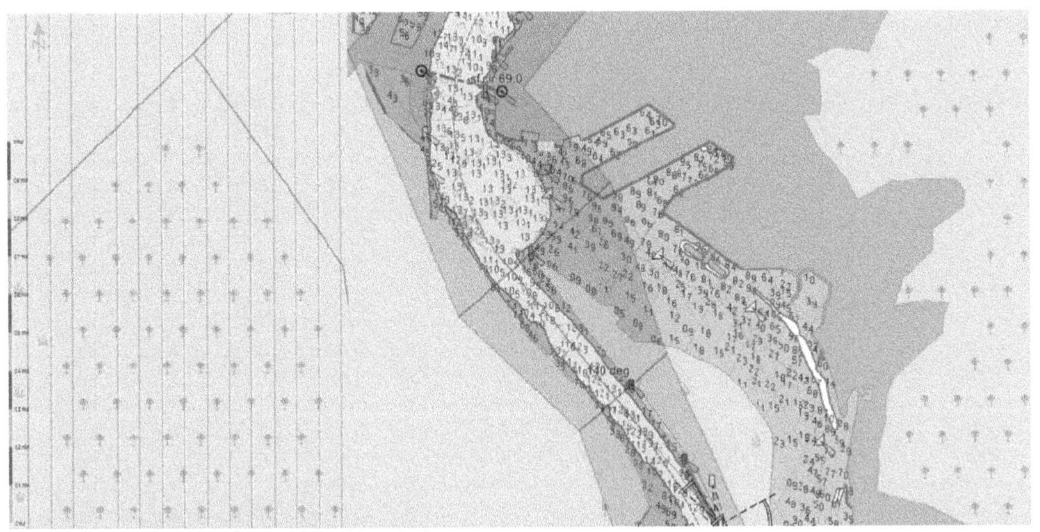

Figure 7. Presentation of sections used for risk analysis.

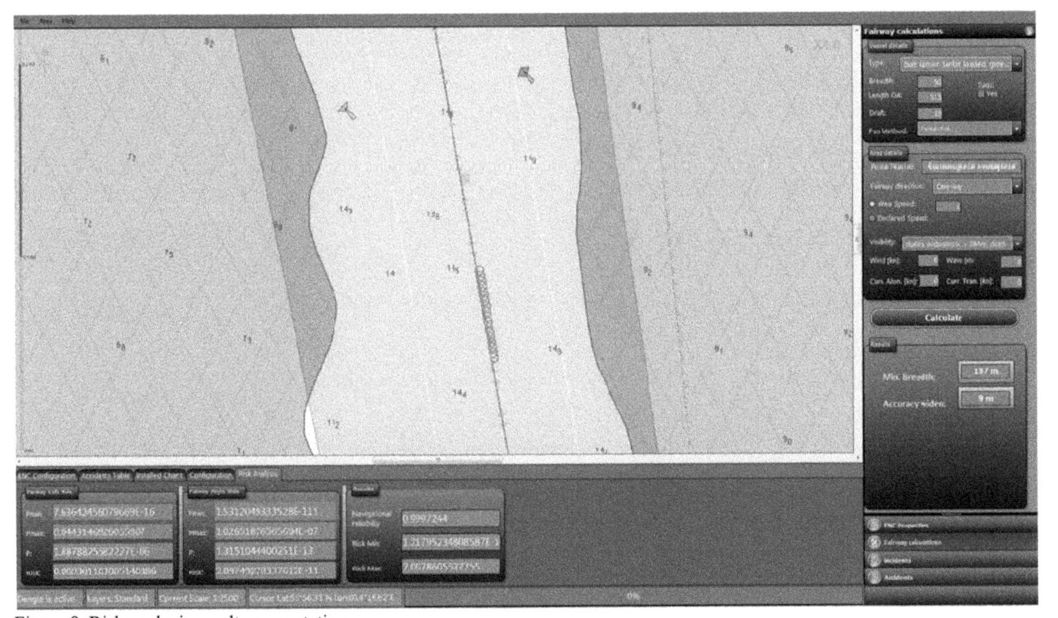

Figure 8. Risk analysis results presentation.

The calculated risk is presented together with graphical representation of safe fairway (Figure 8). Additionally for every 5 meters of fairway the navigational and technical probability is calculated and archived for any other statistical purposes.

As it was mention above indicated dangerous sections are the basis of any other analysis. More over the results given by experts can be easily verified in order to check if they turn out to be useful. The crucial property of the system is its update procedures which can lead to the most reliable system for calculating the risk and what is more important to be useful for designing new and evaluating existing fairways in the light of improving the navigational safety.

REFERENCES

PIANC PTC II (1997): Approach Channels A Guide for Design. Report of Working Group II−30. Supplement to Bulletin No.95, Brussels, Belgium.

W.Feller (2007): An Introduction To Probability Theory and Its Applications, PWN, Warszawa.

S.Gucma (2001): Inżynieria ruchu morskiego, Okrętownictwo i Żegluga Sp. z o. o., Gdańsk.

Traffic Incidents Analysis as a Tool for Improvement of Transport Safety

J. Skorupski
Warsaw University of Technology, Poland

ABSTRACT: Safety of traffic is one of the most important factors to be taken into account during organisation and operation of transport systems. One of the primary sources of information and inspiration in the creation of new, more secure solutions are the events with most serious consequences called accidents (air), serious accidents (rail) or catastrophes (road, sea). The investigation of the causes of these events provides answers to questions concerning the possible ways of modernization of components of the transport system. In this paper, it is proposed to draw more attention to the events of somewhat less severity of the consequences - serious incidents (air), traffic conflicts (road) and incidents (rail, maritime). These events include a broad class of cases in which the number of barriers against accident (crash) failed - but not all of them. Quantitative analysis of these events can provide information about the reliability of safety management system (SMS) components, and thus allow improving transport safety.

1 INTRODUCTION

Transport traffic safety is one of the most important goals of organization, management and supervision systems in different modes of transport. Proposals for new organizational and technical solutions are often the result of the analysis of the causes and circumstances of accidents, especially the most serious, which resulted in numerous fatalities. Examples of major transport accidents that resulted in key changes in safety perception were: an air traffic accident in Tenerife in 1977 (Netherlands Aviation Safety Board 1978), the Estonia ferry disaster in the Baltic Sea (Joint Accident Investigation Commission 1997), catastrophe of Senegalese ferry MV Le Joola off the coast of Gambia (Republique du Senegal 2002), the train disaster in Ufa near Chelyabinsk in Russia in 1989 (Surhone et al. 2010), aircraft collision over Überlingen (Brooker 2008). An example of a solution resulting from the last of these cases was systematic and comprehensive regulation of the use of traffic collision avoidance systems for aviation TCAS (Federal Aviation Administration 2011). A major change introduced into procedures was the rule of absolute subordination to the communicates of TCAS system, even in the case of different air traffic controller command.

Accidents investigation is usually conducted in terms of searching for the reasons and circumstances favourable to these events. These studies are carried out within the existing organizational structures that make use of well-established and legally sanctioned methods and procedures. They are directed at determining the causes of accidents and to make preventive recommendations aimed to eliminate these causes, and indirectly prevent the formation of analogous events in the future.

In transport, there are numerous safety barriers established - technical, procedural, organizational, management - in order to prevent accidents with catastrophic consequences. In this paper a method of traffic incidents analysis is proposed. Of course, incidents (serious incidents) are also examined by the aforementioned institutions. This examination, however, as in the case of accidents is focused on searching for reasons. Analysis of the incidents, however, also allows for exploration of possible scenarios for the development of the incident and checking what other effects it could bring about. This approach allows evaluating and verifying whether in this particular case, no transformation of the incident into accident was the result of hedging activities, or whether it was a coincidence. In the latter case one should suggest preventive recommendations relating to these events that did not actually occurred. In other words, it comes to the belief that set up safety barriers will work forever. Also in slightly different circumstances, such as worse weather conditions, or worse technical condition of the vehicle, etc. The fact that safety barrier worked in a particular incident does not give us a guarantee that it will be in any case.

In this paper it is proposed to analyze serious incidents quantitatively. The proposed approach is illustrated in the example of air traffic. The term "serious air traffic incident" usually involves a very dangerous event in which almost all barriers to protect against accidents have failed (in most cases except one). This allows to attempt to quantify the likelihood of failure of each of the elements of the safety assurance system. Unfortunately, in most cases, we do not have enough data that allows for a statistical determination of the frequency of events that make up the accident scenarios. There are also no measurement methods that can achieve such data. This is due to two reasons. The first is the extraordinary rarity of these events, and, until recently, a lack of public awareness of the need to report events of less important safety consequences. The second reason is very frequent participation of the so called human factor in these events. Analysis of the probability of particular human action or the probability of human error is very uncertain and subjective. The only available method of obtaining real knowledge about such events is the use of experts' opinions. These, obviously, are characterized by a lack of precision and clarity not allowing the use of such methods in the probabilistic analysis.

This paper is divided into five sections. The first contains the introduction to the research problem. The second part deals with the principles and practice of investigating the causes of aviation accidents. The third section shows the process of formation of a serious incident and the essence of traffic incident analysis focused on searching for the quantitative assessment of the effectiveness of the safety barriers. The fourth presents a simple example of a serious incident analysis, explaining the method of analysis of the effectiveness of safety barriers. The fifth section contains summary and presentation of the findings of the research.

2 AIR TRAFFIC ACCIDENTS AND INCIDENTS INVESTIGATION

Polish aviation regulations define three categories of events (Aviation Law 2002):

- accident - an event associated with the operation of the aircraft, which occurred in the presence of people on board, during which any person has suffered at least of serious injuries or aircraft was damaged,
- serious incident - an incident whose circumstances indicate that there was almost an accident (such as a significant violation of the separation between aircraft, without the control of the situation both by the pilot of the aircraft and the controller),
- incident - an event associated with the operation of an aircraft other than an accident, which would adversely affect the safety of operation (e.g. a violation of separation, but with the control of the situation).

Air traffic events investigation is regulated both by international and national regulations: Annex 13 to the Chicago Convention (ICAO 2001) and EU Regulation 996/2010 on the investigation and prevention of accidents in civil aviation (European Union 2010). These documents define the basic principles of accidents and incidents investigations, which include:

- key role of EASA,
- cooperation between committees investigating accidents in different countries,
- the absolute need for reporting incidents,
- recommendations for accident prevention.

The basic legal act of national importance "Aviation Law" in Part VI "Air Navigation" in Chapter 3, "Managing the flight safety and investigating accidents and incidents" regulates the operation of accident investigation committee (Aviation Law 2002). Detailed rules of proceeding are defined in Regulation of the Minister of Transport on accidents and incidents (Minister of Transport 2007).

In recent years, much emphasis is put on a proactive approach to ensuring the safety of air traffic. It is based on preventive reporting of damages and failures, which is obligatory to those who have the ability to detect them before they cause dangerous traffic event.

However, for many years the most important and the most seriously considered are the recommendations issued by the State Commission for Aircraft Accident Investigation (PKBWL) as a result of the study of the reasons for air events. The PKBWL consist of: chairman, two deputy, secretary and members - experts in the field of aviation law, flight training, air traffic, aviation technology maintenance, aircraft construction, and aviation medicine. In some cases additional assistance of experts is necessary, both from the above mentioned areas and from the field of navigation, rescue, meteorology and aeronautical communications.

Generally an air traffic event examination procedure begins with the notification to the accident investigation committee. Then the chairman of the committee shall qualify the event by severity of its effects. Accidents and serious incidents are investigated by PKBWL. Incidents and other events (such as damages or defects detected) examination may be provided by the user of aircraft, air navigation services provider, airport management.

The basic principle of PKBWL operation is to improve the safety of air transport, rather than find and punish the guilty. For this reason, the regulations specify the purpose of the Commission, stating that it does not adjudicate the guilt and responsibility. At the same time law guarantees full security and freedom of testimony for all people co-operating with the Commission (including the participants of the event), by introducing a clause that the release of the results of Commission's investigation can be made only with the consent of the court. And the court while deciding whether to release the results should take into account whether such making information public is more important than the negative consequences that may result for the air transport safety. Additional protection is granted to members of the Commission, which must not be questioned as a witness as to the facts which might reveal the results of investigation.

All these factors make the study of the causes of air events, carried out in accordance with the rigors of this legislation, an extremely effective way to discover the mechanisms that lead to accidents. Preventive recommendations lead to the elimination of these mechanisms and thereby reduce the number of events.

However, analysis of Commission's reports shows that these recommendations are focused on the causes and circumstances leading to the event. In the case of incidents and serious incidents it is implicitly assumed that organizational, technical and human barriers which worked properly and prevented the development of incident into accident, will work properly in other similar situations. This assumption may not be true. One can identify many cases of similar incidents in which a barrier in one of them worked, and in another did not. Analysis of the effectiveness of these key barriers in air traffic incidents enables the quantitative assessment, which is very difficult using other methods.

3 ANALYSIS OF AIR TRAFFIC INCIDENTS

As mentioned in previous sections, this paper proposes an analysis of serious incidents not only in terms of reasons of their occurrence, but rather the probability of their transformation into an accident with consequences of loss of life or serious damage to equipment. In (Skorupski 2012a) a model (in the form of Petri net (Jensen 1997)) of conversion of incident into accident was presented. The actual incident occurred at Chopin Airport in Warsaw, where a simultaneous takeoff of two aircraft on two intersecting runways occurred. Fault tree of this incident is shown in Figure 1.

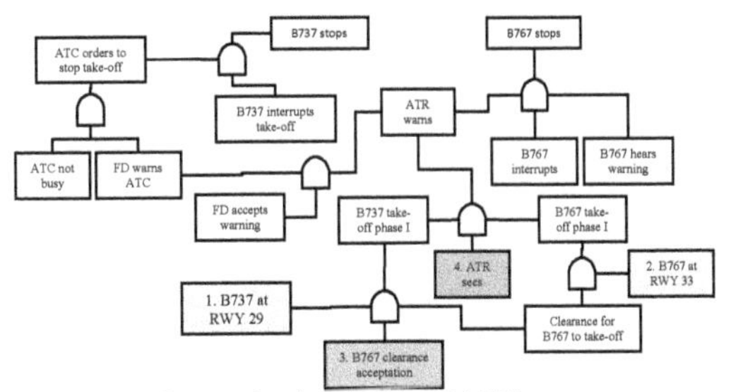

Figure 1. Fault tree of serious incident 344/07.

In this tree, we can distinguish the events that make up the incident. The most important are: occupying at the same time two intersecting runways (events 1 and 2 in Figure 1) and the key event for the creation of an incident - acceptation by the B737 crew, clearance for takeoff issued to other aircraft (event 3). Qualitative analysis of this event was performed by PKBWL. It should be noted it is not involved in quantitative analysis, which means that it does not consider how the proposed preventive recommendations reduce the likelihood of an accident. They are looking for such events (actions, omissions, organizational errors etc.), which elimination would break the causal sequence leading to the incident.

From the point of view of the method of serious incidents analysis, as proposed in this paper, the most important event is marked by number 4 in Figure 1. It represents the most important safety barrier, which protected before the accident (Lower et al. 2013b). Usually, there are several scenarios that lead to the failure of all barriers. These scenarios can be analyzed using event tree method (ETA). An example of the analysis is presented in (Lower et al. 2013a). This analysis allows to determine the probability of conversion of the incident in the accident, but also to quantify the effectiveness of the barrier. In the case of high risk of breaking the barrier, propose of preventive measures in relation to the events that have played a positive role in preventing the accident should be also considered. In this case, those measures should be based on strengthening the effectiveness of the barrier. This is a novel approach, but it seems appropriate and effective. It may give better results in prevention than the standard prophylactic recommendation for reasons of incidents. Especially as one can learn the quantitative characteristics (probability) characterizing such events.

Speaking about the risk of breaking the safety barrier one should understood the combination of probability of breaking it and possible consequences. We can determine the probability by event tree analysis. Determining the consequences may require further analysis of aircraft movements dynamics or use of other methods to estimate the effects of an accident, for example evaluation by experts.

4 EXAMPLE OF TRAFFIC INCIDENT ANALYSIS

In this section serious air traffic incident will be presented and analyzed with the use of Petri nets. The incident happened at Chopin airport in Warsaw. A method described in (Skorupski 2012b) was used during analysis.

4.1 *The course and the circumstances of the incident*

An example concerns the incident No. 291/05, which took place in winter conditions, on 31 December 2005 (Civil Aviation Authority, 2008). There was only one aircraft involved - Boeing 757-200 (B757). During the take-off operation, in acceleration phase, the crew felt a sudden shudder to the left and immediately performed a rejected take-off procedure. The maximum velocity of the aircraft was 70 knots. The crew of another aircraft, waiting for permission to take-off, reported by the radio that they can see flames in the left engine of B757.

After discontinuation of the take-off, the plane has been surveyed. No damages were found. Engine test has also been performed. Its operating parameters were consistent with the standard, resulting in the crew found that the engine is working properly. The crew reported to be ready to take-off again. Due to the large number of operations on that day, expected waiting time for permission to taxi was 45 minutes. In this situation, captain made a consultation with an experienced mechanic, after which he decided to cancel the flight. As a result of this decision, the plane was suspended in operation and two days later a detailed borescope study of left engine was made. Borescopes are commonly used in the visual inspection of aircraft engines, aeroderivative industrial gas turbines, steam turbines, diesel engines and generally in cases where the area to be inspected is inaccessible by other means. The inspection showed serious damage to the compressor blades, which excluded further use of this engine.

The period of suspension in operation of the aircraft was used to carry out the survey of the right engine, which did not show any irregularities so far. During the survey serious damages to the right engine compressor blades were also found. They were of the same nature as in the left engine. This precluded also the right engine from further operation.

4.2 *The causes of the event*

PKBWL investigated this event and determined that the most likely cause of the incident was a collision with a foreign object of a soft nature, pulled into both engines during the take-off operation. As a result, there was a density shock, local air density increase, causing overload and breakage of the rotor blades in both engines. Contributing factors to the creation of such a situation were: winter weather and release of the parking brake while switching aircraft's thrust automaton, which increases the likelihood of snow and ice to be sucked into the engine.

4.3 *Analysis of the causes of the event and activities of the Commission*

As always in such cases, PKBWL was focused on the causes of the incident. The Commission's report shows that its actions were in full effect. The cause was aspiration of snow during take-off. Commission adopted, as always in such situations, preventive recommendations. They consisted of:

- recommendation to use the information about the circumstances of this event for pilots and technical crew training, as well as the personnel responsible for the winter maintenance at the airport,
- consideration of the possibility to change internal procedures and equipment to improve winter maintenance of the airport,
- consideration to improve procedures of reporting technical problems, and their removal by the technical staff of the carrier.

As one can see, preventive recommendations are focused mainly on the elimination of events that led to the incident. In this case, to eliminate the possibility of aspiration of snow into the engine. It should be noted, however, that events affecting the safety of the travelling passengers, took place also in the cockpit of the B757 and during technical inspection. An analysis of the incident, along with the model of transforming it into an accident, using Petri nets is presented below.

4.4 *Model of incident*

The process of transformation of analyzed incident into accident can be modelled with Petri net:

$$S_{INC} = \{P, T, I, O, M_0, \ \tau, X, \Gamma, \ C, G, E, R, r_0\} \qquad (1)$$

where:
P – set of places,
T – set of transitions, $T \cap P = \varnothing$,
I, O, are functions respectively of input and output:
$I, O: T \rightarrow B(P)$, where $B(P)$ is the multiset over the set P, and functions I, O are determined for transition $t \in T$ as:

$t^+ = \{p \in P : I(t,p) > 0\}$ – input set of transition t,

$t^- = \{p \in P : O(t,p) > 0\}$ – output set of transition t,

$M_0 : P \to \mathbb{Z}_+ \times R$ – initial marking,

$\tau : T \times P \to \mathbb{R}_+$ – delay function, specifying static delay $\tau(t)$ of transition t moving tokens to place p,

$X : T \times P \to \mathbb{R}_+$ – random variable, describing random time of realization of traffic event (transition) t leading to traffic situation (place) p,

Γ – nonempty, finite set of colours,

C – function determining what colour tokens can be stored in a given place: $C : P \to \Gamma$,

G – function defining the conditions that must be satisfied for the transition, before it can be fired; these are the expressions containing variables belonging to Γ, for which the evaluation can be made, giving as a result a Boolean value,

E – function describing the so-called weight of arcs, i.e. expressions containing variables of types belonging to Γ, for which the evaluation can be made, giving as a result a multiset over the type of colour assigned to a place that is at the beginning or the end of the arc,

R – set of timestamps (also called time points), closed under the operation of addition $R \subseteq \mathbb{R}$,

r_0 – initial time, $r \in R$.

Petri net model of this incident is shown in Fig. 2.

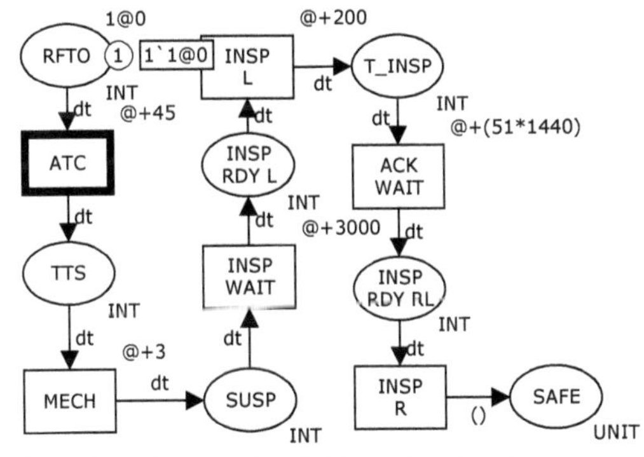

Figure 2. Basic model of air traffic incident 291/05

This model allows tracing the course of the actual incident, without taking into account other events that could change it. Parameters of the model presented in Figure 2 reflect of both the static and dynamic phenomena occurring in the analysed incident.

Designation of places is as follows:

- *RFTO* - aircraft ready for takeoff (p_1),
- *TTS* - aircraft crew knows the time to beginning of taxiing process (p_2),
- *SUSP* - aircraft use is suspended until left engine borescopes inspection (p_3),
- *INSP RDY L* - aircraft is ready for left engine borescopes inspection (p_4),
- *T_INSP* - aircraft starts waiting for admission to flight (p_5),
- *INSP RDY R* - aircraft is ready for right engine inspection (p_6),
- *SAFE* - there is no takeoff before detection of failures in both engines (p_7).

Designations of transitions are as follows:

- ATC - controller's decision to wait for taxiing (t_1),
- MECH - consultation with an experienced mechanic (t_2),
- INSP WAIT - waiting for the borescopes inspection of the left engine (t_3),
- INSP L - left engine borescope examination (t_4),
- ACK WAIT - waiting for authorization to use the aircraft (t_5),
- INSP R - right engine inspection (t_6).

4.5 *Simulation experiments*

By using CPN-Tools software package - a tool for creation and simulation of models using Petri nets, a number of simulation experiments were conducted. Their goal was to:

- find the probability of transformation of the incident into accident, where many elements of the model were treated as random values, but with expected values equal to those actually observed in the incident,
- seek the barrier, which appropriate functioning has prevented the creation of an air accident, and which caused that in fact only a serious incident occurred.

The plan of experiments assumed modification of the model resulting from scenarios of conversion of the incident into an accident. The following scenarios were considered:

1. There is a small traffic at the airport - in this case the time necessary to begin taxiing at the second attempt to take-off would be small and the captain would have decided to begin the take-off procedure instead of consulting the situation with the mechanic.
2. The captain does not choose to consult the problem with an experienced mechanic - being sure that the positive results of left engine inspection ensure safety.
3. Right engine survey is not performed - in case of quick repair of the left engine.

To perform analysis of the above scenarios, it was necessary to modify the basic model in such a way as to take into account the probability of each event, and to reproduce a random durations of events. Both groups of values have the large impact on the probability of conversion from the incident into an accident. Petri net for examination of scenarios 1 to 3 is shown in Figure 3.

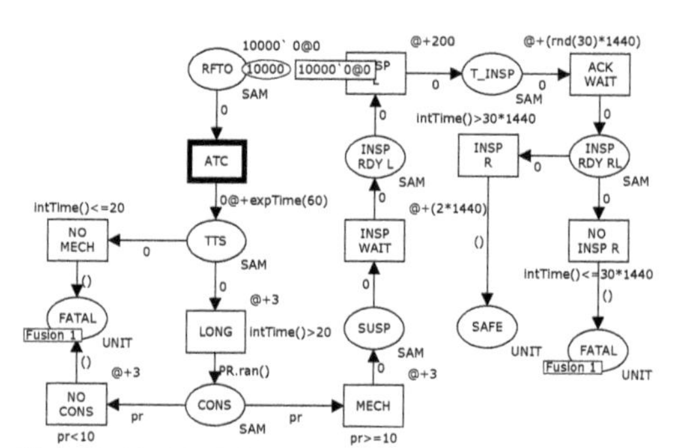
Figure 3. Model of air traffic incident 291/05 for analysis of transformation of incident into accident.

Compared to Figure 2 there are new places in this model:

- CONS - pilot is considering the consultation with a mechanic (p_8),
- FATAL - take-off with inoperative engine was performed, which would be likely to result in an accident (p_9),

There are also additional transitions:

- NO MECH - pilot decides to begin the second take-off without consulting the situation with a mechanic, because of short waiting time to begin taxing (t_7)
- LONG - recognition of the taxi waiting time to be long, that makes the pilot to consider the decision to undertake consultations with the mechanic (t_8),
- NO CONS - captain decides not to take the consultation with the mechanic (t_9),
- NO INSP R - decision not to examine the right engine due to the short waiting time for decision allowing the use of an aircraft (t_{10}).

Analysis of a simple event tree aimed at determination of probabilities of the above scenarios indicates that the total probability of taking off with a damaged engine can be described by the equation:

$$P_k = P_a + (1 - P_a) \cdot P_b + (1 - P_a)(1 - P_b) \cdot P_c \tag{2}$$

where

P_a, P_b and P_c are the probabilities of the scenarios, respectively 1, 2 and 3.

After transformation we obtain

$$P_k = P_a + P_b + P_c - P_a \cdot P_b - P_a \cdot P_c - P_b \cdot P_c + P_a \cdot P_b \cdot P_c \tag{3}$$

Assuming that the second attempt of take-off with severely damaged one or two engines would cause an accident, formula (3) is also a formula for determining the likelihood that analyzed incident becomes an accident.

Simulation experiments required modification of the model shown in Figure 3 in the following way

1 Modification of function $e \in E$ describing an arc between transition t_1 (ATC), and the place p_2 (TTS). The parameter of function *expTime* appearing in this expression, determines the expected time, after which the beginning of taxiing for take-off is possible. In the case of very low traffic, this time will be close to 0; otherwise it can reach values greater than 60 minutes.

2 Probability of consulting the decision with a mechanic is modelled using PR.*ran* function that generates a random value uniformly distributed in the range [0..100]; it can be identified with the values of probability. Depending on the value of generated random variable *pr*, which is the label of the output arcs from place p_8 (CONS), the values of function G for arguments t_2 (MECH) and t_9 (NO_CONS) determine the probability of consultation with a mechanic (transition t_2 MECH) or no consultation (transitions t_9 NO_CONS).

3 Time of suspension in service of the aircraft is represented by the value of function X(ACK_WAIT, INSP_RDY_R). Function *rnd* appearing in this expression determines the random duration of the aircraft remaining in the suspension, expressed in number of days. This random variable is described with a Poisson distribution with a parameter which is also the *rnd* function parameter.

The results of the 10,000 simulation runs show that at accepted values of probability and time constants of the dynamic analysis, the likelihood of continuation of airplane operation with at least one of the engines being defective is 0.62. It should be assumed that such an event would end in the destruction of the engine during the next take-off, and this would result in the inevitable accident with catastrophic consequences. It is therefore the probability of converting the incident into accident determined by simulation, at random values characterizing the events in the formation of the incident. It was adopted that the expected values of random variables characterizing the dynamic events (in time domain) are equal to the values observed during the actual incident.

Sensitivity analysis of the likelihood of conversion of the incident into accident, according to the application of additional measures (in an ad hoc or systemic manner) to eliminate specific threats (scenarios) was also conducted. For example, the relationship between probability P_k and traffic volume with corresponding to this time of waiting to begin taxiing (scenario 1) is shown in Figure 4. Probability of conversion of the incident into accident, depending on the likelihood of resignation from the consultation with the mechanic, which corresponds to a probability of the scenario 2 is shown in Figure 5. Dependence of P_k on the waiting time for authorization for the plane to operate (scenario 3) is shown in Figure 6.

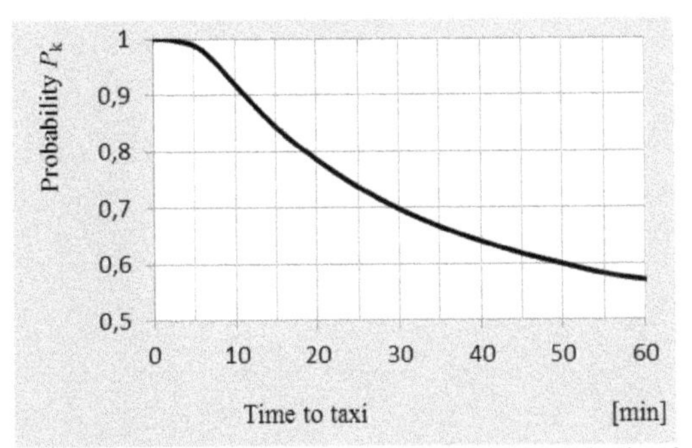

Figure 4. Relation between P_k and estimated time to begin of taxi procedure.

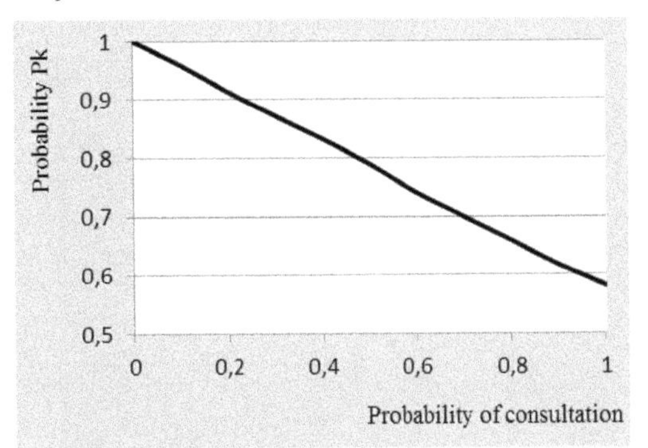

Figure 5. Relation between P_k and probability of consultation with the experienced mechanic.

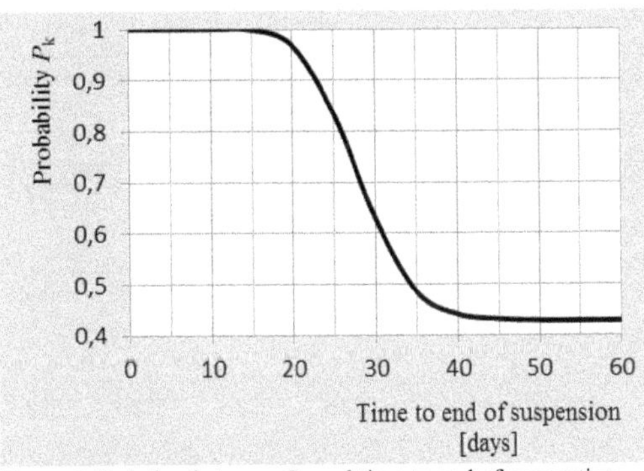

Figure 6. Relation between P_k and time to end of suspention.

5 SUMMARY AND CONCLUSIONS

The analysis of scenarios, transforming the incident into accident clearly shows the potential of analyzing incidents that can be used to improve transport safety. Cited in Section 3, the analysis of incident 344/07 shows the dependence of the possibility of an accident on the actions of ATR 72 aircraft waiting in line for take-off. Its activity in the observation of the neighbourhood area was crucial to avoid an accident. However, this barrier is not permanent. The pilot of this aircraft could be busy with his own take-off procedure and could not observe the situation on the runways. One might then consider the introduction of such a procedural requirement. The analysis of only the reasons for such an incident does not result in this kind of safety recommendations.

These findings are confirmed by the analysis of the incident 291/05, which uses different analytical methods and different types of modelling. According to the author, this suggests that in many similar situations of traffic incidents, also in other modes of transport, it is possible to receive interesting and important results. The essential common element may be a demonstration of effectiveness of the various barriers aimed at preventing accidents with catastrophic consequences. Sample analysis results for the case in question are presented in Section 4.

REFERENCES

Aviation Law 2012. Act of 3 July 2002, *Journal of Laws*, No. 130, item. 1112.

Brooker, P. 2008. The Uberlingen accident: Macro-level safety lessons, *Safety Science* 46 (2008), p. 1483-1508.

Civil Aviation Authority 2008. Statement No. 7 of President of the Office of Civil Aviation from 24 April 2008 on air event No. 291/05, Warsaw. (in Polish)

Joint Accident Investigation Commision of Estonia, Finland and Sweden 1997. Final report on the capsizing on 28 September 1994 in the Baltic Sea of the ro-ro passenger vessel, Edita Ltd, Helsinki.

European Union 2010. Regulation (EU) No 996/2010 of the European Parliament adn of the Council of 20 October 2010 on the investigation and prevention of accidents and incidents in civil aviation and repealing Directive 94/56/EC, Official Journal of the European Union, L 295/35

Federal Aviation Administration 2011. Introduction to TCAS II version 7.1, US Department of Transportation, DOT HQ-111358.

International Civil Aviation Organisation 2001. Aircraft Accident and Incident Investigation, International Standards and Recommended Practices, Annex 13 to the Convention on International Civil Aviation.

Jensen, K. 1997. Coloured Petri Nets. Basic Concepts, Analysis Methods and Practical Use. *Monographs in Theoretical Computer Science*. Springer Verlag.

Lower, M., Magott, J. & Skorupski, J. 2013a. Analysis of Air Traffic Incidents with the Use of Fuzzy Sets, *Reliability of Critical Infrastructure*, Proceedings of the XLI Winter School on Reliability, Szczyrk.

Lower, M., Magott, J. & Skorupski, J. 2013b. Air Traffic Incidents Analysis with the Use of Fuzzy Sets, Lecture Notes in Artificial Intelligence, Springer 2013 (in prep.).

Minister of Transport 2007. Regulation of 18+ January 2007 on accidents and incidents, Journal of Laws, No.35, item 225.

Netherland Aviation Safety Board 1978. Final report of the investigation into the accident with the collision of KLM flight 4805 and Pan American flight 1736 at Tenerife airport, *ICAO Circular 153-AN/56*. Amsterdam.

Republique du Senegal 2002. Raport d'Enquette du Comission d'Enquette technique sur les causes du naufrage du Joola. Dakar.

Skorupski, J. 2012a. Method of analysis of the relation between serious incident and accident in air traffic, *Advances in Safety, Reliability and Risk Management* (ISBN 978-0-415-68379-1), p. 2393-2401, CRC Press Balkema, London 2012
Skorupski, J. 2012b. Modelling of Traffic Incidents in Transport, *TransNav International Journal on Marine Navigation and Safety of Sea Transportation*, vol. 6, No. 3, 2012.
Surhone, L.M., Timpledon M.T. & Marseken S.F. 2010. Ufa Train Disaster, VDM Publishing, 2010.

Vessel Traffic Stream Analysis in Vicinity of The Great Belt Bridge

K. Marcjan & L. Gucma
Maritime University of Szczecin, Poland

A. Voskamp
Delft University of Technology, The Netherlands

ABSTRACT: The Baltic Sea is an enclosed body of water in which an accident may have serious environmental and financial consequences. The entrances to the Baltic Sea are difficult to navigate, because the waters are shallow and the currents are strong. The Great Belt is the largest and the most important strait of Denmark, connecting the Kattegat to the Baltic Sea. Across this strait spans The Great Belt Bridge, which connects Zealand with Funen and is one of the world longest suspension bridges. The work in this paper is based on AIS data in the vicinity of this Bridge, examining the distances between vessels while passing each other. The bridge is a narrow passage in the Strait. As a result, encounters with a small distance appear frequently, which may finally lead to a serious accident.

1 INTRODUCTION

1.1 *Study area*

Most people associate Denmark with the theme park Legoland, Hans Christian Andersen and the Vikings. Meanwhile, between the Danish islands of Zealand and Funen hangs one of the finest bridges Great Belt Bridge (Storbeltsbroen), one of the largest suspension bridges in the world. The suspension bridge over the Great Belt is the longest structure of its kind in Europe, and also the third-longest suspension bridge in the world (compared to the famous Golden Gate Bridge in San Francisco which takes the 11th place on the list). The connection between the largest Danish island Zealand and the island Dual was opened in 1998, after two years of preparation and ten years of construction.

Store Baelt 55°30' N, 10°55'E as it is written in (Sailing Directions 2010) also known as the Great Belt, is the central of three passages connecting the Kattegat to the Baltic Sea. It is the widest and deepest of the three passages and so, the most suitable for deep-draft ships. Depths in the Store Bealt are very irregular. The route through the Great Belt is narrow, forcing vessels into close and sometimes dangerous proximity. Large vessels bound for the Baltic Sea should transit Route T, which leads from Lighted Buoy No. 1 off Skagens, through Store Baelt to the S end of Langelands Bealt.

The area of research described in the paper is divided into three parts as depicted by the red squares in Fig. 1. The central one is the TSS(Traffic Separation Scheme) under the Great Belt East Bridge more precisely as it extends between 55°21.76'N and 55°19.22'N. The other two sectors are chosen in such way that navigation conditions were similar to the system TSS, but increased for vessels maneuvering areas. The second examined reservoir north of the TSS extends between 55°25.63' N 10°56.09' E and 55°21.65' N 11°05.34' E. The last one, south of the TSS extends between 55°19.23' N 10°56.91' E and 55°15.14' N 11°06.13' E. The aim of the study is to check the differences between statistical domains constructed for each of these areas.

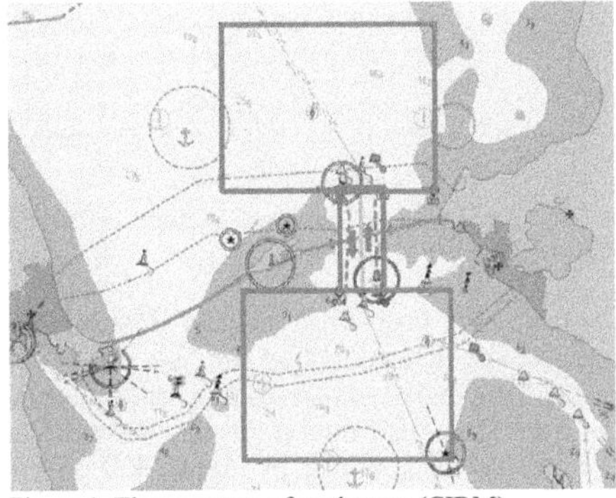

Figure 1. Three sectors of study area (CIRM)

2 GREAT BELT BRIDGE-TRAFFIC ANALYSIS USING AIS.

The AIS is a very good source for the analysis of traffic and the safety of bridges in the area. One may encounter difficulties in GPS satellite signal reception during the transition due to obscuration by the elements of the bridge and there is a possibility of losing the position of the ship. Traffic analyses were carried out during the transition under the Great Belt East Bridge, where there is two-way vessel traffic. AIS recorded unfiltered signal coming from the HELCOM AIS network of data exchange. A total of about 6,500 ships were registered in one direction. The study comprised data from the year 2011. Registered routes of ships are depicted in Figure 2. It can be seen that some of the vessels are moving close to the pillars of the bridge, indicated by black arrows.

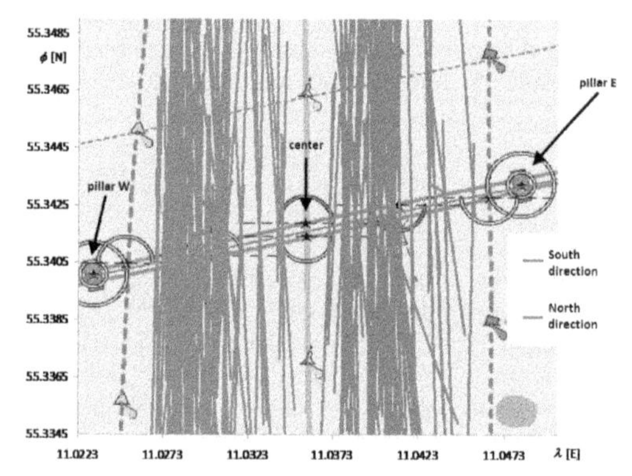

Figure 2. The routes of ships registered in the region of the Great Belt East Bridge.

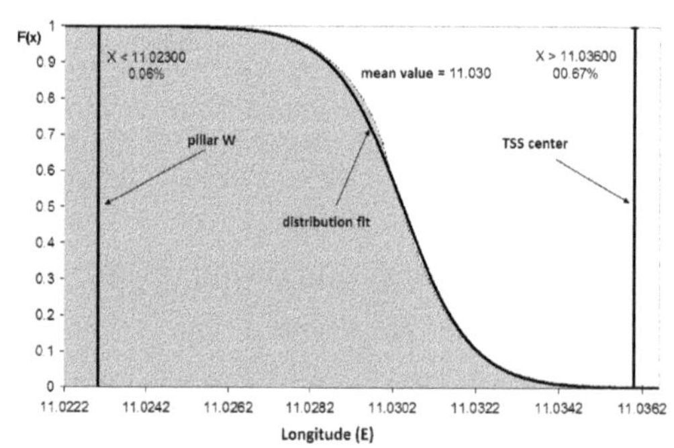

Figure 3. The distribution of positions of ships under the bridge (traffic lane west).

In the next step, probability density distributions were estimated as determined by the opening of the position of ships registered under the bridge during the transition. The distribution of the position of ships moving north is shown in Figure 3. Logistic distribution with parameters a = 11.03 and b = 8.32 showed a good fit to the empirical data (using the Kolmogorov-Smirnov test and χ 2). It can be seen that the probability of a collision with the western support of the bridge is $6 \cdot 10^{-4}$, thus showing good agreement with the literature data. Likelihood to go beyond their own traffic lane and enter the lane for opposing traffic amounts $6.4 \cdot 10^{-3}$, which is higher.

3 COMBINING A STATISTICAL SHIP DOMAIN WITH THE STUDY AREAS

By a ship domain we shall understand a certain area around the ship which the navigator likes to keep clear of other fixed and movable objects. Navigational situation can be considered safe as long as any other vessel or navigational obstacle does not exceed this area. Searching for a safety area around a vessel was started just after the first collision of the ship when people started looking for the reason of that disaster. The concept of the ship domain for the first time was proposed by (Fujii Y. & Tanaka K. 1971).

To determine ships domain shape, area around the vessel has been divided into 8 sectors (every 45 degrees starting from 0 degrees) due to relative bearing at the approaching vessel.

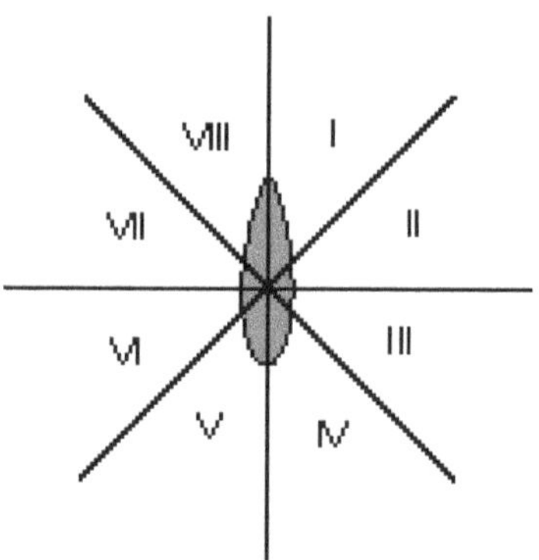

Figure 4. Domains division by sector (Gucma, L. & Marcjan, K. 2012).

Construction of domain shape was held together by linking distances (determined statistically domain boundary in a particular sector) plotted in the middle of each sector.

Data was selected for the period of four months from January to April 2011. AIS data is filtered in such a way that the density of the position is not less than 2 minutes. This allows to determine quite accurately the distance of ships encounters near the bridge, all the more important is that they are within short distances. Unfortunately, this filtering can be the reason for not qualifying (excluding) some of near misses. The total number of CPA (The Closest Point of Approach) during the encounters before filtration qualified during the period of time is 1082, after the filtration the number decreases to 914. In addition, the study rejected the intended meeting ships at short distances or situations associated with the towing and pilot embarkation, the study also did not include fishing vessels. Total number of qualified encounters for the first reservoir TSS under the Great Belt East Bridge was the 433. Figure 5a shows the location of CPA speaking to one of the vessels involved in the encounters.

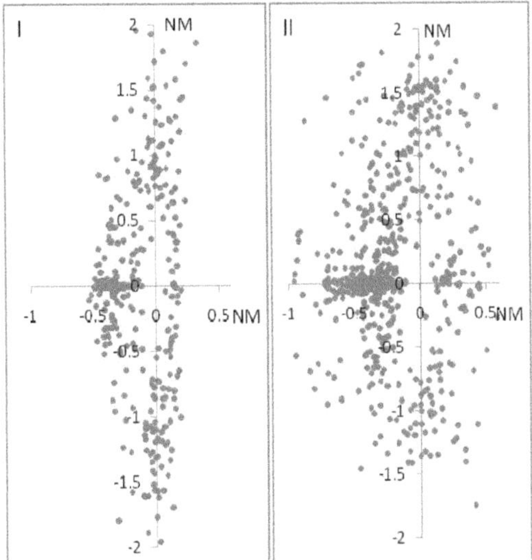

Figure 5a. Relative positions of CPA from bridge area I - TSS under Great Belt East Bridge area, II - area S of the TSS.

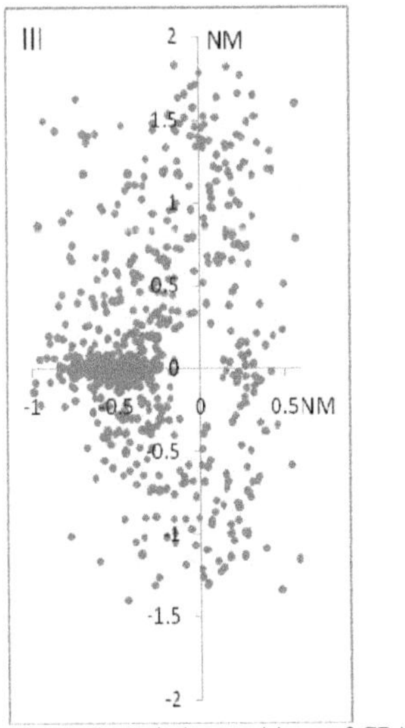

Figure 5b. Relative positions of CPA from bridge area. Study area N of the TSS.

Figure 5b summarizes all charts of relative CPA in selected areas. Due to parallel traffic lanes on opposite directions, many of the encounters attribute to the bearing of 90 degrees. Based on the chart it can be seen that there are sectors where the number of CPA is insufficient and because of that stochastic phenomena are inevitable. Similar problems can be seen in the work of (Ya-lei, R. & Jun-min, M. & Yan, D. & Jian-feng 2012) whose domain sizes based on insufficient data were taken from very short periods of time.

In order to construct the statistical domain around the ship, assumptions are used as described in (Gucma L.&, Marcjan K. 2012). These involve searching for a range constituting a maximum distance over which most of the navigators passes other vessels while encounter on a defined relative bearing. Situations between vessels in a TSS while passing under the bridge because of the parallel courses were divided into head-on and overtaking (COLLREG 1972). After dividing the statistical domain shapes were constructed as following:

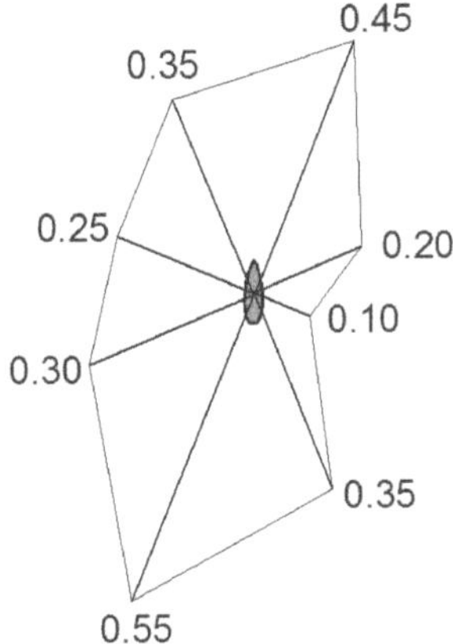

Figure 6a. Overtaking, domain for the area of The East Bridge.

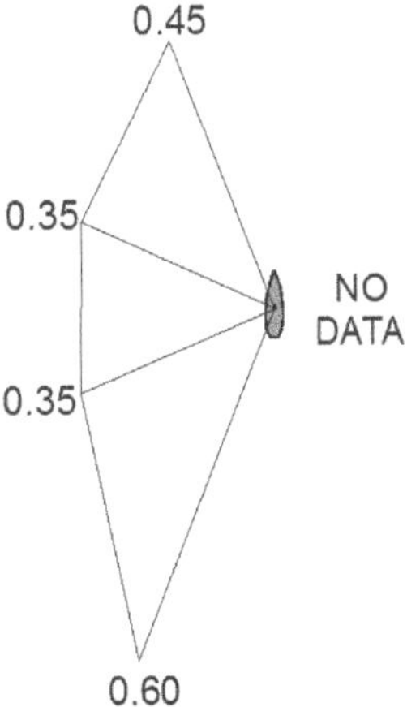

Figure 6b. Head-on situation, domain for the area of The East Bridge.

According to the AIS and shapes made up, statistical domains show that overtaking in the TSS is rare because of the small amount of available space between the bridge pillars or buoys separating traffic lanes. The limits in sectors II-III and VI-VII were determined by a small number of meetings during which the CPA has occurred in the above sector:
sectorII (45^0 - 90^0) - 4encounters,
sectorIII (90^0 - 135^0) - 7encounters,
sectorVI (225^0 - 270^0) - 3encounters,
sectorVII (270^0 - 315^0) - 7encounters.

An interesting approach to the subject of ship domain in TSS was described in (Ervin, v. I 2012).

In other cases, the vessels were approaching the other vessel from the bow sectors I,VIII, or have been catching upon the aft sectors. When it comes to the domain for Head-on situations, due to the fact that ships are moving along designated traffic lanes, the limits were calculated only for sectors V, VI, VII and VIII.

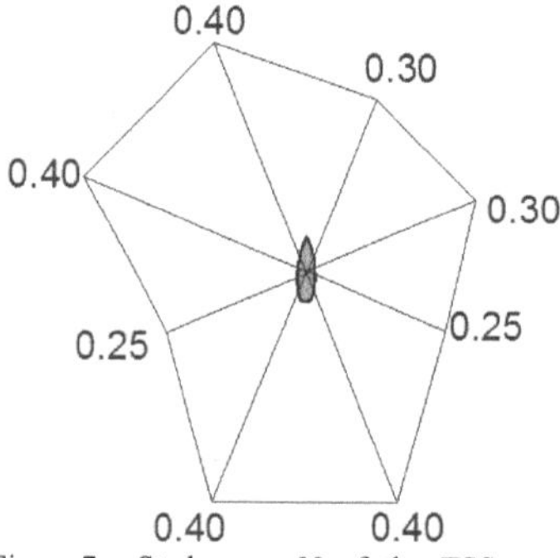

Figure 7a. Study area N of the TSS, overtaking situation domains.

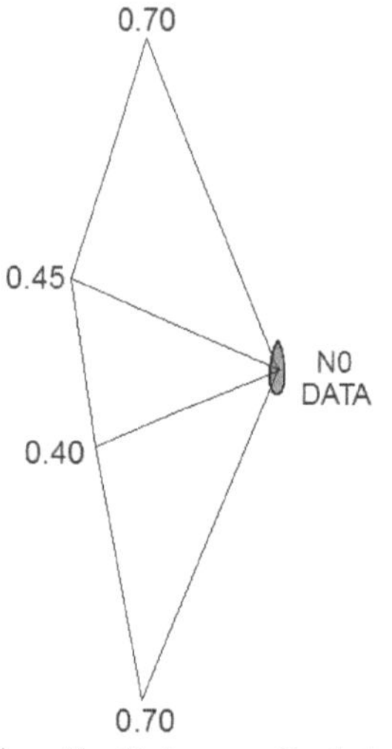

Figure 7b. Study area N of the TSS. Head-on situation domains.

As depicted in Figure 7 and Figure 8, the passing distance between vessels increases with increasing maneuvering area. However, vessels move along a narrow passage because of depth limitations and it causes no head-on situations in sectors I-IV and comparatively small number of crossing situations.

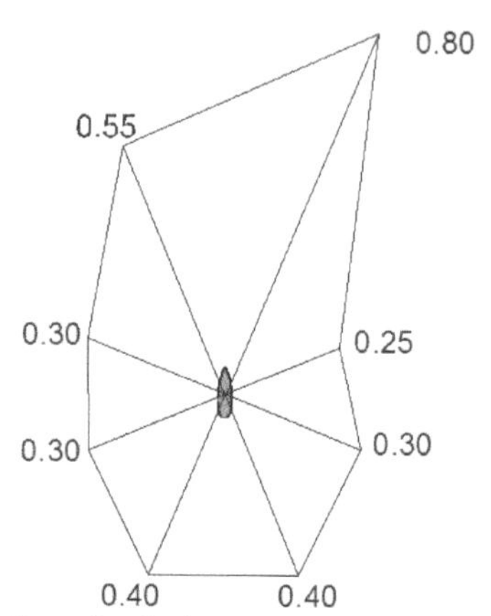

Figure 8a. Study area S of the TSS. Overtaking situation domain.

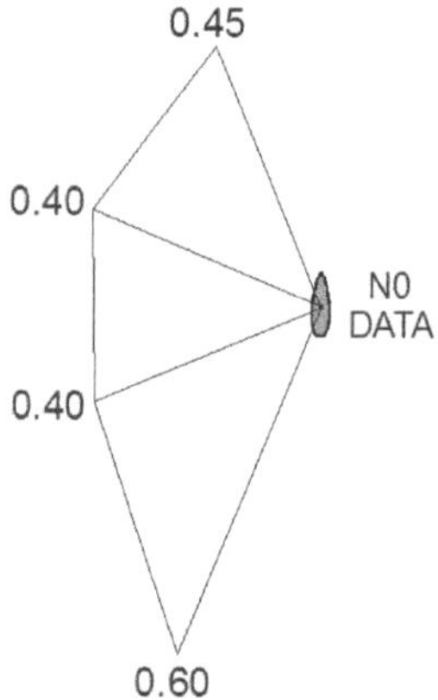

Figure 8b. Study area S of the TSS. Head-on situation domain.

4 CONCLUSIONS

Based on extensive research on the restricted waters of the Great Belt, it can be concluded that the size and shape of a ship domain is closely related to the area for which the domain is calculated. Therefore, working on a domain generalization - ship building domain, which can be applied also to other areas, must take into account a number of factors associated with the limits of the restricted area among others, such as:

- maneuvering space available;
- vessel maneuvering distribution;
- the presence of TSS;

- crossing the traffic lanes;
- limitations of the available depth;
- traffic density.

To build a domain whose shape and size could be universal for more than one restricted water area, one should first examine the relationship between the shape of the above-mentioned factors.

REFERENCES

Ervin, v. I. 2012 Detection of Hazardous Encounters at the North Sea from AIS Data. *The International Workshop on Next Generation of Nautical Traffic Model.* Shanghai, China 32-44.

Fujii Y. & Tanaka K. 1971. Traffic Capacity. *Journal of Navigation.* 24, (543-552).

Gucma, L. & Marcjan, K. 2012. Examination of ships passing distances distribution in the coastal waters in order to build a ship probabilistic domain. *Scientific Journals Maritime University of Szczecin* 32-2/2012: 34-40.

International Regulations for Preventing Collisions at Sea 1972.

Sailing Directions (Enroute). 2010. Baltic Sea (Southern part). National Geospatial-Intelligence, Springfield Virginia.

Ya-lei, R. & Jun-min, M. & Yan, D. & Jian-feng, Z. 2012 Study on Ship Domain Using AIS Data. *The International Workshop on Next Generation of Nautical Traffic Model.* Shanghai, China 51-57.

CIRM Navigation chart.

Chapter 4

Search and Rescue

Search and Rescue of Migrants at Sea

J. Coppens
Maritime Institute, Ghent University, Belgium

ABSTRACT: It is an international legal obligation for States to render assistance to persons in distress at sea. However, a comparable legally binding duty to disembark these rescued persons does not exist in the law of the sea. As a result, these persons – often migrants – can spend weeks on a ship at sea before a State allows them to go ashore. Moreover, in order to cope with the problem of migrants at sea, States sometimes rely on the principles associated with search and rescue at sea as a means of interdicting vessels that could not otherwise lawfully be visited on the high seas. This article analyses whether this practice constitues an abuse of right.

1 INTRODUCTION

The Arab Spring recently highlighted the problem of migrants at sea and the shortcomings of the international legal framework. Indeed, due to the social uprisings in Tunesia and Lybia, thousands of people tried to reach Europe by sea. This is a dangerous journey, as these asylum seekers often travel in unseaworthy vessels. As a result of the Arab Spring, it is estimated more than 1.500 people drowned or went missing while attempting to cross the Mediterranean to reach Europe in 2011.[1] In 2012, UNHCR also expressed its concerns on the loss of life in maritime incidents in the Caribbean among people trying to escape – often in unseaworthy vessels – difficult conditions in Haiti following the 2010 earthquake.[2] For example, in summer 2012, several persons lost their lives in Bahamian and US waters while trying to reach the shores of Florida. US Coast Guard data show that as of December 2011 over 900 people have been found on boats in rescue operations including some 652 Haitians, 146 Cubans and 111 people from the Dominican Republic.[3] These events are a reminder of the extremes that people in difficult situations sometimes resort to. Although no firm statistics exist, it is estimated that hundreds of deaths occur yearly as a result.[4] The international community is aware that this problem has to be tackled as soon as possible in order to prevent further loss of life. This paper will first deal with the search and rescue obligations of flag States and their shipmasters as well as those of coastal States. Secondly, we will take a look at the problem of disembarkation. Indeed, as most rescued persons are asylum seekers, States are reluctant to let them disembark onto their own territory. Recently, there is a regional initiative for the Mediterranean Basin to solve the disembarkation problem. This will be discussed in a next part. Although States have indeed the duty to render assistance to persons in distress, their actual intent here is interdicting a vessel. As interdiction on the high seas is only possible in a limited number of cases, states have thus tried to 'disguise' these interdictions as rescues. The final part will thus focus on the question whether this practice can be considered to be an abuse of right under international law.

---

[1] UNHCR, "Mediterranean Takes Record as Most Deadly Stretch of Water for Refugees and Migrants in 2011", Briefing Note (31 January 2012), available online: <http://www.unhcr.org/4f27e01f9.html>.

[2] UNHCR, "More People are Risking Lives in the Caribbean to Reach Safety", Statement (13 July 2012), available online: <http://www.unhcr.org/refworld/docid/50001c5a2.html>.

[3] US Coast Guard, "US Coast Guard Maritime Migrant Interdictions", available online: <http://www.uscg.mil/hq/cg5/cg531/AMIO/FlowStats/currentstats.asp>.

[4] UNHCR, "More People are Risking Lives in the Caribbean to Reach Safety", Statement (13 July 2012), available online: <http://www.unhcr.org/refworld/docid/50001c5a2.html>.

2 SEARCH AND RESCUE

It is a legal obligation for shipmasters and States under customary international law, as well as under Articles 58(2) and 98(1) LOSC to render assistance to persons in danger of being lost and to proceed with all possible speed to the rescue of persons in distress. Article 98(1) LOSC states: "*Every State shall require the master of a ship flying its flag, in so far as he can do so without serious danger to the ship, the crew or the passengers: (a) to render assistance to any person found at sea in danger of being lost; (b) to proceed with all possible speed to the rescue of persons in distress, if informed of their need of assistance, in so far as such action may reasonably be expected of him; (c) after a collision, to render assistance to the other ship, its crew and its passengers and, where possible, to inform the other ship of the name of his own ship, its port of registry and the nearest port at which it will call.*" As assistance must be given to 'any' person[5], the obligation applies regardless of the persons' nationality or status or the circumstances in which they are found.[6] Therefore, migrants can not be excluded from this.

First of all, there is a duty to render assistance persons in danger of being lost. There is a variety of acts that may constitute assistance, for example to tow the vessel to safety, extricate a grounded vessel, fight a fire aboard a ship, provide food and supplies, embark crewmen aboard to replace the tired or the missing, secure aid or assistance from other nearby ships, or simply stand-by to provide navigational advice.[7] Secondly, there is an obligation to proceed with all possible speed to the rescue of persons in distress. However, the LOSC does not mention what 'distress' is. The actual distress phase is defined by the 1979 International Convention on Maritime Search and Rescue (SAR Convention)[8] – a treaty monitored by the IMO that imputes multi-State coordination of search and rescue systems – as: "*A situation wherein there is reasonable certainty that a person, a vessel or other craft is threatened by grave and imminent danger and requires immediate assistance.*"[9] When exactly a situation is identified as requiring immediate assistance, can be different according to which State is handling the situation. For some States the vessel must really be on the point of sinking.[10] However, the ILC stated that – although a situation of distress may at most include a situation of serious danger – it is not necessarily one that jeopardizes the life of the persons concerned.[11] In contrast, for other States it is sufficient for the vessel to be unseaworthy.[12] MORENO-LAX even suggests that unseaworthiness *per se* entails distress.[13] Nevertheless, every situation is different and whether persons at sea are in distress or not will dependent on the specific circumstances. Therefore, an assessment can only be made on a case-by-case basis. Although the definition of distress is quite vague, this allows shipmasters and States to take all relevant elements into account. Their discretionary power to decide whether persons are in distress or not is regarded as being essential. However, one element that is indisputable, is that the existence of an emergency should not be exclusively dependent on or determined by an actual request for assistance.[14]

Coastal States shall establish adequate and effective search-and-rescue (SAR) services (for example, through the creation of a Rescue Co-ordination Centre or RCC) and, where circumstances so require, cooperate with neighbouring States for this purpose.[15] The basic elements for a search and rescue service are a legal framework, the assignment of a responsible authority, the organization of

[5] LOSC, Art. 98(1).

[6] International Convention for the Safety of Life at Sea (adopted 1 November 1974, entered into force 25 May 1980) 1184 **UNTS** 278 [SOLAS Convention], Chapter V Regulation 33 para. 1.

[7] NORRIS, Martin J., *The Law of Salvage* (Mount Kisco NY: Baker/Voorhis, 1958), 15-31; WILDEBOER, Ina H., *The Brussels Salvage Convention: Its Unifying Effect in England, Germany, Belgium, and The Netherlands* (Leiden: A.W. Sijthoff, 1965), 95; KENNEY, Frederick J. & TASIKAS, Vasilios, "The Tampa Incident: IMO Perspectives and Responses on the Treatment of Persons Rescued at Sea", 12 *Pacific Rim Law & Policy Journal* 143 (2003), 151-152.

[8] International Convention on Maritime Search and Rescue (adopted 27 April 1979, entered into force 22 June 1985) 405 **UNTS** 97 [SAR Convention].

[9] SAR Convention, Annex Chapter 1 para. 1.3.13.

[10] European Commission Proposal for a Council Decision of 27 November 2007 supplementing the Schengen Borders Code as regards the surveillance of the sea external borders in the context of the operational cooperation coordinated by the European Agency for the Management of Operational Cooperation at the External Borders, COM (2009) 658 final, Explanatory Memorandum, para. 2.

[11] ILC, *Yearbook of the International Law Commission* (New York: ILC, 1979), Vol. II, Part II, 135, para. 10, available online: <http://untreaty.un.org/ilc/publications/yearbooks/Ybkvolumes%28e%29/ILC_1979_v2_p2_e.pdf>. Although this definition was given during the discussions on the concept of 'distress' as one of the grounds for excluding wrongfulness with regard to the Draft Articles on State Responsibility, the definition is often being used to describe the situation of distress of persons at sea. See for example: BARNES, Richard A., "Refugee Law at Sea", 53 *International and Comparative Law Quarterly* 47 (2004), 60.

[12] European Commission Proposal for a Council Decision of 27 November 2007 supplementing the Schengen Borders Code as regards the surveillance of the sea external borders in the context of the operational cooperation coordinated by the European Agency for the Management of Operational Cooperation at the External Borders, COM (2009) 658 final, Explanatory Memorandum, para. 2.

[13] MORENO-LAX, Violeta, "Seeking Asylum in the Mediterranean: Against a Fragmentary Reading of EU Member States' Obligations Accruing at Sea", 23 *International Journal of Refugee Law* 174 (2011), 195.

[14] Council Decision (EU) No. 2010/252 of 26 April 2010 supplementing the Schengen Borders Code as regards the surveillance of the sea external borders in the context of the operational cooperation coordinated by the European Agency for the Management of Operational Cooperation at the External Borders, *OJ* L 111/20 of 4 May 2010, Annex Part II para. 1.4.

[15] LOSC, Art. 98(2); SAR Convention, Annex Chapter 2 para. 2.1.1.

available resources, communication facilities, coordination and operational functions and processes to improve the service including planning, domestic and international cooperative relationships and training.[16] In the SAR Convention, rescue is described as "*an operation to retrieve persons in distress, provide for their initial medical or other needs, and deliver them to a place of safety.*"[17]

Until the adoption of the SAR Convention, there was actually no international system covering search and rescue operations. Basically, the world's oceans are divided into 13 search and rescue areas, in each of which the countries concerned have delimited search and rescue regions for which they are responsible.[18] States must ensure that suffcient Search and Rescue Regions (SRR) are established within each sea area. These regions should be contiguous and – as far as practicable – not overlap.[19] Each SRR shall be established by agreement among parties concerned.[20]

In April 2009, the Panamanian flagged Turkish ship *M/V Pinar E* rescued 142 African migrants off the coast of Lampedusa. The ship and the rescued migrants were the subject of an ensuing stand-off between Italy and Malta regarding who would receive the migrants. While Malta insisted that the *M/V Pinar E* would take the migrants to Lampedusa because it was the nearest port to where the stricken boats were found, Italy maintained that the persons were rescued in the Maltase SRR and thus fell under Malta's responsibility. Although Italy finally agreed to allow disembarkation in Sicily, the decision was made exclusively in consideration of the painful humanitarian emergency aboard the cargo ship. Italy made clear that it's acceptance of the migrants must not in any way be understood as a precedent nor as a recognition of Malta's reasons for refusing them.[21] This is one of the many incidents that highlight the lack of cooperation and coordination between SAR services of these two States.

3 DISEMBARKATION

Neither treaty law nor customary international law requires States to let rescued persons disembark onto their territory. Both the International Convention on Safety of Life at Sea (SOLAS Convention)[22] – a treaty seeking to ensure protection of passengers aboard ships in distress through the prevention of situations of distress – and the SAR Convention[23] only provide that States must arrange for the disembarkation of persons rescued at sea as soon as reasonably practicable.[24]

As a result, persons rescued at sea can spend weeks on a ship at sea before a State allows them to go ashore. The case of the *Marine I* provides an example. On 30 January 2007, the Spanish Coast Guard received a distress call from the vessel *Marine I*. It was alleged that over 300 migrants from Guinea were on board. Although the *Marine I* was within the Senegalese SRR, Senegal requested Spain to proceed with a rescue operation, claiming that Senegal did not have the proper means to assist. Because the Mauritanian port of Nouadhibou was closest to the emergency, Senegal also informed Mauritania of the situation. On 4 February, a Spanish maritime rescue tug reached the *Marine I* and provided immediate relief by handing out supplies of water and food. The Spanish government also commenced negotiations with Senegal and Mauritania on the fate of the migrants. On 12 February (two weeks after the distress call), Spain, Senegal and Mauritania finally reached an agreement regarding the passengers. It was reportedly agreed that Spain would pay €650,000, in return for Mauritania allowing the passengers to disembark. Repatriation commenced the day after the migrants had disembarked. Guinea agreed to readmit thirty-five passengers, all of African origin.[25] In total, Spain reported 18,000 irregular arrivals by sea from West Africa that year.[26] The fact that Spain was prepared to pay as much as €650,000 to prevent the disembarkation of 300 migrants shows that some States are reluctant to allow disembarkation of rescued persons onto their territory.

4 REGIONAL MOU FOR THE MEDITERRANEAN ON DISEMBARKATION

At the meeting of the IMO Sub-Committee on Radiocommunications and Search and Rescue

[16] SAR Convention, Annex Chapter 2 para. 2.1.2.

[17] SAR Convention, Annex Chapter 1 para. 1.3.2.

[18] IMO, "SAR Convention", available online: <http://www.imo.org/OurWork/Safety/RadioCommunicationsAndSearchAndRescue/SearchAndRescue/Pages/SARConvention.aspx>.

[19] SAR Convention, Annex Chapter 2 para. 2.1.3.

[20] SAR Convention, Annex Chapter 2 para. 2.1.4.

[21] BBC, "Italy takes in stranded migrants" (20 April 2009), available online: <http://news.bbc.co.uk/2/hi/europe/8007379.stm>; Times of Malta, "Migrants standoff: PM insists Malta did its duty" (20 April 2009), available online: <http://www.timesofmalta.com/articles/view/20090420/local/migrants-standoff-pm-insists-malta-did-its-duty.253648>.

[22] International Convention for the Safety of Life at Sea (adopted 1 November 1974, entered into force 25 May 1980) 1184 **UNTS** 278 [SOLAS Convention].

[23] International Convention on Maritime Search and Rescue (adopted 27 April 1979, entered into force 22 June 1985) 405 **UNTS** 97 [SAR Convention].

[24] SOLAS Convention, Chapter V Regulation 33; SAR Convention, Chapter 3 para. 3.1.9.

[25] WOUTERS, Kees & DEN HEIJER, Maarten, "The Marine I Case: A Comment", 22 *International Journal of Refugee Law* 1 (2010), 2-3.

[26] UNHCR, "All in the Same Boat: The Challenges of Mixed Migration", available online: <http://www.unhcr.org/pages/4a1d406060.html>.

(COMSAR) in March 2010, the United States stated that the discussions between Mediterranean countries concerning rescue and disembarkation of migrants at sea is based on a regional problem requiring a regional solution. However, Italy, Malta and Spain expressed their disappointment that other countries seemingly did not recognize that the problem was much wider than simply a regional one. Indeed, other parts of the world are also confronted with similar difficulties and, even more importantly, ships of all flags are currently involved in the resulting rescue operations. Therefore, the IMO Secretary-General proposed to first develop a pilot project for a regional solution in the Mediterranean. Second, if this project works, it could be applied in other parts of the world.[27] One of the primary concerns of the IMO is the integrity of the search and rescue and, consequentially, the safety of life at sea regime.[28] Therefore, the IMO wants to prevent incidents – which cause loss of life at sea – from recurring.[29] COMSAR launched the idea of developing a pilot project for a regional solution in the Mediterranean in March 2010. On the one hand, the system of rescuing migrants in the Mediterranean Basin has to be improved. On the other hand, these persons also have to be disembarked at a place of safety in accordance with the SAR and SOLAS Conventions.[30] If the project works, it could be extended to other parts of the world experiencing similar situations.[31] Meanwhile, the IMO is even waiting to take steps on the international level – for example amending the Facilitation Convention[32] – until the results of this Regional Agreement are ready.[33]

5 ABUSE OF RIGHT

States sometimes rely on the principles associated with search and rescue at sea as a means of interdicting vessels that could not otherwise lawfully be visited on the high seas.[34] Indeed, for a high seas interdiction to be lawful, it must meet the conditions set out in Article 110 LOSC. For example, in the case *Hirsi Jamaa and Others v. Italy*[35] (2012), Italy submitted that it intercepted the migrant vessel in the context of a rescue on the high seas.[36] The UNCHR already stressed that States should avoid the categorization of interception operations as search and rescue operations, because this can lead to confusion with respect to disembarkation responsibilities.[37] Although States have indeed the duty to render assistance to persons in distress, their actual intent here is interdicting a vessel. As interdiction on the high seas is only possible in a limited number of cases, states have thus tried to 'disguise' these interdictions as rescues. This part will focus on the question whether this practice can be considered to be an abuse of right under international law.

5.1 *The principle of good faith in international law*

Abuse of right is often being linked to the principle of good faith. Bin CHENG – author of the timely publication *General Principles of Law* – believed that good faith eludes *a priori* definition. According to him, the notion can be illustrated by means of international judicial decisions. However, the concept can not be defined.[38] The *legal* concept of good faith entails the *moral* elements of honesty, fairness and reasonableness and therefore it is not easily reducible to precise rules. However, as a legal principle, it must be applied only where there is a legal obligation in question.[39] In the *Nuclear Tests Case*, the International Court of Justice (ICJ) said: "*One of the basic principles governing the creation and performance of legal obligations, whatever their source, is the principle of good faith. Trust and confidence are inherent in international co-operation. ... The very rule of pacta sunt servanda in the law of the treaties is based on good faith.*"[40]

[27] COMSAR, "Report to the Maritime Safety Committee", *IMO Doc.* COMSAR 14/17 (22 March 2010), paras. 10.1-10.26.

[28] MSC, "Report of the Maritime Saftey Committee on its eighty-seventh session", *IMO Doc.* MSC 87/26 (25 May 2010), para. 14.18.

[29] LEG, "Report of the Legal Committee on the work of its ninety-eight session", *IMO Doc.* LEG 98/14 (18 April 2011), para. 13.25.

[30] LEG, "Report of the Legal Committee on the work of its ninety-eight session", *IMO Doc.* LEG 98/14 (18 April 2011), para. 13.25.

[31] COMSAR, "Report to the Maritime Safety Committee", *IMO Doc.* COMSAR 14/17 (22 March 2010), paras. 10.1-10.26.

[32] Convention on Facilitation of International Maritime Traffic (adopted 9 April 1965, entered into force 5 March 1967) 591 *UNTS* 265 [FAL Convention].

[33] COMSAR, "Report to the Maritime Safety Committee", *IMO Doc.* COMSAR 15/16 (25 March 2011), para. 10.3.

[34] KLEIN, Natalie, "International Migration by Sea and Air", in OPESKIN, Brian, PERRUCHOUD, Richard & REDPATH-CROSS, Jillyanne (Eds.), *Foundations of International Migration Law* (Cambridge: Cambridge University Press, 2012), 270.

[35] ECtHR, *Hirsi Jamaa and Others v. Italy*, 23 February 2012, Appl. No. 27765/09 (2012). See also: UNHCR, "UNHCR's oral intervention at the European Court of Human Rights Hearing of the case *Hirsi and Others v. Italy* (Application No. 27765/09), Strasbourg, June 22, 2011", available online: <http://www.unhcr.org/refworld/pdfid/4e0356d42.pdf>.

[36] ECtHR, *Hirsi Jamaa and Others v. Italy*, 23 February 2012, Appl. No. 27765/09 (2012), para. 65.

[37] UNHCR, "The Treatment of Persons Rescued at Sea: Conclusions and Recommendations from Recent Meetings and Expert Round Tables Convened by the Office of the United Nations High Commissioner for Refugees – Report of the Office of the United Nations High Commissioner for Refugees,", *UN Doc.* A/AC.259/17 (11 April 2008), para. 20, available online: <http://www.un.org/Depts/los/consultative_process/consultative_process.htm>.

[38] CHENG, Bin, *General Principles of Law as Applied by International Courts and Tribunals* (London: Steven & Sons Limited, 1953), 105.

[39] O'CONNOR, John F., *Good faith in International Law* (Hants: Dartmouth Publishing Company, 1991), 123.

[40] ICJ, Nuclear Tests Case, *New Zealand v. France*, 20 December 1974, *ICJ Reports* 457 (1974), para. 49. The association of *pacta*

The Vienna Convention on the Law of the Treaties of 1969 (VCLT) codified and progressively developed the customary rules on the law of treaties.[41] Article 26 VCLT says: "*Every treaty in force is binding upon the parties to it and must be performed by them in good faith*". In this respect, the principle of good faith has three functions. First of all, the principle is particularly relevant in relation to the performance of treaties. For example, Article 2(2) United Nations Charter mentions: "*All Members, in order to ensure to all of them the rights and benefits resulting from membership, shall fulfill in good faith the obligations assumed by them in accordance with the present Charter.*"[42] Parties must observe what they have actually agreed to observe. Secondly, the principle has got a function when it comes to the interpretation of a treaty. Article 31(1) VCLT stipulates that a treaty shall be interpreted in good faith in accordance with the ordinary meaning to be given to the terms of the treaty in their context and in the light of its object and purpose. This means that the interpretation of a treaty is first of all based on the actual text or 'plain meaning'. Both in the jurisprudence of the Permanent Court of International Justice (PCIJ) and the ICJ, the principle of *ut res magis valeat quam pereat* has been invoked.[43] This principle entails that if a piece of law seems unclear, one should try to understand it in a way that makes sense of it. However, there are certain limits to this principle. As the Court said in the Interpretation of Peace Treaties Advisory Opinion, the principle can not be applied in a way that would be contrary to the spirit of the treaty.[44] Therefore a Court may take into account – in interpreting a treaty – honesty, fairness and reasonableness. Thirdly, good faith has a function in the process of negotiations for a treaty. International law may invoke specific rules derived from good faith, such as estoppel, which may be applied as appropriate to negotiations.[45]

5.2 *The principle of abuse of right in international law*

In the Dictionnaire de la terminologie du droit international of 1960 abuse of right is defined as: "*Exercise par un Etat d'un droit d'une manière ou dans des circonstances qui font apparaître que cet exercise a été pour cet Etat un moyen indirect de manquer à une obligation internationale lui incombant ou a été effectué dans un but ne correspondant pas à celui en vue duquel ledit droit est reconnu à cet Etat.*"[46] The abuse of right thus refers to a State exercising a right in such a manner or in such circumstances either in a way that avoids an international obligation or for a purpose not corresponding to the purpose for which that right was recognized in favour of that State. This definition of the concept was footed on two judgments of the PCIJ, namely in the *Case concerning Certain German Interests in Polish Upper Silesia*[47] and the *Free Zones Case*[48]. KISS, however, limits the definition to the relations between States. According to him, abuse of rights refers to a State exercising a right either in a way which impedes the enjoyment by other States of their own rights or for an end different from that for which the right was created, to the injury of another State.[49]

Some authors challenge the actual existence of the principle. SCHWARZENBERGER stipulated that the arbitrary or unreasonable exercise of a right is not illegal, but merely an unfriendly act. He therefore rejects the notion that there is a general rule of international customary law prohibiting the abuse of right.[50] The best known proponent of abuse of rights has been Hersch LAUTERPACHT. He stated that the determination of when the exercise of a right becomes abusive must depend on the specific facts of each case, rather than the application of an abstract legislative standard. Abuse of right would occur "*when a State avails itself of its right in an*

sunt servanda with 'faith' was already well marked in the history of ancient Rome. The keeping of treaties and pacts was associated by the Romans with the Goddess Fides, the personification of trust. See: O'CONNOR, John F., *Good faith in International Law* (Hants: Dartmouth Publishing Company, 1991), 17.

[41] Vienna Convention on the Law of Treaties (adopted 23 May 1969, entered into force 27 January 1980) 1155 *UNTS* 331 [VCLT].

[42] Charter of the United Nations (adopted 26 June 1945, entered into force 24 October 1945) 1 *UNTS* XVI.

[43] See for example: PCIJ, Free Zones of Upper Savoy and the District of Gex Case, *France v. Switserland*, 19 August 1929, *PCIJ* Ser. A No. 22 (1929), 13; ICJ, Corfu Channel Case, *United Kingdom of Great Britain and Northern Ireland v. Albania*, 9 April 1949, *ICJ Reports* 4 (1949), 24.

[44] ICJ, Interpretation of Peace Treaties with Bulgaria, Hungary and Romania Advisory Opinion, 30 March 1950, *ICJ Reports* 65 (1950), 229.

[45] O'CONNOR, John F., *Good faith in International Law* (Hants: Dartmouth Publishing Company, 1991), 111.

[46] BASDEVANT, Jules, "Abus de droit", in X., *Dictionnaire de la terminologie du droit international* (Paris : Sirey 1960).

[47] PCIJ, Case concerning Certain German Interests in Polish Upper Silesia, *Germany v. Poland*, 25 August 1925, *PCIJ* Ser. A No. 7 (1926), 30. The Court held: "*Germany undoubtedly retained until the actual transfer of sovereignty the right to dispose of her property, and only a misuse of this right could endow an act of alienation with the character of a breach of the Treaty; such misuse cannot be presumed, and it rests with the party who states that there has been such misuse to prove his statement.*"

[48] PCIJ, Free Zones of Upper Savoy and the District of Gex Case, *France v. Switzerland*, 19 August 1929, *PCIJ* Ser. A/B No. 46 (1932), 167. The Court suggested that if a State attempted to avoid its contractual obligations by resorting to measures having the same effect as the specifically prohibited acts, an abuse of rights would result.

[49] KISS, Alexandre, "Abuse of Rights", in BERNHARDT, Rudolf (Ed.), *Encyclopedia of Public International Law* (Amsterdam: North-Holland, 1992-2003), Vol. I.

[50] SCHWARZENBERGER, Georg & BROWN, Edward D., *A Manual of International Law* (Abingdon: Professional Books, 6th ed. 1976), 119.

arbitrary manner in such a way as to inflict upon another State an injury which cannot be justified by a legitimate consideration of its own advantage."[51] Nevertheless, he acknowledged that this was a relatively ambiguous definition. Before international courts and tribunals, the application of the principle would therefore result in a great deal of discretionary power being granted to judges and arbitrators. He thus promoted some caution when studying this principle.[52]

It is difficult to establish what is supposed to amount to an abuse, as distinct from a harsh but justified use, of a right under international law.[53] However, largely due to its widespread existence in national legal systems, many authors have considered abuse of rights to be part of international law, whether as a general principle of law or as part of customary international law.[54] Abuse of rights first found support in international law when the Advisory Committee of Jurists was drafting Article 38 of the Statute of the PCIJ[55], which identifies sources of international law and which later became Article 38 of the Statute of the ICJ.[56] One of the members of the Committee, referred to the principle "*which forbids the abuse of rights*" as one of the "*general principles of law*".[57] According to him, disputes concerning the right of a coastal state to fix the breadth of its territorial sea are an example of this principle. At that time, there was no international rule defining the outer limit of the territorial sea. Therefore he suggested that the Court be permitted to admit the rules of each State in this respect as "*equally legitimate in so far as they do not encroach on other principles, such for instance, as that of the freedom of the seas.*"[58]

The principle also appears in case law of the PCIJ and the ICJ. When dealing with the right to draw straight baselines in a territorial sea delimitation in the *Fisheries Case*, the ICJ said: "*The base-line has been challenged on the ground that it does not respect the general direction of the coast. It should be observed that, however justified the rule in question may be, it is devoid of any mathematical precision. In order properly to apply the rule, regard must be had for the relation between the deviation complained of and what, according to the terms of the rule, must be regarded as the general direction of the coast. Therefore, one cannot confine oneself to examining one sector of the coast alone, except in a case of manifest abuse; nor can one rely on the impression that may be gathered from a large scale chart of this sector alone.*"[59] Some additional support for the principle may be found in separate and dissenting opinions as well as in international arbitral decisions. Also, a number of States have argued for the applicability of abuse of rights in State-to-State litigation and arbitration.[60]

5.3 *The link between good faith and abuse of right*

It is possible to argue that abuse of right is redundant because it is itself only a more specific expression of a broader principle, namely that of good faith. For example, BIRNIE & BOYLE argue that abuse of right is merely a method of interpreting rules concerning matters such as the duty to negotiate and consult in good faith, or another way of formulating a doctrine of reasonableness or a balancing of interests. Therefore they conclude that the principle does not add anything useful.[61] CHENG similarly writes that the theory of abuse of right is merely an application of good faith to the exercise of rights.[62] Nonetheless, the principle of abuse of right is not redundant since it is – in a small yet important respect – supplemental to the principle of good faith since it provides the threshold at which a lack of good faith gives rise to a violation of international law, with all the attendant consequences.[63]

---

[51] OPPENHEIM, Lassa, *International Law: A Treatise* (London: Longmans, Green & Co., 8th ed. 1955 by LAUTERPACHT, Hersch), 345.

[52] LAUTERPACHT, Hersch, *The Development of International Law by the International Court* (London: Stevens & Sons, 1958), 164; see also LAUTERPACHT, Hersch, "Droit de la paix", 62 *Recueil des Cours* 95 (1937), 342.

[53] SCHWARZENBERGER, Georg & BROWN, Edward D., *A Manual of International Law* (Abingdon: Professional Books, 6th ed.1976), 84.

[54] See generally BYERS, Michael, "Abuse of Rights: An Old Principle, A New Age", 47 *McGill Law Journal* 389 (2002) 389-431; WHITEMAN, Marjorie M., *Digest of International Law* (Washington D.C.: Government Printing Office, 1965), Vol. V, 224-30; KISS, Alexandre, "Abuse of Rights", in BERNHARDT, Rudolf (Ed.), *Encyclopedia of Public International Law* (Amsterdam: North-Holland, 1992-2003), Vol. I.

[55] Statute of the Permanent Court of International Justice (adopted 16 December 1920, entered into force 5 September 1921) 6 *LNTS* 279.

[56] Statute of the International Court of Justice (adopted 16 June 1945, entered into force 24 October 1945) 33 *UNTS* 993 [ICJ Statute], Art. 38(1)(c) which reads: "*The Court, whose function is to decide in accordance with international law such disputes as are submitted to it, shall apply: ... (c) the general principles of law recognized by civilized nations*"; See generally: CHENG, Bin, *General Principles of Law as Applied by International Courts and Tribunals* (London: Stevens & Sons, 1953), 490 p.

[57] X., *Procès-verbaux des séances du comité* (The Hague : Van Langenhuysen Brothers, 1920), 314-315 and 335.

[58] X., *Procès-verbaux des séances du comité* (The Hague, Van Langenhuysen Brothers 1920), 315.

[59] ICJ, Fisheries Case, *United Kingdom v. Norway*, 18 December 1951, *ICJ Reports* 116 (1951), 141-142.

[60] For an extensive overview of the case law see: BYERS, Michael, "Abuse of Rights: An Old Principle, A New Age", 47 *McGill Law Journal* 389 (2002), 389-431.

[61] BIRNIE, Patricia & BOYLE, Alan E., *International Law and the Environment* (Oxford: Clarendon Press, 1992), 126.

[62] CHENG, Bin, *General Principles of Law as Applied by International Courts and Tribunals* (London: Stevens & Sons, 1953), 21.

[63] BYERS, Michael, "Abuse of Rights: An Old Principle, A New Age", 47 *McGill Law Journal* 389 (2002), 411.

5.4 *Good faith and abuse of right in the law of the sea*

Article 300 LOSC stipulates: *"States Parties shall fulfil in good faith the obligations assumed under this Convention and shall exercise the rights, jurisdiction and freedoms recognized in this Convention in a manner which would not constitute an abuse of right"*. The inclusion of this provision provides circumstantial evidence of the acceptability of the doctrine in international law.[64] In the Southern Bluefin Tuna Case, Australia and New Zealand alleged before the International Tribunal for the Law of the Sea (ITLOS) that Japan was "*in breach of its obligations under international law, specifically Articles 64 and 116-119 LOSC, and in relation thereto Article 300 and the precautionary principle which, under international law, must direct any party in the application of those articles.*"[65] Although they cited the provision as a useful guide in interpreting Japan's duties, they did not invoke the principle as the basis of an independent cause of action.[66] When ITLOS issued its decision, the order did not refer to Article 300 LOSC or relied on allegations of abuse of right.[67]

When the controversy moved to the Arbitral Tribunal, Australia and New Zealand again cited Article 300 LOSC.[68] These allegations – based on a treaty that specifically refers to the abuse of right – would seem to provide an ideal situation for the parties to invoke that principle as a separate legal basis for their claims.[69] However, in reality both Australia and New Zealand specifically stated that they were not accusing Japan of an independent breach of an obligation to act in good faith.[70] This reluctance of States to allege an independent breach of the article reflects an awareness of the diplomatic cost a State may pay in making such allegations against another State. It will be difficult to prove that a State is guilty of a substantive breach of the abuse of right principle.[71] The Arbitral Tribunal put forward that the burden of proof on a State making such allegations is very high. It does not exclude, however, that a court or a tribunal might find that the obligations of Article 300 LOSC provide a basis for jurisdiction.[72] We can conclude that, although it is very rare for a provision of this kind to be included in an international treaty[73] and despite the explicit language of the provision, the experience to date suggests that Article 300 LOSC is unlikely to have much practical meaning or effect. In fact, no international tribunal has yet expressly founded liability on such an abuse of right doctrine. However, this can not be completely excluded in the future.

5.5 *Abuse of right and interdictions in disguise*

Interdiction on the high seas is subject to a number of conditions which – if not met – can lead to illegal actions. Therefore, States tend to categorize interdictions on the high seas as rescues. Could these actions constitute an abuse of right under international law, and more specifically Article 300 LOSC? In its commentary on the LOSC, NORDQUIST stipulates that it would appear that the parameters of the notion of abuse of rights as enunciated in Article 300 LOSC are limited to relations between States Parties as defined in Article 1(2) LOSC. He therefore refers to the definition put forward by KISS.[74] Thus, to amount to an abuse of right, these rescues on the high seas have to be carried out in a way which impedes the enjoyment by other States of their own rights or for an end different from that for which the right was created, to the injury of another State.[75] The purpose of these rescues is definitely not consistent with the purpose of the duty to render assistance.

Therefore, we can conclude that this constitutes an end different from that for which the duty was created. The most difficult element is however that

[64] ILUYOMADE, Babatunde O., "The Scope and Content of a Complaint of Abuse of Right in International Law", 16 *Harvard International Law Journal* 47 (1975), 71.

[65] ITLOS, Southern Bluefin Tuna Cases, *Australia and New Zealand v. Japan*, 30 July 1999, Australia's Request for Provisional Measures, *ITLOS Reports* (1999), 25.

[66] HOFF, Paul S. & GOULDING, Michael I., "Japan's Whale Research Program and International Law", 32 *California Western International Law Journal* 151 (2002), 197.

[67] ITLOS, Southern Bluefin Tuna Cases, *Australia and New Zealand v. Japan*, 27 August 1999, *ITLOS Reports* (1999).

[68] Arbitral Tribunal constituted under Annex VII of the United Nations Convention on the Law of the Sea, Southern Bluefin Tuna Cases, *Australia and New Zealand v. Japan*, 15 July 1999, Australia's Statement of Claim and Grounds on Which It Is Based in the Dispute Concerning Bluefin Tuna, paras. 37 and 45.

[69] HOFF, Paul S. & GOULDING, Michael I., "Japan's Whale Research Program and International Law", 32 *California Western International Law Journal* 151 (2002), 198.

[70] Arbitral Tribunal constituted under Annex VII of the United Nations Convention on the Law of the Sea, Southern Bluefin Tuna Cases, *Australia and New Zealand v. Japan*, 31 March 2000, Reply on Jurisdiction of Australia and New Zealand, para. 182.

[71] HOFF, Paul S. & GOULDING, Michael I., "Japan's Whale Research Program and International Law", 32 *California Western International Law Journal* 151 (2002), 198.

[72] Arbitral Tribunal constituted under Annex VII of the United Nations Convention on the Law of the Sea, Southern Bluefin Tuna Cases, *Australia and New Zealand v. Japan*, 4 August 2000, Award on Jurisdiction and Admissibility, para. 64.

[73] NORDQUIST, Myron H. (Ed.), *United Nations Convention on the Law of the Sea 1982: A Commentary* (Dordrecht: Nijhoff, 1985-2003), Vol. V, para. 300.6.

[74] NORDQUIST, Myron H. (Ed.), *United Nations Convention on the Law of the Sea 1982: A Commentary* (Dordrecht: Nijhoff 1985-2003), Vol. V, para. 300.5.

[75] KISS, Alexandre, "Abuse of Rights", in BERNHARDT, Rudolf (Ed.), *Encyclopedia of Public International Law* (Amsterdam: North-Holland, 1992-2003), Vol. I. See also: GOODWIN-GILL, Guy S., "State Responsibility and the 'Good Faith' Obligation in International Law", in FITZMAURICE, Malgosia & SAROOSHI, Dan (Eds.), *Issues of State Responsibility before International Judicial Institutions* (Oxford: Hart Publishing, 2004), 75-104.

this must lead to the injury of another State. As *in casu* we deal with migrant vessels with potential asylum-seekers on board, it is highly unlikely that any State considers the interdictions in disguise as injurious. Nevertheless, we can conclude that the duty to render assistance was not created for the end it is being used sometimes.

6 CONCLUSION

Although it is a legal obligation for shipmasters and States under customary international law, as well as under the LOSC to render assistance to persons in danger of being lost and to proceed with all possible speed to the rescue of persons in distress, there is no comparable duty to disembark these persons. As a result, rescued asylum seekers can spend weeks on a ship before going ashore. Shipmasters are therefore reluctant to rescue migrants at sea.

To cope with the influx of migrants, States sometimes rely on the principles associated with search and rescue at sea as a means of interdicting vessels that could not otherwise lawfully be visited on the high seas.[76] However, this might constitute an abuse of right as the duty to render assistance was not created for the end it is being used sometimes.

REFERENCES

Bailliet, Cecilia, "The Tampa Case and its Impact on Burden Sharing at Sea", 3 Human Rights Rights Quarterly 741 (2003).

Barnes, Richard A., "Refugee Law at Sea", 53 International and Comparative Law Quarterly 47 (2004).

Betts, Alexander, "Soft Law and the Protection of Vulnerable Migrants", 24 Georgetown Immigration Law Journal 533 (2010).

Cheng, Bin, General Principles of Law as Applied by International Courts and Tribunals (London: Steven & Sons Limited, 1953).

Giuffre, Mariagiulia, "Watered-Down Rights on the High Seas: Hirsi Jamaa and Others v Italy", 61 International and Comparative Law Quarterly 728 (2012).

Hammarberg, Thomas, "It is Wrong to Criminalize Migration", 11 European Journal of Migration & Law, 383 (2009).

Klein, Natalie, "International Migration by Sea and Air", in Opeskin, Brian, Perruchoud, Richard & Redpath-Cross, Jillyanne (Eds.), Foundations of International Migration Law (Cambridge: Cambridge University Press, 2012).

Matthew, Penelope, "Australian Refugee Protection in the Wake of Tampa", 96 American Journal of International Law 661 (2002).

Moen, Amy E., "For Those in Peril on the Sea: Search and Rescue under the Law of the Sea Convention", 24 Ocean Yearbook 377 (2010).

Moreno-Lax, Violeta, "The EU Regime on Interdiction, Search and Rescue, and Disembarkation: The Frontex Guidelines for Intervention at Sea", 25 International Journal of Marine & Coastal Law 621 (2010).

Moreno-Lax, Violeta, "Seeking Asylum in the Mediterranean: Against a Fragmentary Reading of EU Member States' Obligations Accruing at Sea", 23 International Journal of Refugee Law 174 (2011).

Oxman, Bernard H., "Human Rights and the Law of the Sea", in Charney, Jonathan I., Anton, Donald K. & O'Connell, Mary Ellen (Eds.), Politics, Values and Functions: International Law in the 21st Century – Essays in Honor of Professor Louis Henkin (The Hague: Kluwer Law International, 1997).

Padrón Armas, et al., "New Proposal for Search and Rescue in the Sea", 5 International Journal on Marine Navigation and Safety of Sea Transportation 121 (2011).

Pallis, Mark, "Obligations of States towards Asylum Seekers at Sea: Interactions and Conflicts Between Legal Regimes", 14 International Journal of Refugee Law 329 (2002).

Papanicolopulu, Irini, "The Law of the Sea Convention: No Place for Persons?", 27 International Journal of Marine & Coastal Law 867 (2012).

Papastavridis, Efthymios, "Interception of Human Beings on the High Seas: A Contemporary Analysis under International Law", 36 Syracuse Journal of International Law and Commerce 145 (2009).

Papastavridis, Efthymios, "Rescuing Migrants at Sea: The Responsibility of States under International Law", Working Paper Series Social Science Research Network (27 September 2011), available online: <http://papers.ssrn.com/sol3/papers.cfm?abstract_id=1934352>.

Ronzitti, Natalino, "Coastal State Jurisdiction over Refugees and Migrants at Sea", in Ando, Niseku, McWhinney, Edward & Wolfrum, Rüdiger (Eds.), Liber Amicorum Judge Shigeru Oda (The Hague: Kluwer Law International, 2002).

Severance, Arthur Alan, "The Duty to Render Assistance in the Satellite Age", 36 California Western International Law Journal 377 (2006).

Treves, Tullio, "Human Rights and the Law of the Sea", 28 Berkeley Journal of International Law 1 (2010).

Wouters, Kees & Den Heijer, Maarten, "The Marine I Case: A Comment", 22 International Journal of Refugee Law 1 (2010).

Trevisanut, Seline, "Search and Rescue Operations in the Mediterranean: Factor of Cooperation or Conflict?", 25 International Journal of Marine & Coastal Law 523 (2010).

---

[76] KLEIN, Natalie, "International Migration by Sea and Air", in OPESKIN, Brian, PERRUCHOUD, Richard & REDPATH-CROSS, Jillyanne (Eds.), *Foundations of International Migration Law* (Cambridge: Cambridge University Press, 2012), 270.

Ergonomics-based Design of a Life-Saving Appliance for Search and Rescue Activities

H.J. Kang
Marine Transportation Research Division, Maritime & Ocean Engineering Research Institute, KIOST, Korea

ABSTRACT: When fatal accidents occur at sea, many of the victims are suddenly plunged into cold water. Because the victims can survive for only a brief time, the required rescue time is important since it is difficult to locate a person in distress during night time and inclement weather at sea. Rescue equipment such as light sticks (lifejacket lights), SARTs (Search and Rescue Transponders) and EPIRBs (Emergency Position Indicating Radio Beacons) are effective rescue devices, but they are expensive, large and limited in function for personal use. For this reason, a new life-saving appliance (LSA) using radar cross-section characteristic has been conceptually proposed. To realize this concept, the service environment must be considered and the functional reliability of the LSA, based on characteristics of the human body, must be guaranteed. In this paper, a conceptual design of the LSA is proposed using an ergonomics-based systems engineering approach.

1 INTRODUCTION

Figure 1 depicts the usage of the conceptually designed personal life-saving appliance (LSA) utilizing radar cross-section characteristics.

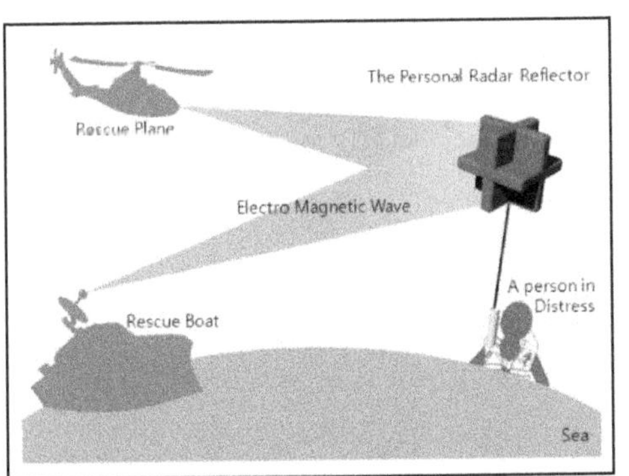

Figure 1. Usage concept of the conceptually designed personal LSA using RCS characteristics (Kang et al., 2008)

By using tri-hedral radar reflectors, the suggested LSA is compatible with other LSAs such as lifejackets, light sticks (lifejacket lights), SARTs and EPIRBs. The suggested LSA has long-lasting usability regardless of daylight and weather conditions (compared to a life stick). It can indicate the location of a person in distress not only for X-band radar but also for other radar bands (compared to SART) with a minimum error of indicating position (compared to EPIRB). In addition, the suggested LSA is structurally simple, lightweight and small and thus ideal for personal use. However, the LSA suggests its own principal usage concept. To effectively realize the concept design, the service environment, including accident scenarios and characteristics of the human body, must be considered.

2 DESIGN METHODOLOGY

Systems engineering as "an interdisciplinary approach and means to enable the realization of successful systems" is an effective, widely accepted method for successful systems development (INCOSE 2004). A systems engineering process is considerably helpful in creating an effective design process for the LSA. The Mil-STD-499 serves as the origin of systems engineering and has developed to ISO/IEC 15288, ANSI/EIA 632 and IEEE 1220(Burtain et al., 1994). Each standard is applicable for the new system development project, as all these standard track system life and level of detail (ISO/IEC, 2002; EIA, 1998; IEEE 1998). However, in this paper, simplified Mil-STD-499 model of 'requirements analysis', 'functional allocation' and 'synthesis' has been applied because the design target is comparatively simple.

3 REQUIREMENRS ANALYSIS

3.1 *Characteristics of a Person in Distress*

While the number of marine accidents is continually decreasing, fatal marine accidents still occur (The International Association of Oil and Gas Producers, 2010). When an accident such as a collision, stranding or fire (explosion) occurs, people on the deck may suddenly fall into the water. Donning lifejackets and running to the shelter deck may be difficult tasks under stress. Sometimes, victims fall into the water over their heads which describes the difficulties of proper actions to face the situation (Shibue et al., 2012). Water is cold in the fall and winter season, and survival time is dramatically restricted. If we assume a water temperature of 10-16° C, a victim can survive at best for 1 to 6 hours. Water conducts heat away from the body 25 times faster than air because it has a greater density (Curtis, 2009).

3.2 *Accident Scenarios*

A distress accident usually occurs when a ship sinks, capsizes or is fatally damaged after a collision, stranding or fire (explosion). In these cases, a ship's behavior and status change rapidly. It is difficult to properly face this emergency situation. In the case of passengers on the deck of a passenger ship, they may suddenly fall into the sea without a lifejacket. In this situation, search and rescue activities will commence after an SOS (save our souls) signal is sent to the rescue team. From the viewpoint of LSA design, all accident cases should be considered on the assumption that a person falls into the water and requires search and rescue activities. For this reason, we must address critical aspects of survivability such as lifejacket implementation, swimming capability, water temperature and search and rescue time. Figure 2 shows various accident scenarios for the LSA design. Based on these accident scenarios, the lethal cases of S1, S2, S3, S4 and S5 should be incorporated into the design to enhance a victim's survivability.

3.3 *Requirements for the LSA Design*

Related requirements must be derived from accident scenarios S1, S2, S3, S4 and S5 to better face these situations. Each requirement is derived from the accident scenarios as shown in Table 1, the requirements should be considered in the functional allocation phase. From a functional viewpoint, the requirements 'R3.' through 'R8' are important. Other requirements should also be considered to be system analysis and control tools for following design phase.

4 FUNCTIONAL ALLOCATION

In the functional allocation phase, a feasibility study of the conceptually designed LSA will be performed; applying the requirements to the LSA design also will be studied. All functional allocation results should meet the requirements. Figure 3 shows the conceptually designed structure of the LSA (Kang et al., 2008). As shown in Figure 3, the LSA consists of a switch (Percussion Device), compressed gas, rolled line and folded radar reflector. The LSA is activated by pushing or pulling the percussion device.

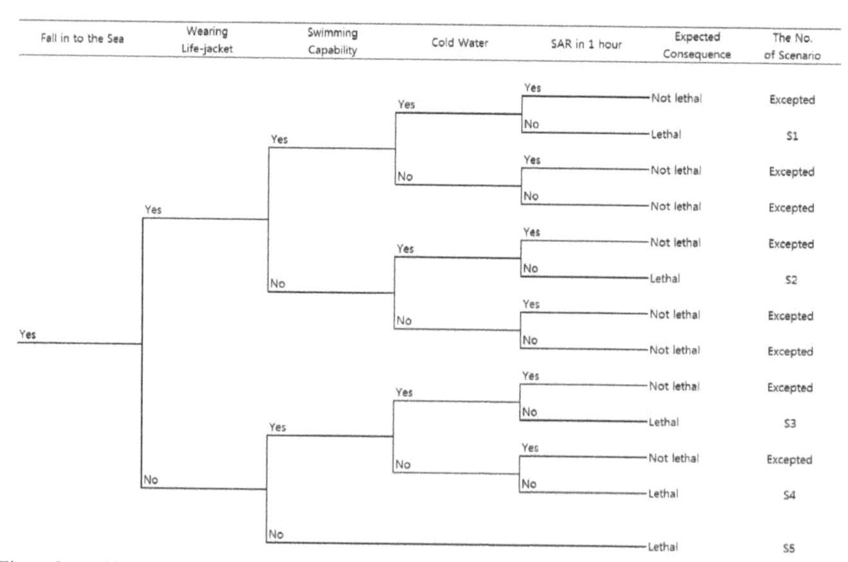

Figure 2. Accident Scenarios Concerned with Person in Water

Table 1. Requirements for the LSA

Related Requirements
R1. Indicate location regardless of weather conditions and time
R2. Support person who not good at swimming
R3. Keep RCS through SAR activity
R4. Life-buoy function for not proficient swimmers
R5. Usability in tough conditions especially in cold water
R6. Usability in the case of malfunction
R7. Size and weight for personnel use
R8. Reasonable price compare to existing LSA
R9. Location recognition (rescuing person in distress in time)
R10 [illegible]
R11 [illegible]
R12 [illegible]
R13 [illegible]
R14 [illegible]
R15 [illegible]
R16 [illegible]

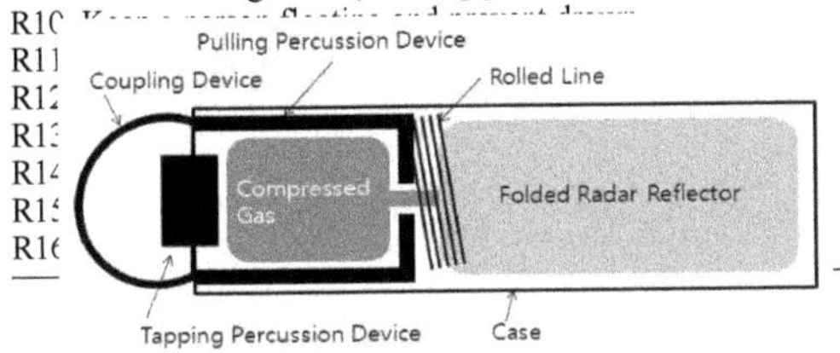

Figure 3. Structure of the Conceptually Designed Personal Radar Reflector (Kang et al., 2008)

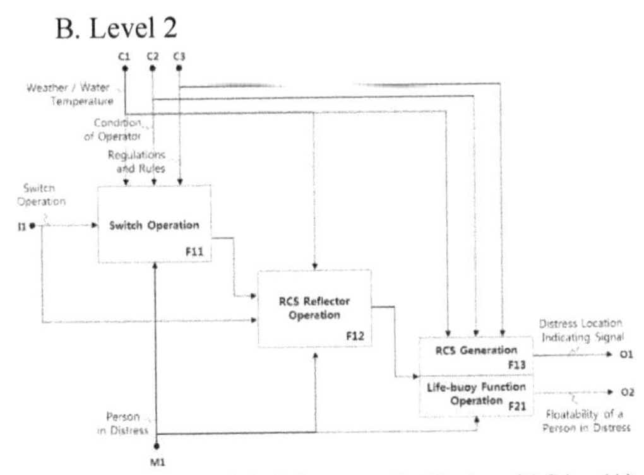

Figure 4. IDEF 0 Model of Conceptually Designed LSA - (A) Level 1; (B) Level 2

For functional allocation, the conceptually designed result was reanalyzed and rearranged by using the Integration Definition for Function Modeling (IDEF 0). IDEF 0 is a function modeling methodology that describes and integrates engineering analysis (Department of Defense 2001). Based on the requirements analysis, the function of the existing LSA device has increased, as shown in Figure 4 (A).

The primary function of the device is increased from 'location indication (R1)' to include 'keeping a person float (R2)'. Figure 4 (B) is a Level 2 IDEF 0 model. Each component of F11, F12 and F13 should be formulated to allow the requirements for the LSA.

4.1 *F11 Switch Operation*

The conceptually designed LSA has tapping and pulling percussion devices. The device assumes that it will be coupled with a lifejacket. To verify the feasibility of the design, we applied two push buttons as percussion devices instead of tapping and pulling the percussion devices. For the comparison, we assumed that a survivor will push the tapping percussion device with a finger or pull the pulling percussion device (coupled with the lifejacket) by gripping the case of the LSA. In each case, the survivor must keep the percussion device pushed or pulled until the radar reflector is entirely spread out. Each design's usability is compared by using Boolean algebra and a 4-variable map, as shown in Figure 5. In Figure 5 (B), the mix of pull and push percussion devices has 2 more usable cases compared to the push only percussion device case of Figure 5 (A).

A.

ZW \ XY	00	01	11	10
00				
01				
11		1	1	1
10				

X Availability of Tapping Device A
Y Availability of Tapping Device A'
Z Availability of Keeping Device A(A') Push
W Availability of Fingers

F = ZW(X+Y)

B.

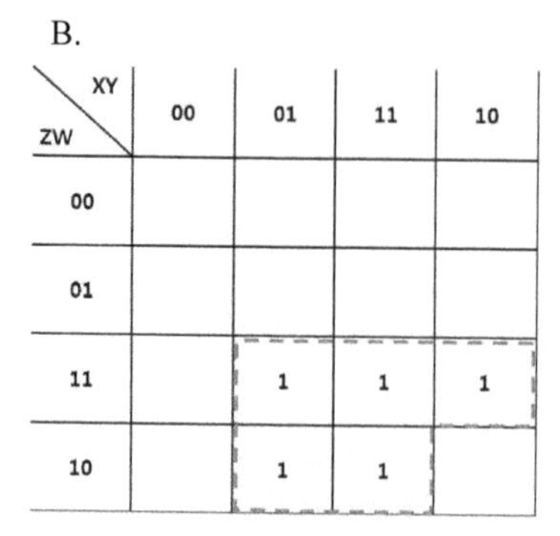

ZW \ XY	00	01	11	10
00				
01				
11		1	1	1
10		1	1	

X Availability of Tapping Device A
Y Availability of Pulling Device B
Z Availability of Keeping Device A(B) Push(Pull)
W Availability of Fingers

F = Z(W(X+Y) + Y)

Figure 5. Comparison of LSA Operation Method – (A) Push Percussion Device only (B) Mix of Pull and Push Percussion Device

4.2 *F12 RCS Reflector Operation*

It is suggested that the LSA use RCS to indicate the location of a survivor. To avoid back scattering from the sea surface and to be detectable from a long distance, the RCS reflector must hang relatively high. For this reason, the inflatable RCS reflector has a balloon shape and uses a compressed gas such as helium. For personal use, the RCS reflector folds and remains in the LSA case until the survivor taps or pulls the percussion device.

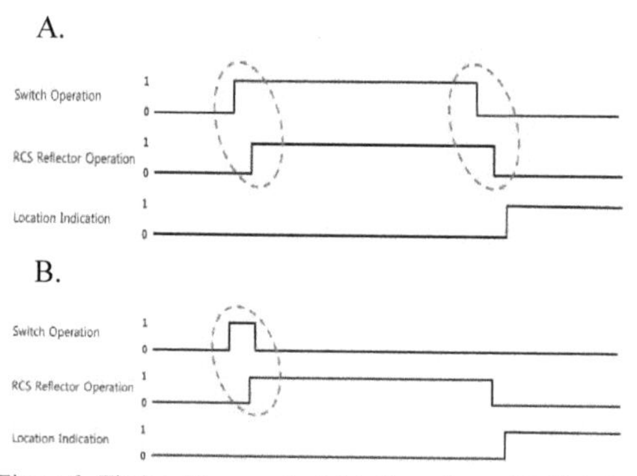

Figure 6. Timing Diagram for LSA Operation - (A) Existing Design; (B) Revised Design

Because the LSA does not automatically deploy (to avoid injury from a malfunction), the survivor must keep the percussion device pushed or pulled until the RCS reflector entirely spreads out, as shown in the timing diagram of Figure 6 (A).In an actual situation, however, it may be difficult to keep the percussion device pushed or pulled because of sea waves, cold water, injury or unconsciousness. To solve this operational problem, the LSA must be operable with one of the pushing or pulling activities shown in Figure 6 (B). A malfunction of the LSA should be functionally avoidable for safe use. Because the LSA has a large percussion device, a malfunction is possible. To avoid malfunctioning, the percussion device should not engage unless the percussion device is pushed or pulled with sufficient force. Additionally, if a person falls into the water, the inflated RCS reflector should easily unthread from the case when the survivor operates the LSA. An emergency measure to override a malfunctioning LSA is also required.

4.3 *F13 RCS Generation*

The conceptually designed LSA's RCS reflector must have a radar cross-section of 10 dBsm according to the SOLAS (Safety of Life at Sea) requirements for avoiding collisions between ships and vessels of less than 15 m length. In the case of personal use, however, the required RCS can vary because SAR activities are typically performed with acute awareness of the situation. Regardless of the judgment for a proper RCS, they can vary with the azimuth (phase) of the RCS reflector, as shown in Figure 7.

A.

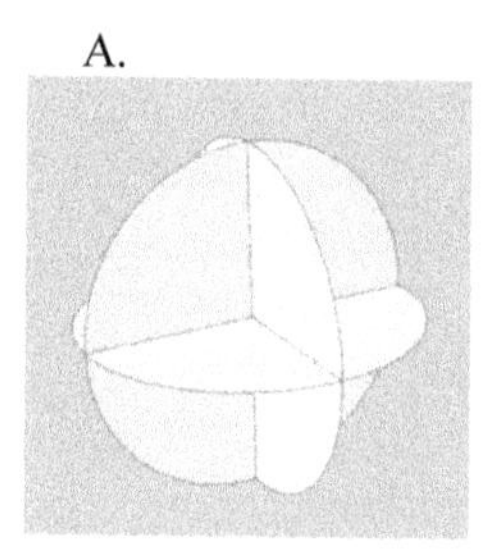

B.

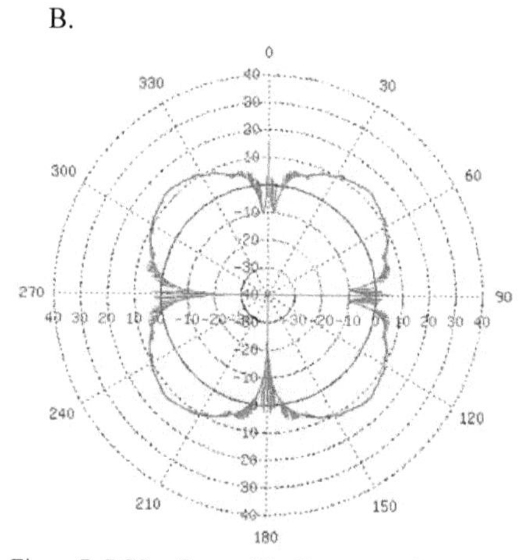

Figure 7. RCS reflector (A); Geometry of RCS reflectors (B). RCS values can vary with the azimuth (phase) (Kang et al., 2008)

To generate the appropriate RCS, the phase of the RCS reflector should be fixed, and its height should be kept consistent. Functionally, design alternatives should be proposed and applied to the LSA to maintain the phase and height of the RCS reflector. In case of RCS generation, a single RCS reflector's value may differ from crowded RCS reflectors' values. For this reason, the required RCS value for the LSA should be studied in the design phase of synthesis.

4.4 *F21 Lifebuoy Function Operation*

To use RCS reflectors as lifebuoys, the characteristics of the human body should be considered. Survivors in cold water should maintain body heat with an appropriate position. However, as mentioned above, it may be difficult to keep this position without a lifejacket. For this reason, when using the RCS reflector as a lifebuoy, the ability to strap the survivor's body in with the RCS reflector should be suggested and applied to the LSA. The ability to personally carry the LSA without a lifejacket should also be suggested and applied to the LSA.

5 SYNTHESIS

5.1 *S11 Switch Operation*

As studied in the functional allocation phase, the realization of the switch operation function (F11) to an actual system requires large tapping and pulling percussion devices for coupling with a lifejacket or clothes.

This has been applied to the LSA design, as shown in Figure 8.

5.2 *S12 RCS reflector operation*

To apply the RCS reflector operation function (F12) to the LSA design, some critical systems have been applied to the LSA. First, a stopper to keep the percussion device in the push or pull state in place has been applied to the LSA, as shown in Figure 8.

The stopper has counterchanged inclines, which are not movable unless one pushes or pulls the percussion device with sufficient force (above normal level). A sodium alginate cover, which melts in water, is also applied to the LSA to facilitate the unthreading of the RCS reflector when the LSA is activated. The combined result allows the LSA to be usable even for an injured survivor who may be losing consciousness. A karabiner-style coupling has been applied to the pulling percussion device for co-usability with the lifejacket and other clothing as shown in Figure 8.

5.3 *S13 RCS Generation*

To reflect the designed RCS value, the RCS reflector has trihedral geometry as shown in Figure 9. The reflector has been designed to interface with X- and S-band radar. The RCS reflector has a 400 mm diameter and a mean value of 7.06/7.68 dBsm for HH and VV, a maximum value of 23.42 dBsm for both HH and VV.

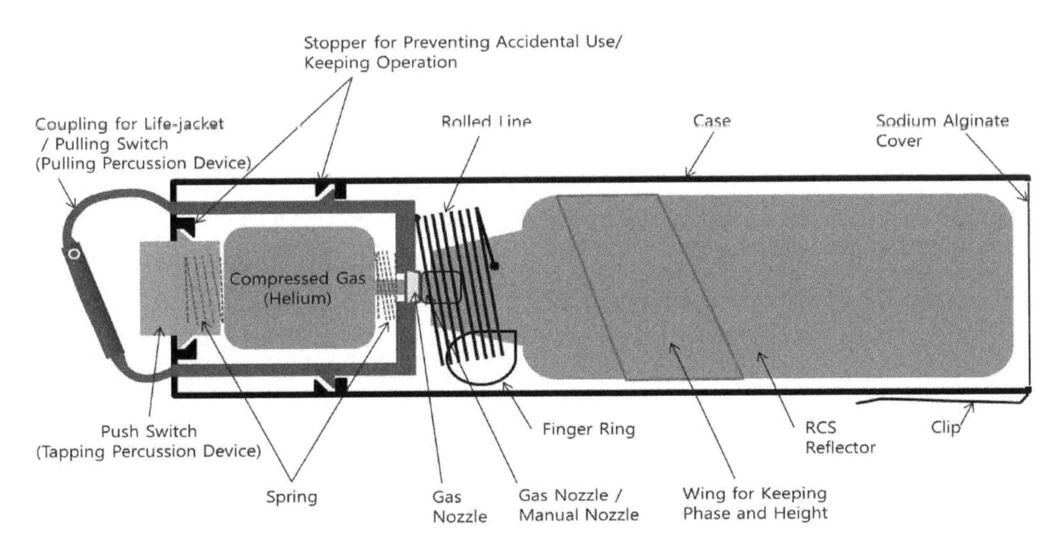

Figure 8. Revised Design Result for F12

A.

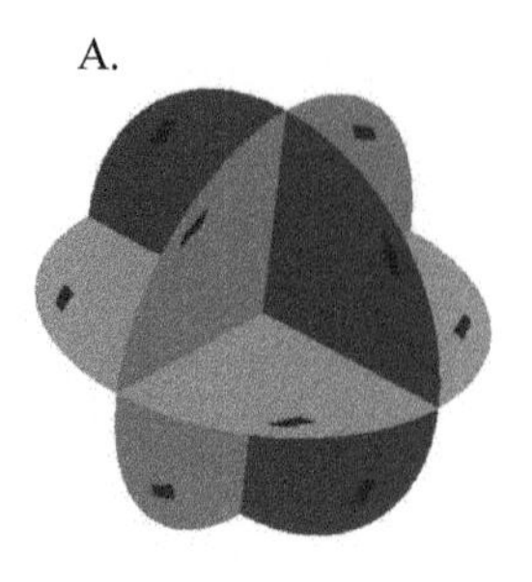

B.

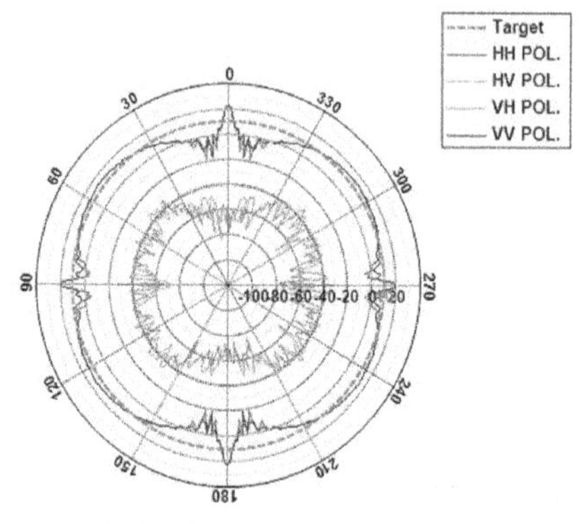

Figure 9. RCS reflector geometry for the LSA - (A) Geometry of RCS reflectors; (B) RCS values with its azimuth (phase) at 10Ghz Frequency with 1 degree step.

5.4 *S21 Lifebuoy Function Operation*

A lifejacket is essential for maintaining body heat with an appropriate position. However, as mentioned earlier, in some cases, a survivor may not have time to deploy a lifejacket. In such a case, the survivor should use the LSA as a lifebuoy. For this scenario, a finger ring, a karabiner-style pulling coupling (percussion) device and a wing strap have been designed and applied to the LSA. As shown in Figure 10, a survivor can use the RCS reflector as a lifebuoy. The LSA can be attached to the lifejacket or clothes with a karabiner-style coupling device, and by using the finger ring, the survivor can hold the RCS reflector. The survivor can affix the RCS reflector to his or her body by using the wing of the RCS reflector.

Figure 10. Revised Design Result for F21-Life-buoy function of the RCS reflector

6 CONCLUSION

The concept of an LSA that uses the characteristics of RCS is highly effective, and the LSA will improve the efficiency of SAR activities. To fully realize the concept, however, service environments and ergonomics should be considered. To achieve this, a systematic and reasonable design process and tools are required. By using suggested design process and tools in this paper, it is possible to consider ergonomics and service environments effectively. The conceptual design results of the LSA design were revised to create more functional reliability while considering the characteristics of a human body in a marine accident. However, the suggested design processes and results of the LSA still have many rooms to fulfill. In addition, some tests in the sea should be done to verify the effectiveness of the design results.

ACKNOWLEDGEMENTS

The contents of this paper are the result of MOERI/KIOST research projects PES 160B and PN65393

REFERENCES

Burtain, R.G., Collier, K.I., 1994. Implementing Mil-STD-499b system engineering on a mature program, *Proc IEEE Natl Aerospace Electron Conf 2*: 1330–1337.

Curtis, R., 2009. Outdoor Action Guide to Hypothermia and Cold Weather Injuries. *United States Search and Rescue Task Force.*

Department of Defense, 2001. Systems engineering fundamentals, *Defense Acquisition University Press*, Fort Belvoir, VA.

EIA, 1998. EIA-632: Processes for engineering a system, *Electronic Industries Alliance.*

IEEE, 1998. IEEE Std 1220: IEEE Standard for Application and Management of the Systems Engineering Process.

INCOSE, 2004. Systems engineering handbook version 2a. *INCOSE*, San Diego.

ISO/IEC 15288, 2002. ISO/IEC 15288 : Systems engineering-system life cycle processes.

Kang, H.J., Lee, D., Shin, J.G., Park, C.S., Park, B.J., Choi, J., 2008. A Study of a Rescue Device for Marine Accidents using Radar Cross Section Characteristics. *Marine Technology Society Journal*: 38-44

Shibue, T., Misumi, R., Hayami, T., Sawai, T. Ohmasa, M., Hirokawa, N., 2012. Falling Behavior Simulation of a Standing Human Body on a Carriage at Experimental Collision, to Simulate a Collision Between a High Speed Ship and a Quay. *ISOPE 2012.*, Rhodos, Greece: 747-752.

The International Association of Oil and Gas Producers., 2010. OGP Risk Assessment Data Directory-Water transport accident statics.

The Signals of Marine Continuous Radar for Operation with SART

V.M. Koshevoy & D.O. Dolzhenko
Odessa National Maritime Academy, Odessa, Ukraine

ABSTRACT: An approach to the selection of periods of the signals ensuring the joint operation of marine continuous radar and search and rescue radar transponders which takes into account the sizes of sea targets is considered. A method for preventing the appearance of false targets on the display of radar is presented. A method of signal generation, allowing to increase their coherent part for improving the energy efficiency on the base of considered processing methods is proposed.

Existing marine radars are working at impulse type and have the disadvantage such as high output power, which adversely affects the health of the crew and the environment. Nowadays a design of marine radar with continuous radiation exists, allowing to reduce significantly the peak power of radiation and thus to improve the working conditions of seafarers and improve environmental safety [1].

For continuous wave (CW) radar it is important to choose the signals providing the main tasks of radiolocation: radio detection and ranging. In addition to detecting and ranging it is important for the safety of navigation to ensure the functioning of the existing search and rescue radar transponders (SART). But the problem is that SART have no compression filter and are designed for pulsed signals. Within the confines of continuous signals this problem may be solved by using the signals with non-uniform radiation power. The structure of type of these signals was considered in [2, 3]. There was also showed a number of drawbacks which follow from the next moment: for ensuring the typical radar ranges it is necessary to use a large number of pulses in the repetition period. It's lead to significant irregularities of the amplitudes in the signal period and a significant increase of its peak factor.

Therefore, this work focuses on the choice of the signals and its processing, which overcome above drawbacks.

Previously in [2, 3] there was proposed a method of forming continuous periodic discrete signals allowing to co-operate CW radar and SART. Such signals have irregular structure, which provides an acceptable gain in peak power compared to the pulse at relatively low repetition period. It's suggested to use them in two modes: quasi mode (as parcels of a certain duration, which consist of pulses) and continuous mode. The misfit of the first mode is the presence of side lobes of a correlation function and this complicates the solution of radar tasks. The second mode provides zero side-lobe level, but it is necessary to use small periods of signals to achieve the desired gain in peak power. However, due to this the problem of ambiguous measure distance to the target appears, but it can be eliminated by using multiple aliquant periods consequentially. There is an additional problem, which consists in the fact that the difference between the marks range on the area of reflected periods can be comparable to the size of the target, which leads to the appearance of false targets in the case of large objects. Further, the proposed method of removing the ambiguity has a problem of energy losses caused by the short coherent length of the signal, because during the processing the accumulation of all periods does not used.

A goal of this article is the selection of periods of such signals which can satisfy certain restrictions with regard to possible sizes of targets and ensure normal operation of radar. In addition, the goal of this paper is to develop algorithms that generate signals for the operation of SART (with certain peak factor other than 1 (about 10 ... 20)), which allows to increase the coherent part with a corresponding slight increasing of peak factor.

The method of forming continuous periodic signals with non-uniform structure which provide the interoperability of CW radar and SART, was proposed in [2, 3]. The signal is:

$$\vec{s} = [s_0 \quad s_1 \quad \dots \quad s_{N-1}] = [\overbrace{a \quad b \quad \dots \quad b}^{N}], \qquad (1)$$

where $\vec{s}$ = vector of complex amplitudes of one period of continuous periodic signal with components s_n. For binary sequences it consists of two values: $s_0 = a$ and $s_1 = s_2 = \dots = s_{N-1} = b$; a and b are the arbitrary complex amplitudes, N = the number of pulses in one period of signal.

An important requirement for CW radar signals is ensuring of minimal level of side lobes of periodic autocorrelation function (PAF) which increases the accuracy of measurement. So let consider the signals with zero level of side lobes of PAF and a, b will be the next:

$$a = |a| \cdot e^{i\varphi_a} = -|a| = -\frac{N-2}{2}, \text{ where } N \geq 2 \qquad (2)$$

$$b = |b| \cdot e^{i\varphi_b} = 1. \qquad (3)$$

Unevenness of these signals is given by the ratio of a to b and when the PAF has zero level of side lobes it equals:

$$\frac{|a|}{|b|} = \frac{N-2}{2}. \qquad (4)$$

The expression (4) links with the peak factor $\aleph$ by the next way

$$\aleph = \frac{s^2_{n\max}}{\left(\sum_{n=0}^{N-1} s_n^2\right)/N}.$$

For signals (2), (3) it is obtained the expression for peak factor

$$\aleph = \begin{cases} \dfrac{N}{\dfrac{(N-2)^2}{4} + (N-1)}, & 2 \leq N < 4; \\ \dfrac{N}{1 + \dfrac{4(N-1)}{(N-2)^2}}, & N \geq 4. \end{cases} \qquad (5)$$

The table 1 allows to trace how the peak factor is changing when the the number of pulses in one period of the signal (2), (3) increases.

Table 1. The dependence of the peak factor $\aleph$ from N.

N	5	10	20	30	40
$\aleph$	1,8	6,4	16,2	26,1	36,1
N	50	60	70	80	100
$\aleph$	46,1	55,9	65,9	75,9	96,0

From the Table 1 and (5) it follows that it is necessary to choose N between 20 ... 50 because for large values of N the signal will have the pulse form and the win in peak power will be lost. Thus the example for $N = 100$ shows that the peak factor for the signal (2), (3) will be the same as for pulse with a period N ($\aleph = 96 \approx 100$, i.e. $\aleph = N$)

However, the usage of such short periods has a problem of ambiguous of range-finding. To overcome this, it can be use the quasiimpulse mode [3]. But it has a drawback - the presence of side lobes of aperiodic correlation function equal to 2 / N. Therefore it's necessary to return to a continuous mode, which ensures zero level of side lobes of PAF. To remove the ambiguity of range-finding the multiple signals with aliquant periods should be used.

For illustration it will be presented the situation with two signals of aliquant periods: $T_{I_1} = N_1 \cdot T_0$ (when $N_1 = 20$) and $T_{I_2} = N_2 \cdot T_0$ (when $N_2 = 23$), where pulse width T_0 ensures standard range resolution, $T_0 = 0{,}33$ mcs. When processing of these signals in a scheme of elimination of ambiguity of range-finding the received marks of range are matching in a certain moment and a target appears on a radar. This is shown on a Fig.1.

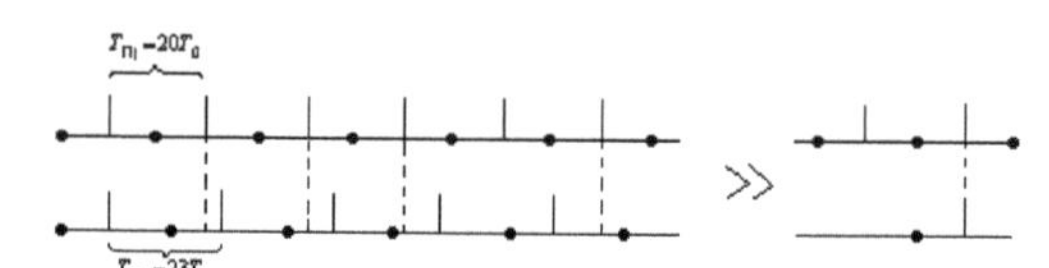

Figure 1. Schematic representation of the reflected signals at different stages.

But if we choose only the prime numbers for periods, that was made up, this is suitable only for the detection of point targets. Due to the fact that each object has its dimensions (a ship in particular is extended object), each can hold multiple range marks in reflection. Therefore, the selected periods of repetition should not only be mutually prime numbers, but also satisfy the following: to provide the difference between the time position of the pulses of two signals not less then n marks (in some cases n takes values 3 ... 4 range marks). If the difference between the values of periods less than n, then in the case of large targets it will lead to the appearance of false marks on a radar display.

For the selection of suitable periods of repetition it is enough to fulfill condition:

$$|l \cdot N_1 - m \cdot N_2| \geq n, \qquad (6)$$

where the variables l and m determine the number of periods for the signals with N_1 and N_2 accordingly; n - the number of elements of the resolution matching the length of the target.

Based on (6), we obtain the following inequality

$$m_{\min} \le \left[\frac{N_1 - n}{\Delta N}\right], \quad (7)$$

where $\Delta N = N_2 - N_1$, $[X]$ - integral part of number X. Accordingly, the minimum period for the radar range is:

$$m_{\min} \cdot N_1 \cdot T_0 = \left[\frac{N_1 - n}{\Delta N}\right] \cdot N_1 \cdot T_0 . \quad (8)$$

Analysis (8) shows that $m_{\min}$ increases with increasing N_1 and with decreasing ΔN. But ΔN can not be less than n. So it can be obtained an expression for $m_{\min}$ (in condition $\Delta N = n$) from (7):

$$m_{\min 1} = \left[\frac{N_1}{n}\right] - 1 . \quad (9)$$

After substituting (9) in (8) it is obvious that for the increasing the resulting period it is necessary to increase N_1.

$$\left(\left[\frac{N_1}{n}\right] - 1\right) \cdot N_1 \cdot T$$

Furthermore, it is necessary to increase N_1 because the short period of signal has a short length of a coherent part and the energy of signal is underutilized.

If for example we choose $N_1 = 43$, then in accordance with the condition $\Delta N = n = 3$: $N_2 = 47$. From (7) it follows:

$$m_{\min} \le \left[\frac{43-3}{4}\right] = 10 .$$

The analysis of the structure of this signal ($N_1 = 43$, $N_2 = 47$) really shows that for the number of periods over 10 the difference between range marks will not exceed 3. For signal $N_1 = 19$, $N_2 = 23$ this value will be:

$$m_{\min} \le \left[\frac{23-3}{4}\right] = 5 .$$

Thus only the part of a resulting signal satisfies the condition: for the case $N_1 = 43$, $N_2 = 47$ - 10 periods, for the case $N_1 = 19$, $N_2 = 23$ - 5 periods. Using of periods longer than the calculated value will lead to the appearance of false targets.

To solve this problem and avoid the appearance of false marks on the radar display, the blanking of the area of period of the reflected signal which leads to the appearance of false targets can be performed. On the radar screen there will be only the reflected signals for which the condition "not less than n range marks between the reflected pulses" fulfills. Once this condition is violated, a special scheme blanking does not skip the rest of the signal.

Selection of the appropriate periods and the application of the scheme of blanking can solve the problem of ambiguity, but it does not eliminate the problem of underutilization of the energy of continuous signals, because of the small periods of repetition. The disadvantage of signals used above is that it can be used only a limited length of the coherent part of these signals (in this case $N_1 = 43$). But its increasing by increasing the number of pulses in the period is unacceptable, because when N increasing the peak factor increases too. Therefore, it's need to increase the coherent part of the signals.

Coherent part of the signal (1) may be increased if the requirement of the zero level of side lobes of PAF will not fulfill. Then let the level of side lobes of PAF be 1/*N*. The signal will have the form:

$$a = -\frac{N-3}{2}, \quad b = 1 . \quad (10)$$

For signal (10) it was obtained the equation for peak factor:

$$\aleph = \begin{cases} \dfrac{N}{\dfrac{(N-3)^2}{4} + (N-1)}, & 2 \le N \le 5; \\[2ex] \dfrac{N}{1 + \dfrac{4(N-1)}{(N-3)^2}}, & N \ge 5. \end{cases} \quad (11)$$

Table 2 shows the peak factor of the signal (10) in the case of an increase N.

Table 2. The dependence of the peak factor $\aleph$ from N.

N	5	10	20	30	40
$\aleph$	1,0	5,8	15,8	25,9	35,9
N	50	60	70	80	100
$\aleph$	45,9	55,9	65,9	75,9	95,9

As follows from (11) and from the Tab.2 with increasing N the values of the peak factor from the Table 2 are less compared to Table 1. Thus for $N = 5$ peak factor of signals (10) is almost two times less than the peak factor of signals (2), (3). Thus, this method allows to increase the coherent part of the signal, but with the increase N peak factor is significantly reduces.

Therefore, to increase the coherent part we can propose a method of synthesis of signals based on known property of element wise multiplication of signals with mutually prime periods, and as a result their correlation functions are multiplying [4, 5].

The product of two periodic signals with mutually prime periods

$$\vec{s}^{(1)} = [\overbrace{a^{(1)} \quad b^{(1)} \quad \ldots \quad b^{(1)}}^{N_1}]$$

and

$$\vec{s}^{(2)} = \overbrace{[a^{(2)} \quad b^{(2)} \quad \ldots \quad b^{(2)}]}^{N_2}$$

forms a new sequence:

$$\vec{s} = \overbrace{[a^{(1)} \cdot a^{(2)} \quad b^{(1)} \cdot b^{(2)} \quad \ldots \quad b^{(1)} \cdot a^{(2)} \ldots a^{(1)} \cdot b^{(2)}]}^{N_1 \cdot N_2} \quad (12)$$

In resulting period $\grave{O}_{\grave{I}.\eth} = N_1 \cdot N_2 \cdot T_0$ the PAF has a zero level of side lobes.

For (12) it was obtained an expression for calculating the peak factor:

$$\aleph = \frac{N_{11} \cdot N_{12} \cdot \max\left[\left(\frac{N_{11}-2}{2}\right)^2 \cdot \left(\frac{N_{12}-2}{2}\right)^2 ; \left(\frac{N_{11}-2}{2}\right)^2 ; \left(\frac{N_{12}-2}{2}\right)^2 ; 1\right]}{\left(\frac{N_{11}-2}{2}\right)^2 \cdot \left(\frac{N_{12}-2}{2}\right)^2 + (N_{12}-1) \cdot \left(\frac{N_{11}-2}{2}\right)^2 + (N_{11}-1) \cdot \left(\frac{N_{12}-2}{2}\right)^2 + N_{11} \cdot N_{12} - N_{11} - N_{12} + 1} \quad (13)$$

For values $N_{1,2} \geq 4$ (13) takes form:

$$\aleph = \frac{N_{11} \cdot N_{12}}{1 + \frac{N_{11}-1}{\left(\frac{N_{11}-2}{2}\right)^2} + \frac{N_{12}-1}{\left(\frac{N_{12}-2}{2}\right)^2} + \frac{N_{11} \cdot N_{12} - N_{11} - N_{12} + 1}{\left(\frac{N_{11}-2}{2}\right)^2 \cdot \left(\frac{N_{12}-2}{2}\right)^2}} \quad (14)$$

To evaluate the possibilities of this approach let calculate the peak factor for signal $N_1 = 19$: according to (5) $\aleph \approx 15$. If we take $N_1 = N_{11} \cdot N_{12} = 21 \cdot 4 = 84$ then according to (14) $\aleph \approx 17$. Obviously, the length of the coherent part of the second signal is much higher (in 4.5 times) than the same value of the peak factor, which indicates the effectiveness of the proposed method of increasing the coherent part.

The research showed that using the product of three or more signals greater gain of the peak factor can be received with increasing the coherent part of the signal.

The ambiguity function of the conceived sequence was also calculated [6]:

$$\chi(k,l) = \sum_{n=0}^{N-1} s_n^* \cdot s_{(n+k)} \cdot e^{i\frac{2\pi nl}{4N}}, \quad (15)$$

where k is a discrete interval of time and l is discrete interval of frequency with step $\Delta f = 1/4NT_0$.

Under there are some cross-sections of the ambiguity function of waveform (12), (2), (3) (when $N = 21 \cdot 4 = 84$):

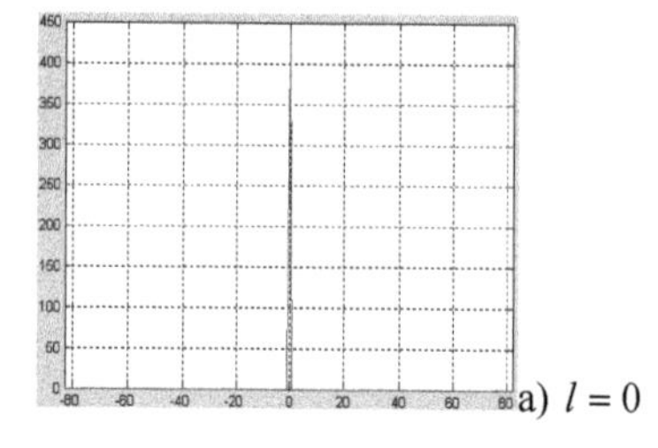

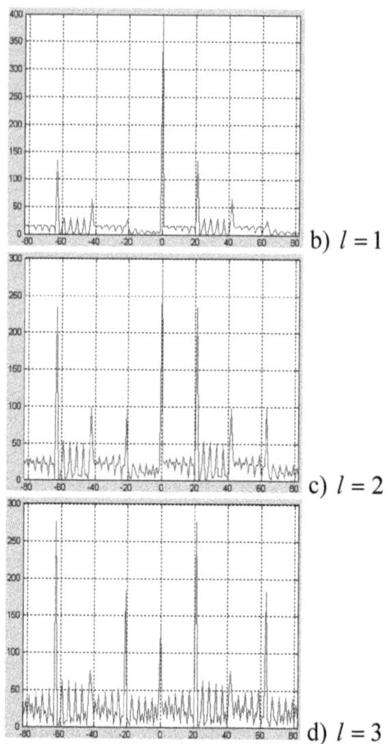

Figure 2. The cross-sections of ambiguity function of waveform (12) ($N = 84$)

Thus, the method of selection of periods of signals taking into account the sizes of targets was proposed. Such signals provide a joint operation of CW radar and existing SART. There were presented several approaches to the selection of signals that allow to increase the coherent part with a relatively low increase of the values of peak factor.

REFERENCES

[1] Levanon Nadav & Eli Mozeson. 2004. Radar signals, *J. Wiley*: 411 p. NJ.

[2] V. M. Koshevyy, D. O. Dolzhenko. The Synthesis of Periodic Sequences with Given Correlation Properties // Proc. of IEEE East–West Design & Test Symposium (EWDTS'11), Sevastopol, Ukraine, September 9–12, 2011. – P. 341 – 344.

[3] Koshevyy V. M., Dolzhenko D.O. 2012. The selection of signals that ensure the possibility of joint operation of continuous radar and existing SART [in Russian] // Судовождение. – № 21. Odessa: ONMA. – p. 124 -128.

[4] M. B. Sverdlik. 1975. Optimal Discrete Signals [in Russian]. *Sov. Radio*, Moscow.

[5] Titsworth, Robert C. Optimum and Minimax Sequences, *Proc. of Int. Telem. Conf.*, London, 23 – 27 September. 1963. UK.

[6] Shirman Y. D. 1970. Theoretical fundamentals of radar [in Russian], *Sov. Radio*, Moscow.

Risk Analysis on Dutch Search and Rescue Capacity on the North Sea

Y. Koldenhof
Maritime Research Institute Netherlands (MARIN), Wageningen, Netherlands

C. van der Tak
Maritime Traffic and Safety Consultant (MTSC), Heelsum, Netherlands

ABSTRACT: The Ministry of Infrastructure and Environment is responsible for and coordinates the Search and Rescue (SAR) operations on the Netherlands continental shelf. In this frame work MARIN was requested to help in optimizing their task by using a risk assessment. The work consisted of three main steps.
First, the request for SAR capacity by ships in distress has been established by determining the probability of nautical incidents spread over the geographical area of responsibility, using the SAMSON model (Safety Assessment Model for Shipping and Offshore in the North Sea). This delivers the number of Potential Persons at Risk (PPaR).The second step was to establish the available SAR capacity; the lifeboats of the KNRM (Royal Dutch Rescue Society) and aviation support resources of different ministries and industry, all with own locations, rescue capacities and response times.
Finally, by combining the SAR request and the SAR capacity for each geographical grid cell, the percentage of the PPaR that could be rescued has been calculated. The model was used to determine the effect of different scenarios in which the weather conditions and the duration of the incidents were varied. Finally, the overall capability of the SAR fleet has been determined by summarizing the results of these scenarios. The objective was the optimal matching of the available SAR units on the demand of SAR and to optimize the effect of new investments in SAR helicopters.

1 INTRODUCTION

In 2012 it was again sadly proven, with the collision of the Baltic Ace with the Corvus J that serious accidents with loss of life can occur at sea. Therefore it is necessary and essential to be prepared. The Ministry of Infrastructure and Environment is responsible for and coordinates the Search and Rescue (SAR) operations on the Netherlands continental shelf. The Netherlands Coastguard performs this task for the ministry using various resources e.g. the KNRM. In this frame work MARIN (Maritime Research Institute Netherlands) was requested in 2004 to help in optimizing their task using a risk assessment.

1.1 *Objective*

The main objective of the study was to develop a tool that can quantify the effectiveness of the SAR fleet under predefined weather conditions. Included herein is the probability and geographical spread of nautical incidents that lead to a certain number of PPaRs This approach differs from the usual method to determine the need and the capacity of SAR independent of the locations and likelihood of incidents.

1.2 *General approach*

The general approach of the study consisted of three main steps. First the request for SAR capacity has been established by determining the probability of nautical incidents and the distribution of the number of so-called Potential Persons at Risk (PPaR) resulting from these incidents spread over the geographical area of responsibility. Secondly the available SAR capacity, now and for future scenarios has been determined. Finally, by combining the SAR request and the SAR fleet for each geographical grid cell, the percentage of the PPaR that could be rescued has been calculated. By introducing various (future) scenarios for the SAR capacity the influence of these scenarios, also in different weather conditions and the duration of the incidents could be determined. Finally, the overall effectiveness of the SAR fleet has been determined by summarizing the results of these scenarios

2 REQUEST FOR SAR CAPACITY

2.1 *Potential persons at risk*

To determine the number of people that needs to be rescued the term "Potential Persons at Risk" or PPaR is introduced. A "Potential Person at Risk" is a person on board a ship that is involved in an accident that could lead to the sinking of the ship; collision, contact with a fixed object, hull failure or fire/explosion. This is not necessary the expected number of people that will end up in the water, but it is an indication of the number of people for which a SAR-action is launched.

The expected number of Potential Persons at Risk (PPaR) with the corresponding frequencies is normative for the SAR-capacity. Therefore the determination of these frequencies is subject of this paper.

2.2 *SAMSON*

The expected number of nautical accidents that results in a SAR request has been determined using the SAMSON model. SAMSON stands for "Safety Assessment Model for Shipping and Offshore on the North Sea", and has been developed and evolved over the last 30 years. The model is a collection of several mathematical models, ship traffic data, statistical accident data and other information such as ship characteristics, see Koldenhof 2009 and Glansdorp 2010. The model has been developed for the Dutch Authorities and further evolved within many commercial and EU-funded projects.

SAMSON consists of three main parts. First a Maritime Traffic System, that consists of all characteristics of the study area, including environmental information, a detailed traffic database, characteristics of the various ships, e.g. the number of people on board and other layout issues e.g. shallow water or fixed objects at sea.

The second part of the model contains the mathematical models that determine the frequencies of the various nautical accidents. First the possible dangerous situations are determined based on the defined Maritime Traffic System. For example the expected number of encounter between sailing ships. Secondly this number of expected possible dangerous situations is multiplied by a so-called CasualtyRate (CasRat). This CasRat indicates the probability that a dangerous situation actually ends up in an accident, for example the probability that an encounter between two ships results in a collision. This CasRat is different for each accident type, ship type and ship size and is based on worldwide accidents database statistics (IHS Fairplay 1990-2010).

The third main part of the model is the determination of the possible consequences of the accidents. For this study this means the probability that a ship will sink after an accident and the expected number of people on board. In this part also the possible risk reducing measures, in this case SAR, can be included.

3 AVAILABLE SAR CAPACITY

Helicopters and lifeboats from the KNRM can be used in Search and Rescue actions. In case of large incidents both helicopters and lifeboats can be deployed whereby the strong points of both types of equipments can be used in the most optimal manner.

An advantage of using a helicopter is its high speed so it can reach distant places faster. The disadvantage is the fact that there are only few helicopters available and that they operate from fixed locations, which means relatively large distances towards the incident. Also a helicopter has a relatively small capacity to take on board people.

The advantage of lifeboats is that they are located along the whole coastline and they have a relatively small response time of 10 minutes. They can arrive relatively quickly at the scene of the incident. Only in case of incidents far away from the coast a helicopter will arrive sooner.

For the study different SAR-scenarios are determined together with the responsible organisations. Three different scenarios with helicopters are considered, all in conjunction with the lifeboats of the KNRM. The helicopters (existing and new investments) have different rescue capacities and different response times for working days, weekend days, day and night. The airbase for all helicopters is Den Helder, located at northwest point of the Netherlands. Table 1 and Table 2 show an overview of the rescue capacity, speed and response time of the three types of helicopters used in the study. (bft = Beaufort class of wind; endurance is the maximum flying time without refuelling, response time is the time after first call to take off of the SAR unit)

Table 1. Overview of the rescue capacity and speed of the different type of helicopters (bft = Beaufort class for wind)

Helicopter type	Capacity [person]	Speed [kn] 0-7 Bft	$\geq$ 8Bft	Endurance [hour]
Lynx	4	135	100	2
S61	27	110	75	4.5
NH-90	16	155	120	4.5

Table 2. Overview of the response time of the different helicopter types during weekdays (Monday - Friday)

Helicopter type	Response time during weekdays [minute]			
	Work-Time	Day-light	Darkness <2100h	>2100h
Lynx	20	45	60	60
S61	15	15	15	60
NH-90	15	45	60	60

The first scenario represented the situation of 2007, using the Lynx and S61 helicopter. The second scenario gives the situation when the smaller helicopter is replaced by a new, slightly larger one, so using NH-90 and the S61. And the final scenarios modelled the situation that the large helicopter S61 is not longer deployed by the Coastguard, so only using the new NH-90 helicopter.

There are 5 types of lifeboats taken into account in the study. An overview of the capacity, speed and response time is given in Table 3.

Table 3. Overview of different type of lifeboats used in the study.

Name	Number available at sea	Speed [kn]	Capacity	Response time [min]
Arie Visser	8	32	120	10
Johannes Frederik	4	32	90	10
Valentijn	12	32	50	10
Harder	1	32	20	10
Altantic	9	32	12	10

4 RESULTS

The results of the additional model can also be divided into to three main parts, first the results of the SAMSON model on the request for SAR capacity, second the available SAR capacity based on the different SAR-scenarios, and finally the combined results regarding the effectiveness of the SAR-scenarios.

4.1 *Results SAMSON; request for SAR-capacity*

The result of the calculations with the SAMSON model is the expected number of incidents per year per accident type, ship type and ship size on, in this case, the Dutch Continental Shelf.

Based on an analysis of the reports of the KNRM only the following incidents are taken into account:

- Collision between sailing ships
- Contact with a fixed object (platform)
- Sinking of a ship
- Hull failure
- Fire Explosion

Table 4 shows the expected number of accidents per accident type with an expected number of PPaR above a certain value per year. In total almost 4 accidents with a SAR-request are expected per year. The expected frequency for very large incidents, with more than 500 potential persons at risk will occur once every 51 years. And accidents with more than 1200 PPaR are expected to occur once every 396 year, this still means that these accidents can happen and that one needs to be prepared. Further the probabilities of some incidents depend on the weather conditions.

For the further calculations is it important to have a geographical overview of the expected number of PPaR. In figure 1 the probability of an incident with more than 40 PPaR is given.

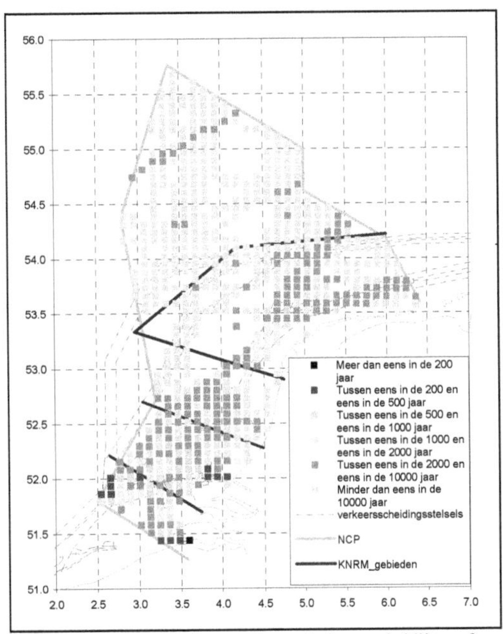

Figure 1. Geographical overview of the probability of an incident with more than 40 PPaR.

Table 4. Expected total number of incidents with at least the given number of PPaR per year, based on the traffic of 2004.

PPaR	Number of total incidents per year	Once every year an incident
>0	3.953	0.3
>10	1.877	0.5
>40	0.1588	6
>100	0.0750	13
>200	0.0458	22
>500	0.0195	51
>800	0.0131	76
>1200	0.0027	369

The results of this step, the calculation of the request for SAR-capacity by ships in distress is independent from the SAR-scenario that is chosen at a later stage. The number of accidents and the geographical spread of the accidents have been determined for normal and storm conditions because the availability of the SAR is partly dependent on these two conditions.

4.2 *Available rescue capacity*

The calculations have been performed for three helicopter scenarios. An important factor in the available capacity after a certain period of time in a certain grid cell is the response time of the different

SAR units, thus when the incident occurs and under which environmental conditions. In total the calculations have been performed for 288 cases:
- 3 helicopter scenarios;
- 4 times after the reporting of the incident (1, 1.5, 2 and 3 hours)
- 8 response times for the helicopters
- 2 weather conditions (0-7 Bft and storm)
- 3 speed settings for the lifeboats (0-7 Bft and two for storm (sea along and sea against)

Figure 2 shows the available rescue capability one hour after the incident is reported for scenario 1 (Lynx and S61), during working days and working hours. A dark blue grid cell means that 1 hour after the reporting of an incident, SAR units with a summarized rescue capacity of 20-30 can be at that location, thus enough to rescue all crew of most ships, however, far from sufficient to rescue passengers and crew of a ferry. Close to the coast the summarized rescue capacity amounts many hundreds, thanks the many lifeboats of the KNRM along the whole coast.

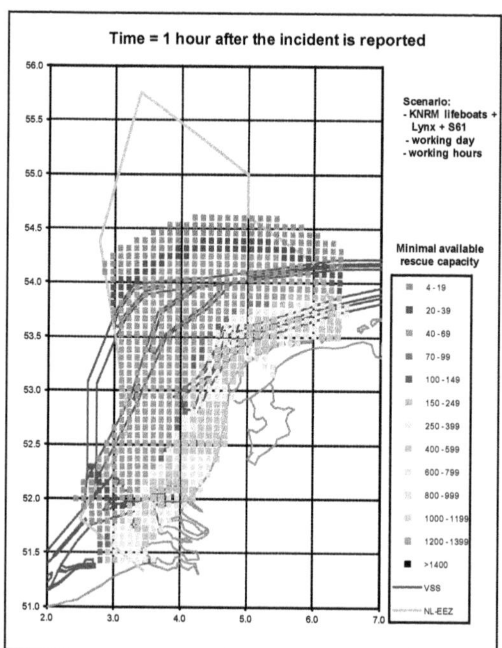

Figure 2. Geographical overview of available rescue capacity one hour after the incident is reported for scenario 1 (Lynx, S61 and lifeboats), during working days and working hours (VSS = Traffic Separation Schemes, NL EEZ = Dutch Exclusive Economic Zone)

4.3 *Successful SAR-action*

With the SAMSON model the expected number of PPaRs has been determined for each grid cell (8x8km). The probability has been determined for 14 different PPaR classes, with also for each cell the average number of PPaR.

Next to the request for SAR capacity also the available SAR-capacity after x hours for each grid cell has been calculated. Combining both results for each grid cell and time after the accident can lead to two outcomes:

1. The available rescue capacity in a grid cell exceeds the number of PPaR. This means that all potential persons at risk can be rescued and that the rescue operation is 100% successful.
2. The available rescue capacity in a grid cell is less than the number of PPaR. This means that not all persons at risk can be rescued. Assumed is that the number of PPaR that is rescued is equal to the available rescue capacity. Because not all PPaR are rescued the incident will still count as an incident with possible fatal ending.

The calculations have been performed for 1, 1.5, 2 and 3 hours after the incident and it is assumed that the capacity for each SAR-unit is only used once. The use of own rescue equipment on board the ships and assistance of other passing ships is not taken into account.

Furthermore the term "rescued after x hour" in this study means that after x hours the rescue capacity is present at the location. So the question whether a person is still alive after x hour in the water or the time it takes to get people out of the water is not taken into account. Also the probability that a person can be successfully taken out of the water is not included.

In table 5 the results for scenario 1 (Lynx, S61 and KNRM boats) is shown. The table shows the percentage of "successful" rescue operations and the percentage of PPaR's rescued, one hour after the incidents has been reported and during working days. This means that for 59.4% of all incidents on the Dutch part of the North Sea where 41 till 70 persons will be at risk there will be sufficient SAR-capacity available after one hour, given scenario 1. And for 82.2% of all PPaRs' involved this category of incidents (41-70 PPaR) there will be SAR-capacity available after one hour.

Table 5. Overview of results of the scenario 1 (Lynx, S61 and lifeboats), during working days and working hours. one hour after the incident has been reported (corresponding to figure 1)

PPaR	% of incidents where SAR-capacity > PPaR	% of PPaR "covered" by available SAR-capacity
1-5	99.3%	99.3%
6-20	93.5%	94.7%
21-40	82.2%	91.1%
41-70	59.4%	82.2%
71-100	60.8%	75.2%
101-150	58.7%	70.6%
151-250	87.4%	91.7%
251-400	94.7%	95.4%
401-600	14.6%	44.5%
601-800	3.0%	34.3%
801-1000	0.0%	25.2%
1001-1200	0.0%	18.7%
1201-1400	0.0%	17.1%
> 1400	0.0%	6.5%

The percentage of the incidents for which the available SAR capacity is larger that the number of PPaR, increase again for the PPaR classes 151-250 and 251-400, because ships with these number of passengers are mainly sailing west of the Dutch coast and not in the north point of the Dutch sea area thus on shorter distance of the coast. This is the strength of the method, namely that it combines the requests (probability and locations of incidents) for SAR with the supply of SAR.

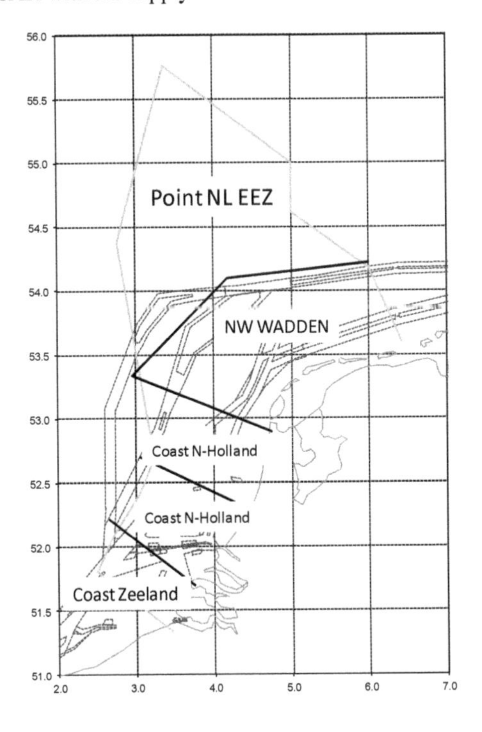

Figure 3. Geographical overview the different areas at the Dutch EEZ

Thus the results depend on the location where the incident occurs and the helicopter scenario and weather conditions. Therefore similar results are determined for 4 different areas along the coastline and one location far away from the coast, and for the different scenarios. The location of the different areas is shown in Figure 3.

Finally all results are combined in different graphs. Figure 4 shows for the three scenarios the percentage of "successful" incidents (vertical axis) over all PPaR-classes, weather conditions and response times for a number of hours after the incident is reported (horizontal axis). The different colours are the results for the different areas. It is clearly visible that the percentage of incidents where sufficient SAR-capacity is available for the area far from the coast (the red-line, point NL EEZ) is significantly less than for the other areas.

The second graph shows the results when no lifeboats of the KNRM are used. In this case the results for the different areas are different, this is caused by the fact that the helicopters are located at one fixed location and the life boats from the KNRM are located all along the whole coast. The results only taken into account the lifeboats are given in the last graph, as expected the percentage of "successful" actions in the area further from the coast is less than for the area closer to the coast, also compared to the scenarios where the lifeboats are combined with the helicopters. This means the use of helicopters is especially useful for a quicker response in these areas in the North.

Also the percentage of PPaRs rescued after a given hour is combined in different graphs, for the different scenarios. These results are shown in figure 5.

The conclusions that can be drawn for these graphs (in figure 5) are similar with the conclusions regarding the number of "successful" actions. A difference is the fact that none of the lines reached the 100%, this means that even after 3 hours still not 100% of all PPaR are "rescued". This is the result for the expected incidents with a very large number of PPaR, e.g. the larger cruise ships or ferries. Also in the calculations it is assumed that the available rescue capacity is only used once. In case of large incident this will not be the case. When a large cruise vessel is in trouble the lifeboats of the KNRM will return to the site of the incident after they have taken persons to the shore

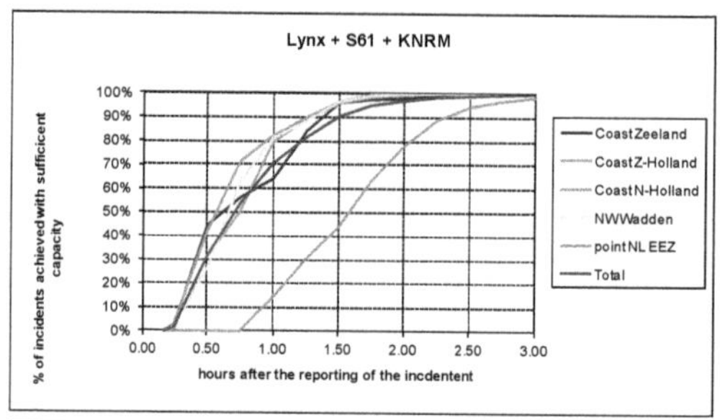

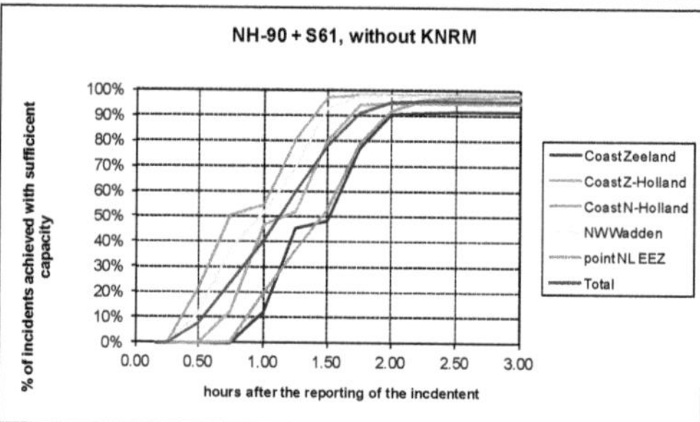

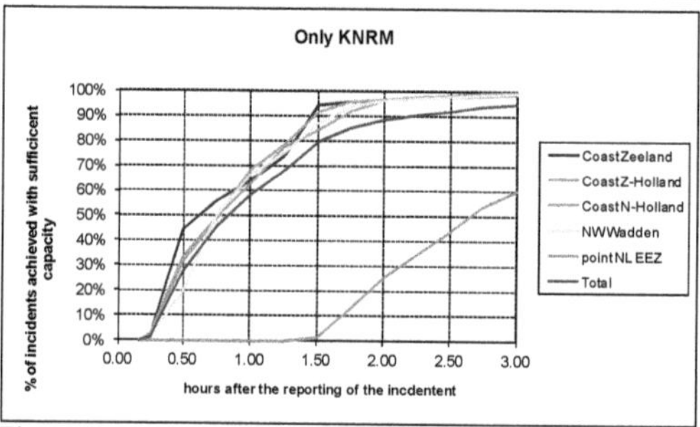

Figure 4. Overview of the percentage of "successful" SAR-actions for three different scenarios, summarized over all conditions (point NL EEZ, is the sea area far up north of the Dutch EEZ)

Another conclusion that can be drawn from the graphs is that when only helicopters are used, so no lifeboats, the total percentage of "rescued" PPaR will always be less than 75%. This means helicopters alone are not sufficient.

4.4 *Implementations of the results*

The results given in the previous paragraph have also been determined for the different weather conditions, this means the effectiveness of the SAR-operations for the different scenarios under normal and storm conditions. Also the response time of the different SAR-units has been set to different values to see the result of that on the effectiveness of the operations. Using all these different "settings" it can be easily seen what the effect of the different scenarios will be.

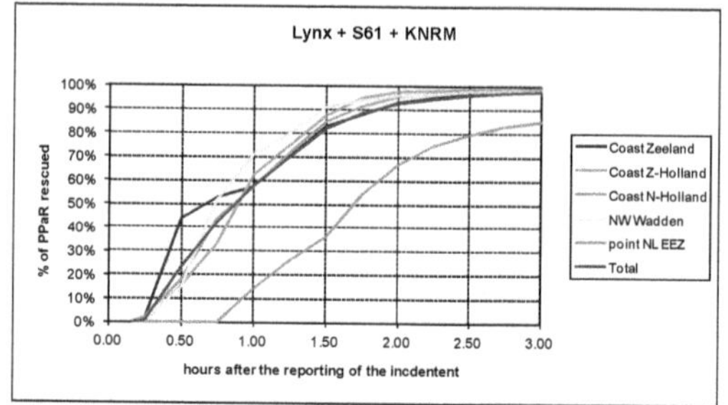

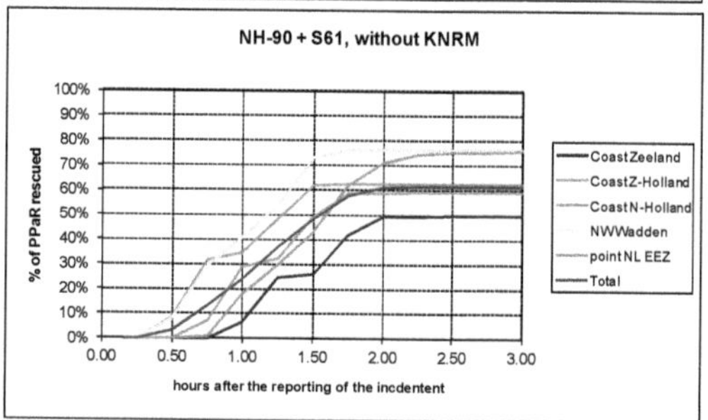

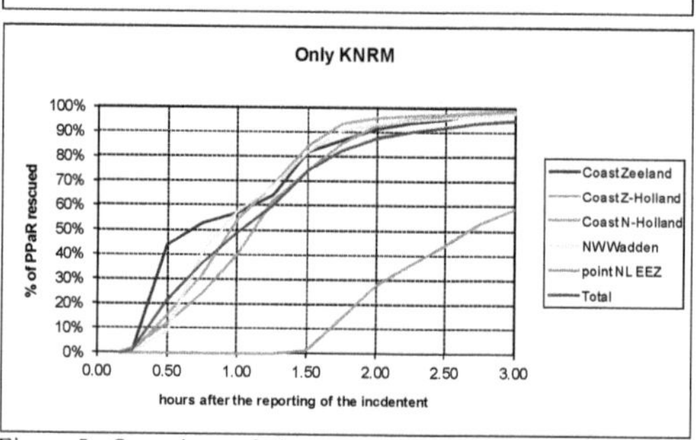

Figure 5. Overview of the percentage of PPaR rescued for three different scenarios, summarized over all conditions

And this was found to be very useful information for the Ministry of Infrastructure and Environment and the Coastguard. The results are used in the decision making process for the contracting of different parties. Eventually the results are used "Nota Maritieme Noodhulp" (Memorandum Maritime help in case of distress) in which the general policy and views regarding all type of rescue operations on the Dutch part of the North Sea is set.

4.5 *Validation Dutch Continental Shelf*

The results of the SAMSON-calculations have been compared with the observed number of accidents on the Dutch Continental Shell. In the period 1993-2003 the average number of observed accidents was 24.7, for the period 2000 – 2003 the average number was 19.5 (Koldenhof,Y 2005). This means an overall decrease of 21% over 4 years. When this same percentage is used the expected total number of relevant accidents in 2004 would have been 16.8.

The total expected number calculated by the model was 16.3, this means 3% lower.

5 CONCLUSIONS

Based on the results of the SAMSON-model and the inventory of the SAR-capacity at different locations, under different weather conditions and for different scenarios, it is very good possible to indicate the effectiveness of the different SAR-scenarios. And based on the new addition to the SAMSOM-model it is possible to see the effect of different settings, concerning for example the type of SAR-units and the response time.

This new tool has proven to be very useful for the Ministry of Infrastructure and Environment and the Coastguard in the decision and policy making process.

6 DISCUSSION

The limitations of the approach are the fact that study only focussed on the total SAR-capacity that could be at the location of the incident after a certain time period. The method does not take into account the real effectiveness of the SAR-action at the scene. How many people can really get on board and how many people are still alive after being in the water for several hours? These are some questions that are not answered in this study.

There have been many discussions with the Dutch Coastguard and KNRM about these questions to maybe determine "factors" for the possible effectiveness of the SAR-operations. However it was found that there are too many uncertain factors. Also looking at the "results" of different rescue operations gave to little information to determine a good overall quantitative factor for the effectiveness of an operation.

Finally the study provided sufficient insight for the authorities, in the possible request for SAR-capacity after a ship is in distress and the availability of the SAR-units, to help them in establishing policy.

REFERENCE

Koldenhof, Y., van der Tak, C., Glansdorp, C.C. Risk Awareness; a model to calculate the risk of a ship dynamically, Marine Traffic Engineering Conference 2009, Sweden

Glansdorp, C.C., van der Tak, C., Koldenhof, Y., The MarNIS RISK Concept, IALA conference 2010, Cape Town

Koldenhof, Y 2005. Verkeersveiligheid op de Noordzee 2000-2003. MARIN, 20402.620

Koldenhof, Y & van der Tak, C 2006. Risicoanalyse Search and Rescue, Noordzee, inventarisatie drenkelingen. MARIN, 20921.620

Koldenhof, Y & van derTak, C 2007. Risicoanalyse Search and Rescue, Noordzee, effect reddingscapaciteit. MARIN, 21647.620

The Operational Black Sea Delta Regional Exercise on Oil Spill Preparedness and Search and Rescue – GEODELTA 2011

A. Gegenava
LEPL Maritime Transport Agency, Batumi, Georgia

I. Sharabidze
Batumi State Maritime Academy, Batumi, Georgia

ABSTRACT: This paper presents the Operational Black Sea Delta Regional Exercise on Oil Spill Preparedness and Search and Rescue – GEODELTA 2011. Exercise was conducted on a regional scale on 14-16 September 2011 in Batumi, near of the river "Chakvis -Tskali", located on the Eastern Black Sea coast of Georgia. The exercise composed of all relevant emergency response components including marine pollution preparedness and response, Search and Rescue (SAR), with a scenario based on collision of an oil tanker and a Ro-Ro vessel, 7.5 nautical miles off the Batumi coast. Having as main objective the enhancement of the cooperation between the national authorities with responsibilities in coordinating and conducting search and rescue activities in accordance with the Georgian Law of Marine Rescue Service, as well as to test the collaboration under the Protocol on cooperation in combating pollution of the Black Sea by oil and other harmful substances in emergency situations to the 1992 Convention on the protection of the Black Sea against pollution (The Bucharest Convention), the exercise was developed by the relevant Georgian institutions, under the coordination of the Maritime Transport Agency (MTA), in close cooperation with the Black Sea Commission Permanent Secretariat (BSC PS), and with the participation of Bulgaria, Romania, Turkey and the Ukraine. Turkey and European Maritime Safety Agency (EMSA) contributed the exercise by special type of vessels, equipment and trained staff. Other participants - Black Sea littoral states, International Maritime Organization (IMO), Oil Spill Response Regional Initiative (OSPRI), Oil Spill Response and Oil Company's and International tanker Owners Pollution Federation (ITOPF) - participated with evaluators and observers. All aspects related to SAR operations, off-shore and on-shore oil pollution response were planned by Maritime Rescue Coordination Centre (MRCC) Georgia, MTA, Ministry of Economy and Sustainable Development of Georgia. This paper presents the outlines of the exercise GEODELTA 2011 with its elements and describes the process of implementation and inter-governmental cooperation in the Black Sea region. The related preparatory activities and lessons learned from the Exercise GEODELTA 2011 are also presented. GEODELTA 2011 was a full test of the Black Sea Contingency Plan (BSCP) and the mobilization and deployment of oil pollution combating resources.

1 INTRODUCTION

The Bucharest Convention signed in 1992, establishes the legal base for the protection of the marine environment of the Black Sea ratified by its six littoral States in 1994. Basic framework of the Bucharest Convention includes joint action in the case of accidents resulting in marine pollution. Article IX of Bucharest Convention specifically indicates to take necessary measures and cooperate in cases of grave and imminent danger to the marine environment of the Black Sea or to the coast of one or more of the Parties due to presence of massive quantitates of oil or other harmful substances resulting from accidental causes (Anon., 1992). The Emergency Response Protocol to the Bucharest Convention addresses cooperation mechanisms to make intervention in emergency situations for cases of major pollutions caused by oil or other harmful substances resulting from shipping accidents and the related operational procedures are given in the BSCP. This regional framework complements International Convention on Oil Pollution Preparedness, Response and Co-operation 1990 (OPRC Convention 1990), which is currently ratified by littoral States. The Black Sea Commission's Advisory Group on Environmental Safety Aspects of Shipping (AG ESAS), in 2005 established a detailed regional exercise programme, designed to implement and test the BSCP. There are five types of exercises:

- BLACK SEA ALPHA: Synthetic or Table-top Exercise;
- BLACK SEA BRAVO: Alarm or Communication Exercise;
- BLACK SEA CHARLIE: Equipment Deployment Exercise;
- BLACK SEA DELTA: Operational Exercise;
- BLACK SEA ECHO: State-of-the-art Exercise.

The first delta exercise within this programme was made bi-lateral in collaboration with Russian Federation and Republic of Turkey on 14-15 June 2006. The second but more comprehensive one, held in Turkey in September 2007 with the participation Bulgaria, Romania, Ukraine, Russian Federation and Georgia, covered the major oil spill exercise also incorporating SAR aspects, it was called as SULH 2007.

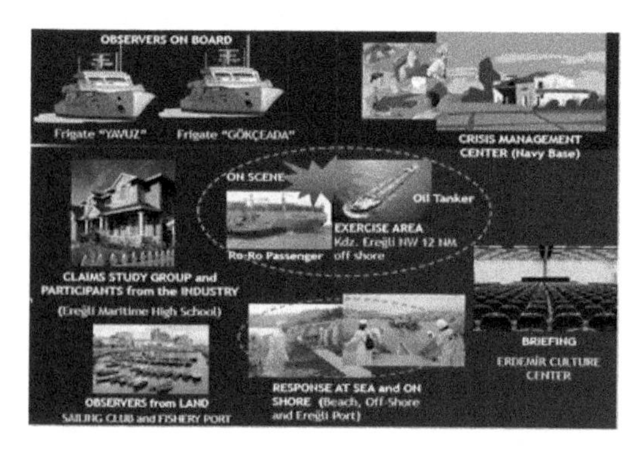

The third delta exercise was held in Romania in August 2009 with the participation all Black Sea Littoral countries and it was called RODELTA 2009.

GEODELTA 2011- fourth operational Black Sea delta regional exercise extended over three days and involved all the littoral States and the other relevant stakeholders including the port operating and oil spill response private and State companies and the oil and shipping industries.

Exercise – GEODELTA 2011 was conducted under the national legislation, with the special provisions of the:

- Comission on the Protection of the Black Sea Against Pollution, and its substantial support from the dedicated pilot project MONINFO - (Environmental Monitoring of the Black Sea Basin: Monitoring and Information Systems for Reducing Oil Pollution) and Guidelines for oil spill exercises developed under the scope of MONINFO as a new annex of the BSCP;
- AG ESAS, that in its last 19 meeting (November 2010) planed for further actions in the BS DELTA-Georgia preparation.

The analysis of the guidelines for oil spill exercises under the BSCP and trainings held earlier show that organization, planning and conduction of such frame training is a very high responsibility and having the aim to test the cooperation among the Black Sea littoral states, to use all the state and private sectors and the whole system of maritime infrastructure.

2 IMPLEMENTATION OF THE EXERCISE

The regional aspects of exercise were established during Steering Committee Meetings. First meeting was held in Istanbul, Turkey in March 2011 in the BSC PS. The Second Steering Committee Meeting was held in Istanbul, Turkey, on 8-10th June, 2011, on the MONINFO Project 2nd Stakeholders and 20th ESAS AG Meetings, supported by BSC PS. At the national level, working group meetings for exercise planning were regularly held. Every working group meeting was held in MTA, headed by the Director of MTA and participating with representatives of Georgian Maritime Infrastructure State and private sectors.

Exercise – GEODELTA 2011 held on 14-16, September 2011. On the 14th of September some preparing activities were held, like tests communications and Claims and compensations working group meetings, Opening ceremony Exercise – GEODELTA 2011 and Short briefing.

Date of the operational exercise - 15, September 2011.

All aspects related to SAR operations, off-shore and on-shore oil pollution response were planned by MRCC Georgia, and all aspects related to land response were planned by MTA, Ministry of Economy and Sustainable Development of Georgia. MRCC received some indications for objectives establishing and pollution response planning from OSPRI and ITOPF.

Four Black Sea States participated in the exercise that was coordinated by Maritime Transport Agency, of Georgia in close cooperation with relevant Georgian Authorities and the BSC PS. Governmental and/or industry owned response units from Georgia, Turkey and EMSA participated in the exercise together with representatives of OSPRI and ITOPF.

3 PARTICIPATING CONTRACTING PARTIES

3.1 *Exercise participants*

Ministry of Economy and Sustainable Development of Georgia, Civil Aviation Agency, Government of Ajara, Batumi Townhall; MTA, MRCC Georgia; Operational Group – Duty officers; Radio Operators; OSR Group; Batumi, Poti and Kulevi Harbour Masters; (handling/operation, security consultant, Seamen, Drivers etc.); Emergency Sub-Contractors (Ambulance, Firefighting Service, Emergency Situation Service, Police, Civil Contractor); External organizations – Ministry of Environment; Ministry of Energy and Natural Resources – Black Sea Office; Emergency Department of MIA; National Agency of Environment – Black sea Monitoring Division; Police; Georgian Coast Guard Department; Georgian State Hydrographic Service; Ministry of Finance of Georgia - Revenue Service Department of Economic Boarder Protection; Lifeguards of Adjara; Municipality of Kobuleti, Batumi Sea Port Ltd; APM terminals "Poti", Black Sea Terminal – Kulevi; Black Sea Pilot Service, Maritime Agency SALVO MARINE, SEACOR RESPONCE Ltd; Batumi Oil Terminal Ltd; Scientific Research Company "GAMA" Ltd;

BSC PS; Black Sea countries: Turkey, Environment Protection Services (MEKE LTD, MARE LTD, SEAGULL LTD); Turkish Coast Guard; Bulgaria; Romania; Midia Marine Terminal; EMSA; OSPRI; ITOPF; Sea Alarm Foundation Ltd; Oil Spill Response Limited; LLP Tengizchevroil; Caspian Region Chevron; Chevron Eurasia.

3.2 *Resources and OSR equipment*

Table 1. OSR equipment offshore

№	Title	Quantity
1.	Containerized offshore Boom (2 diesel engine)	200 m.
2.	Inflatable Light Boom ILB 750	400 m.
3.	Oil transfer Pump 32 cub.m. LPP 6	5 units
4.	Air Fans for inflation of the booms	6 units
5.	Mini Weir Skimmer LWS 500	4 units
6.	Anchors for use with Fast Current Boom	10 units
7.	Absorbent Pad	20 pack
8.	Absorbent Boom	100 m.
9.	Motorboat of Black Sea Service - "BUNEBA-5"	crew 6
10.	Helicopter Ministry of Defense	1 pilot, 1 observer
11.	Tug - "GONIO"	crew 5
12.	Tug - "KAPITAN T. FAGAVA"	crew 3
13.	Oil Recovery Vessel - "AISI"	crew 4
14.	Vessel "GIDROGRAF 81"	crew 6
15.	Firefighting Tug - "PATRIOT"	crew 13
16.	Boom deployment boat "GAGRA"	crew 2
17.	High Speed Coast Guard Boat	3 units
18.	Oil Response Team	6
19.	EMSA Vessel "GSP ORION"	crew 20
20.	Turkish Coastal Directorate Tug "GEMI KURTARAN"	crew 22
21.	Turkish Coast Guard Boat "TCSG-95"	----

Table 2. OSR equipment on the shore line

№	Title	Quantity
SEACOR Ltd		
1.	Booms BSB 550/10	200 m.
2.	Fas tank	1 unit
3.	Oil transfer Pump RBS-5	1 unit
4.	Air Fans for inflation of the booms	1 unit
5.	Brush Skimmer RBS-5	1 unit
	DISC Skimmer KOMARA 20	1 unit
6.	Anchors for use with Fast Current Boom	1 unit
7.	Absorbent Pad	200 m.
8.	Absorbent Boom	100 m.
9.	Boom Deployment Boat	crew 2
10.	Oil Response Team4	
Batumi Oil Terminal Ltd		
1.	Booms LAMOUR BEACH BOOM	100 m.
2.	Collapsible Tanks LST TSC 11.4	2 units
3.	Oil transfer Pumps 32 cub.m. LPP 6HA/C75	2 units
4.	Air Fans for inflation of the booms	1 unit
5.	Skimmer MINIMAX 12	2 units
6.	Anchors for use with Fast Current Boom	4 units
7.	Absorbent Pad	200 u-its
8.	Absorbent Boom	100 m
9.	Oil Response Team	6
Emergency Situation Department of MIA		
1.	OSR Trailer 1 unit	
2.	Collapsible Tanks LST TSC 11.4 cub.m.	1 unit
3.	Skimmer C-14D	1 units
4.	Absorbent Pad	200 u-its
5.	Absorbent Boom	100 m.
6.	Diver	1 person
7.	Rubber Boat	1 unit
8.	OSR Team	6
LIFEGUARDS		
1.	Water Lifeguards Brigade	4
MARE Sea Cleaning Services Inc. Turkey		
1.	Inflatable Boom/Containment Boom / For Off Shore, PU	200 m

2.	Backpack Air Blower	1 pcs
3.	Anchor Set (Anchor, Rope, Buoy, Chain)	4 pcs
4.	Sorbent boom –absorption boom/ 20 cm. dia.	100 m
5.	Sorbent pad-absorption sheet / 40x50 cm sizes	1 box
6.	Zodiac Boat	1 pcs
7.	Diving Equipment	2 set
8.	Equipment for close the damage (branda)	1 pcs
9.	Stand Equipment	7 pcs
10.	OSR Team	9
MEKE MARINE ENVIRONMENTAL PROTECTON SERVICES LTD, Turkey		
1.	EROS OFFSHORE BOOM 150 cm	50 m
2.	BACKPACK AİR BLOWER	1 pcs
3.	SOLİD FLATATİON BOOM	50 m
4.	ANCHOR SET	4 pcs
5.	PORTABLE TANK (15 m³)	1 pcs
6.	SORBENT BOOM (20 cm x 3 m)	96 m
7.	SORBENT PAD (40cm x 50cm)	200pcs
8.	STAND EQUİPMENT	2 pcs
9.	OSR Team	9
SEAGULL Environmental Protection Service Ltd, Turkey		
1.	Fence boom/containment boom/ For near shore, PVC	50 m
2.	Inflatable boom /containment boom / For off shore, PU	50 m
3.	Backpack Air Blower	1 pcs
4.	Anchor Set (Anchor, Rope)	1 set
5.	Sorbent boom –absorption boom / 13 cm.dia.x300 cm. length	2 pockets
	Sorbent boom-absorption boom/ 20 cm.dia.X300 cm. length	2 pockets
	Sorbent pad-absorption sheet/ 40x50 cm sizes	10 poc-es
6.	Oil skimmer/ used to remove oil from The water.	1 Set
7.	Floating storage tank/ for storage on the Sea Cap.15 cbm.	1 Set
8.	Stand Equipment	1 pcs
9.	OSR Team	5

3.3 *Vessels in "Marine causality" (simulation)*

- M/T "NORD WIND" – Batumi Oil Terminal Ltd, Real Name- Tug "Tamara 1";
- Ro-Ro Vessel ANNA MARIA – Batumi Oil Terminal Ltd, Real Name-Tug "Tamara 2".

4 RUNNING AND FINALIZATION OF EXERCISE

4.1 *Scenario GEODELTA 2011*

2011, September 15 at 11:00 local time (GMT+4) Maritime accidents have happened (Maritime Casualty). Tanker "NORD WIND" which went out of Batumi port oil terminal (course 354°) was loaded Azeri Crude oil. In position 41° 45' N; 041° 38' E, offshore 7, 5 nautical miles the Ro-Ro vessel "ANNA MARIA" (course 174°), collided with the tanker "NORD WIND". As a result of collision, tanker "NORD WIND" has a serious tear, crude oil spills into the sea (Pollution Incident). One crew member of Ro-Ro vessel «ANNA MARIA" have fallen overboard and another was seriously injured. DISTRESS signal is received by MRCC Georgia in Batumi from both ships at VHF channel 16 and collision position determined from VHF DSC. The National Marine Oil Spill contingency plan is activated, all SAR actions being co-ordinated by MRCC Georgia according to national procedures. National On-Scene Commander is a head of MRCC. To conduct operations at sea/on shore National On-Scene Commander designates Local On-Scene Commander/s. The helicopter and Georgian Coast guard high-speed boats nominated for aerial assessments of spilled area, ditched on the sea, requesting SAR assistance; MRCC coordinated SAR activities at sea.

After observation and calculation it was investigated that up to 250 t crude oil was spilled, which corresponded to Tier 3 spillage liquidation. Competent National Authority (CNA) with Emergency Response Centre e(MRCC-OPRC) established 6 working groups:

- The environment impact assessment and area monitoring group;
- The logistics and actions planning group;
- The financial-procurements-permits group;
- The communications group;
- The claims and compensations group;
- The public information communications group.

4.2 *Objectives of the exercise*

Purposes of the exercises according to their importance were:

- To conduct the operation safely and environmentally responsible manner;
- To test company response capability for this type of emergency and identify escalation for tier 3 response;
- To test effectiveness and competency of emergency group for tier 3 response:
- To monitor the spilled area by helicopter and high-speed motorboats.
- To response SAR operation;
- To assist damaged tanker;
- To deploy booms and skimmers;
- To determine spillage size, quantity and direction;
- To identify short term strategy, potential for escalation, protection of sensitive areas – River "Chakvis-Tskali", others;
- To start Immediately the containment of oil slick and commence recovery;
- To run BS WEB TRACK and GIS, obtain weather forecast and collect oil sample;
- To prepare the waste disposal plan – involve environmental IMT officer;
- To send Pollution Report to Black Sea countries;
- To test communication means.

4.3 *Exercise running*

Respond significant oil spill in the open sea, rescue operation "Man over Board" and assist the damaged tanker sending the three high-speed coast guard boats from Batumi and Poti to rescue the "Man over board" and monitoring of the spilled area by helicopter. The following sources were sent to the spillage area: Batumi Port Crane-"CHERNOMORETS-9", Tug "Captain T. Pagava", Oil recovery vessel "AISI", Poti Port Firefighting Tug "PATRIOT", Hydrographic service vessel "GIDROGRAF 81" to put down and up equipments;

Turkish Coast Guard Boat "TCSG-95" was sent to the area of spillage with the purpose of marine security and safety operation.

After the observation and calculation investigation showed that, up to 250 t crude oil was spilled, which corresponded to Tier 3 spillage liquidation. Wind direction was NW and there was danger that spilled oil could reach the offshore and places where the river "Chakvis-Tskali" connects with the sea. CAN with MRCC-OPRC mutually decided that with only the national resources, oil spill could not be liquidated, it was necessary to ask international assistance from Black Sea littoral countries and from EMSA. BSCP was activated. MRCC Georgia sent POLREP - Pollution Report of the MRCC Black Sea countries. Georgia, as a leading country required assistance from them.

Additional sources EMSA Vessel "GSP ORION" and Turkish Coastal Directorate Tug "GEMI KURTARAN" were sent to the spillage area.

At the estuary of river "Chakvis-Tskali" and nearby shoreline blocking with booms, protection of shore line, spilled oil collection and transfer to the safe place.

The following information was processed from GIS/OSIS to assist OS response planning and operations:

- Satellite Oil Service Report from ITU-CSCRS.

- Shoreline modeling satellite picture of the river "Chakvis-Tskali".

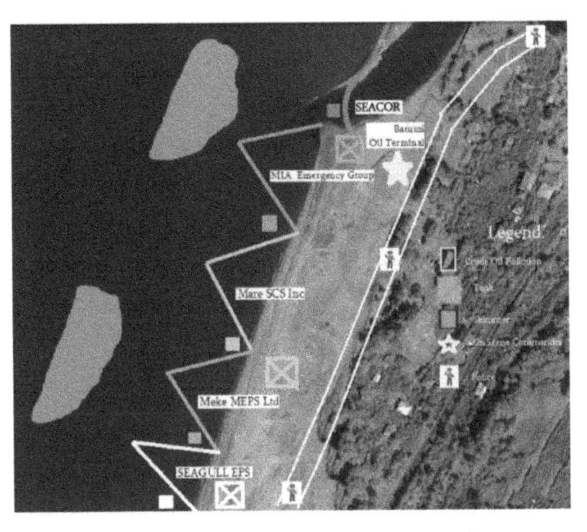

- A potential spill point was identified on the aerial photo and offshore diagram.

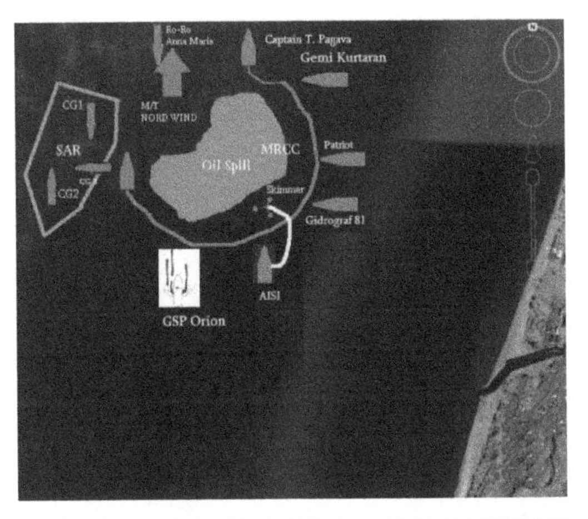

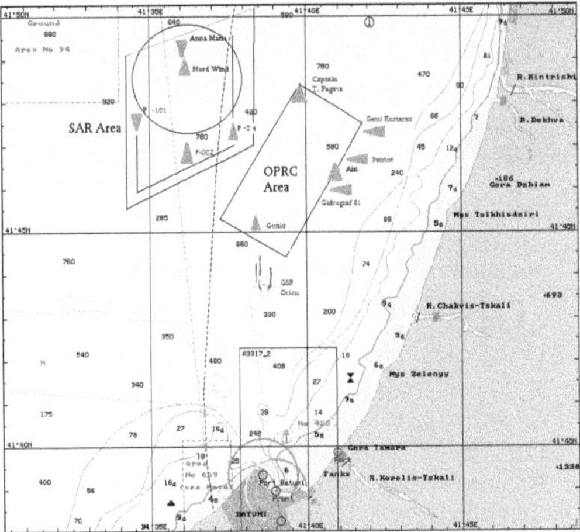

- Sea current direction/speed was identified and spill modeling was conducted by BS WEB TRACK.

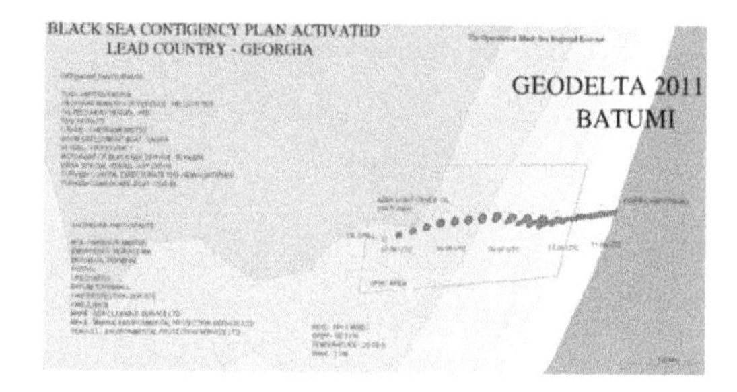

4.4 *Tasks for exercises*

CNA:

- To coordinate the processes of managing the exercises;
- To establish the intervention decision, scheme of communications and decision flow;
- To identify any additional national support for intervention;
- To approve the waste management;
- To approve the international assistance (asking for assistance and conditions imposed) and to facilitate it (custom clearances and fees exceptions);
- To establish administrative and financial rules of intervention;
- To inform the national authorities and the public about the situation;
- To survey the recovery of damages from the polluter.

National On-Scene Commander - head of MRCC:

- To receive the alert signal, quickly assess the situation and spread information to CNA also during the response operations;
- Based upon the level of response called for, the LOSC had the Emergency Response Team (ERT), the Black Sea service, the Port Authorities of Poti and Batumi, the Hydrometeorology Department, the Coast Guard, the Oil Terminal in Batumi briefed and ready to mobilize to site.
- Implementation of strategic decisions – SAR Operation, Liquidation of Oil Spill, protection of Shore Line, Coordination of spilled oil collection and transfer to the safe place, Coordination of response of operation team – Duty officer and Radio operator. Logistic and action planning, establishment of financial and purchasing, public and media relation Teams;
- The NOSC was ultimately responsible to keep accurate records of spill related costs as back up documentation for compensation claims. In the case of spills from vessel operations the NOSC might wish to liaise with the tanker owner and his P&I club to provide advice on financial aspects of clean up measures in relation to pollution damage. Access to legal services may also be required;

Offshore Participants:

- Batumi Harbor Master-Offshore Local On-Scene Commander - Coordinating SAR and OSR Operation Offshore;
- Oil response Team, Boom Deployment Boat "GAGRA"- Deploying of Booms and Equipment, OSR operation;
- Boat "BUNEBA-5"- Oil Spill Monitoring;
- Helicopter - Aerial Surveillance of SAR Operation and Oil Spill;
- Crane-"CHERNOMORETS-9", Tug "KAPITAN T. FAGAVA", Vessel "AISI"- Deploying and travelling of booms to the place of accident, OSR operation, Transferring of collected oil to the Oil Refinery;
- Firefighting Tug-"PATRIOT"- Operations on Firefighting and Spoiled fields;
- Georgian Coast Guard 3 Boats-P-101, P-24, P-002 - SAR Operation, Security and Safety provision in place of accident;

- Vessel "GIDROGRAF 81" Deploying of equipment;
- EMSA Vessel "GSP ORION"- OSR operations, Sweeping Arms;
- Turkish Coastal Directorate Special Tug "GEMI KURTARAN"-Firefighting, OSR operations;
- Turkish Coast Guard Boat "TCSG-95"- Security and Safety provision in place of accident.

Shore Line Participants:
- Poti Harbor Master-Onshore Local On-Scene Commander – Coordinating of blocking the estuary of the river "Chakvis-Tskali" and adjacent shore line with booms, protection of shore line, spilled oil collection and transfer to the safe place;
- SEACOR RESPONSE, Emergency Service of MIA Ajara, Batumi Oil Terminal – Blocking and Deploying of Booms and equipment at the estuary of the river "Chakvis-Tskali" and collecting of crude oil at the adjacent shore line;
- Lifeguards - Providing the lifesaving operations shore line;
- Batumi Town hall - Providing transport shore line;
- Fire protection and Emergency Situation Service - Firefighting Operations;
- Ambulance - First Aid On Shore;
- Kobuleti Municipality - Providing shore line readiness for exercises;
- MARE Sea Cleaning Services Inc., MEKE Marine Environmental Protection Services Ltd, SEAGULL Environmental Protection Service Ltd - Blocking the adjacent shore line of the river "Chakvis-Tskali" with booms, collecting of crude oil at the adjacent shore line;
- Patrol - Controlling of the adjacent areas, the territory of exercise was closed to any person, which did not participate in it.

5 THE MOST SUCCESSFUL EVALUATIONS MADE BY OBSERVERS, EVALAUTORS AND PARTICIPANTS

The biggest success is that the GEODELTA 2011 Exercise became a reality, in spite of all the factors impeding the organization of an event of such a scale; Whole organization of the event (incl. logistics) was excellent; Permanent telecommunication with the bodies from Government, Customs, Boarder Police and the main participants of the exercise; MSA participation; OSPRI and ITOPF participation; Cooperation between Georgia and Turkey was really successful in terms of sharing equipment and supporting clearing activities; Provision of a participant list with contact details and information on their status and location was excellent; The organization of facilities, space and logistics for the Emergency Response Centre, shoreline activities and offshore deployments was excellent; There appeared to be excellent communication between the command centre and the field activities, including the TV monitors to allow continuous live viewing.

6 CONCLUSIONS

1. The GEODELTA 2011 Exercise was a very good example for the collaboration between different national organizations/institutions involved, as well as for a good co-operation on international level.
2. Events like GEODELTA 2011 are great opportunity to stress the attention of the politicians and different level decision-makers on the seriousness of the problems, related to major oil pollution incidents, as well as on the

importance of further planning and some investments in the fields of pollution preparedness, response and co-operation.

3 It is recommended that personnel delegated to work in the command center consider additional oil spill incident management training. This will not only highlight the importance of communication (both internally and externally) but will demonstrate techniques used to carry this out. Ensuring the command staff fully understands the contingency plan, lines of communication and responsibilities means that effective decision making can made in an efficient manner. Not only will this training identify any weaknesses in the management process, it will ensure staff are more confident to adapt to unforeseen events or 'injects'.

REFERENCES

Anon., 1992. The Bucharest Convention "The Convention on the Protection of the Black Sea against Pollution", April 1992. Bucharest, Romania;

Anon., 2003. Black Sea Contingency Plan to The Protocol On Cooperation In Combating Pollution of the Black Sea By Oil and Other Harmful Substances in Emergency Situations. www.blacksea-commisson.org;

Anon., 2007. National Marine Oil Spill Contingency Plan. March 2007, Tbilisi, Georgia;

Arsenie P., Hanzu-Pazara R.: Human Errors and Oil Pollution from Tankers. TransNav - International Journal on Marine Navigation and Safety of Sea Transportation, Vol. 2, No. 4, pp. 409-413, 2008;

Gucma L., Goryczko E.: The Implementation of Oil Spill Costs Model in the Southern Baltic Sea Area to Assess the Possible Losses Due to Ships Collisions. TransNav - International Journal on Marine Navigation and Safety of Sea Transportation, Vol. 2, No. 4, pp. 405-407, 2008;

Anon. Steering Committee Meetings. Istanbul, Turkey, March 2011 in the BSC PS. www.blacksea-commisson.org;

Anon. Steering Committee Meetings. Istanbul, Turkey, on 8-10th June, 2011, on the MONINFO Project. www.blacksea-commisson.org;

GEODELTA 2011 Exercise Report. 20th ESAS AG Meetings, Istanbul, Turkey, on 11-13th October, 2011. www.blacksea-commisson.org;

ITOPF, Information services / data and statistics. ITOPF. http://www.itopf.com/information-services/data-and-statistics/statistics/index.html#no, 2011.

Chapter 5

Meteorological Aspects and Weather Condition

Operational Enhancement of Numerical Weather Prediction with Data from Real-time Satellite Images

Ł. Markiewicz, A. Chybicki, K. Drypczewski, K. Bruniecki & J. Dąbrowski
Gdansk University of Technology, Gdansk, Poland

ABSTRACT: Numerical weather prediction (NWP) is a rapidly expanding field of science, which is related to meteorology, remote sensing and computer science. Authors present methods of enhancing WRF EMS (Weather Research and Forecast Environmental Modeling System) weather prediction system using data from satellites equipped with AMSU sensor (Advanced Microwave Sounding Unit). The data is acquired with Department of Geoinformatics' ground receiver station (1.5 metre HRPT/MetOp). Aforementioned improvement is based on real-time updates of initialization and boundary conditions dataset of the NWP system. Conclusions and advantages of proposed solution are presented in the paper.

1 INTRODUCTION

The numerical weather prediction (NWP) is an advanced software application that utilizes mathematical models of the atmosphere and the hydrosphere for preparing weather predictions based on current weather data. Due to rapid development of fields of software data processing, remote sensing and parallel computing techniques, numerical modelling became a major tool for studying atmosphere dynamics and is utilized in various industrial applications, such as inland and marine navigation (Krata 2012), sustainable energy forecasting and others (Weintrit 2012).

Numerical models are based on the physical laws that govern the temporal evolution of the flow and express the mass, momentum and energy of the fluid. These computations are performed by algorithms that use independent variables (input), i.e. initial conditions needed for model initialization, along with many parameters that define the physical and numerical conditions. The models' dependent variables (output) consist of the temporal sequence of meteorological fields created during the integration process. They also include many derived quantities that one may wish to compute from that sequence.

In this context, one of most challenging tasks for operational NWP systems is optimal and efficient observational data assimilation. In the paper, authors present method of enhancing one of the NWP models, namely Weather Research and Forecasting Environmental Modeling System (WRF EMS), which is based on utilizing satellite meteorological data acquired from Department of Geoinformatics' imagery data receiver ground station. Proposed improvement is based on real-time updates of initialization and boundary conditions datasets of the NWP system. This paper presents the advantages of proposed solution.

2 NUMERICAL WEATHER PREDICTIONS

2.1 *Basic equations*

During numerical weather simulations the values of all the meteorological parameters are determined by the initialization data, which describes current weather conditions over the given area. The state of the atmosphere, aforementioned values and their mutual interactions are all reduced to the form of mathematical equations, which are solved using numerical methods. This enables the prediction of future atmospheric states in addition to time and location of various weather phenomenon (e.g. rainfall, snowfall, storm, fog, tornado, dew). Majority of these equations are based on the momentum equation (1), the energy conservation (thermodynamic) equation (2), the mass conservation (continuity) equation (3), the water vapour conservation equation (4) and the equation of state (5).

$$\frac{dV_3}{dt} = -2\Omega \times V_3 - \frac{1}{\rho}\nabla_3 p - \nabla_3 \Phi + F, \tag{1}$$

$$\frac{dT}{dt} = \frac{R}{C_p}\frac{T}{p}\frac{dp}{dt} + \frac{Q}{C_p}, \tag{2}$$

$$\frac{d\rho}{dt} = -\rho \nabla_{3\bullet} V_3, \tag{3}$$

$$\frac{dq}{dt} = M, \tag{4}$$

$$p = \rho R T, \tag{5}$$

where: V_3 = wind 3-dimensional velocity vector; Ω = Earth angular velocity vector; ρ = air density; ∇_3 = scalar function gradient of 3-dimensional vector; p = air pressure; Φ – Earth gravitational field potential; F = torque force; R = gas constant; C_p = heat capacity (at constant pressure); T = absolute temperature; Q = heat in unit of mass and time; $\nabla_{3\bullet}$ = divergence operator of vector field; q = relative humidity; M = water vapour mass.

Additionally, several simplifications can be made taking into account the order of magnitude of various terms to be considered. For instance, if it is assumed that the horizontal speeds are far greater than the vertical and the area of the atmospheric changes' horizontal scale exceeds 10km the momentum and thermodynamic equations can be reduced to (6) and (7).

$$\frac{dV}{dt} = -fk \times V - \frac{RT}{p}\nabla p, \tag{6}$$

$$\frac{dT}{dt} = \frac{R}{C_p}\frac{T}{p}\frac{dp}{dt}, \tag{7}$$

where: V = horizontal wind velocity; f = Coriolis parameter; k = unit vector (vertical); ∇p = pressure gradient.

The scale requirement makes it impossible to use these expressions to simulate convection - it allows inertia-gravity waves, but not sound waves because of the filtering effect. Nonetheless, the reduced equations are the foundation of most models used for numerical weather prediction.

Moreover, this set of equations can be simplified by considering the atmosphere as a fluid of limited depth in which density and the vertical distribution of horizontal velocity are constant (Jean Coiffier 2011). Owing to the second condition that the wind is also constant in the vertical, the vertical influx of the air masses is no longer existent in a model. When taking aforementioned assumptions into account the atmospheric state on a given area is described by the horizontal wind vector and the surface height (where the pressure is zero). Furthermore the pressure can be identified as the geopotential gradient of the atmosphere's free surface. This allows obtaining continuity equation from the conservation of mass for a fluid column of constant density, thickness and base area. Altogether, these equations and presumptions enables the description of change in velocity of a fluid column along with the change in the geopotential the free surface. It is known as the Saint-Venant system, which was originally created for the river water motion research. It is widely used in NWP systems because it permits the easy assessment of the numerical methods' attributes before employing them.

In many weather prediction models the vertical pressure coordinate, which is strictly surface dependent, is used. It enables the simplification of the continuity equation and the whole model, by describing the atmosphere as a group of layers separated by surfaces of the same pressure. To solve the problem of the lower layers, which intersect the higher areas of the surface (i.e. hills, mountains), the sigma coordinate was proposed (9).

$$\mu = \pi_s - \pi_t, \tag{8}$$

$$\sigma = \frac{(\pi - \pi_t)}{\mu}, \tag{9}$$

where: π_s = pressure at surface level; π = pressure at given height; π_t = pressure at the top of the modelled layer; $\boldsymbol{\sigma}$ = vertical coordinate.

It is also known as the normalized pressure coordinate, because it is equal to 0 at the upper boundary of the domain and equals 1 at the bottom (where the pressure is zero). When using sigma coordinate, the atmosphere can be represented as a group of layers of the equal sigma value, which embrace the surface at its higher areas, no longer intersecting them.

2.2 *Numerical weather prediction suite*

An atmospheric numerical prediction model is the main component in a much larger system (numerical prediction suite), which comprises of many various implemented processes that make operational weather forecasts possible. Among these processes such elements as the acquisition of meteorological data, the objective analysis, forecasting based on one of the atmospheric models, the determination of model run parameters, and finally, results dissemination and visualization can be distinguished.

Though the quality of weather forecast obtained in a purely deterministic way is mainly related to the accuracy with which the initial state is defined, major effort has been made to make the best of all available assimilated observations. Among many observation techniques two groups, namely in-situ observations and remote sensing observations, provide basic input for NWP systems.

In-situ observations network comprises of the network of not uniformly distributed fixed-location stations that measure the main weather parameters at given reference time coordinated in Universal Time Clock (UTC). Most of the network stations make measurements every hour, however, some of them (mainly upper air weather stations) make vertical soundings every 12 to 24 hours. This observation network is supplemented by a set of measurements made at fixed time by ships at sea, buoys and on board airliners.

The measurements of the Earth global parameters are made using remote sensing techniques. Apart from images of Earth and cloud cover provided by meteorological satellites, radiation measurements over a wide range of spectral channels are also available. Basically, meteorological satellites can be divided into two groups: sun-synchronous polar orbiting satellites travelling in a low orbit (about 800km over ground) providing relatively high-resolution images in daily routines, and geostationary satellites travelling in equatorial orbit remaining relatively motionless to the Earth. Unlike polar-orbiting satellite, geostationary satellites provide measurements for the same zone of the Earth at high frequencies, i.e. each 15 minutes for MeteoSat.

2.3 *Boundary conditions and initialization dataset*

Due to the fact that forecasts for ranges not exceeding 48 hours (over a limited geographical areas) may be functional on a relatively small domain, compared with the sphere, so called Limited Area Models (LAM) have become significant tools in many weather services. When integrating a limited area model, it is necessary to prescribe the values of the fields on the boundary of the working domain by interpolating in space and time the forecasts from another model operating over a larger domain at lower resolution. Although LAM models omit important atmospheric processes such as the effects of gravity waves, experience shows that the gain from using a greater horizontal resolution exceeds the degradation from perturbations introduced on the lateral boundaries of the LAM model domain.

Since WRF EMS model is running as a LAM model, lateral and boundary conditions datasets need to be imported from larger domain running model. In this context Global Forecast System (GFS), containing a global computer model and variational analysis, administrated by National Oceanic and Atmospheric Administration (NOAA) is utilized. The resolution of GFS model varies in each part of the model: horizontally, it divides the surface of the earth into 35 or 70 kilometre grid squares; vertically, it divides the atmosphere into 64 layers and temporally, it produces a forecast for every 3rd hour for the first 192 hours. The GFS model output is freely available in the public domain for variety of national and commercial weather services.

GFS model data is served as files in General Regularly-distributed Information in Binary form (GRIB, also known as GRIdded Binary) format which is a general purpose, bit-oriented data exchange format, that is well suited for transmitting large volumes of data. It is widely used in meteorological and climate research for storing weather (and forecast) information. GRIB format was approved by World Meteorological Organization (WMO) (Guide To GRIB). There were three versions of the file format: versions 0, 1 and 2. Currently version 0 is deprecated and version 2 is most commonly used in meteorological or climate projects. GRIB data is also self-describing, meaning that the information needed for file reading is present within the file. GRIB file contains collection of messages (term record is also used), each of them holds the gridded data for one parameter at a given time and at only one level (BADC GRIB Documentation). Messages may contain sub-messages which allows creating hierarchical and ordered structures.

2.4 *Data assimilation*

In order to take advantage of the various information, acquired during data acquisition process, the quality of obtained information needs to be represented mathematically. Variational data assimilation is a technique that measures quality of the data basing on error statistics obtained by combination of repeating analysis and measurements. In this context, data assimilation is a relatively complex process and the effective implementation has been in the interest of considerable scientific investment and requires very substantial computing capabilities. Basically, data assimilations schemes are based on modifying an earlier forecast as background field at grid points using a weighted sum of differences between background field and measurements at fixed locations or statistical approaches.

Among many approaches for data assimilations, techniques of successive correction method and statistical interpolation by the least squares approach made the most significant improvement in this area of research over past decades.

Enhancement of aforementioned methods relies on the least squares estimation (sometimes referred as Best Linear Unbiased Estimator - BLUE) of the analysed state that aims to minimize the cost function considered as variational problem (Gandin 1963, Lorenc 1981, Daley 1991).

In 1985 Lewis and Derber proposed the new solution by seeking to minimize an objective criterion defined as a function of the problem, later called 3D variational approach (3DVar). Minimalization relates to a quadratic function quantifying the deviations from the available information combined from observation and the background (previous analysis) weighted by their respective standard deviations. Enhancement of 3DVar approach was presented in four-dimensional assimilation scheme (4DVar), in which the objective function to be minimized, quantifies the distance between the models trajectory and measurements within the assimilation time interval.

3 WRF EMS NUMERICAL WEATHER PREDICTION SYSTEM

The Weather Research and Forecast Environmental Modeling System is a complete, numerical weather prediction package that incorporates dynamical cores from both the National Center for Atmospheric Research (NCAR) Advanced Research WRF (ARW) (Klemp et al. 2007) and the National Center for Environmental Predictions (NCEP) non-hydrostatic mesoscale model (NMM) (Janjic et al. 2001) as one integrated forecast system. All the capabilities of the NCEP and NCAR WRF packages, the installation, configuration, and execution of the cores have been greatly simplified in order to encourage to use of them throughout the operational forecasting by universities, private companies or research communities (Michalakes et al. 2004).

Nearly every element of an operational NWP system has been integrated into the WRF EMS, including the acquisition and processing of initialization data, model execution, output data processing, file migration and archiving. Real-time forecasting operations are enhanced through the use of an automated process that integrates various fail-over options, the synchronous post processing and distribution of forecast files.

The system includes pre-compiled binaries optimized for 32- and 64-bit Linux systems running in shared or distributed memory environments. The MPICH2 executables are also included for running on local clusters across multiple workstations. The WRF EMS is designed to give the users flexibility in configuring and running NWP simulations, whether it is for local "offline" research or real-time forecasting purposes. It also allows for the acquisition of multiple initialization data sets via Network File System (NFS), FTP and HTTP. The post processing software (WRF-post) supports wide variety of display software including Advanced Weather Interactive Processing System (AWIPS), BUFSKIT, NCAR Command Language (NCL), Grid Analysis and Display System (GrADS), GEMPAK, NCEP Advanced Weather Interactive Processing System (NAWIPS) and Network Common Data Form (netCDF). The WRF-post can process forecast fields on 81 different pressure levels from 10 to 1025mb.

3.1 *WRF EMS data processing diagram*

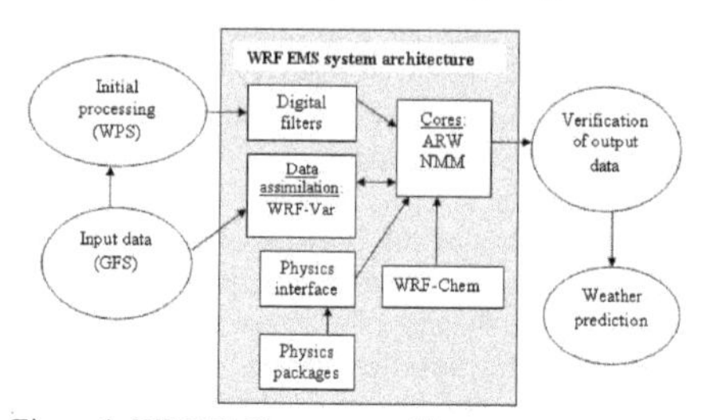

Figure 1. WRF EMS system architecture.

WRF EMS system architecture is modular and consists of many software components (Fig. 1). They were created in joint cooperation by various scientific institutes, which specialize in atmospheric and weather research (e.g. NOAA, NCEP, NCAR). These segments, by sharing data at various stages of the simulation process, are capable of creating a complete weather prediction using given parameters (temporal, atmospheric, etc.) for a predetermined area.

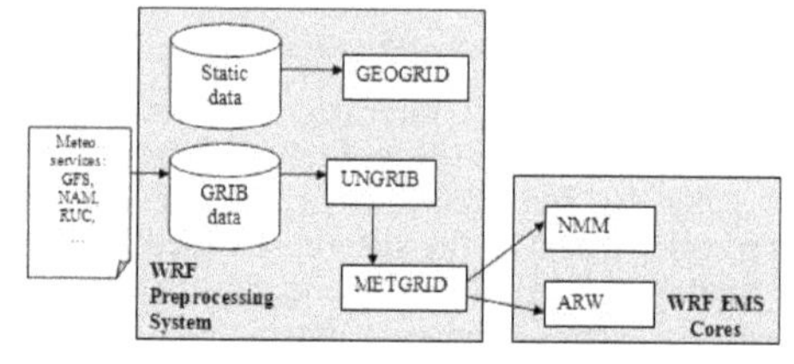

Figure 2. WPS module.

Initial processing of the input data is made in the WRF Preprocessing System (WPS) module and is divided into 5 phases (Fig. 2). Predefined domain (configured prior to the system run) is copied into the system in a form of static data. It determines the prediction process' area and its geographical projection. This domain is used to create GEOGRID files. Initialization data is downloaded (in compressed GRIB format) from the worldwide meteorological services (i.e. GFS, North America Mesoscale Model - NAM etc.) and uncompressed by UNGRIB module. The whole process is completed when uncompressed data is copied into METGRID file. It stores all the needed weather parameters. Finally METGRID file is sent to one of the cores (set of algorithms) that handles the subsequent steps of NWP process.

3.2 *WRF EMS configuration and data dissemination*

There are number of variables that determine model initialization and model computing process, including parameters of physical modelling, processing grid parameters, input and output data format and processing environment (processing cluster configuration). The instance of WRF EMS system that operates in the Department of Geoinformatics utilizes multi-node Ubuntu cluster based on 4 multicore workstations running in MPICH architecture. The model generates short-term 48-hour forecasts 4 times a day in 4-km spatial grid.

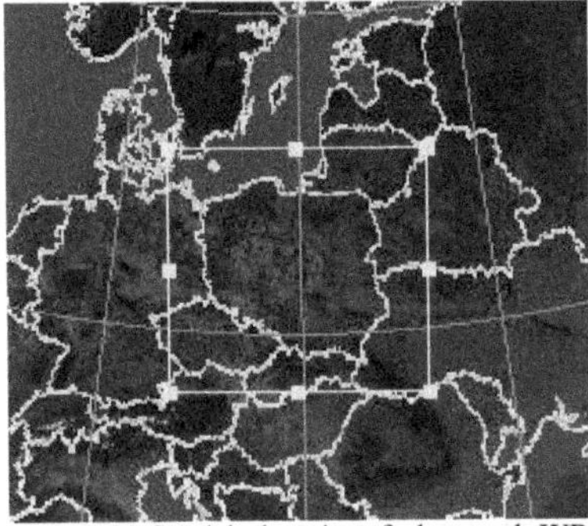

Figure 3. Spatial domain of the used WRF EMS system instance.

Spatial domain covers the territory of Poland as shown in Figure 3. The results of predictions are presented via Web portal as weather maps, diagrams, icons, descriptions and are available at http://www.weathersense.pl (Chybicki et al. 2011).

4 SATELLITE DATA ASSIMILATION

There is a number of Earth observing meteorological satellites that make their data available by direct broadcast service. Direct broadcast approach demands that the end-user is equipped with satellite ground station capable of receiving signal from satellite transmitter. In this research project authors utilized a ground station (1.5 metre HRPT/MetOp), which is located on the roof of the Electronics, Telecommunication and Informatics faculty building (Fig. 4).

The received signal consists of the High Resolution Picture Transmission (HRPT) stream of digital data where the real time Earth observations are multiplexed.

4.1 *HRPT-MetOp groundstation*

The Gdansk University of Technology (GUT) is operating 1.5 meter HRPT/MetOp satellite groundstation from the year 2009. The groundstation is capable of obtaining data from the Advanced Very High Resolution Radiometer (AVHRR) which is a major sensor on board of NOAA-* and MetOp-A/B satellites.

Figure 4. Department's HRPT-MetOp ground station.

The Advanced Microwave Sounding Unit – A1 (AMSU-A1) sensor is deployed on board the NOAA-* satellites as well as on the European MetOp-A, -B and -C.

The data from the AMSU is not natively supported by the station software, however authors of this paper have developed AMSU data acquisition software module.

The main advantages of using the data from the ground station are:

- the negligible time to delivery for the end-user (almost immediately after the reception from HRPT),
- accessibility to the Internet and other third-party service providers.

For the future research it is essential that the HRPT/MetOp-A ground station is adapted and upgraded into the National Polar-orbiting Operational Environmental Satellite System Preparatory Project (NPP/NPOESS) compliance mode as well as for receiving the data from the Chinese FY2 meteorological satellite system, which is equipped with sensors essential in meteorology and weather forecasting.

4.2 *HRPT stream/AMSU data*

The geometry of the AMSU sensor is of the whisk broom type and is presented in the Figure 5. Two versions of the sensor are available: a 15-channel (AMSU-A version) or 5-channel (AMSU-B) microwave radiometer.

These sensors are used for measuring global temperature and humidity profiles (T/Q profiles). AMSU sensors are also capable of verifying water existence in all forms (i.e. snow/ice coverage, rains and vapour) in the atmospheric layers by microwave radiation analysis (NOAA AMSU-A Guide).

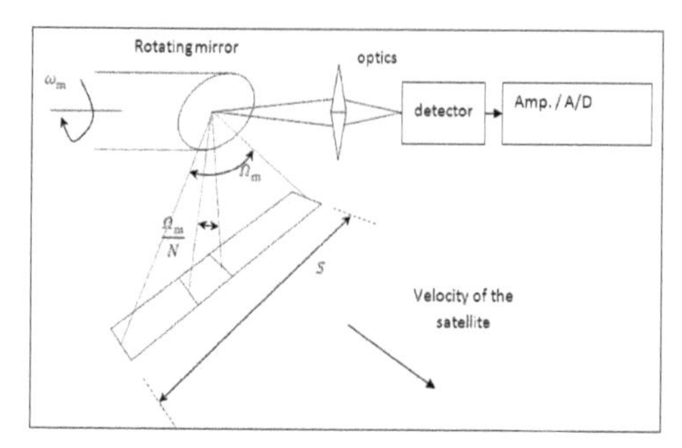

Figure 5. AMSU sensor geometry.

Cloud and ozone properties can be also determined by mutually using AMSU and other sensors. This kind of data is extensively used in numerical weather prediction and climate studies.

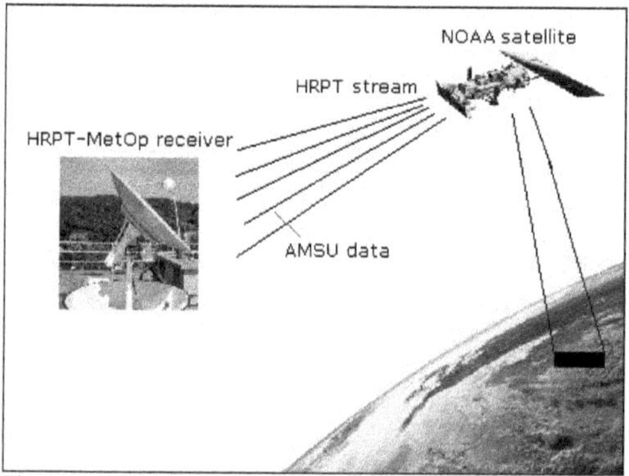

Figure 6. NOAA HRPT system diagram.

AMSU data is a part of the HRPT data stream. It is relayed to the end-users using split-phased S-band transmission with time-division multiplexing, with rate of about 660 kbps (NOAA AMSU-B Guide). Transmission power is 6.35W (38.03 dBm) and it occurs on three frequencies: 1698/1707 MHz (primary), 1702.5 MHz (secondary). Data resolution is 1.1km, which is sufficient for short- and long-term weather prediction (NOAA KLM User Guide).

The HRPT data stream is divided into major frames, each of them incorporates 3 minor frames. Data from the AMSU sensors is located in the third minor frame. Minor frame's data is separated using 10-bit words, as shown in Table 1.

Table 1. AMSU frame data description.

Data segment	Number of bits
Frame sync	60
ID	20
Time Code	40
Telemetry	100
Calibration target view	300
Space data	500
Synchronization data	10
Data words	5200
Spare words	1270
Earth data	102400
Auxiliary sync	1000

Each of the 520 data words consists of module identification and device flags (temperature values of mixer, amplifier, reflector and board, local oscillator monitor current, bridge voltage, processor and pixel data verification). Last two bits are used for parity and inverted bit checks.

HRPT stream contains many channels with data and headers from various satellite sensors. Earth data, the most significant data segment, stores all the samples from each channel (sample 1 – channel 1, sample 1 – channel 2 etc.).

ID's first word (second word is empty) determines minor frame number and the source satellite. It also includes the most significant bit (MSB) and least significant bit (LSB) for frame stability check, outcome of which in also located in ID.

Next segment describes time code, which identifies day and millisecond count. Telemetry and calibration target view are used for sensor ramp calibration. All of the frame/auxiliary synchronization and spare words have predefined bit values.

Spare words consist of three sets of data from every major frame; each has information about three minor frames. First minor frame includes five minor TIROS information processor (TIP) frames, second has five frames of spare data and third has five frames of AMSU data from AMSU information processor (AIP).

AMSU digital format synchronization is divided into 8-second pulses. Synchronization data stores information about sensor synchronization and MSB/LSB bits. Not only Earth data is collected - space data segment stores values of the space scan in channels 1 to 5.

AMSU telemetry contains all the radiometric data acquired in each sensor scan, along with the information about device self-adjustments done on orbit (i.e. reflector and scene positions, sensor state, sensor power and result of the cold/worm calibrations).

4.3 *Raw data processing*

The system, developed by the authors, processes the raw AMSU data in three important modules. First is responsible for georeferencing of the AMSU data. Georeferencing is done using the whisk broom sensor geometry, precise time from the telemetry inside the HRPT and the satellite parameters: position, altitude and attitude, which are based on the orbital parameters from Two Line Element data set (satellite) and Simplified Perturbation Model (orbit prediction). For the georeferencing purposes authors assumed the simplified Earth ellipsoidal shape (WGS84 standard). After georeferencing, raw data is calibrated in the second module. The calibration, in the authors' system module, is done for every spectral band accordingly to the additional information from the warm and cold calibration of the satellite's sensor.

The third module, developed by the authors, is responsible for transition from radiative to temperature and humidity profiles (T/Q). As for now, this submodule processes the data from the AMSU in order to calculate the atmospheric temperature profile from the Earth's surface to the upper stratosphere, at approximately 2-millibar pressure altitude (48 km). The humidity, as well as other products like precipitation, surface measurements, snow cover, sea-ice concentration and soil moisture are planned to be added in the future.

The temperature profiles are generated inside the module with the appropriate weighting functions. The AMSU coefficients are produced by Remote Sensing Systems (RSS) (MSU and AMSU Data Products). Over the years, RSS have received support for the development of this dataset from a number of sources, including NOAA's Office of Global Programs, NOAA's Climate Program Office, and NOAA's Climate Data Record Program. Production of the current dataset (version 3.3) is supported by NOAA's Climate Data Record Program, while improvements to the methods used are currently supported by NASA's Earth Science Division (part of the Science Mission Directorate). The weighing functions are based on the brightness temperatures' components related to the specific altitude as shown on the Figure 7. The vertical weighting function describes the relative contribution of microwave radiation emitted by a layer in the atmosphere. It is the total intensity measured above the atmosphere by the satellite.

Each of the Temperature Middle Troposphere (TMT), Temperature Troposphere–Stratosphere (TTS), Temperature Lower Stratosphere (TLS) and C10-C14 correspond to a specific channel of the AMSU-A1 sensor. Temperature Lower Troposphere (TLT) and Temperature Total Troposphere (TTT) are more complicated and include differencing and weighting of multiple channels (Carl 2009).

Precisely the TLT is constructed by calculating a weighted difference between C5 measurements from near limb views and measurements from the same channel taken closer to nadir. This extrapolates the C5 measurements made lower in the troposphere and removes most of the stratospheric influence. TTT is a multichannel combination created by calculating a linear combination of TMT and TLS. TTT equals 1.1*TMT minus 0.1*TLS. Such combination reduces the influence of lower stratosphere (Fu 2005).

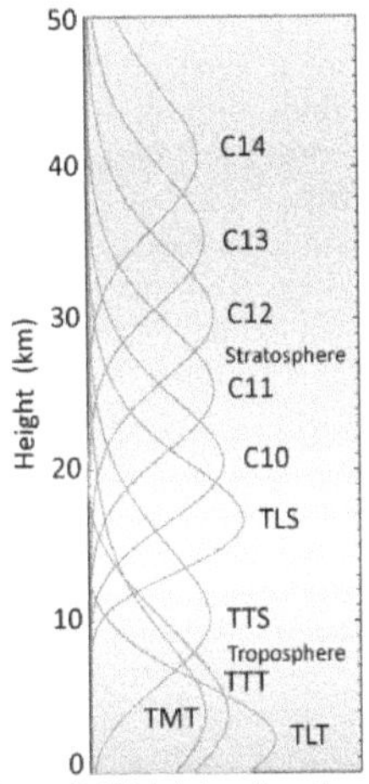

Figure 7. Weighting function for each RSS product.

4.4 *Assimilation of acquired data to WRF EMS system*

In the described solution WRF EMS boundary and condition data sets are taken from GFS service. Aforementioned data is a series of GRIB files that need to be merged with acquired T/Q profiles. There are several applications available that allow reading, editing and writing a binary GRIB version 2 files. The most popular of them are: NOAA Climate Prediction Center (CPC) command line tool wgrib2 (wgrib2 documentation), toolset for working on climate and NWP data Climate Data Operators (CDO) (cdo documentation) and the European Centre for Medium-Range Weather Forecasts (ECMWF) GRIB API (GRIB API documentation). Authors have chosen wgrib2.

Process of assimilation of the GFS and T/Q data is divided into three steps. In the first phase the parameters, which will be merged with T/Q profiles, are converted from GRIB format to text files. Created files contain only the values of one chosen parameter at each of the grid points. Additionally the entire GRIB file is also converted into the text file. Thereafter, previously created files of each variable are used to determine the position of given parameter in the main text file. Found grid values are then assimilated with the data from the T/Q profiles. Last step is a creation of a new GRIB file containing the updated values. To keep the GFS format, boundary and condition files' structure is used (with the modified values) to create a new GRIB file.

Updated boundary and condition dataset is then loaded into WRF EMS weather simulation process by the EMS_PREP routine. GRIB files do not have to be identical with the GFS files in a manner of variables' order and other characteristics, because WRF EMS enables different configurations for every (standard or custom) dataset. NWP system uses input files to create V-table, configuration file that determines which variables will be used for model initialization, so the process can be easily adapted for a specific GRIB file.

Authors have been able to edit GRIB files to change specific variables' values accordingly to the AMSU sensor data downloaded using the department's HRPT/MetOp ground station. Additionally, by creating custom configuration, this new datasets have been successfully used to run weather prediction processes in WRF EMS. Further research will include operational verification of the NWP process enhancement methods described in this paper.

5 SUMMARY

Preliminary results of AMSU satellite data assimilation to WRF EMS system and modification methods to initialization datasets were presented in the paper.

Preliminary results show that the fusion of AMSU data retrieved from HRPT stream in real-time is possible to implement in operational weather forecasting. However, additional research needs to be done due to the rapid development of novel technologies (in the field of satellite observation) that affects the AMSU instrument – for example AMSU will be operationally replaced by the Advanced Technology Microwave Sounder (ATMS), which was already launched on October 28, 2011.

REFERENCES

BADC GRIB Documentation, source: http://badc.nerc.ac.uk/help/formats/grib/

cdo documentation, source: https://code.zmaw.de/projects/cdo

GRIB API documentation, source: https://software.ecmwf.int/wiki/display/GRIB/Home

Guide To GRIB, source: http://www.wmo.int/

MSU and AMSU Data Products; Remote Sensing Systems; source: http://www.ssmi.com

NOAA AMSU-A Guide, source: http://www2.ncdc.noaa.gov/docs/klm/html/c3/sec3-3.htm

NOAA AMSU-B Guide, source: http://www2.ncdc.noaa.gov/docs/klm/html/c3/sec3-4.htm

NOAA KLM User Guide, source: http://www.ncdc.noaa.gov/oa/pod-guide/ncdc/docs/klm/html/c4/sec4-1.htm

Wgrib2 documentation, source: http://www.cpc.ncep.noaa.gov/products/wesley/wgrib2/

Chybicki, A. & Lubniewski, Z. & Kaminski, L. & Bruniecki, K. & Markiewicz, L., 2011, Numerical Weather Prediction – data fusion to GIS systems and potential applications. GIS *Odyssey, ISBN HIZ 953-6129-32-9,* 2011

Coiffier, J., 2011. *Fundamentals of Numerical Weather Prediction*. ISBN 978-1-107-00103-9

Daley R., 1991. *Atmospheric Data Analysis*. Cambridge: Cambridge University Press.

Fu, Q. & Johanson C. M., 2005. Satellite-Derived Vertical Dependence of Tropospheric Temperature Trends, *Geophysical Research Letters*, 32, L10703

Gandin L.S., 1963. *Objective Analysis of Meteorological Fields*, Lenignrad, Gidromet Isdaty, Jerusalem: Israel Program for Scientific Translations (1965)

Janjic, Z. I. & Gerrity J. P. & Jr. & S. Nickovic, An Alternative Approach to Nonhydrostatic Modeling. *Monthly Weather Review. Vol. 129.* 1164-1178,2001

Klemp, J. B. & Skamarock W. C. & Dudhia J., 2007. Conservative split-explicit time integration methods for the compressible nonhydrostatic equations. *Monthly Weather Review*.

Krata P. & Szlapczynska J., March 2012, Weather Hazards Avoidance in Modelling Safety of Motor-Driven Ship for Multicriteria Weather Routing, *TransNav, the International Journal of Marine Navigation and Safety of Sea Transportation*, Vol. 6, No. 1

Lewis J. M. & Derber J. C., 1992. *The use of adjoint equations to solver a variational adjustment problem with advective constraints.* Tellus, 37A, 309-22

Lorenz A., 1981. A global three dimensional multivariate statistical interpolation scheme. *Monthly Weather Review*. 109, 701-21

Mears C. A. & Wentz F. J., 2009. Construction of the RSS V3.2 lower tropospheric dataset from the MSU and AMSU microwave sounders. *Journal of Atmospheric and Oceanic Technology*. 26, 1493-1509.

Michalakes J. & Dudhia J. & Gill D., & Henderson T. & Klemp J. & Skamrock W. & Wang W. & Mozdzynski G. (ed.), 25-29 October 2004. The Weather Research and Forecast Model: Software Architecture and Performance. *Proceedings of the 11th ECMWF Workshop on the Use of High Performance Computing In Meteorology*. Reading, U.K.

Weintrit A. & Neumann T. & Formela K., December 2012, Some Problems of the Offshore Wind Farms in Poland, *TransNav, the International Journal of Marine Navigation and Safety of Sea Transportation*, Vol. 6, No. 4

Analysis of the Prevailing Weather Conditions Criteria to Evaluate the Adoption of a Future ECA in the Mediterranean Sea

M. Castells, F.X. Martínez de Osés & J.J. Usabiaga
Universitat Politecnica de Catalunya (UPC), BarcelonaTech, Spain

ABSTRACT: Appendix III of MARPOL´s Annex VI sets out the criteria and procedures for designating an emission control area (ECA).This criteria includes: a clear delineation of the proposed ECA; land and sea areas at risk; emission quantification and impact assessment; prevailing weather conditions; data on marine traffic; and land based measures concurrent with the ECA adoption. This paper analyses one of these parameters: prevailing weather conditions to evaluate the adoption of a future ECA in the Mediterranean Sea. Results will demonstrate how marine emissions will impact the sea and land ecology in the Mediterranean area.

1 INTRODUCTION

There is a growing voice calling for an ECA in the Mediterranean area, claiming significant damages to the environment, crops and health; produced by emissions from shipping both in Mediterranean Sea coastal countries as well as further in shore. The last ship emissions inventory for the Mediterranean developed by Entec UK limited in 2007 appointed that intra-European movements, i.e. Short Sea Shipping (SSS), contributed in 2005 significantly to emissions in the Mediterranean Sea, as 38% of the fuel consumed corresponded to intra-European movements (16% domestic and 22% between EU countries).

Building on the statistics of "Maritime transport statistics – short sea shipping of goods" published by the Eurostat (Eurostat, 2012), short sea shipping traffic volumes in the Mediterranean are already recovering from the downturn suffered due to the current economic crisis. Containerized and RoRo cargo which in 2010 represented 29.4% of the total short sea shipping volumes in the Mediterranean are emerging strong, registering highest traffics shares ever.

The International Maritime Organization has adopted the mandatory installation of Automatic Identification System (AIS) requirements. Based on the AIS data analysis and processing, data of traffic around the world and in the Mediterranean can be analyzed (Xiang, 2012).

Figure 1 clearly depict the high vessel density areas around the world and the main traffic lanes connecting the economic centers in Asia, Africa, Europe and the Americas crossing the Atlantic, Pacific and Indian Oceans. As can be seen, the Mediterranean sees lots of ship traffic.

Table 1. Short sea shipping cargo volumes (in percentage) in the Mediterranean, 2006-2010. Source: Own, based in Eurostat

Cargo (%)	2006	2007	2008	2009	2010
Liquid bulk goods	51.0	50.3	49.8	49.8	48.6
Dry bulk goods	16.6	15.8	15.6	18.0	15.7
Large containers	16.2	16.7	17.4	18.0	18.3
Ro-Ro (self propelled units)	5.3	5.3	6.0	4.8	7.1
Ro-Ro (non –self propelled units)	4.6	5.4	3.9	3.9	4.0
Other	6.4	6.4	7.3	5.5	6.3
Total (million tons)	564	584	589	563	570

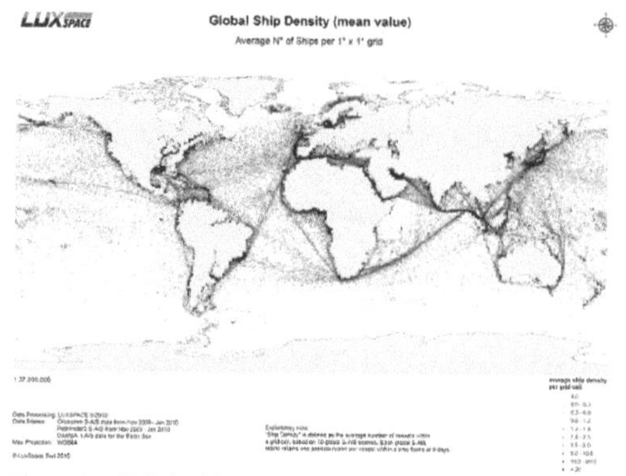

Figure 1. Global ship density map of all class A vessels. Source (Eiden, 2010)

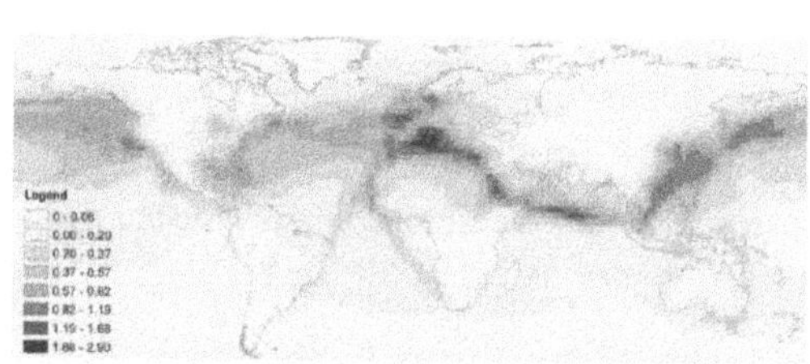

Figure 2. Concentrations of PM2.5 with ICOADS data in micrograms per cubic meter. Source: Winebrake, et al. 2009

On the other hand, as can be seen in Figure 2, the Mediterranean ship emissions appear important probably due to the more shipping (and ship emissions) along coastal routes frequented by containerships along coastal shipping, particularly in the Europe and Mediterranean areas.

2 REGULATION

The International Maritime Organization (IMO, 2009) commenced a debate on the reduction of emissions to air form ships in the 1980's. MARPOL's Annex VI was published in 1998 and came into force May 19, 2005, and was revised in 2008. The main changes were as follows:

Reduce the global cap on sulphur content in fuel oil to 3.50% (effective January 2012), then progressively down to 0.50 % (effective January 2020)

Reduce limits applicable in Sulphur Emission Control Areas (SECAs) to 1.00% sulphur content (effective July 2010), being further reduced to 0.10% (effective January 2015)

Reduce limits on nitrogen oxide (NOx) emissions from marine engines, with the most stringent controls on so-called "Tier III" engines (i.e., those installed on ships after January 2016) operating in Emission Control Areas.

Appendix III defines the criteria and procedures for designation of an Emission Control Area (ECA). An ECA should be considered by adoption by the IMO is supported by a demonstrated need to prevent, reduce and control emissions of SOx, NOx and particulate matter (PM) from ships.

The proposal considering criteria for designation of an ECA shall include:

1 A clear delineation of the proposed area of application, along with a reference chart on which the areas is marked;
2 The type or types of emission(s) that is or are being proposed for control;
3 A description of the human populations and environmental areas at risk from the impacts of ship emissions
4 Emission quantification and impact assessment
5 Prevailing meteorological conditions
6 Data on marine traffic
7 Land based measures
8 The relative costs of reducing emissions from ships when compared with land-based controls and the economic impacts on shipping engaged in international trade.

Existing Emission Control Areas include:

- Baltic Sea (SOx, adopted: 1997 / entered into force: 2005)
- North Sea (SOx, 2005/2006)
- North American ECA, including most of US and Canadian EEZ (NOx & SOx, 2010/2012).
- US Caribbean ECA, including Puerto Rico and the US Virgin Islands (NOx & SOx, 2011/2014).

Experts expect proposals for other ECAs to be submitted to the Marine Environment Protection Committee (MEPC) in the near future. Most likely candidates are the coastal waters of Mexico and Japan. Norway is also expected to propose an ECA for its coastal waters in the Norwegian Sea, which would be the first ECA in Arctic waters. Proposals for ECAs in the Mediterranean Sea and Straits of Malacca are also expected; however, it will likely be years before it is feasible to meet ECA requirements in these highly trafficked areas.

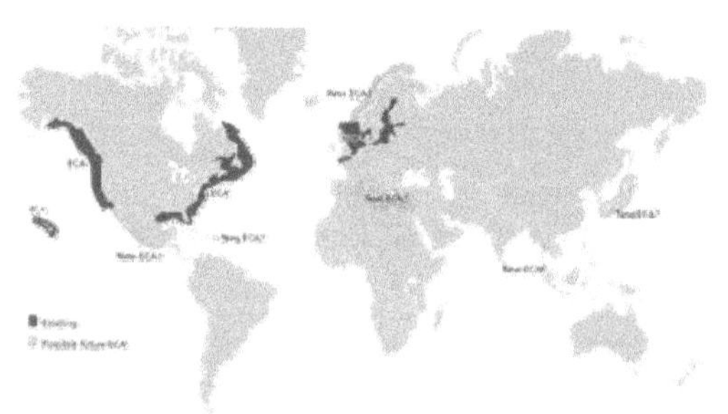

Figure 3. DNV's map of current and possible ECAs in the future. Source: Det Norske Veritas (DNV)

This paper will analyze relevant information pertaining to weather conditions to demonstrate how marine emissions will impact the sea and land ecology at risk in the Mediterranean.

3 METHODOLOGY

To evaluate prevailing meteorological conditions criteria, relevant information pertaining to weather conditions and, particularly, wind patterns in the Mediterranean area is required to demonstrate how marine emissions will impact the sea and land ecology at risk.

Descriptions of climatic conditions which are found in aids to navigation (Pilots, Routing Charts, and Pilot's Charts) make use of average values of meteorological elements. Such an approach follows the methods applied in classical climatology where climate is treated as "a mean state of atmosphere in a many year period" (Fredynus, 2012). The frequency of occurrence of types of adverse or unfavourable weather conditions is extremely important for determinate marine emissions at land.

Predominately onshore winds result in greater pollution on land. Moreover, topographical, geological, oceanographically, morphological or any other conditions that could lead to increased probability of higher localized air pollution or levels of acidification. Mountainous regions inshore impacted by onshore winds can lead to intensification of air pollution and acid deposition. The situation in the Mediterranean is more complex with varied geography and more complex wind conditions that are more localized and seasonal.

This paper will focus on predominant wind pattern in the Mediterranean to demonstrate areas at risk. Detailed weather conditions are held by a number of organizations, for instance: www.puertos.es, www.idromare.it, www.eurometeo.com and Mediterranean Pilot Volumes of the Admiralty charts and publications.

3.1 *Mediterranean climate*

The Mediterranean climate use to be known due to its mild and wet winters and dry and hot summers. This climate even being very similar to the one of California, central Chile, South Africa or SW of Australia; is due to the topographic close of the sea basin.

We can define two main questions: First of all, in general terms, from October and during the winter season, the high pressure of Azores is reduced in extension and it affords the Atlantic lows to reach southernmost areas. In this season, the Northern Eurasia countries are cooled quickly, whilst the rest of the Mediterranean remains warmer. This means that the Polar and Arctic air masses in its advance to South, will suffer a very intense convection. This situation drives to a Ciclogenesis episode, mainly in the Gulf of Genoa, South of Ionian Sea and around Cyprus Island.

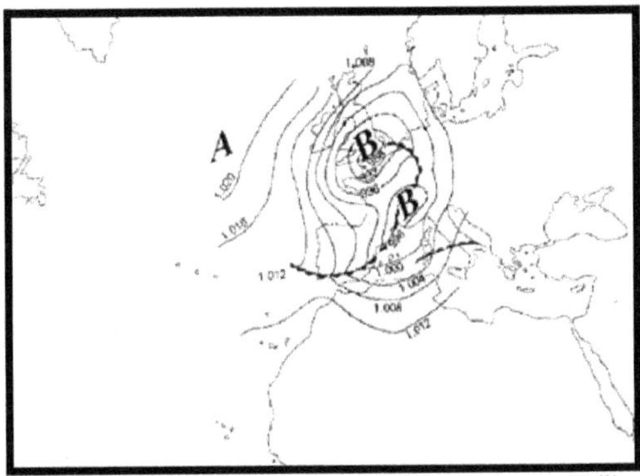

Figure 4. North situation in the Gulf of Genoa. Source: Sailing Directions Nr. 1 Costa Oriental de la Península Ibérica. Instituto Hidrogáfico de la Marina. Ministerio de Defensa. 1990.

On the other hand, to understand the Mediterranean climate, the local winds must be understood. Some of them have a synoptic effect like the Mistral, whilst some others are mainly local. The Mistral (NW), blowing down the Rhone valley, empties into the Gulf of Lion and usually appears with a pattern characterized by the position of a low pressure centred in the Balkans, close to the centre of the high pressure the Azores, together generating a pattern of situation of North winds in French territory. The topographic barriers will later address it as a North West.

During the winter, the wind effect is reinforced by the cooler snow coverage of that area, causing the downward flow of air from the Alps and the Massif Central (France) (see figure 4). The gravitational collapse of cold air and therefore, dense towards the bottom of the valleys, is called katabatic wind. Which is a dry wind, cold, down the valley of the Rhone, and when combined with a strong pattern of synoptic wind, can produce wind speeds of up to 80-85 knots in the vicinity of the Rhone delta. Between December and May, occurs an average of 26 days during which the Mistral can blow at 33 knots or more, with a slight peak of activity between March and April (11 days between the two). These winds are falling rapidly when they penetrate into the sea, but occasionally can be extended to the island of Malta or North Africa. The danger for navigation is actually the rise in a short time: high seas. This phenomenon that occurs mainly in March, when the significant wave height in the southern part of the Gulf of Lion, reaches 2 meters, one of the highest among statisticians in the Mediterranean.

The same effects are associated with the katabatic Bora, which is a NE wind, which blows on the Eastern shore of the Adriatic Northern winter occurs where violent storms and gusts at times up to 100 knots. These conditions are intensified when the Mediterranean Low pressure is well developed and maintained a high pressure lingering over Europe. In particular behind a cold front moving southeast over the Adriatic effect, is reinforced by katabatic descent of cold air from the mountains Dalmatians. All these features well defined throughout the year, we find unexpected effects that do pose a greater hazard to navigation.

Along the Eastern coast of Spain, we find the East wind and gales, which are part of the change as well as the local winds during the winter half of the year. The first characteristic of these storms pattern NE and ENE, passing squalls associated with the Mediterranean between France and Algeria may cause seas really hard when the wind settles and finds considerable NE fetch, showed in figure 5 & 6.

These storms are more frequent and dangerous especially in autumn and spring. The storm characterized by strong winds from the SW, across the Strait of Gibraltar, up the Spanish coast depressions associated with advancing from the late fall to early spring. The main danger is triggered violent storms and electrical appliance involved. The transition from prevailing winds between the two halves of the year completely alters the character of the local winds. In the absence of ciclogenesis, local wind patterns are dominated by the following: where the wind comes from and what happens when it blows.

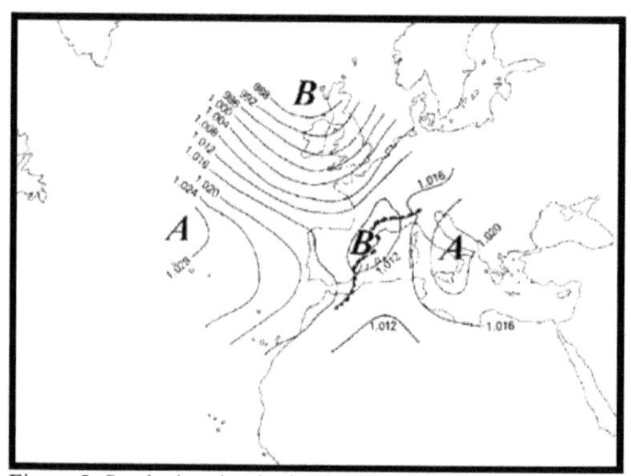

Figure 5. South situation in the Gulf of Genoa. Source: Sailing Directions Nr. 1 Costa Oriental de la Península Ibérica. Instituto Hidrogáfico de la Marina. Ministerio de Defensa. 1990.

For example, the Sirocco is the warm wind SW associated with the advance of a depression moving east direction across the Mediterranean (see figure 7), being most common in the spring because the subtropical high pressure moves north. Since it has warm, dry and full of dust from the Sahara, crossing the Mediterranean when it absorbs large amounts of moisture, makes it an unpleasantly warm and wet air. The most obvious consequence of the Khamsin is its emergence as almost spontaneously without previous evidence, of a narrow channel of wind posing a threat to coastal terminals with open basins, where large ships could be docked (big Bulk carriers, Tankers or Containerships). Due to that channel wind lifts. Another problem for navigation is when the air passes over the colder waters in the Northern Mediterranean in the spring or early fall, quickly forming a dense fog.

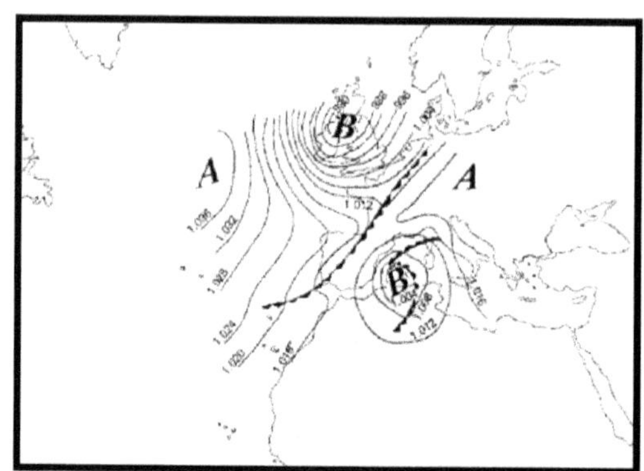

Figure 6. Typical Algeria Low in the Gulf of Genoa. Source: Sailing Directions Nr. 1 Costa Oriental de la Península Ibérica. Instituto Hidrogáfico de la Marina. Ministerio de Defensa. 1990

But perhaps the most famous summer winds are Etesian (the Meltemi Turkish) blowing from the NE and NW Aegean Sea. These are the consequence of a low pressure system extending from NW Anatolia to India, formed by the intense warn down suffered in the region. These winds reach maximum intensity during the month of August when hovering around 15 knots in the southern Aegean, resulting in variable weather conditions, which benefit the coastal regions moderating temperature. Occasionally they can be associated with violent storms, produce winds and sudden squalls (known as Bourini in Greece), causing considerable damage to local navigation.

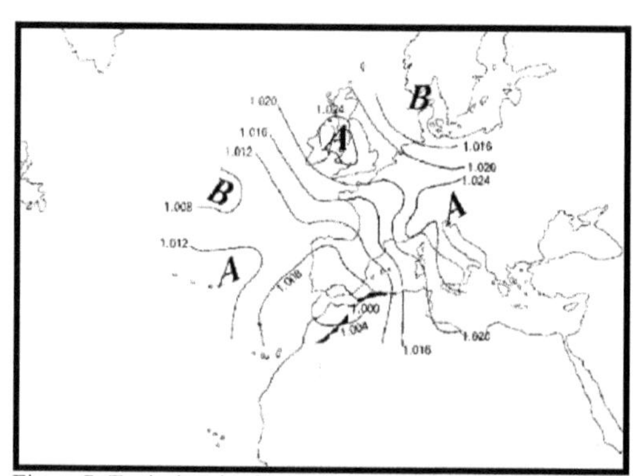

Figure 7. Typical path of a Mediterranean Low in the Strait of Gibraltar. Source: Sailing Directions Nr. 1 Costa Oriental de la Península Ibérica. Instituto Hidrogáfico de la Marina. Ministerio de Defensa. 1990.

4 CONCLUSIONS

This paper has analyzed one of the parameters to evaluate the adoption of a future ECA in the Mediterranean Sea: the prevailing weather conditions.

Obtained analysis demonstrates that the weather pattern in the basin of the Mediterranean is affected by many differing systems and is quietly unpredictable being quick to change and often very different at two places only a short distance apart. Due to the high surrounding land masses and the latitude, the climate can at times be extreme but on the average it is very pleasant especially in the summer months. As big conclusions, we can confirm that excluding the coastal breezes that can carry the smoke among ten to twelve miles inside the shore when it blows from sea, the Mediterranean is characterized by strong Northerly winds in winter time within the Gulf of Lyon, Adriatic Sea and Aegean sea, that carry the smoke to the south coast of the basin. In the other hand, we can face opposite situations when lows travel across the Mediterranean, bringing the air masses from South to the European coasts, as the cases of Sirocco, Ghibli or Khamsin

Further research is needed to evaluate the rest of criteria to designate a future ECA in the Mediterranean Sea. The emission quantification and impact assessment and land based measures concurrent with the ECA adoption criteria has been evaluated in previous work (Usabiaga et al, 2012).

In this respect an ECA proposal by a Mediterranean country is all the more essential keeping in mind that since its proposal, around five years is needed until its adoption. On the contrary by 2017 road transport will have swept away maritime transport.

REFERENCES

[1] Admiralty Charts and Publications. Mediterranean Pilot. Volumes I to VI.

[2] Eiden G. et al, 2010. Performance of AIS sensors in space - PASTA-MARE project final report executive summary

[3] European Commission, 1998. Directive 1998/70/EC of the European parliament and of the council. Official Journal of the European Union.

[4] European Commission, 2005. Directive 2005/33/EC of the European parliament and of the council. Official Journal of the European Union.

[5] European Commission, 2009. Directive 2009/30/EC of the European parliament and of the council. Official Journal of the European Union.

[6] European Union, 2011. EU transport in figures. Statistical pocketbook 2011. 22 p.

[7] Eurostat, 2012. Maritime transport statistics – short sea shipping of goods. Available from internet: <http://epp.eurostat.ec.europa.eu/statistics_explained/index.php/Maritime_transport_statistics_-_short_sea_shipping_of_goods>.

[8] Ferdynus, J, 2012. Polish Seaports – Unfavorable Weather Conditions for Port Operation (Applying Methods of Complex Climatology for Data Formation to be Used by Seafaring). TransNav, the International Journal on Marine Navigation and Safety of Sea Transportation, Volume 6 Number, p 131-139.

[9] International Maritime Organization (IMO), 2009. Revised MARPOL Annex VI – Regulations for the prevention of air pollution from ships and NOx Technical Code. London.

[10] Meech R., 2008. Designation of the Mediterranean Sea as a SOx Emission Control Area (SECA) under MARPOL Annex VI. SAFEMED project, task 3.7.

[11] Official Spanish ports portal. Available from internet: www.puertos.es

[12] Usabiaga J.J, Martínez de Osés F.X., Castells M, 2012. ASSESSMENT FOR POSSIBLE FUTURE ECA ADOPTION IN THE MEDITERRANEAN AREA (Short Sea Shipping vs. Road Transport). International Conference on Traffic and Transport Engineering. Belgrade

[13] Wall C., 2007. Ships Emissions Inventory – Mediterranean Sea. Entec UK limited.

[14] Website of general Mediterranean weather. Available from internet: www.eurometeo

[15] Website of Istitute Superiore per la Protezione e la Ricerca Ambientale. Available from internet: www.idromare.it

[16] Winebrake, J. J.; Corbett, J. J.; Green, E. H.; Lauer, A.; Eyring, V., 2009. Mitigating the Health Impacts of Pollution from International Shipping: An Assessment of Low-Sulfur Fuel Mandates. Environmental Science and Technology, 43(13), p.4776-4782.

[17] Xiang, Z et al, 2012. Applied Research of Route Similarity Analysis Based on Association Rules. TransNav, the International Journal on Marine Navigation and Safety of Sea Transportation, Volume 6 Number 2, p 181-185.

Monitoring of Ice Conditions in the Gulf of Riga Using Micro Class Unmanned Aerial Systems

I. Lešinskis
Latvian Maritime Academy, Riga, Latvia

A. Pavlovičs
Transport Accident and Incident Investigation Bureau of Latvia, Riga, Latvia

ABSTRACT: The process of Global Warming enables researchers to conduct maritime operations in the new Polar Regions. Due to dynamic process of ice movement, occasionally ships and off-shore installations are exposed to unfavorable situations. As the result of this, officials of maritime industry have to arrange safe operational fulfillment, concerning valuable and sound information about the ice conditions at sea.
This study describes the potential use of the Unmanned Aerial Systems (UASs), since ice conditions recognition has been evaluated and tested practically. The particular experiment has been conducted by means of UAS, worked out at Riga Technical University and equipped with the electromotor, thus considerably minimizing the ecological impact, compared with unmanned vehicles equipped with combustion propulsion.
The outcome of experiment is practical proof, that UAS can be effectively implemented, however with minor limitations, for local ice conditions reconnaissance and surveillance at sea.

1 ICE NAVIGATION AND EFFICIENCY OF INFORMATION

Almost every particular Mariner faces the practical necessity to navigate thru the ice-fields. Ice navigation requires specific knowledge and, moreover, specific experience as for the crew of ship, as for the ship's technical settings and ultimate construction. In order to arrange the winter maritime traffic, there should be involved different authorities and services. One of the most essential services in this case is a regularly and efficiently updated information and data about the ice conditions and situation. The data on ice conditions at sea can be in form of assumed forecast or/and factually collected, real data.

Nowadays, blocks of real information on ice cover at sea can be originated from different sources: ships on sites, coastal surveillance and weather stations, satellites, and aircraft (Lešinskis & Pavlovičs 2011).

The main source of data on the state of the ice cover the freezing seas is a satellite remote sensing. Information concerning ice may come from satellites: NOAA, Fengyun and EOS (Terra, Aqua), RADAR-SAT1, Envisat (Frolov et al. 2012)

The information from satellites in details isn't sufficient enough. For instance, the 1 pics of picture that receive Latvian Environment, Geology and Meteorology Centre from NOAA cover the territory 1km X 1km. This means that resolution of picture is bigger than 1km.

Ice situation related to information from satellites delivers the general understanding if the ice is being presented there or not. The limited resolution of a satellite image must be supplemented with information about ice fields structure and conditions from other source.

Additional information about ice condition at sea may be received from coastal observation station and ships at sea. For compilation of the ice picture along the Latvian coast line there are used 9 coastal stations.

The Information from ships might be received when ships are physically within ice covered areas and usually this is information from ice breakers.

The observational capabilities of a ship at sea and a coastal station are quite limited, and even within the good weather and visibility they do not go beyond the sight of apparent horizon. At normal weather conditions a distance D in nautical miles of apparent horizon visibility can be calculated by formula:

$$D = 2.08\sqrt{h} \quad (1)$$

where, h = height of observer from sea level in meters.

Distance D in kilometers can be calculated by formula:

$$D = 3,56\sqrt{h} \quad (2)$$

The height of a ship's bridge differs from ship to ship; depending on the ship size it could be from 3 to 25meters in average, but anyway, common observational capabilities of average ship are counted to be up to 10 nautical miles within favorable weather.

Due to very low land scope along the Latvian coast line the distance of observation from the observation distance are not better than from ships at sea.

The ''ad-hoc'' interviewing of Latvian Coast Guard ships Commanding Officers and Master of icebreaker ''Varma" has proved, that factual knowledge about current ice situation can be acquired just within the range about 3-5 nautical miles from around the ship.

In case of poor visibility, when visual ship's outlook becomes useless, ships' Masters can estimate the ice condition by use of radars. Nevertheless, taking into consideration the flat tangential distribution of the radar impulse along the terrain, the distances of credible data, depicting whether waters are free of ice or heavily covered by packed-ice, do not exceed 3-5 nautical miles (House 2006). The distance of detection s is dependable not only by height of radar antenna and electromagnetic wave form but by type of ice (Table 1).

Additionally, when ice situation becomes complicated, it is quite unlikely to get any relevant ice information from other ships, due to general avoidance of marine traffic of such the areas.

Table 1. Ice Features and Shipborne Radar detection

Ice Features	Shipborne detection range (nautical miles)
Large icebergs	15-20
Small Growlers	2
Bergy bits	3
concentrated pack ice	3
hummocked ice	3
Leads trough static ice	0.25

The most sufficient information may be provide by ice patrol aviation, but organize such patrols in all areas where vessels operated are very costly. Additionally it necessary to take into accounts that aviation operations in Polar areas are very dangerous because of bad weather conditions (Lešinskis&Pavlovičs2011).

2 CAPABILITIES OF UNMANNED AVIATION

Due to rapid developments of UAVs worldwide, there is opportunity to deploy them for needs of ice surveillance and reconnaissance. The UAV might be used as information source for the center of local monitoring (Frolov, 2012)

Direct UAV surveillance and visualization of ice conditions at sea provides the real-time capture of ice situation. Use of UAVs for these purposes massively decreases associated risks and functional expenses (Maslanik et al. 2002).

The consumer of this information might be a ship in marine area, which has to use optimal and best passage, once the ice field is expected to be penetrated.

Types of unmanned air vehicles (UAV) for ice operations are dependable by the tasks. For ice survey might be used different types of UAV and UAS. In order to be cost effective from one side and be able to fulfill the provisional tasks and minimize impact on ecology from other side is possible to use three types of UAV:

- Small fix wing stroke engine unmanned aircraft with capacity to operate of 10 to 12 hours and payloads approximately 4.5 to 5.5 kg. This type needs special prepared runway or very powered catapult and large sized arresting gears
- Micro-class fix wing electric engine unmanned aircraft with capacity to operate from 20 minutes to several hours. A payload is about 1-2 kg. Might be launched from hand or small size catapult. For landing need some space or arrested gears.
- Micro unmanned rotorcraft with electric engine usually can operate no more than one hour with payload less than 1 kg.This type does not require any extra installations for take-off or landing.

In general there are three types area where might be used UAS:

1 Ice conditions surveillance in the large water basins
2 Ice conditions surveillance in medium range marine areas
3 Ice observation at close range around the vessel or coastal station

The basic methods of UAS deployment as the means of ice conditions surveillance are as follows:

1 Perspective surveillance;
2 Detailed surveillance;
3 Pre-planned aerial photo surveying.

The methods are being implemented depending on missions, particular situation and location where the UAV is being launched from (Lešinskis et al. 2012).

The relation between area, type of UAV and methods of deployment are showed in Tab. 2. The analysis of Tab.2, as well as UAV take-off and landing procedures, we can draw a conclusion on most efficient patterns of operation:

- Small UAS deployment for observation of large water basins, launched from deliberately equipped airfields;
- Micro fix wing UAS deployments for short constrained water basins observations, launched from short and small launching spots; these spots might be Naval Sea Surveillance System's Sites along the coastline, or vessels with the sufficient length of open deck - for instance, icebreakers.
- Rotor wing UAS for local short range observations, when even smallest launching spots are not available, thus small sized ships are most feasible for this role.

Table 2. Area's and tasks for deferent type of UAS

Size of area	Perspective surveillance	Detailed surveillance	pre-planned aerial photo surveying
Ice conditions surveillance in the large aquatorium as bigger Baltic Sea gulfs (Gulf of Finland, Gulf of Riga, Bothnia Bay)	Small UAS	Small UAS	Small UAS
Ice conditions surveillance in close sea areas (approaches to the ports)	Small UAS & Micro fix wing UAS	Small UAS & Micro fix wing UAS	Small UAS
Ice observation around the vessel or coastal station	Micro fix wing UAS & Rotorcraft	Micro fix wing UAS & Rotorcraft	Micro fix wing UAS

In the process of flight and collecting of visual photo and video data on and ice-openings are being surveyed, with the aim to clarify their sizing and dimensions. Additionally, the more detailed data on ice thickness, ultimate ''breed of ice'' and other related ones are being acquired. The thickness of ice packs could be estimated and assumed by mutual comparison of visual and video data, when dedicated data library is being dynamically created and updated, in order to extract provisional ice-thickness from visual images.

The different source shows that the color and form of one year ice are related to the thickness of ice (Pavlovskis&Kautiņš 1994 Ning at el, 2009).

The analysis of available data reveals, that a correlation between the form of ice pack, as appeared on site, and ice thickness could be directly related with the color and shading (Tab.3). Thinner ice renders darker outlook of an ice pack. This is a general conformity observed not only in Baltic Sea basin, but also in Antarctic, and is the direct characteristic of ice albedo.

New ice has low albedo, which increases gradually as the ice thickens, but then jumps exponentially when it acquires even just a thin snow cover. Therefore, the main determinant of area averaged albedo is ice concentration (Brand et al. 2005).

Table 3. Forms and Thickness of ice (Pavlovskis&Kautiņš 1994)

Class of ice	color	Thickness in cm
Dark nilas	Very dark	≤5
Light nilas	dark	5-10
Grey ice	grey	10-15
Grey-white ice	Light grey	15-30
White ice	white	30-70

The analysis of different UAS capabilities to be selected for the means of ice conditions monitoring, shows that UAS implementation pattern depends on the basic method of deployment, the size of covered water space, dimensions and technical equipment of the launching spot. Stroke engine propelled UASs, equipped with the optical means of observation are more feasible for ice conditions monitoring in Baltic Sea, spacious Riga Bay, Bothnian Bay, Finnish Bay. Areas of a port access might be more efficiently observed from the Micro class UASs, particularly launched from the icebreakers. Rotorcraft UAS are good means of ice control around the ship or coastal surveillance facilities.

3 CAPABILITIES OF THE MICRO CLASS UASS

Taking into consideration the accumulated experience and data on the operational capabilities of UASs to be used for ice conditions monitoring, there is completed comparative analysis of available Micro class UASs. One of the most suitable, up to date, system is the Riga Technical University developed micro class UAS.

The implementation of micro-class UAS as the means of control of maritime environment is dependable on the set of parameters and technical characteristics of an aircraft, flight control equipment, sensors, environmental and hydrometeorology factors, local and international regulatory procedures and requirements.

The efficiency of a UAS to be used for ice conditions monitoring depends on every particular setting and ratio between the sweeping observation range and the square-area of designated area.

The Facts and details on the UAS, used for the ice conditions reconnaissance are as follows:

Airborne time - 45 minutes
Cruise speed - 45-90 км /h
Effective range of flight control system - 3км
Flight height margins 58m – 500 м

Acceptable wind - 8 м\s

Air temperature (for optimal capacity of electrical battery) from -7°C to +30°C. Temperatures below zero decrease the endurance of battery in linear manner.

The maximum observation range of a sea surface is being calculated taking in consideration parameters of: distance of the horizon; sensors resolution capability; air transparency factors

Distance of the horizon is being calculated by formula (1).

Resolution capability of sensors has been determined experimentally in an on-line mode, with the deployed UAS equipment. The experiment has been completed in Riga Technical University Laboratory. The experimental conditions are as follows: the distances (black and white line of the 3 and 8 mm wide) have been measured by observation digital camera with the configured resolution zoom 10 X. Average effective resolution has been defined as 1'3.

The discrimination of an object by UAS is being described by the formula:

$$d = \frac{l \cdot 57{,}3^0}{\alpha}, \quad (3)$$

where l = lenght of an object, α = angular discrimination: capability to visually discriminate a distant object, expressed in degrees of arc.

More convenient measurement of an angular discrimination is being expressed in arc minutes, so formula (3) could be expressed as:

$$d = \frac{l \cdot 3438'}{\alpha} \quad (4)$$

The basic factors of the air transparency are fog, low ceiling and precipitations. Meteorological conditions, affecting the distance of a horizon, and their periodicity in Riga Bay, when the ice appearance is likely, are described in Tab.4

Table 4. Meteorological conditions in Riga Bay

Months	Average Cloud cover (0-10	Days with the precipitations	Days with the fog
January	6-7	9	3
February	6	7	3
March	6	8	3
April	5	8	2

The rectilinear UAV flight provides the square-area coverage (Fig.1) of an observed marine area, being calculated by formula:

$$S = (V \cdot \frac{t}{2}) \cdot 2D_{max} + \frac{\pi D_{max}^{2}}{2} \quad (5)$$

Where S-square area coverage; V=cruise speed of UAV flight; t=airbome time; D_{max}=maximum sweeping range, depending on the minimal values of three factors: distance of the horizon, resolution capability of sensors, air transparency factors.

In order to extract practical calculations, we should exclude the air transparency factors, due to their extreme irregular patterns, and to define parameters as follows:

H – 500м

l- 50 м

t- 0.5 hour

V- 60 км/h

D- Distance of the horizon in kilometers, if H – 500м by formula (2)

D = 79,6 км

The distance of object discrimination l = 10 м (a split in ice-plate) and α =1,3′ has been defined in accordance with the formula (4):

d = 26.4 км

Due to the assumption that $d < D$, observed square-area S is being calculated by formula (5):

S = 1094 км²

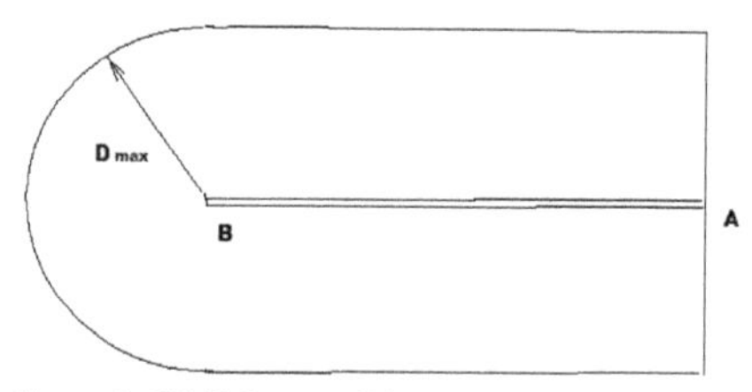

Figure 1. UAV Ice conditions surveillance in close sea areas. A- take-off and landing position. B- Position of turn. D_{max}- observation distance

4 THE EXPERIMENT

The aim of the experiment was practical effort to deploy micro class UAV in ice conditions recognizance role, as well as to check up the feasibility of its deployment methods. Secondary purpose of the experiment was to work out practically launching (take off) and landing (recovery) procedures of UAV in beach zone.

The experiment has been conducted within the weather conditions as described in the table (Table 5), and UAV's technical performance has been limited by relatively strong wind.

The final outcome of experiment is the practical proof, that micro-class UAV can be effectively used for the purposes of ice-conditions surveillance.

In first Phase of experiment there was used the method of Perspective Surveillance, when aircraft executed a flight in transversal plane of imaginary ship's general course (Fig. 2).

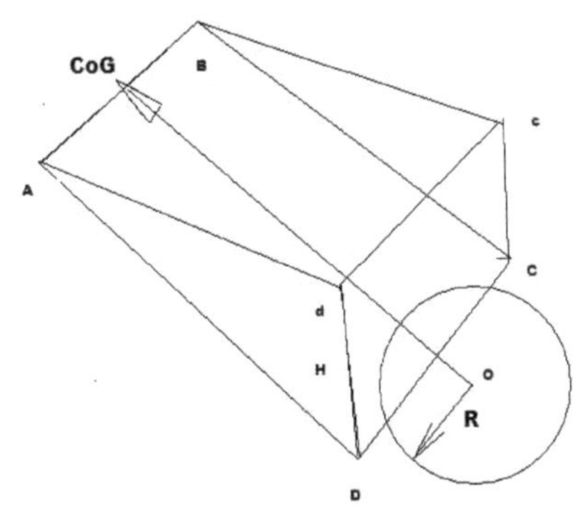

Figure 2. Perspective Surveillance. O – Position of ship; R- Area of observation from bridge; CoG- General Course of ship; dc- UAV flight path; ABCD- Area of observation from UAV

During the first flying Phase there were observed tightened ice-packs, snow drifts, and there was visually visible ice-free water space in a remote viewing perspective (Fig.4).

Figure 3. Perspective observation from UAV in Riga Bay

Second flight Phase has been executed for aim of data rectifying on ice-conditions in accordance with the Detailed Surveillance method's procedures. Overall ice structure has been estimated more detailed in visual view (Fig.5).

The procedures of UAV landing on the beach were complicated due to slanting gusty wind. Taking in consideration quite limited practical experience of flight operator to work in such the conditions, the final overall results of experiment could be assessed as successful.

Table5. Meteorology conditions during the experiment

visibility	9 km
Clouds Ceiling	1000 – 1500 feets
Winds	N-6 M/s
Temperature	-4.0^{0}C
Pressure	1031.0 mb

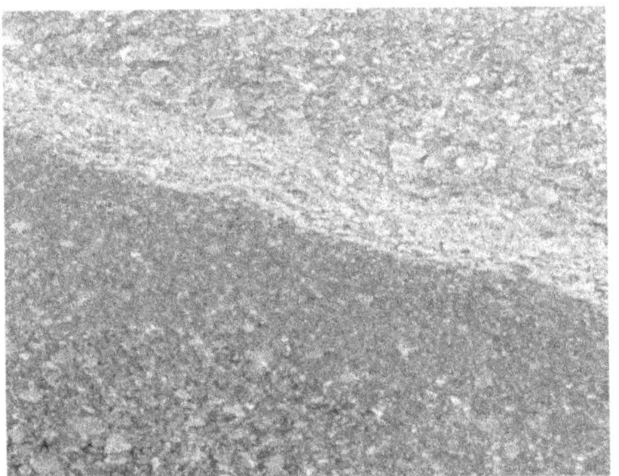

Figure 4. Method of Detailed surveillance with zoom 2x at H= 155m

Due to extreme negative temperature the battery working power has been decreased considerably, so some 59 % from overall battery's power has been spent out. Flight's average speed was 55 km/h on the 160 meters altitude. Ice congelation of the apparatus has not been observed. Gusts of wind heavily influenced the stability of flight in horizontal plane.

Figure 5. Method of Detailed surveillance with zoom 5x at H= 155m

5 ANALYSIS OF UAV DEPLOYMENT AND THE RESULTS OF AN EXPERIMENT

All data from particular UAV flight are being processed by specific software GNU GIMP (Genuinely Not Unix Image Processor) (Inoue at el. 2008), which is suitable for a variety of image manipulation tasks, including photo retouching, image composition, and image analysis. GIMP has many capabilities. It can be used as a simple paint program, an expert quality photo retouching program, an online batch processing system, a mass production image renderer, an image format converter, etc. Every visual color image can be processed and quantified by special GIMP featured capabilities of ''layering'', ''thresholds'' and ''dynamic masking". The task of processing images from UAV particular flight is to calculate the visual mutual correlations by colors, contrast and brightness of images, in order to create and support Data Base for given flight circumstances.

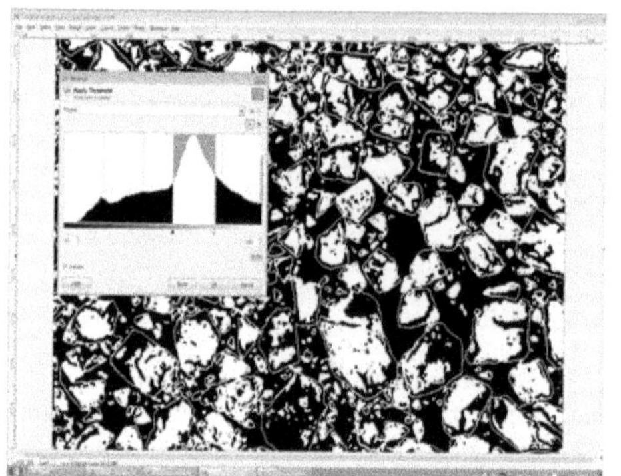

Figure 6. Example of GIMP software histogram: manually selected most clear from ice areas on image from UAV (white color areas) with the designation of threshold

6 CONCLUSIONS

The final conclusions might be deducted as follows:

1 UAVs (UASs) can be efficiently used for the shipping route planning in complex situation of ice cover; as from big-sized as from a small ship.
2 UASs do not require complicated infrastructure and many operators to operate with, thus being a cost effective project.
3 The flight patterns of UAVs depend on the size of area for observation and type of UAV (propulsion and capabilities).
4 The provisional mathematical tools can be efficiently deployed for UAVs flight and data parameters: sensors' resolution, sweeping (search) ranges and others.
5 The efficient image processing software, GNU GIMP and others can be efficiently used for the digital processing of images and creation of ice patterns data base.
6 UASs have the huge technology potential in matters of sensors, propulsion systems, data processing features, and operational ranges, to be implemented in numerous tasks at sea.

REFERENCES

Brand, E., R., Warren, D., S., Worby, P., A., Grenfell, C., T. Surface Albedo of the Antarctic Sea Ice Zone. Journal of Claimate, Volume 18, 1 September 2005. American Meteorological Society 2005. p. 3606-3622.

Frolov, I.Ye.,Mironov, Ye.U., Zubakin, G.K., Gudoshnikov, Yu.P., Yulin, A.V., Smirnov, V.G. &Buzin I.V. 2012 Ice Management – From the Concept to Realization. International Journal on Marine Navigation and Safety of Sea Transportation. Volume 6 Number 2 June 2012. p 215-220

House, D. 2006 Navigation for Masters. London.

Inoue, J., Curry, J., A., Maslanik, J., A. 2008 Application of Aerosondes to Melt-Pond Observations over Arctic Sea: Notes and Correspondence, American Meteorological Society, February 2008.

Lešinskis, I., Pavlovičs, A. 2011The Aspects of Implementation of Unmanned Aerial Vehicles for Ice Situation Awareness In Maritime Traffic. Proceedings of 15th International Conference Transport Means 2011. Kaunas. p 65-68.

Lešinskis, I., Urbahs, A.,Pavlovičs,A., Pertovs, V. 2012.Monitoring of Ice Conditions at Sea Using Micro Class Unmanned Aerial Vehicles.Proceedings of 14th International Conference „Maritime Transport and Infrastructure-2012" Riga.p 57-63.

Maslanik, J., Curry,J., Drobot, S., Holland, G. 2002 Observations of Sea Ice using a Low-cost Unpiloted Aerial Vehicle. Ice in the Environment: Proceedings of the 16th IAHR International Symposium on Ice. Dunedin, New Zealand, 2nd–6th December 2002- available in internet 2012.03.26:http://www.riverice.ualberta.ca/IAHR%20Proc/16th%20Ice%20Symp%20Dunedin%202002/Volume%203/84.pdf

Ning, L., Xie, F., Gu,W., XU,Y., HUANG,S., YUAN , S., CUI,W. , LEVY.,J. 2009 Using remote sensing to estimate sea ice thickness in the Bohai Sea, China based on ice type. International Journal of Remote Sensing Vol. 30, No. 17, 10 September 2009, p 4539–4552

Pavlovskis, G., Kautiņš, A. 1994 Ledus veidu un formu albums. LJA, Rīga ''The album of ice patterns and forms . Latvian Maritime Academy (in Latvian)'', Rīga.

Global Warming and Its Impact on Arctic Navigation: The Northern Sea Route Shipping Season 2012

E. Franckx
Faculty of Law and Criminology, Vrije Universiteit Brussel, Belgium

ABSTRACT: Major changes have taken place in the Arctic during the last couple of decades, political as well as physical. The cumulative effect of the end of the Cold War and climate change has finally made it possible for foreign ships to seriously start considering the Northeast Passage as a possible alternative to the traditional shipping routes making use of the Suez Canal or going around the Cape of Good Hope. Since 2010 foreign ships have finally started to make use of this alternative route, proposed at different time periods by Soviet or Russian authorities, to ship goods from the Atlantic to the Pacific, or *vice versa*. Despite the fact that the number of such ship movements, as well as the tonnage of cargo transported by them, have both been increasing at a very fast rate ever since, these figures are nevertheless dwarfed by the number of ships and quantity of goods using more southern routes. It remains to be seen whether the recent changes in Russian legislation, which will become operative as of the shipping season of 2013, will be able to really open up this shipping route north of the Eurasian continent.

1 INTRODUCTION

The Arctic has recently been catapulted from the back burner of international attention to the forefront of the global agenda. The dissolution of the Soviet Union, but also climate change seem to be responsible for a marked heating up, not only of its physical environment, but also of the political tensions concerning the exact legal regime to be applied there. While the universal 1982 United Nations Convention on the Law of the Sea (1982 Convention) has been accepted by the five Arctic rim countries as the applicable legal framework, on the regional level these same states have tried to ward off possible external interference as much as possible by means of the Ilulissat Declaration of 2008 (2008 Ilulissat Declaration).

Shipping has been at the heart of these developments, especially since the physical opening up of the two possible sea routes connecting the Atlantic and the Pacific Oceans. If 2007 was characterized by the most extensive summer melt ever since satellite measurements started in 1979, records have been broken every year since: 2008 was the first year both passages opened up simultaneously, 2009 broke the ice thickness record, itself surpassed again in 2010, while the sea ice extent of 2011 fell just short of the exceptional year 2007, but with far less favourable weather conditions this time. Finally 2012 broke the 2007 summer melt record.

Of the two main gate holders, Canada and Russia, the latter will in all likelihood play first fiddle in the rise of Arctic shipping. The physical characteristics of the Canadian archipelago (waters contained by a myriad of islands; thickest ice of the whole Arctic) make for exceptionally complex sea ice conditions that will constitute hazards to navigation for many years to come. It is therefore expected that the Northeast Passage, or the Northern Sea Route as the Russians prefer to call it, will develop first (Hassol 2004: 84-85).

It will be argued that the increased interest in promoting Arctic navigation, as evidenced by a net increase in the number of voyages and the tonnage of goods transported through the Northern Sea Route during the 2012 shipping season, will stimulate its growth, which since the beginning of this decade has been increasing at an impressive rate. Nevertheless, these developments are to be placed in their proper context. If one compares these figures not only with the number of ships and tonnage that used the more traditional shipping routes between the Atlantic and the Pacific Ocean, but even with the tonnage shipped through the Northern Sea Route during the Soviet period, it becomes clear that these absolute figures of growth during the last couple of years loose much of their lustre. Brief mention will finally be made of changes in Russian legislation since the summer of 2012, but which will only be applicable for the next shipping season in 2013.

This paper will only focus on the use of the Northern Sea Route as a through passage for international commerce, for it is here that its present use is different from past practices. Only those movements of goods on foreign flagged vessels starting in the Atlantic and reaching the Pacific Ocean, or *vice versa*, will be of interest, excluding that way the goods transhipped within Russia or from Russia to other countries. It will start out with a preliminary remark on terminology (Part 2) before highlighting the recent developments in ice conditions (Part 3) that lie at the heart of the increased through traffic in the Northern Sea Route (Part 4). Before drawing conclusions (Part 6), a few words will be said on the recent legislative initiatives taken by the Russian Federation to foster further international use of this route (Part 5).

2 TERMINOLOGY

Even though Soviet and Russian authors always emphasized the difference between the Northeast Passage and the Northern Sea Route, the practical importance of such distinction was less than clear for these authors normally created a *flou artistique* that would at times allow the two concepts to become very similar indeed (Franckx & Boone 2012b). It is clear that the essence of the geographical scope of the Northern Sea Route consisted of the waters of the Kara, Laptev, East Siberian and the Russian part of the Chukchi Sea up to a distance of 200 nautical miles, *i.e.* within the limits of the Russian exclusive economic zone.[77] However, whether this special regime also applied to the eastern waters of the Barents Sea and the northern waters of the Bering Sea, where somewhat similar ice conditions were to be found, remained unclear. Moreover, because navigation through such waters very much depends on local ice conditions, it was argued that at times the route followed by vessels while transiting the Northern Sea Route might well have to exceed the 200 nautical miles limit. But if all of the maritime boundaries of the Northern Sea Route concept were so permeable, to the west, the north as well as the east, the difference with the Northeast Passage became negligible in fact.

Today, some clarification has been provided by newly enacted legislation, as will be mentioned in Part 5 below, suggesting that the above boundaries have become more fixed and, as a consequence, also less permeable.

3 ICE CONDITIONS

Ever since the very exceptional summer of 2007, where the decrease in ice cover had surpassed the former lowest ever by 25 % of the former lowest ever, namely in 2005, and by almost 40 % when compared to the climatological average (Comiso et al. 2008: 6), during six consecutive years now the extent of summer ice has been the lowest since satellite measurement started in 1979. The table below place these figures in perspective.

Table 1.

Year	Minimum ice extent (km²)	Date
2007	4.17	September 18
2008	4.59	September 20
2009	5.13	September 13
2010	4.63	September 21
2011	4.33	September 11
2012	3.41	September 16
1979-2000 average	6.70	September 13
1979-2010 average	6.14	September 15

*Adapted from: Arctic Sea Ice Extent Settles at Record Seasonal Minimum, *Arctic Sea Ice News & Analysis*, 19 September 2012 (available at <http://nsidc.org/arcticseaicenews/2012/09/>).

This table also indicates that 2012 dethroned the summer of 2007 to take first place, with a further diminution of 760.000 km², *i.e.* an area larger than Germany, Poland and Austria combined. These favourable ice conditions certainly had a positive influence on the shipping season 2012 through the Northern Sea Route. The shipping season started earlier than normal. Its most northerly physical as well as legal choke point when following the coastal route, namely Vil'kitskii Strait (Franckx 1988), was devoid of fast ice about a month earlier than in 2010 for instance (An Early Opening of the Northern Sea Route? 2012). But by the end of August, only nine vessels had used the route mainly because of unfavourable ice condition in the East Siberian Sea that slowed down traffic (An Early Opening of the Northern Sea Route? 2012). The traffic picked up again towards the end of the season breaking new records, as will be seen in the next part.

4 THE 2012 SHIPPING SEASON

The idea of using the Northern Sea Route for international shipping is not new. It was for the first time suggested by the Soviet Minister of Merchant Marine in 1967 that foreign shippers could make use of this route upon payment of certain fees for the use

[77] The basis for this national regulation of navigation is to be found in Art. 234 of the 1982 Convention entitled "Ice-covered areas".

of icebreaker assistance. But this offer appears to have been silently withdrawn in order not to be seen by their Arab friends as providing an alternative shipping route in the wake of the crisis surrounding the Suez Canal (Armstrong 1970: 123). It took twenty more years before M. Gorbachev repeated a similar offer during his Murmansk speech of 1 October 1987: "Through the Arctic runs the shortest sea route from Europe to the Far East, the Pacific Ocean. I believe, depending on the evolution of the normalization of international relations, that we could open the Northern Sea Route for foreign shipping subject to the use of our icebreaker pilotage" (Izvestiia). Soon afterwards, however, the Soviet Union disappeared from the political map of the world, and once again the offer was never taken up by foreign shippers to move goods from the Atlantic to the Pacific Ocean or *vice versa*. Neither the *Tiksi* in 1989, which was a Soviet flagged vessel chartered for hard currency at the time (Franckx 1988), nor the *Astrolabe* in 1991, doing a test run without goods (Franckx 1992), nor the *Uikku* in 1997, a Finnish flagged vessel hired to move Russian oil from Murmansk to Pevek (The *M/t Uikku* Navigated the Entire Northern Sea Route in 1997), can be considered examples of foreign flagged vessels having transported goods between Asia and Europe. These ships fall consequently outside of the framework of the present article.

The first foreign flagged vessel shipping 41,000 tons of iron ore from Kirkenes, Norway, to China, without even visiting a Russian port, was the *Nordic Barents* (see Figure 1) during the shipping season 2010. This vessel is an ice-class 1A bulk carrier flying the flag of Hong Kong and owned by the Norway-based Tschudi Shipping Company. The ship was escorted by Russian nuclear icebreakers and completed the Northern Sea Route in nine days. It was the first and only foreign ship during the shipping season 2010 that transported goods from Europe to Asia (Franckx & Boone 2012a).

Figure 1.
* *Nordic Barents*. Picture provided by Marine Traffic (available at <http://www.marinetraffic.com/>).

During the shipping season 2011, two foreign vessels made such a through passage. The first in the season was the *Vladimir Tikhonov* (see Figure 2). This ice-class 1A vessel, flying the Liberian flag and belonging to the Sovcomflot Group, transported 120,000 tons of gas condensate from Honningsvaag, Norway, to Man Ta Phut in Thailand. The shipment was ordered by the Novatek Company, the largest independent gas producer in Russia. The ship was escorted by Russian nuclear icebreakers along the route and completed the Northern Sea Route in seven days and a half. This was a Suezmax tanker, which was about 100 meters longer, 20 meters wider and with a maximum draught of 6 more meters when compared to the *Nordic Barents* that made the voyage the year before (see Table 2). It set the record of having been the longest, widest and deepest vessel ever to sail the Northern Sea Route. This caused particular difficulties for the coastal route, which passes along the Sannikov Straits through the New Siberian Islands and can only accommodate ships with a draught of less than 10 meters. A new route consequently had to be followed located north of the just mentioned islands (see Figure 4). Also the beam causes a particular challenge for the Russian nuclear icebreakers. Having a beam themselves of only 30 meters, it implies that at least two icebreakers are required to break the ice in such a manner as to provide a sufficiently wide path for the ship behind.

Figure 2.
* *Vladimir Tikhonov*. Picture provided by Marine Traffic (available at <http://www.marinetraffic.com/>).

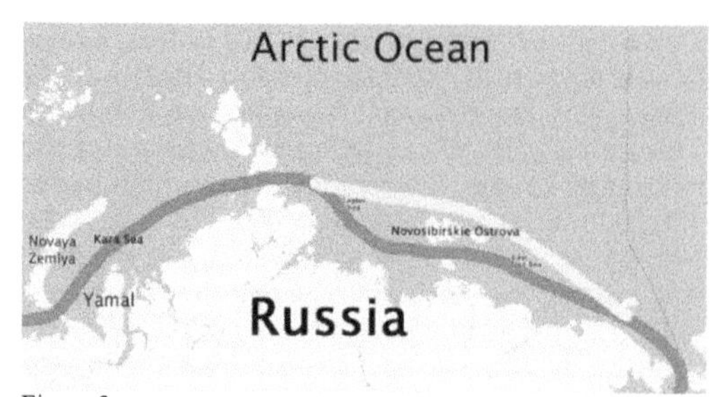

Figure 3.
* Picture provided by the Barents Observer (available at <http://barentsobserver.com/en/briefs/new-pathway-along-northern-sea-route/>).

The second vessel making the first passage through the Northern Sea Route from east to west during the shipping season 2011 was the *Perseverance* (see Figure 4). This is a Belgian owned ice-class 1A tanker flying the flag of Singapore, which sailed from Yosu, South Korea, to Le Havre in France delivering a shipment of 64,000 tons of kerosene. This shipment was in fact one of the return voyages for the multiple shipments ordered by the Novatek Company for the delivery of gas condensate from Russia to Asia. It took the vessel eight days to make the east-west transit of the Northern Sea Route in the company of Russian nuclear icebreakers. It set a record by sailing the Northern Sea Route three times in one single season, of which the first voyage of the 2011 shipping season by leaving Murmansk on 29 June of that year, again a first ever (Franckx 2013).

Figure 4.
* *Perseverance*. Picture provided by Marine Traffic (available at <http://www.marinetraffic.com/>).

This finally brings us to the most recent shipping season 2012, which only just ended. The present account is therefore only a first impression, for one will have to wait for the synthesis reports normally provided later during the year by the Russian Ministry of Transport, in charge of the Northern Sea Route, or Rosatomflot, in charge of the fleet of nuclear icebreakers. For the time being, only five transits seem to fit the present enumeration.

Figure 5.
* *Stena Poseidon*. Picture provided by Marine Traffic (available at <http://www.marinetraffic.com/>).

The first vessel to pass the Northern Sea Route in 2012 from east to west without visiting a Russian port was the *Stena Poseidon* (see Figure 5), flying the flag of Finland. The ice-class 1A vessel, which was escorted by Russian nuclear icebreakers, delivered 66,416 tons of kerosene from Yeosu, South Korea, to Provoo in Finland. On its return voyage the vessel took gas condensate on board in Mursmansk and delivered it to China, but this is of no particular interest here since it concerns a delivery from Russia to another country.

The second vessel to make a similar voyage was the *Marika* (see Figure 6), flying the flag of Norway. This is also an ice-class 1A vessel which very much resembles the *Stena Poseidon*. Like the latter, moreover, it delivered 66,462 tons of kerosene between South Korea and Finland and had a similar return voyage, which does not need to be mentioned here. The ship was escorted by Russian nuclear icebreakers.

Figure 6.
* *Marika*. Picture provided by Marine Traffic (available at <http://www.marinetraffic.com/>).

The third ship to make a similar voyage was the *Palva* (see Figure 7), flying the flag of Finland. This vessel resembles the two other ships having made previous through transits of the Northern Sea Route in 2012, not only as type of vessel (an ice-class 1A vessel of similar dimensions) but also as to the details of cargo transported (about 60,000 tons of kerosene from Yeosu to Provoo with Russian icebreaker assistance). The only distinguishing factor of importance for the present paper is that the return voyage this time went from Finland to South Korea with gas condensate. This return voyage, in other words, constitutes a fourth crossing of the Northern Sea Route that year for trade between the Atlantic and the Pacific, or *vice versa* (South Korea Fuels Finnish Aviation Via Arctic Route 2012).

Figure 7.
* *Palva*. Picture provided by Marine Traffic (available at <http://www.marinetraffic.com/>).

The last vessel to be mentioned in this listing is of a totally different nature. It concerns the *Ob River* (see Figure 8), the first liquefied natural gas (LNG) ship ever to have crossed the Northern Sea Route. The ship, flying the flag of the Marshall Islands and being slightly larger than the *Vladimir Tikhonov*, was assisted by Russian nuclear icebreakers at two occasions during the shipping season 2012. Once in ballast from Japan to France, and subsequently from Norway to Japan transporting a LNG cargo of 66.342 tons. This is so far the longest vessel having made the crossing, not only of foreign vessels making the through passage as illustrated in Table 2, but also in general. Because only the west-east crossing needs to be counted here, this brings the overall total for 2012 to five crossings.

Figure 8.
* *Ob River*. Picture provided by Marine Traffic (available at <http://www.marinetraffic.com/>).

Table 2.

Name	Length	Beam	Dwt	Max. Draught
Nordic Barents	190 m	31 m	43,732 t	10.20 m
Vladimir Tikhonov	281 m	50 m	162,397 t	16.50 m
Perseverance	228 m	32 m	73,788 t	9.50 m
Stena Poseidon	228 m	32 m	74,927 t	8.60 m
Marika	229 m	32 m	74,996 t	11.80 m
Palva	229 m	33 m	74,999 t	9.80 m
Ob River	288 m	44 m	84,682 t	11.00 m

* Chronological listing of ships having sailed the Northern Sea Route based on information provided by Marine Traffic (available at <http://www.marinetraffic.com/>).

5 NEW LEGISLATION

The increased interest in promoting Arctic navigation, as evidenced by a general analysis of the 2012 shipping season along the Eurasian continent, stands in stark contrast to the applicable Russian legal framework during that shipping season, which in essence dates back to the 1990s, a time when this route had not yet been used for international commercial trade linking the Atlantic and Pacific Oceans (Franckx 2010). This flaw has recently been remedied. First, on 28 July 2012 President Putin signed the Federal Law on the Introduction of Changes to Certain Legislative Acts of the Russian Federation Related to the Governmental Regulation of Merchant Shipping in the Water Areas of the Northern Sea Route. This federal law sets out the main principles, such as the centralization of competences in the Administration of the Northern Sea Route, increased government funding of the necessary infrastructure and, last but not least, a further clarification of the Northern Sea Route concept, which has been clearly limited in the west, excluding the Barents Sea, in the east, excluding the Bering Sea, and in the north, excluding maritime areas beyond 200 nautical miles (Franckx 2013).

Second, early 2013 the Regulations for Navigation on the Seaways of the Northern Sea Route, dating back to 1990, have been replaced by the Order of the Minister of Transport of 17 January 2013 on the Adoption of the Rules of Navigation in the Water Areas of the Northern Sea Route. It might suffice here to mention that these new rules contain no less than 74 articles compared to the 12 articles contained in the 1990 Regulations. Besides some restatements, such as the authorization requirement, the 2013 rules also contain a number of new provisions in order to further attract foreign shipping, such as the strict time limits to be respected by the Administration of the Northern Sea Route when considering requests for authorization, the obligation to justify refusals, the requirement for icebreakers to fly the Russian flag, and a schematic indication of which ships are allowed where and when in the different water areas of the Northern Sea

Route, just to name a few. Finally, as already indicated in the federal law, a further confirmation is to be found that the icebreaker fees need to be in relation to a number of listed parameters. But again, further enactments will be necessary to flesh out the details of the latter provision.

Because these novel enactments listed above will become operational for the first time during the shipping season 2013, the Russian authorities moreover officially installed the Administration of the Northern Sea Route on 28 January 2013 in Moscow. This location was preferred over Arkhangelsk and Murmansk which both wanted to host this institution (Moscow to Rule the Northern Sea Route 2013).

6 CONCLUSIONS

If one takes the official Russian statistics indicating the overall increase in the number of cargo and ships transported by means of the Northern Sea Route, one ends up with the following table:

Table 3.

Year	Cargo shipped through NSR	Voyages with cargo
2010	111,000 tons	4
2011	834,931 tons	34
2012	1,261,545 tons	46

* Based on 46 Vessels Through Northern Sea Route, *Barents Observer*, 23 November 2012 (available at <http://barentsobserver.com/en/arctic/2012/11/46-vessels-through-northern-sea-route-23-11>).

No matter how impressive this table may look at first glance, it should always be remembered that about 14,000 vessels pass through the Suez Canal each year. The same observation applies of course with respect to the tonnage transhipped. But even within the restricted framework of the Northern Sea Route, the 1.2 million tons -- be they a 53 % increase from last year's tonnage -- still do not compare to the 6 million tons shipped along the same route in 1987.

If one applies a similar breakdown specifically to foreign flagged shipping delivering cargo between Europe and Asia, or *vice versa*, as analysed in the present article, the table looks as follows:

Table 4.

Year	Cargo shipped through NSR	Voyages with cargo
2010	41,000 tons	1
2011	164,000 tons	2
2012	320,000 tons	5

* Based on research conducted by the author. The figures concerning the cargo shipped through the Northern Sea Route are only approximate.

Again, one reading of Table 4 could be that there has been a constant increase in cargo as well as in number of voyages, and this by a factor 2 or more. But at the same time the small numbers involved should make it clear that the Northern Sea Route is not yet a viable alternative which foreign shippers naturally envisage when trading between Europe and Asia. So far all the shipments moreover concerned natural resources (oil, gas and ore) found in this northern region and needed to be shipped south. The shipments of kerosene from Asia to Europe should in this respect rather be viewed as return voyages for the much more important tonnage of Russian Arctic mineral resources delivered to Asia. It remains therefore to be seen whether the recent legislative changes will finally trigger the increase of foreign flagged shipping in the Northern Sea Route that the Soviet and Russian policymakers have at times envisaged.

BIBLIOGRAPHY

Literature (in alphabetical order)

Armstrong, T. 1970. The Northern Sea Route in 1967. *Inter-Nord* 11: 123-124.

Comiso, J.C., Parkinson, C.L., Gersten, R. & Stock, L. 2008. Accelerated Decline in the Arctic Sea Ice Cover. *Geophysical Research Letters* 35 (L01703): 1-6.

Franckx, E. 1988. Non-Soviet Shipping in the Northeast Passage, and the Legal Status of Proliv Vil'kitskogo. *Polar Record* 24 (151): 269-276.

Franckx, E. 1991. New Developments in the North-east Passage. *International Journal of Estuarine and Coastal Law* 6(1): 33-40.

Franckx, E. 1992. The Soviet Maritime Arctic, Summer 1991: A Western Account. *Journal of Transnational Law and Policy* 1(1): 131-149.

Franckx, E. 2010. The Legal Regime of Navigation in the Russian Arctic. *Journal of Transnational Law and Policy* 18(2): 327-342 (available at <http://www.law.fsu.edu/journals/transnational/vol18_2/franckx.pdf>).

Franckx, E. & Boone, L. 2012a. New Developments in the Arctic: Protecting the Marine Environment from Increased Shipping. In M.H. Nordquist, J.N. Moore, A.H. Soons & H.-S. Kim (eds.), *The Law of the Sea Convention: US Accession and Globalization*. Leiden, Martinus Nijhoff Publishers: 178-205.

Franckx, E. & Boone, L. 2012b. The Northeast Passage and the Northern Sea Route: Unity in Division? In S. Choo (ed.), *The 18th International Seminar on Sea Names: Asian and European Perspectives (7-9 March 2012, Brussels, Belgium)*: 63-69 (available at <http://koreaipm.hubweb.net/seanames/files/2012_18th/Papers_18th_Seminar_on_Seanames.pdf>).

Franckx, E. 2013. The Shape of Things to Come: The Russian Federation and the Northern Sea Route in 2011, *Yearbook of Polar Law* 5. Forthcoming

Hassol, S.J. (ed.) 2004. *Impacts of Warming Arctic: Arctic Climate Impact Assessment*, Cambridge, Cambridge University Press.

Newspapers & Other On-line Publications (in chronological order)

Izvestiia, 2 October 1987, p. 1, 3, col. 7. Translation by the author.

The *M/t Uikku* Navigated the Entire Northern Sea Route in 1997, *Suomen Merimies-Unioni SMU ry, s.d.* (available at <http://www.smu.fi/in_english/communication/articles/the-m-t-uikku-navigated-the-enti/>).

An Early Opening of the Northern Sea Route?, *Arctic Sea Ice Blog*, 9 June 2012 (available at <http://neven1.typepad.com/blog/2012/06/an-early-opening-of-the-northern-sea-route.html>).

Slow Start on the Northern Sea Route , *Barents Observer*, 27 August 2012, (available at <http://barentsobserver.com/en/arctic/slow-start-northern-sea-route-27-08>).

Arctic Sea Ice Extent Settles at Record Seasonal Minimum, *Arctic Sea Ice News & Analysis*, 19 September 2012 (available at <http://nsidc.org/arcticseaicenews/2012/09/>).

South Korea Fuels Finnish Aviation Via Arctic Route, *Barents Observer*, 25 October 2012 (available at <http://barentsobserver.com/en/arctic/south-korea-fuels-finnish-aviation-arctic-route-25-10>).

46 Vessels Through Northern Sea Route, *Barents Observer*, 23 November 2012 (available at <http://barentsobserver.com/en/arctic/2012/11/46-vessels-through-northern-sea-route-23-11>).

Moscow to Rule the Northern Sea Route, *Barents Observer*, 2 January 2013 (available at <http://barentsobserver.com/en/arctic/2013/01/moscow-rule-northern-sea-route-02-01 >).

International Legal Instruments (in chronological order)

1982 Convention. United Nations Convention on the Law of the Sea. Multilateral convention, 10 December 1982, *United Nations Treaty Series*, vol. 1833, pp. 397-581. This convention entered into force on 16 November 1994 (available at <www.un.org/Depts/los/convention_agreements/texts/unclos/unclos_e.pdf>). At the time of writing there were 164 states party to the convention, as well as the European Union.

2008 Ilulissat Declaration. This declaration was adopted at Ilulissat, Greenland, by the Ministers of Foreign Affairs of Canada, Denmark, Norway, the Russian Federation and the United States of America, 28 May 2008, *International Legal Materials*, vol. 48, 2008, pp. 382-383 (available at <arctic-council.org/filearchive/Ilulissat-declaration.pdf>).

Russian Legal Enactments (in chronological order)

Federal Law on the Introduction of Changes to Certain Legislative Acts of the Russian Federation Related to the Governmental Regulation of Merchant Shipping in the Water Areas of the Northern Sea Route, 28 July 2012 (available at <http://kremlin.ru/acts/16075>). Translation by the author.

Order of the Minister of Transport on the Adoption of the Rules of Navigation in the Water Areas of the Northern Sea Route, 17 January 2013 (available at <http://www.mintrans.ru/documents/detail.php?ELEMENT_ID=19481 >). Translation by the author.

Unloading Operations on the Fast Ice in the Region of Yamal Peninsula as the Part of Transportation Operations in the Russian Western Arctic

A.A. Skutin, N.V. Kubyshkin, G.K. Zubakin & Yu.P. Gudoshnikov
State Scientific Center of the Russian Federation the Arctic and Antarctic Research Institute, St. Petersburg, Russia

ABSTRACT: Paper provides brief retrospective of cargo operations on the Yamal peninsula. It describes main stages of work, types of specialized hydrometeorological support used by AARI nowadays in practice in order to make assistance at unloading of vessels on the fast ice in winter both on eastern and western shores of the Yamal effective and to provide safety of works.

1 ICE COVER OF NATURAL WATER BODIES AS A PART OF TRANSPORTATION INFRASTRUCTURE

Ice cover of seas, lakes and rivers has always represented an obstacle to navigation, so the impact of ice on the shore, onshore and offshore constructions is the threat that must be considered when choosing the location and design of hydraulic constructions. However, people living in areas with freezing water bodies have long accustomed to use ice cover to cut the distances while traveling and transporting goods across water bodies not resorting to the help of water crafts. As vehicles evolved and in the late XIX – early XX century rail transport, wheeled transport and crawler-type vehicles came into use (the weight of which together with the cargo comprised tons and tens of tons) there appeared a need for specialized engineering support of ice crossings that would combine the knowledge of hydrometeorological regime of water bodies and technical approaches to building roads on the ice cover.

As for Russia, one of the first attempts of engineering support of complex ice crossing was winter railroad laid over the ice of Lake Baikal, with a length of 45 km, built in 1904 (Peschansky, 1967). Another interesting example of good arrangement of transportation operations on ice is the history of St. Petersburg tram which ways went over ice of the river Neva in winter periods through 1895–1910. But, without a doubt, the most impressive experience in the history of ice crossings has been "Road of Life" in Leningrad, winter route of 85 km length across the frozen Lake Ladoga, that during its operation throughout two winter seasons of 1941/42 and 1942/43 allowed to transport more than 500 000 tons of cargo, not counting people evacuated from the besieged city.

In the second half of the XX century, with the growth of the network of hydrometeorological stations in the Arctic and appearance of research stations in the Antarctic, the sea fast ice was often used to unload ships on and to deliver cargo to the shore. This experience has been taken into consideration in Soviet Union in 1970's and 80's at the elaboration of regulations and directions on unloading operations on the fast ice for vessels, hydrographic and meteorological services, organizations and expeditions. These documents, practically unchanged, are widely used nowadays in the Russian maritime practice.

2 UNLOADING OPERATIONS ON THE YAMAL PENINSULA

In order to explore gas deposits discovered on the Yamal Peninsula in 1960-70's (including giant Bovanenkovskoe field, Kharasaveiskoye gas condensate field, etc.) it was required to organize regular delivery of construction materials and equipment to the peninsula, which had no rail- and auto-roads and where the river and the coastal navigation from the nearest river port of Salekhard situated on the Ob River, is limited to 70-110 days of ice-free period.

Cargo traffic to Yamal can be significantly increased if vessels are forwarded along the Northern Sea Route. Support of icebreakers allowed to use the route Arkhangelsk / Murmansk – the west coast of the Yamal Peninsula during almost the whole year. The main limiting factor was organization of activities on unloading of vessels. During the ice-free period, unloading was carried out while vessels were at anchorage with the use of barges. However, because of the constant wave surge near the shore, the process of unloading lengthened. Aside from that, the efficiency unloading operations in summer reduced greatly due to the intensive sedimentation of sand on the cargo (including expensive large diameter pipes) in the storage sites near the coast line.

During the ice season, unloading could be performed on the fast ice, which in February, at the stage of the medium first-year ice reached isobaths of 10-20 m. This allowed to anchor vessels with a draft of 9-12 m in to the fast ice and use vehicles to transport cargo across the ice from the vessel to the shore.

Practice of winter-spring unloading operations in the site of the Yamal coast, situated between cape Kharasavey and shoal near cape Burunniy, called "Yamal operations" started back in March-April 1976 with experimental passage of diesel-electric vessel "Pavel Ponomarev" convoyed by nuclear icebreaker "Lenin". April 11, 1976, "Pavel Ponomarev" reached the recommended point and unloading on the shore across the fast ice began. It was the first time in the history of exploration of the Northern Sea Route, that Arctic navigation in the southwestern Kara Sea started so early. The very first passage proved advantages of unloading operations in winter against summer ones performed while vessels are at anchorage, by cutting usual for the last ones terms more than three times (Tchilingarov, 1979).

From the late 1970s until the mid-1990s, unloading operations on the fast ice in the Kharasavey area were performed regularly. Construction materials, machinery, equipment and fuel had been delivered to Yamal; in turn, from Yamal, gas condensate and waste metal had been transported. In the high days of Yamal operations up to 100 000 tons of cargo per winter were transported.

At that time, special hydrometeorological support of unloading operations was provided by Amderma administration for hydrometeorology with the help of specialists from Arctic and Antarctic research institution (AARI), which by that time had acquired vast experience of unloading on the fast ice at early terms in the Arctic (Novaya Zemlya, Franz Joseph Land archipelago). Hydrometeorological support of Yamal operations included: determining the optimal timing of their performance, regular monitoring of ice conditions along the route of navigation in the southwestern Kara Sea and that of the fast ice state immediately in Kharasavey area, composing long-term and short-term weather and ice forecasts, elaboration of recommendations on the choice of unloading sites with regard to safe depths and ice thickness, elaboration of recommendations on building ice roads, monitoring of ice state and roads during cargo operations. Engineering preparations during building of ice roads on the fast ice and cargo transportation across the ice were provided by the Kara Oil and Gas Expedition that possessed a premises and large machinery base in the Kharasavey settlement. In the mid-1990s, due to economic reasons Yamal unloading operations practically ceased.

3 UNLOADING ON THE FAST ICE OF THE YAMAL PENINSULA IN 2007-2012

Resumption of winter navigation to the west coast of the Yamal peninsula was caused by plans of "Gazprom" to explore the deposits of the peninsula as fast as it is possible, and in the first place – Bovanenkovskoye deposit. This task demanded to increase cargo traffic dramatically, what could have been achieved only by the growth of the sea transportation and unloading of cargo on the fast ice. Ordering parties of new Yamal operations were construction organizations – contractors of Yamal gas production department LLC "Gazprom dobycha Nadym". Initiator of unloading operations in winter 2007 was CJCS "Trest Yamalstroygazdobycha". During navigation of 2008 there were already three companies that asked for providing assistance at unloading during winter-spring period: LLC "Stroigazconsulting", CJCS "Trest Yamalstroygazdobycha" and LLC "Severgazmontazh".

Delivery of general cargo was carried out by vessels of Northern Shipping Company, that of fuel – by tankers of Murmansk Shipping Company. Conveyance of vessels in ice was performed with help of nuclear icebreakers "50 Let Pobedy", "Vaigach" and "Taimyr". Berthing of vessels into the fast ice was assisted by shallow-draft nuclear icebreakers "Vaigach" (2007), "Taimyr" (2009) and diesel-electric icebreaker "Kapitan Nikolaev" (2008). Specialized hydrometeorological support of winter-spring unloading operations was provided by AARI experts. Engineering preparations for ice roads and unloading sites were made by crews of mechanical operations of the companies-receivers of cargo.

In April–May 2007, 4 general cargo vessels with deadweight of 6 800 tons and one tanker with deadweight of 15 748 tons were discharged across the fast ice in the region of Kharasavey. In April–May 2008, five general cargo vessels and 2 tankers were discharged, in March–May 2009 –two general cargo vessels and three tankers. After 2009, winter and spring unloading operations in Kharasavey area ceased and railway Obskaya–Bovanenkovo, opened in 2010, took over most part of the cargo traffic related to Bovanenkovo deposit.

Starting 2010, OCJC "Novatek" has intensified their activity on the Yamal peninsula aimed at building LNG plant, which is meant mainly for the South Tambeyskoye gas field. As a site of the plant construction serves Sabetta – settlement in the northern part of the Gulf of Ob. In the result of this, scientific-operational group of AARI (i.e. ice experts group), supporting winter navigation, transferred their activities from the west coast of Yamal to the northeast one in to the Gulf of Ob. Ordering party of winter-spring navigation activities is affiliated company of "Novatek" – OJSC "Yamal LNG", that owns a license on exploration of the South Tambeyskoye field. In April 2011, general cargo vessel "Kapitan Danilkin" (type CA-15 Super, deadweight – 22 850 tons) of the Murmansk Shipping Company was unloaded on the fast ice in the region of the Sabetta settlement. In the spring 2012, "Kapitan Danilkin" delivered cargo to Sabetta again, though this time unloading site was moved 30 kilometers southward of the Cape Poruy due to abnormally mild winter and the lack of ice of sufficient thickness at the depths where the vessel can safely operate in the Sabetta region.

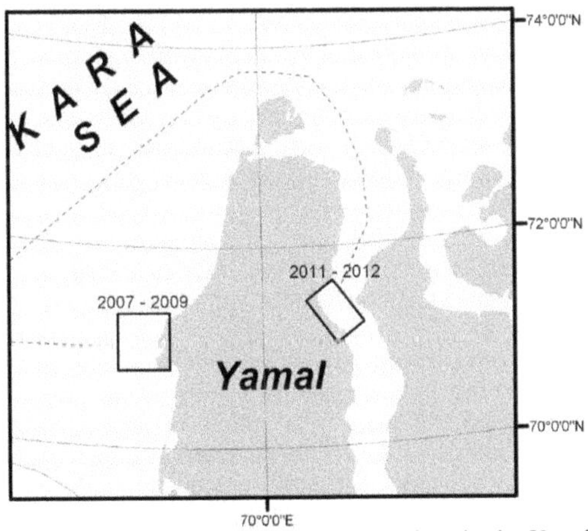

Figure 1. The scheme of the cargo operations in the Yamal peninsula area. The regions of Kharasavey (at the west coast) and Sabetta (east coast) of Yamal are marked.

During winter navigation in 2011 and 2012 berthing of vessels into the fast ice was assisted by a shallow-draft nuclear icebreaker "Vaigach". In the nearest few years, it is expected, that cargo traffic will significantly rise due to the beginning of the large-scale construction works.

4 SPECIALIZED HYDROMETEOROLOGICAL ASSISTANCE IN THE COURSE OF UNLOADING OPERATIONS ON THE FAST ICE

The success of the cargo operations involving unloading on the fast ice largely depends on the quality of hydrometeorological support. While most tasks associated with the arrangement and performance of the winter-spring navigation are solved by the logistics departments of the ordering parties with the help of transportation and building companies, specialized hydrometeorological assistance during ice navigation is provided by ice experts from of hydrometeorological organizations. In general this assistance practically didn't undergo any changes neither through 1970-90's, nor in recent years. Insignificant differences are connected with wider possibilities of satellite monitoring and improvement of methods of field observations on the ice characteristics available in the present time. Below are given main tasks of specialized assistance as they are implemented nowadays.

Monitoring of ice conditions in the region of unloading starts in a period of stable appearance of ice and includes collecting and analysis of satellite images of low resolution and meteorological data from the areas neighboring the area of interest.

Long-term forecasting of evolution of ice processes in the region of unloading is done basing on the existing information about the ice regime of the area and data of the ongoing monitoring. Long-term forecasts are used at the preliminary stage of planning of navigation and determining its optimal timing. As a rule, the preliminary forecast is composed in December – January. In subsequent months updated forecasts released if necessary.

Order, obtaining and interpretation of high resolution satellite images for the area of unloading are performed immediately before the departure of the ice experts group to the area of study. Usually radar images that don't depend on the cloud cover over the water area are ordered. High-resolution images and navigation charts allow to define depths of the sea, to which the fast ice extends, choose areas that are most promising to have level sites of unloading, where vessels can be berthed. Based on high-resolution images, preliminary conclusion on how sufficient the state of the fast ice is for the arrangement of transportation and unloading operations.

Ice reconnaissance is carried out by ice experts group after their arrival to the area of unloading. Depending on ice conditions, reconnaissance can be performed using a helicopter or during the survey with the use of snowmobiles.

In the course of reconnaissance ice conditions are charted, sites suitable for unloading are defined; routes of transporting the cargo to the shore across the ice are marked off. As for unloading sites, these are large areas of the level ice at the depths sufficient for a loaded vessel to approach (usually 10-12 m), with the ice thickness of required carrying capacity. Horizontal dimensions of the site should be sufficient for the vessel to go into it for the entire hull length and for cargo vehicles to maneuver beside the vessel. Number of unloading sites depends on how many vessels it is planned to discharge unloaded, particularly if simultaneously. According to the results of ice reconnaissance, conclusion on principal possibility of unloading vessels on the fast ice is issued.

Choice of cargo transportation routes across the ice and measurements along them. As unloading sites were defined, future ice roads are marked off. At defining the route, the main requirement ice experts group considers is a minimal length, but at the same time road should round sites high hummocks and cross ice barriers, ridges and tidal cracks at the safest points. Ice gauging survey of all the routes, including assessment of the ice thickness, snow depth and sea depth at increments of at least 100 m is performed. Number of routes depends on the possibilities of the mechanical equipment (bulldozers etc.) to clear off ice roads. For example, if the width of cleared of snow and hummocks road comprises at least 50 meters, then one ice road is enough. If the width is 20-30 meters, separate roads for traffic from the shore to the vessel and backwards are prepared. If the width is 8-10 meters, there are usually at the least 3 routes are prepared: one from the shore to the vessel for vehicles without cargo and two routes from the vessel to the shore for loaded vehicles. During the operation, roads for loaded vehicles are used by turns (e.g., once a two days each) in order to give the ice "rest" after heavy and continuous loading. Routes, measured lengthwise, are marked with flags, especially the dangerous sites (such as cracks, areas with lesser ice thickness, etc.). According to the results of measurements, ice experts group provides recommendations and directions for staff, maintaining coastal mechanical equipment on engineering preparations and arrangement of ice roads and unloading areas. Ice roads are cleared by bulldozers and then leveled by graders. Dangerous cracks are overlapped with metal or wooden decks. During all the time of operation, roads are kept clear by means of auger wheel scraper. Quality of ice roads is mainly what the speed of unloading and safety of cargo transportation depend on.

Evaluation of the carrying capacity of the ice cover. Carrying capacity of the ice cover is determined in accordance with (RD 31.41.21-90, RD 31.82.07-88) based on measurements of thickness and salinity of ice and air temperature. In order to additionally control the state of ice, visual assessment of its structure (texture of ice), measurements of temperature, density and flexural strength of ice are carried out. Ultimate and recommended loads (plus safety factor) for ice, safe staying time, safe speed and traffic intervals are calculated. To increase the carrying capacity of ice, method of natural freezing from below was used. According to this method, ice roads were cleared of snow in advance, in 0.5-1 months before the arrival of vessels, which caused intensive growth of thickness of the snowless ice.

Short-term weather forecasting. Continuous work of people and machinery on ice requires constant support with weather forecasts. To improve the quality and reliability of forecasts the ice experts group carried out a set of standard meteorological observations. Weather reports were sent daily to forecasters of AARI, who gave 3-days forecasts.

Advisory assistance to captains of icebreakers and cargo vessels on the berthing into the fast ice. As the vessel caravan approaches, ice experts group contacts vessels and gives advices to captains of icebreakers and cargo vessels on how to approach the unloading site and on all procedures at berthing the vessel for unloading. For it was easier for captains to keep a direction, temporary leading marks are made on ice of machinery and things available. In the night, lights of snowmobiles serve as leading marks. Before the start of unloading works, captain of the cargo vessel and representative of the consignee are given a set of documents stipulated by the rules of safety at performance of cargo operations on fast ice (chart of ice conditions, layout of roads, results of measurements, calculations of the ice carrying capacity, schedule of the machinery motion traffic across the ice).

5 EFFICIENCY OF THE WINTER-SPRING CARGO OPERATIONS

Winter and spring navigation involving unloading on the fast ice large amount of cargo require coordination of work of a variety of organizations and services, the precise compliance with the schedule of works. Efficiency of navigation depends on the timely delivery of cargo to the ports of shipment, prompt charter of the vessel, icebreaker support, forehand- and of high quality preparation unloading sites and ice roads.

Logistics activities start beforehand, at the early stages of the fast ice formation and one cannot be completely sure that the ice conditions will favor unloading on ice. In these conditions, well-organized specialized hydrometeorological assistance allows to reduce risks and to plan all stages of navigation more efficiently. E.g., in 2009, the fast ice in the region of Kharasavey was unexpectedly torn off in February right before the arrival of ice experts group to the region of works. Nevertheless, unloading was successfully performed on the found 10 km southward site of the fast ice that survived the storm due a large amount of stamukhas. Another example took place in 2012, when due to abnormally mild winter ice conditions in the Sabetta region didn't make it for unloading vessel on ice. After the analysis of satellite images and several ice reconnaissance flights a suitable for unloading site of the fast ice was found 30 km southward of the settlement, and unloading, though with great difficulty, was carried out.

At normal ice conditions and properly organized (beforehand) preparations for unloading operations, the speed of unloading of vessels to the shore across the fast ice comprises over 1000 tons / day, which is comparable to the speed of loading and unloading operations in the port. At shallows, both at the western and the eastern coast of Yamal 'shoulder' of cargo transportation across the ice has a length of 3-6 km. The mentioned speed was achieved by arrangement of twenty-four hour work and simultaneous usage of at least 2 vessel's cranes. To do this, it is preferable, that the ship was able to go into the fast ice so that would have been possible to unload across both boards.

Figure 2. Unloading the general cargo vessel "Ivan Ryabov" in the spring of 2008, Kharasavey area.

An important factor affecting the speed and safety of cargo operations is experienced personnel serving the coastal machinery that is skilled at works on ice and sufficient amount of special equipment (light and medium bulldozers, graders, augers, autocranes, trucks). Thus the above described experience of the cargo operations on the fast ice and approaches the AARI uses make this kind of activities possible in the severe conditions of both Arctic and Antarctic.

ACKNOWLEDGEMENTS

Authors would like to thank the reviewers of this article and Dr. Igor V. Buzin (AARI) for assistance on the final stage of the article preparation.

REFERENCES

1. Peschansky, I.S. 1967 Glaciology and ice technology. Leningrad: Hydrometeorology publishing house: 462.
2. Tchilingarov, A.N. 1979 Scientific-operational hydrometeorological support of winter-spring navigation to the Yamal peninsula. VDNKh (National Economy Achievement Exhibition) USSR. Hall "Hydrometerological service of USSR". Hydrometeoizdat: 12p.
3. Safety rules while performing the observation and works at network of Hydrometeorology National Committee. Leningrad: Hydrometeoizdat, 1983. –220p.
4. RD 31.41.21-90 Collection of work technological documentation on the performance of works at unloading vessels across fast landed ice.
5. RD 31.82.07-88 Typical instruction on the labor safety during unloading operations on the beach and on the fast ice.

Chapter 6

Inland, Sea-River, Personal and Car Navigation Systems

The Method of the Navigation Data Fusion in Inland Navigation

A. Lisaj
Maritime University of Szczecin, Poland

ABSTRACT: The paper presents the method of the navigation data fusion in the inland navigation. In the inland and sea-river vessels navigation data fusion between multisensory integration components developed reliable situations in processing data information for the needs of the navigation traffic and remote transmission. The paper described the method of the navigation data fusion on the base numerical multisensory filter. The multisensory data fusion will made the possibilities and accuracy of vessel movement in inland and sea-river navigation. The right verification of the information will enable safety in the vessel traffic.

1 INTRODUCTION

In the vessel's navigation traffic there are flowing and processing of navigational information coming from external and internal sources.

One of the important internal source of the navigation information is radar, from which come information on course, position and own speed [1,5].

One of the important external navigation data source is Inland AIS -from which information on the state vector of the watched target is transmitted, including its position and speed.

Vessel have two sources for the display of own and other vessels positions: radar and inland AIS [2,5].

Situation to have different track of the same object is unacceptable for watching officer.

Inland AIS and radar tracks should not be displayed separately. The integration radar and transponder information becomes a typical problem of track fusion. Both information sources should be used to estimation of target position and movement [4,5,6].

One of the solutions is to use multisensory Kalman filter.

2 THE METHOD OF THE RADAR AND INLAND AIS DATA FUSION

2.1 *The mathematical model of the multisensory Kalman filter*

The mathematical model of the measurement situation consists of two equations[7,8]:

1 Equation 1 – model of the process describes the watched target.
2 Equation 2 – model of the measurement describes the measurement made thereon.

$\mathbf{x}(t+1) = \mathbf{\Phi}(t)*\mathbf{x}(t)+\mathbf{w}(t)$ process model

$\mathbf{z}(t+1) = \mathbf{H}(t+1)*\mathbf{x}(t+1)+\mathbf{v}(t+1)$ measurement model

$\mathbf{x}(t/t)$	old object estimate
$\mathbf{x}(t+1/t)$	prediction of the new object state
$\mathbf{z}(t+1)$	new measurement
$\mathbf{z}(t+1/t)$	prediction of new measurement
$w(t)$	process noise
$v(t+1)$	measurement noise
$\mathbf{\Phi}(t)$	transition matrix
$\mathbf{H}(t+1)$	measurement matrix

1. Initialisation:

Φ (t) = const = **Φ**, **Q**(t) = const = **Q**, (process)

H(t+1) = const = **H**, **R**(t) = const = **R**, (measurement)

Q – covariance matrix of process noise

R - covariance matrix of measurement noise

2. Covariance matrix of object state prediction error:

P(t+1/t)= **Φ** (t)***P**(t/t)* **Φ** ^{T}(t)+ **Q**(t),

3. Kalman amplification:

K(t+1)= **P**(t+1/t)* **H**(t+1)* [**H**^{T}(t+1)* **P**(t+1/t) **H**^{T}(t+1)+**R**(t+1)]$^{-1}$

4. Covariance matrix of object state estimation error:

P(t+1/t+1) = [**I** - **K**(t+1)* **H**(t+1)] **P**(t+1/t)
I – identity matrix

5. Object state prediction:

x(t+1/t) = **Φ** (t)***x**(t/t)

6. Measuring prediction:

z(t+1/t) = **H**(t+1)***x**(t+1/t)

7. New measurement:

z(t+1)

8. Measurement prediction error:

Δz (t+1/t) = **z**(t+1) – **z**(t+1/t)

9. New estimate of object state:

x(t+1/t+1) = **x**(t+1/t) + **K**(t+1) * **Δz**(t+1/t)

10. t=t+1; leap to point 2

In the case when the object is described with a non-linear output equation, the mathematical model can be applied which is presented below.
Equation of the object's state during data fusion:

$$x(t+1) = \Phi x(t) + \Gamma w(t)$$

$$y_i(t) = H_i x(t) + v_i(t) \quad i = 1, 2, \ldots l,$$

where:
$x(t)$, $y_i(t)$ state vectors at the system's input and output,
w(t), v_i(t) disturbances of system state,
Φ, H_i, Γ fixed matrices of suitable dimensions.

We assume that:

- initial state x_0 is a random variable of normal distribution $N(0, \sigma_x)$,
- disturbance of the system's input and output state, *w(t)* i *v_i(t)*, t =1,2,....,l. is Gaussian white noise of zero mean value and covariance matrices Q and R_i.

$$E[v_i(t)v_i^T(k)] = 0, \quad i \neq j; \ \forall t, k,$$

$$E\left\{\begin{bmatrix} w(t) \\ vi(t) \end{bmatrix} \begin{bmatrix} w^T(k) & v_i^T(k) \end{bmatrix}\right\} = \begin{bmatrix} Q & 0 \\ 0 & R_i \end{bmatrix} \delta_{tk}$$

where:
E – expected value,
variables *w(k)* and *v_i(k)* are transposed vectors,
δ_{tk} –Kronecker delta function.

2.2 *Algorithm of the navigation data fusion*

Fusion of l data set of sensors consists in the fact that [7]:

$$\hat{X}_0(t \setminus t) = \overline{A}1(t)\hat{X}1(t \setminus t) + \ldots.. + \overline{A}l(t)\hat{X}l(t \setminus t)$$

where:
$A_i(t), i = 1,2,\ldots,l$, are fusion matrices,
$X_0(t \setminus t)$ are values estimated from particular data sets
1,2,.......l, indexes resulting directly from the from filtration process.

Fusion matrix is calculated from the formula[7,9]:

$$\overline{A} = \Sigma^{-1}e(e^T\Sigma^{-1}e)^{-1}$$

where:
$\overline{A} = [\overline{A}1, \overline{A}2, \ldots\ldots \overline{A}l]^T$ matrix A consists of blocks $A1, A2, \ldots\ldots Al$, each block having dimensions $n \times n$, where n is the number of variables entering the object state vector;
$e = [I_1, \ldots.. I_l]^T$ matrix e consists of blocks $I_1, \ldots.. I_l$, where each block I is a unit matrix on the main diagonal of dimensions.

$$e = \begin{bmatrix} 1_1 & 0 \\ 0 & 1_1 \\ \ldots\ldots\ldots & \\ \ddots & \\ \ldots\ldots\ldots & \\ 1_n & 0 \\ 0 & 1_n \end{bmatrix} nxl,$$

MatrixΣ is a diagonal matrix composed of diagonal matrix elements of covariance P_{ij}.

$$\Sigma = (\overline{P_{ij}})nlxnl \text{ where } \overline{P}_{ij}, i, j = 1,2,\ldots l$$

$$\Sigma = \begin{bmatrix} Pij & .... & Pij \\ ... & ... & ... \\ Pij & ... & Pij \end{bmatrix}_{lxl}$$

$$Pij = \begin{bmatrix} P_{11} & P_{12} & .... & .... & .... & P_{1n} \\ P_{21} & P_{22} & .... & .... & .... & P_{2n} \\ .... & .... & .... & .... & .... & ... \\ .... & ..... & .... & .... & .... & ... \\ .... & ..... & ..... & .... & .... & ..... \\ P_{n1} & P_{n2} & ..... & .... & .... & P_{nn} \end{bmatrix}_{nxn}$$

Covariance matrix is calculated from the formula:

$$P_{ij}(t+1\backslash t+1) = [I_n - K_i(t+1)H_i]*[\Phi P_{ij}(t\backslash t)\Phi^T + \Gamma Q \Gamma^T]*[I_n - K_j(t+1)H_j]^T$$

where:
$P_{ij}(t\backslash t)$, $i,j = 1,2....l$ are matrices of mutual error covariances after filtration.

Initial values $P_{ij}(0\backslash 0)=P_0$, where P_0 is the assessment of accuracy of data set fusion calculated from the formula:

$$P_0 = (e^T \Sigma^{-1} e)^{-1}$$

MatrixΣ is a diagonal matrix composed of diagonal matrix elements of covariance P_{ij}.

$$\Sigma = (\overline{P_{ij}})nlxnl \text{ where } \overline{P}_{ij}, i, j = 1,2,...l$$

Covariance matrix is calculated from the formula:

$$P_{ij}(t+1\backslash t+1) = [I_n - K_i(t+1)H_i]*[\Phi P_{ij}(t\backslash t)\Phi^T + \Gamma Q \Gamma^T]*[I_n - K_j(t+1)H_j]^T$$

where:
$P_{ij}(t\backslash t)$, $i,j = 1,2....l$ are matrices of mutual error covariances after filtration.

2.3 *Examples of the research results for data fusion Inland AIS and radar [1,3,5]:*

Navigational data-position x,y of the own vessel receiving from Inland AIS and radar using in the data fusion process are showing in the figures 1.

The working of the Kalman filter consists in prediction and correction.

The filter makes the prediction on the basis of the previous calculation step, and the corrective component arises as a result of multiplying the matrix of filter amplification through the difference of measurement vector and the estimated vector [5,6,8].

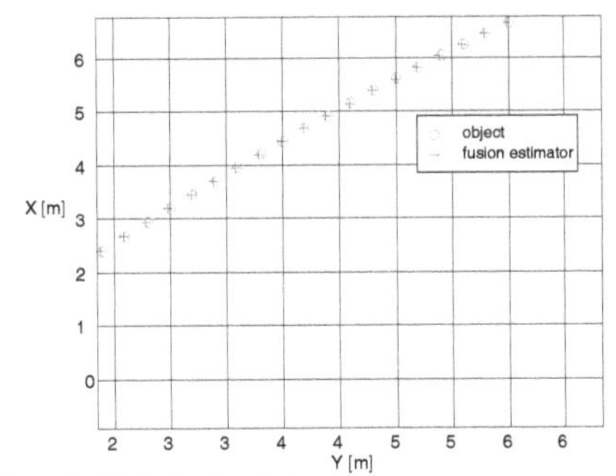

Figure 1. AIS-Radar data fusion

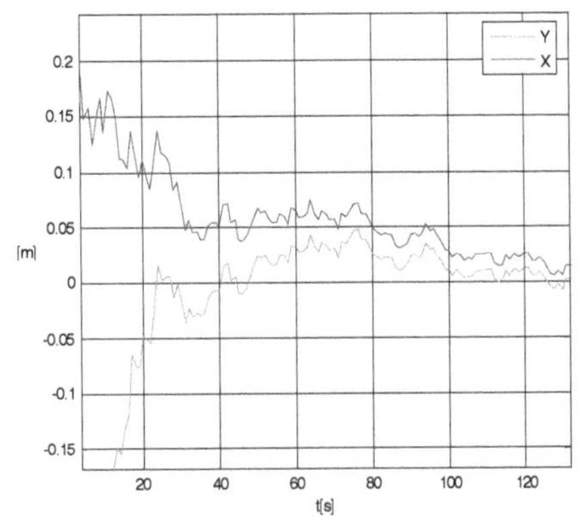

Figure 2. Estimation error of the x,y position after data fusion.

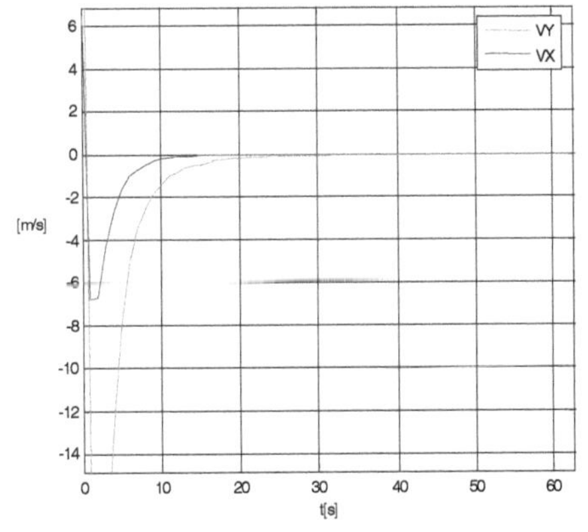

Figure 3. Estimation error of the speed vector Vx and Vy .

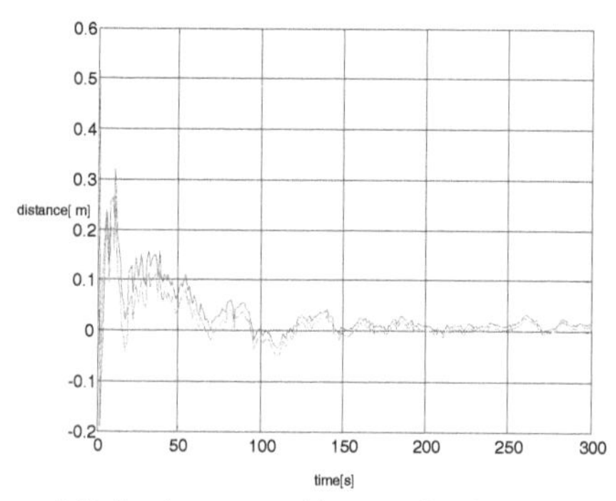

Figure 4. Estimation error multisensory data fusion filter.

3 CONCLUSIONS

The following paper presents the method of the navigation data fusion in inland navigation.

The Kalman multisensory filter is one of the solutions for the estimate of the state vector in navigation data fusion. The accuracy multisensory filter are presented on figure 4.

The inland AIS system from which information on the state vector of the watched target is transmitted, including its position and speed.

The internal source of information is radar, from which come information on course, position and own speed [2,3,5].

The important elements characterising inland AIS are:

- access to identification data of vessels beyond the range of the device – easier communication between vessels,
- data obtained on vessel traffic are updated every few seconds,
- very fast detection of other vessels' manoeuvres,
- high accuracy of data concerning position,
- wider range than the radar's,
- no false echoes,
- no occurrence of echo fading and losing them from tracking as happens with the radar.

Fast and accurate determination of the encountered object's movement vector is the key problem in inland shipping [5,7.]

The radar's limitations:

- limited accuracy;
- limited detection of echoes caused by objects' hiding (turns, bridges, flora, other objects);
- occurrence of false echoes;
- all data calculated by the radar are shown with some delay.
- the shift is also caused by the fact that the vessel's position from AIS is the position of its antenna from the GPS system, and the radar echo indicates the echo's radiolocational centre.

The fusion of data inland AIS and the radar will made the possibilities and accuracy of vessel movement in inland and sea-river navigation.

Fusion of inland AIS echoes and the radar[5,6]:

- AIS objects react faster to course alterations than in the case of the radar;
- indications of objects' positions by both devices may differ;
- AIS does not have dead sectors;
- information from an AIS receiver may be untrue due to irregular updating;
- radar echo and AIS echo are seen on the radar screen;

The right verification of the information will enable safety in vessel manouvering and will ease the whole sea-river transport process to the watch officer.

BIBLIOGRAPHIC

Chen R., Thoran F. and Ventura-Traveset J. – "Access to the EGNOS signal in Space" – GPS Solutions (2003), vol.7 No.1..

Brian Konesky, Dynamic Data Acquisition for GPS (Global Positioning System) : AutoGPS. 2006

Andrzej Lisaj, EGNOS - Przetwarzanie Informacji Nawigacyjnej, International Conference EXPOL-SHIP Szczecin 2004.

Spilker J and Parkinson B. – "Global Positioning System" – Theory and Application, USA 1996.

Stateczny A., AIS and RADAR Data Fusion for Maritime Navigation. Zeszyty Naukowe AM w Szczecinie. 2004.

Stateczny A., Nowoczesne metody nawigacji w żegludze śródlądowej z wykorzystaniem Inland ECDIS. Roczniki Geomatyki t.2/2004.

Shu-li S.: Multi-sensor optimal information fusion Kalman filters with application, Aerospace Science and Technology 2003.

Zieliński T., Cyfrowe przetwarzanie sygnałów. WKiŁ W-wa, 2005

PER Estimation of AIS in Inland Rivers based on Three Dimensional Ray Tracking

F. Ma, X.M. Chu & C.G. Liu
Intelligent Transportation System Research Center, Wuhan University of Technology
Engineering Research Center of Transportation Safety (Ministry of Education)

ABSTRACT: The Automatic Identification System (AIS) is an important maritime safety device, which is populous in inland rivers. Compared with that in open sea, the Package Error Rate (PER) of AIS in inland river has increased sharply due to its complex environment. With the help of hardware in loop simulation, it is possible to make statistical calculation on the PER under a given field strength and describe the data by quadratic rational fraction. Meanwhile, in the three dimensional software environments, the signal field strength is able to be calculated by the ray tracking method, which exhausts all the possible propagation paths, including direct way, reflection, diffractions, and the other medium attenuation matters. Beyond that, in the model, the propagation geography information in inland rivers is required to be simplified in some way, or the computation of the ray tracking is too hard to get. The paper set the Changjiang Wuhan channel as the field testing region, and all the deviations are less than 5% in sunny weather, which proves the method accurate and effective.

1 INTRODUCTION

The Automatic Identification System, or in short AIS, is a rising safety equipment in maritime navigation and positioning, and originally designed for the military use, working on the frequency 161.975 MHz and 162.025 MHz, transmitting power 12.5 W and 2 W (K. Naus et al. 2007). The AIS modulate the data at the speed of 9600 bps, and the mode is GMSK (Guassian Minimum Shift Keying), which means high efficiency (Yan Xinping et al. 2010). The AIS terminals in 20 nautical miles share the 2 frequency in the TDMA (Time Division Multiple Access), which divides one minute into 2250 pieces as the channels, supports them exchanging the name, call sign, ID number, size, speed, vector, ROT and so on. As a highly integrated wireless equipment, the AIS covers the dead zones of RADAR, making the navigation much more reliable. In 2006, the AIS is suggested to be a fundamental equipment in newly built ships by IMO (Tao Linmin et al. 2004).

When the AIS is adopted in inland-rivers, the AIS becomes a maritime management device, more than a navigation tool, but also face some new problems (Wang Feizhou et al. 2001). First, the inland river would cause observable attenuation on the AIS signal, which would make the AIS data link less reliable than that on the open sea. As well as that, the AIS data package contains the CRC checksum, but no error tolerance, which means any bit error, would make the whole package invalid. Just like the other RF system, the descending is usually caused by the attenuation of signal. So, it is necessary to analyze the relationship between the AIS signal field strength and PER, and do the calculation about the precise attenuation in certain environment, which contains the suburb area, mountains, and other dimensions.

2 THE RELATIONSHIP BETWEEN PER AND THE SIGNAL FIELD STRENGTH

In the research of RF system, the simulation is very common, which includes the software simulation and hardware-in-loop simulation. The software simulation is more popular in two areas, one is the performance study in specific hardware circuit, such as the ADS suite from Aglient, the Pspice from ORCAD, which is designed for improve the products and lower the development cost (Dzvonkovskaya A et al. 2010).

The software mentioned are very limited in the AIS study, for the AIS usually covers a large area which is more 80 km, contains varieties of dimensions, even two completely same terminals would show different performance in very small change in environment. The simulation focus on the micro view of AIS terminal is meaningless. Meanwhile, the inland-river environment stands for the complex landforms, the GSM and CDMA simulation tools are unable to describe the un-continuous transmission medium. So the AIS performance is unable to be studied by the software simulation, the hardware-in-loop simulation may be the key (Grasso R 2010).

The hardware-in-loop simulation introduces the real target into the process, which use the real transmitter as the simulating model, combines the hardware and software. In the RF studies, this kind of simulation is very common in the military areas, such as the radars, the guided weapons. The most outstanding characteristic of this simulation, is it can simulate very complex transmission process.

The AIS is very suitable to use the hardware-in-loop simulation, for it is just the system which contains very complex transmission environment.

The structure of the simulation is designed as Figure 1, which includes the software and hardware. The hardware contains the signal source, attenuation simulator, interference simulator, and receiver. And the software would control all the hardware to simulate the attenuation and interference in specific environment.

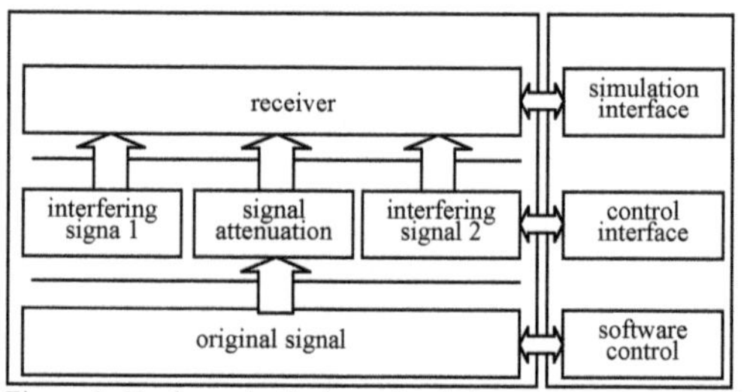

Figure 1. Hardware-in-the-loop simulation system structure of radio frequency

The real hardware-in-loop AIS simulation platform is designed as Figure 2. The signal source is developed by the STM32 MCU and CMX7042 DSP, which produces AIS baseband signal; the Aglient 8920A will modulate the baseband into AIS working frequency in setting field strength; the other 3 8920A will produce the interference signal; all the signal would be combined and sent into the AIS receiver, AIS basestation SAAB R40; and the software will do the statics on how many packages have been missed.

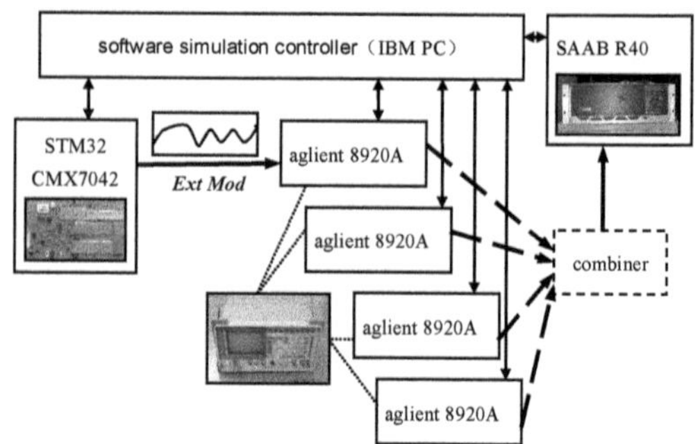

Figure 2. The real hardware-in-loop AIS simulation platform

With the help of the platform, we found that, the relationship between field strength and PER would be decribed as equation 1,

$$y=\frac{2.764\times10^{-3}\cdot x'^{2}+1.863\times10'^{-2}\cdot x'+2.821\times10^{-2}}{7.532\times10^{-3}\cdot x'^{2}+1.548\times10^{-1}\cdot x'+1} \quad (1)$$

where,

$x'=x+99$,

The x stands for the field strength, and the y stands for the PER.

3 FIELD STRENGTH ESTIMATION

3.1 *Ray tracking model*

After the discussion about the relationship between field strength and PER, the filed strength prediction becomes the problem. When in the complex environment, the ray tracking model is a very high efficient tool.

There are three conditions in wireless signal field strength prorogation, which are direct waves, diffractions, and reflections. The multiply will be the final field strength. According to this, the ray tracking method aims to do the exhaustive calculations about direct wave, diffractions, and reflections, and set up attenuation algorithmic model for each condition, to get the real field strength of test spot.

In fact, all the wireless signals propagate under the rule of Maxwell's equations. But the solution to the Maxwell's equation is very difficult to get in practical applications, as the wavelength, paths will cause heavy computations. To simplify the equations, it is required that the wavelength should be far less than the size of barriers in the ray tracking method, and we can assume that the wireless signals' wavelength are close to 0 and the mediums are homogeneous, and ignore most of the electromagnetic wave characteristics. Therefore, the field strength will be shown as below.

$$\vec{E}=\vec{E}_{0}e^{-jk\varphi} \quad (2)$$

$$\vec{H} = \vec{H}_0 e^{-jk\varphi} \quad (3)$$

where:

$k = 2\pi / \lambda$, Vacuum propagation coefficient

$\vec{E}_0 = \vec{E}_0(x, y, z)$, Spatial position real function

$\vec{H}_0 = \vec{H}_0(x, y, z)$, Spatial position real function

The $\vec{E}_0$, $\vec{H}_0$ and $k \cdot \varphi$ change with the shaft direction gradually, and the wavelength is assumed to be 0, so the change in the cross-section is negligible.

Under the assumption, it is reasonable to analyze the wireless wave reflections and infractions with the geometrical optics. When all the propagation paths turn to be known, the field strength will be the multiply of the $\vec{E}$. The reflections and diffractions will be discussed below.

3.2 *Reflection model*

On the surface of perfect medium, there is no loss in the wireless signal reflections. When the signal is propagated to the boundary of different mediums, the reflection and refraction will appear. In Figure 3., the full line stands for a boundary, the signal is transmitted from the point O, and the incident of angle is θ, the reflection angle is θ_R, refraction angle is θ_r, and the dotted line stands for the spreading of field strength.

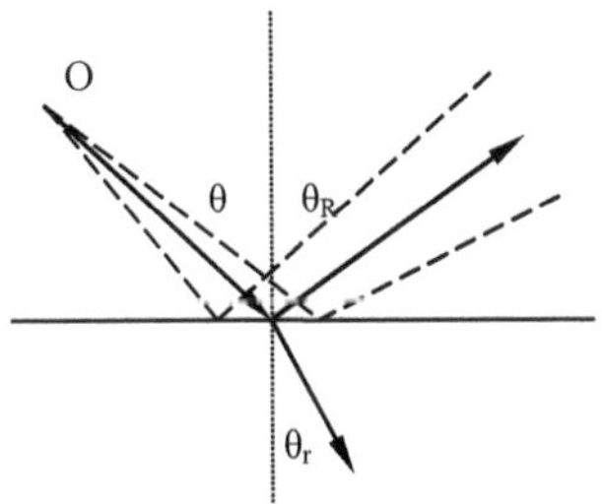

Figure 3. The reflection and refraction

In the reflection angle θ_R, the variable R will be the key to solve the field strength attenuation, where:

$$R_{\perp} = \frac{\cos\theta - \sqrt{\varepsilon_r - \sin^2\theta}}{\cos\theta + \sqrt{\varepsilon_r - \sin^2\theta}} \cdot \eta \quad (4)$$

$$R_{\parallel} = \frac{\varepsilon_r \cos\theta - \sqrt{\varepsilon_r - \sin^2\theta}}{\varepsilon_r \cos\theta + \sqrt{\varepsilon_r - \sin^2\theta}} \cdot \eta \quad (5)$$

ε_r, Relative dielectric constant
η, Reflection efficiency.

In the reflection, there will be some energy consumed by the mediums surface. For AIS, it will be the concrete; we use η to stand for the energy rate that can be reflected.

3.3 *Diffraction model*

The typical diffraction will be shown as below in Figure 4. The wireless signal will propagate to an intersection of two edges, 0 and n. The incident angle is α_1, the reflection angel is α_2, and the diffraction will spread from the intersection.

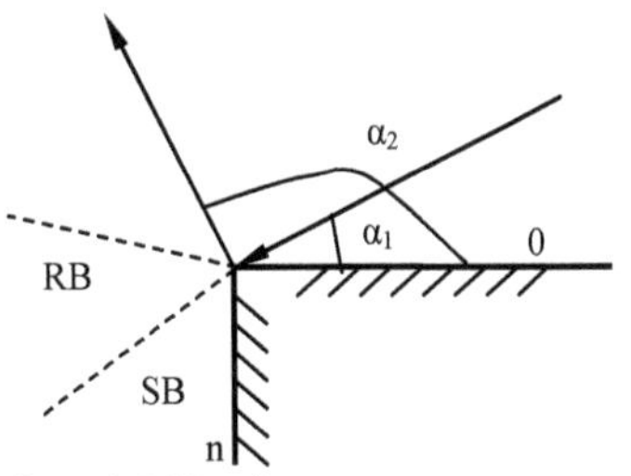

Figure 4. Diffraction

Comparing with the reflections, the diffraction is much more complex, and the analytic geometry is unable to describe. By now, the UTD theory is the only way to make the analysis, the simplification is listed as below. It is necessary to introduce the transition function $f(x)$ to describe the gradual change of the diffraction wireless signal, and use R variable to describe two different edges. According to the equations (4) and (5), there are two value of R when parallel or vertical.

The diffraction parameter D can be calculated by the heuristic equation below:

$$D = [D^{(1)} + R_0 R_n D^{(2)} + R_0 D^{(3)} + R_n D^{(4)}] \cdot \eta \quad (6)$$

η is the same meaning as the reflection. Transition function is shown as follows.

$$D^{(i)} = \frac{-e^{-j\pi/4}}{2n\sqrt{2\pi k}} ctg\gamma^{(i)} F(2k \ln^2 \sin 2\gamma^{(i)})$$

where:

$$\gamma^{(1)} = [\pi - (\alpha_2 - \alpha_1)] / 2n$$

$$\gamma^{(2)} = [\pi + (\alpha_2 - \alpha_1)] / 2n$$

$$\gamma^{(3)} = [\pi - (\alpha_2 + \alpha_1)] / 2n$$

$$\gamma^{(4)} = [\pi + (\alpha_2 + \alpha_1)] / 2n$$

$$F(X) = 2j\sqrt{X} \exp(jX) \int_{\sqrt{X}}^{\infty} \exp(-j\tau^2) d\tau ,$$

3.4 *Field strength estimation model*

Based on the equations (2) and (3), it is easy to set up the estimation model to the AIS signal field strength when we know all the propagation paths, reflection parameter R and diffraction parameter D (Hao Ruijing et al. 2007).

Direct propagation,

$$E_{LOS} = E_0 \frac{e^{-jkd}}{d} \tag{7}$$

Reflection,

$$E_R = E_0 R \frac{e^{-jk(s_1+s_2)}}{s_1 + s_2} \tag{8}$$

Diffraction,

$$E_D = \frac{E_0}{s_1} D \sqrt{\frac{s_1}{s_2(s_1+s_2)}} e^{-jk(s_1+s_2)} \tag{9}$$

where:

$k = 2\pi / \lambda$ Vacuum propagation coefficient

E0 the origin transmit power level,

d direct propagation path distance,

s_1 start point to reflections or diffractions point distance

s_2 reflections or diffractions point to receive point distance,

R reflections parameter, according to (4) or (5)

D diffractions parameter, according to (6)

when the propagation includes more than one path, it needs to use the equations above to do the summation, just as equation (10).

$$E_{total} = \sum_i E_i \tag{10}$$

The path loss will be solved as equation (11)

$$L = 20\lg\left(\frac{\lambda}{4\pi G_t G_r} \frac{|E_{total}|}{|E_0|}\right) \tag{11}$$

where:

G_r Receive antenna gain

G_t Transmit antenna gain

λ Wavelength

In fact, the wireless signal will attenuate markedly in each reflection and diffraction. Therefore, it is logical to ignore the signals which have reflected or diffracted more than specific times, to make the calculation much easier.

As the situation in Figure 5, the signal propagation between A and B in some inland river channel has a lot of paths. The banks are just like the mirrors, the edge K would form A' image of A, and the K1 and K2 edges would form B' and B'' images of B, the A' and B', B'' is reachable, so there are 2 different paths by reflection with the help of K, K1, K2. The diffraction is similar to this. With all the equations above and accumulate all the paths, the prediction of field strength became possible.

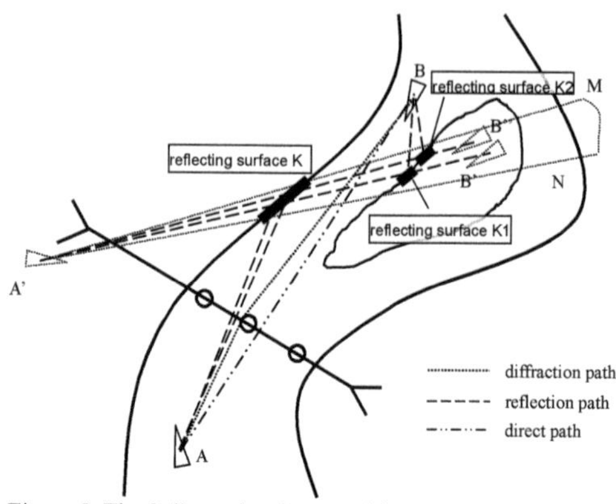

Figure 5. The 3 dimensional ray tracking

4 PER ESTIMATION AND RESULT

Based on the 2 dimensional ray tracking, the 3 dimensional ray tracking is also available, as the Figure 5, between A and B, the propagation paths would be A to C, then B, also A to E, then B. All we should do is exhausting all the paths in define channel.

The A was setting to transmitting at 1W, and the distance between A and B is 4.7 km, the testing time was 6:00 to 18:00, did the statics on every 1 hour, and the weather is sunny, foggy in the morning.

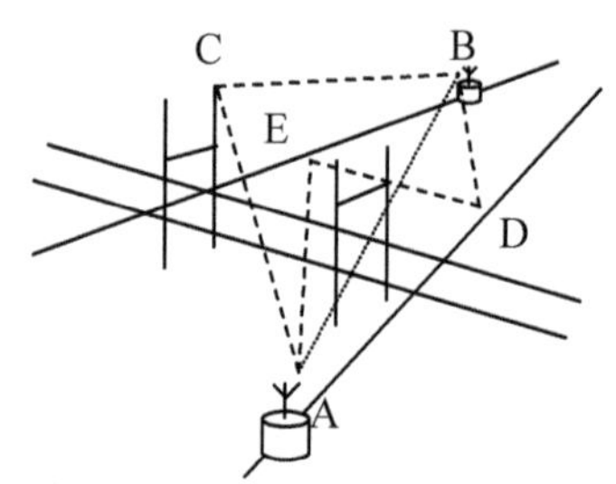

Figure 6. a dimensions structure of the field testing area

The Figure 6 shows a dimensions structure of the field testing area. And the A point field testing is shown in Figure 6. In 12 hours testing, the PER prediction and test result are shown in Table 1.

Table 1. The PER prediction and test result

Time (h)	1st	2nd	3rd	4th	5th	6th	7th	8th	9th	10th	11th	12th
Estimation				26.1								
Test (%)	45.8	44.2	37.2	27.2	27.2	26.5	28.5	26.0	25.1	29.0	21.9	21.0

In the testing, we could infer that, in the good weather, after the foggy, the estimation is very close to the test value.

5 CONCLUSION

The AIS is spreading in the inland rivers, and become an essential part of modern maritime management, the performance of AIS is an inevitable problem. The hardware-in-loop simulation shows the relationship between PER and field strength, meanwhile the three dimension ray tracking mode will calculate the attenuation, so the PER estimation become possible, and also been proved. In the further study, the ray tracking model needs to be improve to adapt the weather changing.

REFERENCES

K. Naus, A. Makar, J. Apanowicz. Usage AIS Data for Analyzing Ship's Motion Intensity. the International Journal on Marine Navigation and Safety of Sea Transportation, 200, 1(3): 237-242.

Yan Xinping, MA Feng. Key Technology of Collecting Traffic Flow on the Yangtze River in Real-Time. Navigation of China, 2010, 6(33):41-45.

Tao Limin, XU Changru. Study of the AIS Netting. Marine Technology, 2004(4) : 31-33.

Wang Feizhou.Comprehensive Assessment of AIS Performance. Navigation of China, 2001,(1) : 14-17.

Dzvonkovskaya A, Rohling H. HF Radar Performance Analysis Based on AIS Ship Information. Page:1239-1244.

Grasso R, Mirra S, Baldacci A, Horstmann J, Coffin M, Jarvis M. Performance assessment of a mathematical morphology ship detection algorithm for SAR images through comparison with AIS data,2009:602-607.

Hao Ruichang, HAO Ruidong, YANG Jian. Research on capacity planning of GSM-R radio network. Journal of Chongqing University of Posts and Telecommunications (Natural Science), 2007,19(2) : 177-180.

Analysis of River – Sea Transport in the Direction of the Danube – Black Sea and the Danube - Rhine River - River Main

S. Šoškić, Z. Đekić & M. Kresojević
Military Academy, Belgrade, Serbia

ABSTRACT: Advantages of river sea transport in the Republic of Serbia in the basin of the river Danube. Development of the river sea transport is directly related to the characteristics of waterway network which enables navigation between river basins and seas. Inland waterways in Serbia belong to the Danube navigational system. There are two navigational directions: Danube East and Danube West. River-sea transport can be directly established in these directions between the Serbian international ports on the Danube (Apatin, Novi Sad, Belgrade, Pančevo, Smederevo and Prahovo). Development of the Pan-European Corridor VII and the River Danube-Black Sea and Danube-Rhine-Main region would allow not only in Serbia but also in neighbouring states and all over Europe connection with the sea. This should be one of the strategic objectives of the Serbian state. In this paper the technical and exploitation characteristics of the river Danube will be analyzed in terms of the navigation of various river and sea vessels with different exploitation characteristics.

1 INTRODUCTION

Advantages of river transport compared to other modes of transport as well as the growing need for the development of river transport and connection with the surrounding seas raise the question of the possibility of connecting the Danube River with the sea ports in the region. Characteristics of the navigation route of the Danube River, the distance of the harbours from other important ports in the region as well as the characteristics of the ports on the Danube River in Serbia are the basis for more complex implementation of river sea transport along the pan-European corridors. In order to get a complete picture of the river sea transport systems, which can be used in direction the Danube - Black Sea and the Danube - Rhine - Main and connection with other river systems, this paper also contains the characteristics of ports on the Danube River in the Republic of Serbia from the perspective of improved infrastructure and new modern systems for classification of the system of waterways. Further more are shown boats (river-sea) which are currently operating on the direction of the Danube - Black Sea. It is also given the review of the ship that could be used for navigation conditions on the Danube. Graphics of the amount of goods transported on the Danube are given in the direction of the Danube - Black Sea and the Danube - Rhine - Main.

2 CURRENT STATUS AND CHARACTERISTICS OF WATERWAYS IN SERBIA WITH A FOCUS ON THE DANUBE

Two European transport corridors pass through Serbia – the overland Corridor X and the Corridor VII linking the 10 European countries which have exits to the navigable parts of the River Danube.

From the Hungarian border to Belgrade, the Danube flows almost parallel to the highway and railway line route in Europe, which makes the area extremely valuable and important for economic and every other development. Danube ports in the Republic of Serbia belong to the middle course of the Danube, while ports in Prahovo belongs to the sector of the lower Danube. Engineering work on cutting sharp curves, the closure of side branches and the establishment of banks were carried out downstream by entering the Danube River in Serbia at the 1433rd km of its flow all the way to the mouth of the river Drava. Downstream from the mouth of the river Drava, river bed of the Danube is much broader with several sharp bends unsuitable for two-way navigation especially: from the 1375th to the

1372nd km, at the 1370th km, at the 1331st km, from the 1286th to the 1284th km and from the 1256th to the 1252nd km. Cutting the curve near the place Mohovo, river is prone to seasonal reductions in flow by about 8 km and it is characterized by the newly formed pan rocky bottom, high speed water and small width of the waterway.

For the ships that sail upstream and pass through the Mohovski channel prevail the water level in Novi Sad, so when the water level is 0 max Draft is 160cm, while for the ships that sail downstream, at water level 0 in Vukovar max Draft is 180cm.

Characteristic shoals of the river Danube between Serbian-Hungarian border and Belgrade or the confluence of the Sava and Danube TMC 1170 are: 1429-1425 km, 1401 km, 1397-1395 km, 1391-1389 km, 1382.5 to 1381 km, 1349 -1347 km, 1287-1283 km, 1276-1274 km, 12646-1264 km, 1248-1244 km, 1241-1238 km, 1230-1228 km and 1202-1197 km. Despite the large number of shallows in this sector in line with the recommendations of the Danube Commission, the sailing conditions at low levels are fulfilled:

- The minimum width of the waterway for straight sections is 180m and 200 m for curves.
- Minimum depth for the specified width is at least 2.5 m,
- Minimum radii of curvature are 1000 m.

Sector of the Danube from Belgrade to Sipa is under the impact of slow river flow due to the construction of hydropower and navigation system Đerdap I (The Iron Gate). The Iron Gate was built at the 943rd km of the river Danube. Two stage navigation locks were also secured, 310 m long and 34 m wide, with 5 m depth lock on the right side and 4.5 m on the left.

Given that the lower Danube is typically low character, with plenty of layers which form shoals and sandbanks, navigation is regulated by several measuring points. The most suitable water meter is in place Djurdje I at 0 cm, the greatest draft permitted is 200 cm.

As we mentioned above, the most appropriate places with water meter station for analysis of the possible draft ships are in Vukovar, for the middle Danube, and Djurdje, for the lower Danube. Using the comparative analysis we will get the approximate ***number*** of ***sailing*** hours per ***day*** for river-sea vessels throughout the year. In accordance with the characteristics of the Danube, we observe periods with no restrictions to navigation and with limitations period primarily due to water levels. The first period is in the range of 1st January till 30th June when navigation is possible without restrictions and the second period July-December when it is possible to limit the draft ships. It is found in the lower Danube for water meter station in Djurdje that there are no shipping restrictions for 294 days a year. The same analysis for water meter station in Vukovar was found that there are no shipping restrictions for 237 days a year.

2.1 *The basic characteristics of the river-sea vessels*

The river-sea vessels are provided for both inland and coastal waterways conditions. In order to successfully respond to the navigation requirements, these ships must have high manoeuvrability characteristics (for mastering shallows, going through the locks etc.) and the limited amount of upgrades due to the passages through bridges in case of high water levels.

There are types of river-sea vessels:

- For waves up to 3.5 m height and the distance from the coast to 50 nautical miles,
- For waves up to 6 m height and distance from the coast to 50 nautical miles for open seas, or 100 nautical miles for closed seas.

In this paper we are going to discuss about the possibility of the use of the vessels whose basic structural and operational characteristics are shown in Table 1

Table 1. General vessel characteristics

General characteristics	1557[78]	1557[79]	1553[80]
Length (m)	114,20	132,60	119,90
Largest between lpp (m)	110,50	128,60	115,20
Maximum Draft (m)	3,50	3,65	3,50
Length (m)	13,22	16,90	13,40
Maximum Computational length (m)	13,00	16,50	13,00
Height of the hull (m)	5,50	5,50	5,80
Maximum height of the vessel (m)	15,70	15,90	13,00
Registered Capacity (t)	2700	4875	2700
Boat Speed in deep water (kn)	10,60	11,00	10,90
Storage capacity (m^3)	4300	6740	3556-fuel 1895-dry cargo
Navigation autonomy (day)	10	7	10

The analysis of the navigation conditions on the Danube near the place Djurdje showed the following facts:

- limited draft period is expected to begin on 18th September
- limited draft period is expected to end on 16th November
- limited draft period is expected to last for 67 days,
- medium possible size of vessels is TSR = 529.09 cm, with a standard deviation that is equal to s = 181.73 cm which gives the interval of possible drafts for river-sea vessels Tmin= 347.36 cm and Tmax=710.82 cm,

[78] projected ship model
[79] model of tanker ship
[80] model of combo ship

- the expected number of days per year for sailing without draft restrictions is 365-67 = 298 days

The analysis of the navigation conditions on the Danube near the middle sector of Vukovar established the following facts:

- limited draft period is expected to begin on 14th January and 14th September
- limited draft period is expected to end on 5th March and 12th December
- limited draft period is expected to last for 130 days,
- medium possible size vessels is TSR = 425.56 cm, with a standard deviation that is equal to s=150.54 cm which gives the interval of possible drafts for river-sea vessels Tmin=272.02 cm and Tmax=576.10 cm,
- the expected number of days in the year for sailing without draft restrictions is 365-130 = 235 days

(Source: The seminar held at the Traffic Engineering, University of Belgrade, 2002.)

2.2 *The Danube River*

The Danube river is an important part of the pan-European Corridor VII and strategic relationship with Europe and Eurasia, which should stimulate the development of trade, tourism and services in Serbia. Out of the total navigable length of the Danube (the total of 2580 km), 22.8% lies in the territory of Serbia. After the NATO bombing in 1999, when the bridges in Novi Sad were bombed and demolished, the navigability of this River was very limited. Apart from this, there are still some destroyed and sunk ships and unexploded landmines from the World War II in the river bed.

Sunken obstacles and bottlenecks are the real difficulties for navigation on the Danube. Bottlenecks on the Danube usually appear on the 1430th km and 1250th km. There are 18 identified bottlenecks on that section (180 km long), mostly sharp bends and narrow cross section. The most difficult bottlenecks are Apatin - where a large section is narrow, Vermelj - with sharp bend near Vermelj, Petreša and near Staklari where the radius of curvature is smaller than the absolute minimum (750m).

Table 2. Total length of navigable waterways Serbia (km) at medium water level for vessels with a total capacity (source: Transport by inland waterways, Traffic Engineering, Zoran Radmilović)

WATER WAY	Total length of Waterways (km)	Total length of the waterway (km) for vessels with following load				
		150 t	400 t	650 t	1000 t	over 1500 t
DANUBE	588	588	588	588	588	588
HS DTD	600	342	321	321	-	-
TOTAL	1188	930	909	909	588	588

On the part of the waterway of the Danube River, which passes through Serbia dimensions of the waterway are determined in relation to authoritative levels (high and low levels of the river navigation)

- width of the waterway towards B = 180 m
- width of the waterway in a curve B = 120 m
- The depth below the low-level navigation (EN or NPN[81]) h = 2.5 m
- The minimum radius of curvature R = 1000 m (Exceptionally is allowed radius of at least 750 m)
- The lowest point in the navigation under the bridge:
  - High above the navigation level H = 9.0 m[82]
  - In the secondary part of the course H = 10.0 m
- width of the navigation bridge openings at low level
  - The part of the confluence of the river Drava, at least L = 80.0 m
  - Downstream from the confluence of the Drava river, at least L = 150.0 m
- The height of the power line voltage of 110 kV
  - High above the level of the vessel B = 19 m (For every kV further than 110 kV height is increased by 1 cm)
- For phone and power low voltage lines above the high-level waterway H=16.5m
- Dimensions of the lock chamber
  - Usable length Commerce L = 150.0 m
  - Usable width of Commerce L_k = 310,0 m
  - Minimum depth at the threshold B_K = 34,0 m

The height of the waterway of the Danube, which is under the impact of the Hydroelectric Power Station "Đerdap" (aka the Iron Gate) was determined in relation to the slow water level:

- Depth below the lowest slow water level h = 3.5 m
- Height of buildings, above the waterway, was determined in relation to the most slow-water elevations.

2.3 *Danube–Tisa–Danube Canal (DTD)*

Table 3. The Navigable canal Danube-Tisa-Danube Canal (source: Transport by inland waterways, Traffic Engineering, Zoran Radmilović)

The Danube-Tisa-Danube canal	The total channel length (km) for vessels with following load			Navi-gable depth (m)	Total length (m)	Width at bottom (m)
	1000 t	500 t	200 t			
TOTAL	337,60	215,30	45,80	1,5-2,15	598,70	7-42-

[81] EN-international sign for low level sailing adopted by the Danube Commission in Budapest, or an abbreviation of the French word étiage navigable (source: Transport by inland waterways, Traffic Engineering, Zoran Radmilović)

[82] Before the destruction of bridges in Novi Sad, 1999. year (source: Transport by inland waterways, Traffic Engineering, Zoran Radmilović)

Table 4. Ports Characteristic on the river Danube in the territory of the Republic of Serbia (source: www.dunavskastrategija.rs)

PORT	km	Coast	total area (m²)	Number of pools	water depth (m)	Number of anchoring	Winter shelter
Apatin	1.401	left	46.830	2		10	yes
Bogojevo	1.367	left	-	-	-	-	-
Bačka Palanka	1.295	left	1.168.000	1	3	1	yes
Beočin	1.269	right	-	-	-	-	-
Novi Sad	1.254	left	350.000	6	4	36	yes
Belgrade	1.168	right	1.000.000	1	(min) 4	NA	yes
Pančevo	1.153	left	2.400.000	1	5	9	yes
Smederevo (old port)	1.116	right	-	1	-	2	-
Smederevo (new port)	1.111	right	31.000	1	8	2	yes
Kladovo	933	right	-	-	-	-	-
Prahovo	861	right	70.473	1	4-5	7	yes

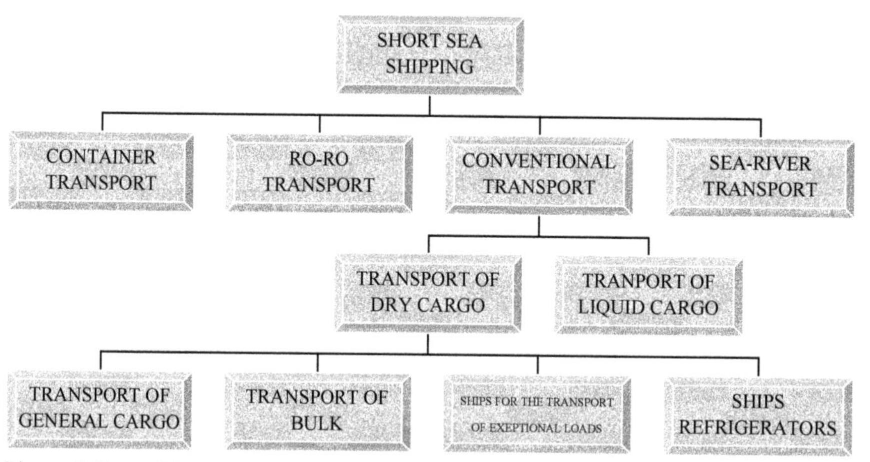

Diagram 1. Types of short sea transport (source: http://www.shortsea.nl/main/attachements/VENICE100608INTEGRATED.pdf)

Hydro system Danube-Tisa-Danube canal is a unique network that connects the Danube and the Tisa in Vojvodina. It represents a hydro system for internal water drainage, irrigation, flood control, water supply, disposal of waters, sailing, tourism, fishing and the cultivation of forests. The DTD hydro system with natural and partially reconstructed streams is 960 km long of which 600 km are navigable. Its network connects 80 villages in Vojvodina. There are 23 gates in the system, five safety constitution, followed by 15 locks and five more that are no longer in operative, the five major pumping stations and 86 bridges (64 road bridges, 21 rail bridges and a pedestrian bridge).

Most of the locks are operational, but revitalisation is more than needed. Although daily maintenance of almost every part of the DTD canal is performed and it seems that it's taken care of the locks, it is clear that a larger revitalisation is not done in recent years to the 5 locks (bottleneck Kučura Lock, Lock Vrbas, Serbian Miletić Lock, Botoš lock and Kajtašovo Lock). The General Master Plan for Inland Water Transport shows that the current situation in the Serbian water transport system requires urgent actions with a view to a faster recovery of transport, which will result in safe navigation, rehabilitation of the network of inland waterways and port development. (General Master Plan for Transport in Serbia, EAR, Belgrade, 2009)

3 CHARACTERISTICS OF PORTS ON THE DANUBE IN SERBIA

According to data from the Danube Commission on the Danube River and its tributaries there are 91 ports for commercial purposes of which 11 are in Serbia. There are eight ports of international importance on the Danube in Serbia: Apatin, Bogojevo, Bačka Palanka, Novi Sad, Belgrade[83], Pančevo, Smederevo and Prahovo.

River ports have adequate capacity for current needs, but the equipment is old and inefficient. Only Port "Belgrade" has conditions for container transport, but neither of these ports is equipped with RO-RO terminals. Port capacities, due to lack of transhipment of goods, are used on average about 30%. Only ports in Pančevo and Belgrade have **container terminals**. The basic information on the features of ports in the Republic of Serbia on the Danube River is presented in Table 4.

[83] maritime (river-sea port)

4 SHORT COASTAL TRANSPORT

A short coastal transport, "short sea shipping" means the movement of cargo and passengers by sea between ports situated in Europe in a geographical sense, or between those ports and ports situated in countries outside of Europe having a coastline on the enclosed sea borders bordering Europe.

The goal and intention of the EU is to stop the anticipated 50% increase in traffic of heavy goods vehicles, to divert part of the vehicle and marine waterways and thus to balance the division of traffic on all forms of transport.

Short coastal transport is divided into:

- container transport,
- RO-RO transport,
- sea - river transport and
- conventional transport.

4.1 *Short sea model with application to the intermodal transportation system*

Short coastal transport allows the use of different levels of transport. The largest used transportation is a combination of ship and truck. In Figure 2 is presented intermodal transport between two ports with a combination of ship-to-truck, inland-time supply and vice versa.

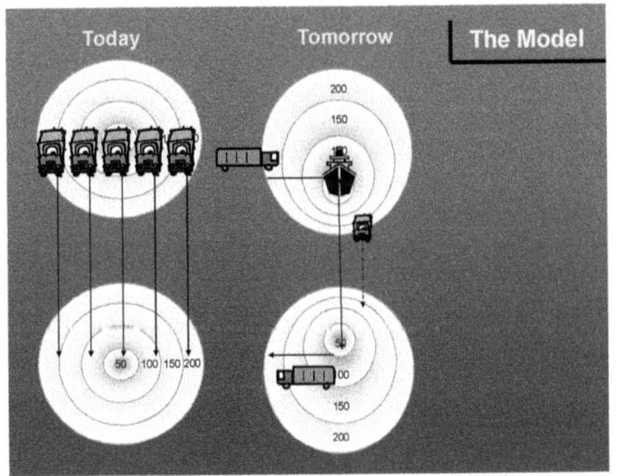

Figure 1a. Short Sea shipping (a- model), source: www.mass.gov/Agov3/docs/Short%20Sea%20Shipping.ppt

Namely, if there are two cities in which certain goods must be carried out and if the distance between these cities is 50, 100 and 200 km, 5 trucks should be hired for each distance nowadays and for each transport should be engaged manpower, time and money. It takes some time to carry the goods and return. Nowadays, goods are mostly transported in this way.

Trucks for transport to a nearby airport can be used to save time and reduce costs and then all of the goods in several tours can be transported to certain destinations where they will wait for other trucks.

Apart from the given model and its specifications, it is also presented how the model could be expanded from small (short haul) to long-distance port (long haul). Figure is related to the cities New Bradford, Bayonne and Jacksonville.

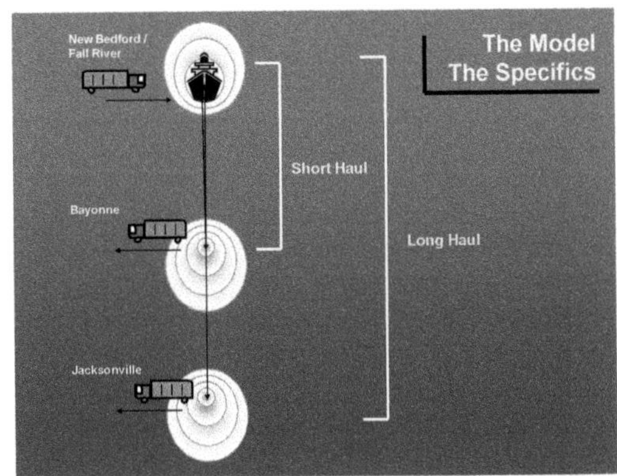

Figure 2b. Short Sea Shipping (b - of implementation of the model in the case of intermodal transport, ship-to-truck between the port), a source: www.mass.gov/Agov3/docs/Short%20Sea%20Shipping.ppt

4.2 *River and sea transport*

River and sea transport are observed and studied as a huge transportation technology system that represents the organizational and material factors focused on quality cargo transportation of certain technical means at specific cost of capital and labour.

Today, there is an increasing need for inland waterways connected with the so-called "sea highways" ("Motorway of the sea") in order to save time and achieve higher profits. Given that transport is the cheapest and that waterways volume of transport is increased in the past few years, so it is assumed that in the future this type of transport will be developed rapidly.

Main advantages of river transport on the Danube River would be the following:

- **economy**, which is reflected in the amount of transported cargo ship and lifetime of external and infrastructure costs.
- **minimum drive energy consumption**, which is reflected in the balance of power savings in ship-transported cargo units;
- **minimum amount of material for the construction of vehicles per ton of transported cargo**, which is reflected in the materials for the construction of ships;
- **safety of navigation**, which is reflected in lower traffic density in comparison to other forms of ground transportation;

- **ecologically most appropriate mode of transportation of cargo**, which is reflected in the slight percentage of water pollution and air pollution in relation to other forms of land transport in Serbia. With the fact that the ships can transport the large amount of cargo, emission of goods transported per unit weight is reduced.
- **minimum requirements for the land** is reflected in projects that would allow the occupation of the land for inland waterways.

(dark)	possible navigation by cleaning and reclamation of river flow
(white)	construction of a new system of canals and dams for transversion of basin
(grey)	needed expansion and regulation of the riverbed

Figure 3. Overview of the possible directions of river transport Danube-Morava-Vardar-Thessalonica (Source: *Open questions and perspektive cooperation with countries in Danube region*, dr. Edita Stojić-Karanovic)

The new government in Serbia has revived the idea of Charlemagne, 13 centuries old, for Europe to overcome the connectivity of rivers from northwest to southeast. Digging a canal to connect the Danube, Morava and Vardar waterway from the Mediterranean to northern Europe should be reduced by 1,200 km compared to the current route through the Black Sea. $ 17 billion is needed for this venture that the optimistic forecasts could be completed within eight years. The idea of building the channel is not new. In the second half of the 19th century a company for waterways construction was operating in the USA. One of the directors of the company was a Serb Igor Eugen Lazarević Hrebeljanović. This company has commissioned a project in Belgrade Danube - Morava - Vardar - Aegean Sea.

The first project of this channel-to-order of the company previously mentioned, was made by Nikola Stamenković, a professor at the University of Belgrade. This project determined, in a professional and creative way, the main route of canal which has survived to this day. The project has been in the Serbian language and it was made public in 1900. and a design solution is published in English 1932.

The main disadvantages that characterize the use of river transport are as follows:

- **unfavourable meteorological and hydrological conditions (ice, wind, fog, low and high water) and the overlap of these phenomena** related to seasonality (the appearance of high and low water levels, ice stagnation and strong winds) and permanent faults (changes in the sector of navigation);
- **limited geographic spread**, which is related to the natural spatial distribution and trends of inland waterways;
- **level of quality of transport services**, which is reflected in the confidence of transportation, speed, ability to transport "door to door", security, flexibility, durability and energy efficiency;

River and sea transport can be divided in several ways:

- the mode of transportation,
- the type of load,
- the amount of cargo and more.

The most important is the way of transportation. River and sea transport can be divided in:

- river-sea transportation of cargo handling from the sea in boats and vice versa, in special places or ports at the mouth and
- river-marine system without handling of cargo at the mouths of the rivers into the sea or ocean.

4.3 *The river-sea system with cargo handling from the sea in river vessels and vice versa, in special places or ports at the confluence*

River-sea transportation of cargo handling at the confluence of three main technologies include following elements:

- Sea transport ship
- River transport ship
- Harbourmaster funds along with reloading equipment and fleet moorings

Comparative analysis of navigational characteristics of the Danube and structural characteristics of the river of sea vessels it was determined that only the depth of its sectors and in certain parts of the year, appears as a limiting factor. It can also be found that part of the Danube River downstream of Mohovski channel has significantly better navigational capabilities in regards to the part of the river upstream of Mohovski channel. One concludes that sea vessels can sail loaded in the ports on the Danube in the Republic of Serbia (Prahovo, Smederevo, Belgrade Pančevo and Novi Sad) wit maximum draft and the biggest registered payload for nearly 300 days a year.

Modern ships for basic technology transport can be divided into:

- ship to transport pallets (pallet carrier);
- <u>ship to transport containers (container ship);</u>
- <u>ship for horizontal transhipment Ro-Ro (Roll-on/Roll-of ship);</u>
- multi-purpose vessel (Lo-Lo/multi purpose Lo-Lo ship) and Ro-Lo
- tank barges ship (barge carriers)

5 INTEGRATED RIVER-SEA TRANSPORT SYSTEM

5.1 *River transport ship*

Connecting maritime capacity of the river resources in a single integrated transport system is achieved with transport barges technology of larger overseas ships.

Technology of transport barges for inland water transport, with sea transport services, has led to the merger with overseas shipments with inland waterway transport in a single integrated intercontinental scale. Technical basis of river-sea intermodal transportation consists of three elements:
- cargo barges as a unit;
- mother ship (carrier barges) and
- reloading equipment and accessories.

6 CASE STUDY OF MERCHANTABILITY RIVER-MARITIME TRASPORT

6.1 *The "Ilde"*

Ilde project (Improvement of intermodal links to the Danube Estuary) is a contemporary project that aims to promote, develop cooperation and increase prosperity countries in the Danube basin (Bulgaria, Romania, Hungary and Serbia).

The project is designed to provide safe navigation and a larger vision of connecting cities in the Danube basin.

According to research navigable vehicles cause the lowest external costs of only 0,28 € per t-km and the largest external costs of up to 2,00 € per t-km which is primarily due to the large number of traffic accidents, noise and high concentration of CO2 :

In addition to the above costs in analyzing intermodal routes the cost of infrastructure (construction and maintenance) are used.

Already known values are taken for analysis and when these distances are included the following values are obtained. Values are calculated for the transport of goods of 300t. The values are obtained when the cost of the graph is multiplied by 3 (100t * 3 * distance).

Table 5 provides the exact look on a savings of goods transport by inland waterways to other modes of transport. However, if the distance is increased, cost of transport inland waterways will also be increased which is not in order.

Internal costs for transportation of goods by truck is reflected in the following parameters:
- Fuel (40L - 100km highway, 1litre is 1 €)
- salary workers (250 € for each)
- fee (about € 100 per country)
- 5t truck transports,
- the train carrying 20 tons and
- boat carrying 300t

For the transport of goods of 300t the following costs are:

For our country, river traffic on the Danube river is interesting, in the shorter distance with the transhipment port in terminals. For such short distances calculations are as follows.

For a distance of 1100 km the time of loading and unloading (with 2 container cranes) is estimated:

Table 5. Cost of external costs for the above-mentioned distance

GENT-BUDIMPEŠTA			GENT - BELGRADE			GENT – KONSTANCA			
ROAD	RAIL	IWW	ROAD	RAIL	IWW	ROAD	RAIL	IWW	SSS
10.370€	10.566€	4319€	13017€	12366€	5001€	17970€	17848€	7700€	10.000€

Table 6. Total costs for the above-mentioned distance (source: *www.ildeproject.com*)

GENT-BUDIMPEŠTA			GENT – BELGRADE			GENT – KONSTANCA			
ROAD	RAIL	IWW	ROAD	RAIL	IWW	ROAD	RAIL	IWW	SSS
26370€	12166€	5319€	29217€	13966€	6101€	34970€	19448€	9200€	13000€

Table 7. Calculation of time for loading and unloading at terminals, (takes a day to carry 300 TEU) (source: *www.ildeproject.com*)

LOADING AND UNLOADING	TIME
the unloading	9 hours
the loading	9 hours
Total time	18 hours

The result for a week carriage of goods by inland is as follows:

Table 8. Calculation of transport and time costs

TOTAL AMOUNT OF TRANSPORT	2100 TEU
TRANSPORTATION COSTS	250 € per one TEU
TIME	108 hours
TRANSFER TIME IN PORTS	18 hours

Generally, container volumes in Serbia are very small. There are only few Serbian exports and the degree of containerization in Serbia is generally very low.

Since 2001, exports have been growing strongly. Despite this growth, trade between Flanders and Serbia remains modestly. The industrial sectors with good development potential for Flemish-Serbian trade are:

- Agribusiness
- Manufactured goods
- Automotive equipment and components (Cars, tractors, tires) and
- Wood products

Due to the lack of interviews in Serbia, it is difficult to draw a potential map for Flemish-Serbian trade. But it seems that, similar to the results gained by interviewed companies in Hungary, the most valuable connections between Flanders and Serbia seem to be container transport by SSS between Flanders and the Adriatic ports, as well as Bulk and Container transport by SSS between Flanders and the Adriatic ports.

The Croatian and Slovenian ports (Koper, Rijeka) would capture the most part of the business if Short Sea Shipping connections would be improved between. Currently sea connections exist between Antwerp and Bar (Montenegro) once a week and are ensured by Maersk and Hapag Lloyd. Regarding rail and IWT connections, it is expected that container terminals in Hungary (Budapest, Szeged, Port of Baja) and Croatia (Zagreb) would profit from a growth in Flemish-Serbian trade because of the limited amount of the Serbian terminal infrastructure.

It appears that shippers prefer the ports of Koper, Rijeka, and Thessalonica rather than port of Bar. For Rijeka, they seem to use mainly trucks, whereas Koper and Bar have good connection with Belgrade by rail that can be improved and increased in the future. However, interesting information and cost indications could be gained regarding an inland navigation shuttle service on the Danube between Constanta and Belgrade. This service is operated by the ILDE project partner from Serbia, Yugoagent.

Yugoagent operates two barges between Constanta and Belgrade. Three more barges can be deployed if extra capacity is needed. Normally one sailing per week in both directions is organised. The barges are classic bulk barges (not cellular) with a capacity from 1300 up to 2000 tons or 80 TEU. Flat-rack containers for heavy and bulky semi-finished goods and out of gauge cargo are used to transport building materials.

Each year an average of 2.000 full TEUs and 1.000 empty TEUs are transported by barge. Imports (Constanta – Belgrade) accounts for 60% and Exports (Belgrade – Constanta) accounts for the remaining 40% of the average total freight. In general the shuttle runs only between Belgrade and Constanta. However, upon request, the ports of Pancevo, Novi-Sad in Serbia and the ports of Ruse and Turnu Severin in Bulgaria can be served as well.

Table 9. Barge Service Constanta – Belgrade (Source: Yugoagent, 2007)

Constanta/Belgrade	20′container	40′container
THC	€ 75.00	€ 130.00
Customs Transit	€ 50.00	€ 50.00
Full container	€ 300.00	€ 430.00
Empty return	€ 50.00	€ 100.00
Documents on border	€ 20.00	€ 20.00
THC Belgrade	€ 65.00	€ 85.00
Total	€ 560.00	€ 815.00

7 CONCLUSION:

According the statistics water transport in Serbia is only 4.7% of the total traffic which shows that we are much behind the EU countries, where the percentage of river transport in relation to the total traffic is 15%. Table 10 presents a comparative review of transport in the period from 2000 to 2006. year.

Table 10. Comparative review of transport (source: Economic Bulletin)

COMPARATIVE REVIEW OF TRANSPORT			
YEAR	RIVER (000 t)	ROAD (000 t)	RAILWAY in stations (000 t)
2000.	3 729	3 900	14 146
2001.	3 609	3 300	11 839
2002.	3 796	3 200	11 947
2003.	2 664	2 400	12 879
2004.	3 295	1 800	14 513
2005.	6 360	3 100	14 219
I-VI 2006.	2 085	1 519	6 680

In any case, there are many combined river transportation options, especially when it comes to the application of modern technology of combined transport, as well as current information-management technologies that are already in use worldwide.

Today only Ukrainian shipping company operates on the inland waterways. The possibility of introducing river-sea transport of goods and combined transport system increase the possibility of introduction of new ships that would sail on the inland waterways.

River and sea transport on the Danube, in accordance with the guidelines of the development of river-sea transport, tend to further develop and improve services. Unfortunately, the expected volume of use of this form of transport is difficult to predict given the economic development of the country. After the renovation of ports and port harbor it can be expected an increase in transport on the Danube river. In consideration of the port on the Danube in the Republic of Serbia (Prahovo, Smederevo, Belgrade, Pančevo and Novi Sad) fluvial marine ships can sail almost 300 days a year loaded with a maximum registrated load capacity.

REFERENCES

[1] „River - maritime traffic opportunity development on route Danube-sea“, Zoran Radmilovic, Vladeta Čolić, Jasna Muškatirović, Slobodan Domandžić, Zlatko Hrle, Radovan Zobenica, Vladimir Škiljaica, Katarina Vukadinović, Faculty of transport and traffic engineering, University of Belgrade, 1998.

[2] „Traffic od inland waterways“, dr. Zoran Radmilović, Faculty of transport and traffic engineering, University of Belgrade 2007.

[3] „Statistical pocketbook of Serbia“, Statistical Office of the Republic of Serbia, Belgrade 2004.

[4] „Statistical pocketbook of Serbia“, Statistical Office of the Republic of Serbia, Belgrade 2009.

[5] „Statistical pocketbook of Serbia“, Statistical Office of the Republic of Serbia, Belgrade 2012.

[6] „Ports Perfomanse in European Intermodal Transport, A comparative study of the development in two intermodal notes Port of Rostock“, Master program, Ulrike Reimer, 2009.

[7] „Container Terminals in river ports“, Milosav Georgijević, Nenad Zrnić, Belgrade 2006.

[8] „Sea - river shipping competitiveness ans its geographical market area for the Rhône-Saône corridor“, Lopez Charles, Journal of Transport Geography, Lyon 2006

[9] „National and Regional Transportation Plans“, Inland Navigation Europe Vicksburg 2002.

[10] „Inland waterway Transport of containerized cargo: from infancy to a fully - fledged transport mode“, Theo Notteboom, Spain 2007.

[11] „An innovative approach for sustannable intermodal transport“, Cumhur Atilgan, October 2005., Istanbul

[12] „Short sea shipping opportunities for the Pyrenean cargo flows“, F. Xavier Martinez and J. Olivella

[13] „The Danube region of Serbia“, Milica Vračarić, Novi Sad 2009.

[14] „LNG as fuel for ships“ in short sea shipping“, Norwegian marine technology research institute

[15] „The role of short sea shipping in European vehicle logistics“, Short sea conference, Brudges 2007, Paul Kyprianov, Grimaldi Group

[16] „Intermodeship, Future Market Opportunities“, Basel 2006., 1°Rromit cluster workshop

[17] „A multicriteria redesign of the midship section of an intermodal RO-RO ship“, Alan Klanac, Finland 2004.

[18] „The TSL(trans-sea-lifter) system:new aspects in SSS“, competitive short sea and river sea transoptation technology, 2005., Hemann J. Jansson, Navtec

[19] „Development of integrated door to door shortsea services“, Sander van 't Verlaat

[20] „Open questions and perspektive cooperation with countries in Danube region “, dr. Edita Stojić-Karanovic, Scientific work, Belgrade 2002,

[21] www.ildeproject.com

[22] www.dunavskastrategija.rs

Study of the Usage of Car Navigation System and Navigational Information to Assist Coastal Navigational Safety

S. Shiotani
Kobe University, Organization of Advanced Science and Technology, Japan

S. Ryu & X. Gao
Kobe University, Maritime Sciences, Japan

ABSTRACT: A variety of information is important to ensure safe navigation and prevent marine disasters. Collisions are the primary cause of marine disasters, particularly in congested seaports. Marine disasters tend to occur in coastal areas, particularly in shallow water. In these regions, effective provision of navigational information is extremely important. Our objective is to propose a methodology integrating a car navigation system using geographical information system (GIS) with traditional marine navigation systems to prevent such disasters.

1 INTRODUCTION

Over time, ship hull structures and motor functions have been improved and various advanced navigation devices have been developed to ensure the safety of marine vessels (Imazu et al. 2003). However, according to statistics from Japan Coast Guard, the number of marine disasters involving vessels has not declined significantly during the past 10 years (Japan Coast Guard 2007). Moreover, the increasing number of vessels may result in an increase in the number of marine disasters.

The authors have conducted a study of effective and appropriate provision of a variety of navigational information with the aim of improving the ease of navigation and the safety of marine travel. Navigational information currently being utilized is based on paper charts or an electronic chart display and in-formation system(Wakabayashi et al. 2002). Our methodology involves a realistic display of the land-scape that ship operators would see from the bridge and overlaying a variety of critical navigational in-formation onto this display (Shiotani et al. 2011; 2011). Our methodology is based on a two- and three- dimensional geographical information system (GIS) (Kawasaki 2006; Sadohara 2005). A three-dimensional chart display is a new concept for the provision of navigational information. Car navigation systems already utilize three-dimensional displays to assist drivers; however, no such device has been sufficiently developed for marine traffic.

The objective of this study is to propose a new and improved navigational information provision methodology to supplement traditional charts using a car navigation system. In recent years, car navigation systems have been developing rapidly and their performance has improved dramatically. We aim to determine the effectiveness and the problems associated with utilizing car navigation systems for marine vessels and to further develop the provision of navigational information.

2 CAR NAVIGATION SYSTEM USAGE

2.1 *Sample vessel and test device*

In our study, a car navigation system, portable automatic identification system (AIS), and video camera were loaded on a small vessel to investigate the performance and effectiveness of our proposed methodology.

The AIS was used to capture the position of the vessel and the video camera filmed the adjacent land space.

Muko Maru is a training vessel that belongs to Kobe University's Graduate School of Maritime Sciences. Figure 1 shows the Muko Maru, and Table 1 indicates the boat's primary specifications. The Muko Maru is a pleasure craft and is usually used for operational training or for surveying the coastal area.

Figure 1. Sample Vessel "Muko Maru"

Figure 2 shows the car navigation system. We used Sanyo's Gorilla SSD portable navigation system (NV-SD760FT).

Figure 3 shows the simplified AIS display system. We stored the received signals from the antenna on a computer and used this information to determine the rhumb line.

Table 1. Primary specifications of Muko Maru

Item	Mukomaru
Lpp (m)	9.33
B (m)	2.54
D (m)	0.89
Δ (ton)	3.4
V (knot)	30
Pw (HP)	270

Figure 2. Car navigation system used for our test

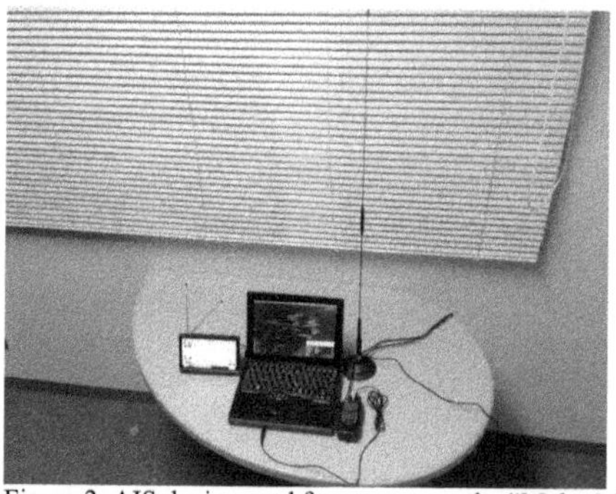

Figure 3. AIS device used for our test on the "Muko Maru"

2.2 *Car navigation system performance on the road*

To understand the car navigation system's performance and basic functions, we deployed the system in a vehicle and performed a road test.

Figure 4 (a) shows the car navigation system's two-dimensional display. The triangle indicates the location and direction of the vehicle. The display indicates detailed building shape and also provides the names of primary buildings. The name of the road is displayed in the lower left corner of the screen. The inset shows the route options at the next intersection with lights. In addition, this type of car navigation system provides other helpful information such as the location of bridges and various geographical features.

Figure 4 (b) shows the car navigation system's three-dimensional display. The primary buildings that can be seen from the driver's seat are displayed, which makes checking the vehicle's current location easier. The identification of bus stops also facilitates determining the vehicle's location. However, the buildings are shown as rectangular parallelepipeds and their colors, windows, and detailed shapes are obscure and not realistic. If the vehicle is on a road that is familiar to the driver, the display would be understandable. However, if it is an unfamiliar road, the obscure information would be less understandable. Therefore, it is assumed that more realistic and more detailed information would be necessary in unfamiliar environments.

(a) Two-dimensional display

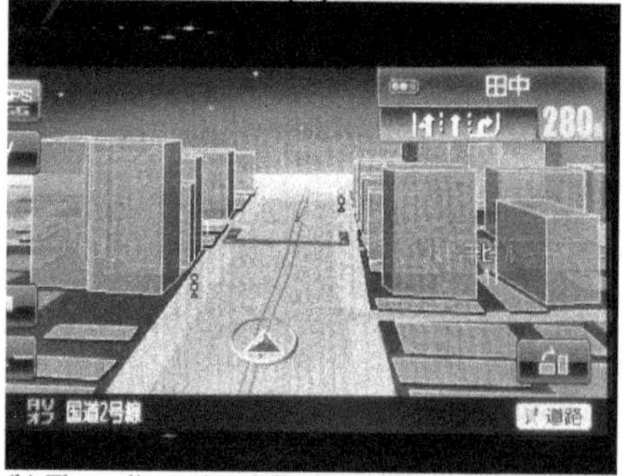

(b) Three-dimensional display

Figure 4. Generic car navigation system displays

There is clearly room for improvement of car navigation systems. Furthermore, two- and three-dimensional information displays are highly useful features for car navigation systems.

3 MARINE NAVIGATION SYSTEM USING A CAR NAVIGATION SYSTEM

To utilize car navigation technology for the sea, it must be limited to a certain distance from the coast.

It is most effective when utilized at very close proximity to the coast. More specifically, it is suitable for sailing in areas where vessel congestion is high and the geography is complicated such as a bay or channel.

Figure 5 shows the Muko Maru's traced route. We overlaid the ship location data from the AIS onto a satellite image from Google Earth. The Muko Maru departed from the pond at the Fukae Campus of Kobe University's Graduate School of Maritime Sciences and sailed westward along the coast. The ship made a U-turn at Port Island in Kobe Port and followed the same course east. After passing Fukae, it sailed along the Ashiya Canal and made a U-turn at the mouth of the Shukugawa River prior to returning.

In Figure 5, Rokko Island is in the center, Port Island is in the lower left corner, and the mouth of the Shukugawa River is in the upper right corner.

The Muko Maru sailed at approximately 6 knots, with a rotational frequency of approximately 1,000 rpm. Because the Muko Maru is a pleasure boat, it is only capable of travelling at a maximum of 30 knots. At a higher speed, there was significant vertical motion of porpoising that made it difficult to record stable images.

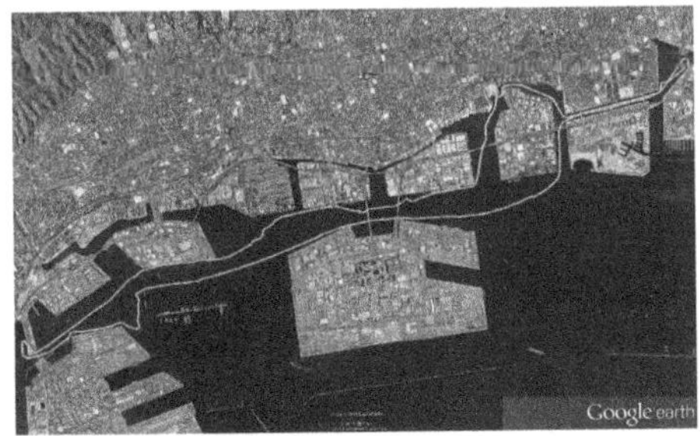

Figure 5. Route of the Muko Maru

Figures 6 (a) and (b) show still images from the video taken while sailing near Rokko Bridge. Figure 6 (a) shows Rokko Bridge and Figure 6 (b) shows waterside warehouses. Shipping/transportation companies maintain docks on both sides of Rokko Bridge. The width of the waterway is approximately 300 m.

(a) Rokko Bridge to Rokko Island

(b) Waterside warehouses

Figure 6. Video images of the coastal area around Rokko Bridge

Figure 7 shows the images displayed on the car navigation system while sailing close to Rokko Bridge. The triangle in the bottom center indicates the location of the ship. The thin line coming out of the triangle indicates the sailing course. Figures 7 (a) and (b) show the car navigation system's two-dimensional display and three-dimensional display around Rokko Bridge, respectively. In addition, the names of the bridges and primary locations on land are shown, which would make it easier for ship operators to determine their current location. In addition, the coastal area is clearly indicated. Therefore, the geographical features are clearly represented, enabling the ship operators to recognize the immediate sailing environment. Although Figure 7 (a) indicates primary buildings in several locations, the total number of buildings is less and many buildings are omitted, which is similar to the case in Figure 4 (b). We attributed this to a general lack of public roadways in the area, and therefore, detailed information would not be relevant to a car-based navigation system.

Furthermore, Rokko Bridge is identified as a road shown in Figure 7 (b). For a vehicle operator driving over a bridge, the detailed appearance of the bridge is not always necessary information; however, for ship operators, bridges and piers are important navigational details.

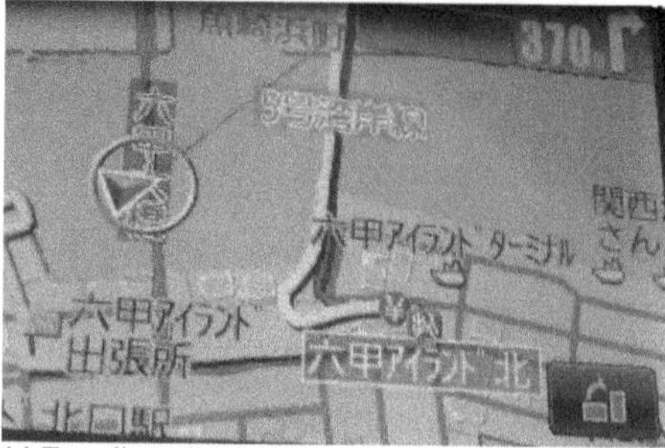

(a) Two-dimensional display

(b) Three-dimensional display
Figure 7. Displays on the car navigation system around Rokko Bridge

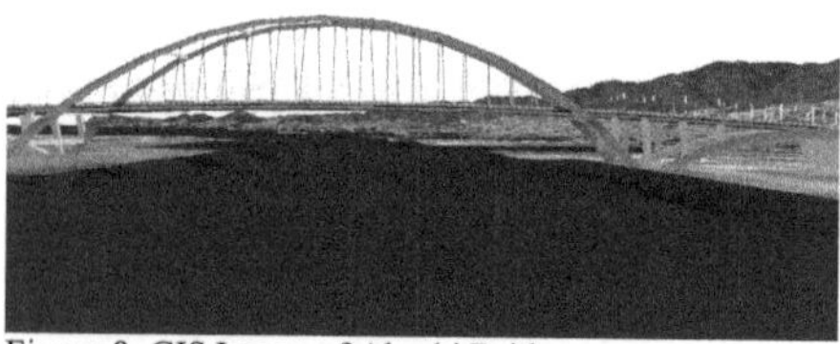
Figure 8. GIS Image of Akashi Bridge

Car navigation systems identify undersea tunnels as roads; however, if they are indicated as such when used for marine navigation, they could be mistakenly identified as bridges. Therefore, proper identification of undersea tunnels is unnecessary information for marine traffic.

Figure 8 shows a GIS image of Rokko Bridge using a navigation simulator that we have developed. A three-dimensional model was created using SketchUp software that is compatible with ArcGIS. Then, we overlaid the image on a satellite image on ArcGIS. Using GIS with our system, it is possible to overlay bridges, buildings, and navigational aids separately over the geography shown in a satellite image. If we could overlay a three-dimensional image of a bridge over a satellite image using a car navigation system, it would be beneficial to marine navigation.

However, Figure 8 does not indicate any navigational aids, such as lighthouses and buoys, primary targets, recommended navigation courses, water depths, sea bottom quality, or other necessary navigational information. Such information is unnecessary for road traffic but important for marine traffic.

Next, we compared a car navigation system with an electronic marine chart. Figure 9 shows the electronic chart of the coastal zone in the course taken by the Muko Maru. It only shows a few pieces of navigational information, such as geography, water depths, lighthouses, and lighted buoys. Therefore, the amount of information that an electronic chart can provide is limited. Normally, ship operators receive basic navigational information from paper charts. In an electronic chart, the detailed information for each navigational aid, such as name, color, and character of each aid, can be provided via text on the right side of the screen upon clicking each aid's icon. If all of the navigational information is provided simultaneously on such a small screen, it would be too complicated to understand.

After we compared a car navigation system and an electronic chart, we found almost no difference other than the amount of navigational information provided. Moreover, the car navigation system indicates more names of locations and buildings, and therefore, is easier to understand.

The navigational information provision system that we have developed can indicate information regarding weather and other marine phenomena. However, currently, car navigation systems are not capable of providing such information. A chart only shows the average tidal flow rates. Therefore, using our GIS would be more useful.

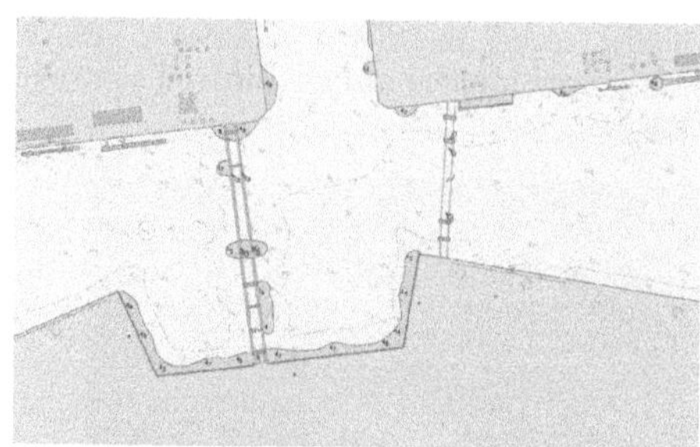
Figure 9. Electronic chart display of sailing area

Figure 10 shows the car navigation system's three-dimensional display while sailing around the Ashiya Canal. Figure 11 shows the front-view image taken by the video camera. The video image is almost the same as the view that ship operators see. However, the visual field is narrow and views behind the buildings are obstructed. A car navigation system makes it possible to observe all of the landscape including areas behind obstructed views. Therefore, it would be easy to comprehend what is going on ahead of the ship. Furthermore, the primary buildings are displayed in three dimensions and the names of the locations are easy to understand. Therefore, it is more comprehensive than the information provided by charts. The width of the Ashiya Canal is approximately 100 m that is very narrow. There are many houses near the coast and the coastal roads are well organized. Consequently, the car navigation system was able to provide detailed road and traffic information.

Figure 10. Car navigation display of the Ashiya Canal

Figure 11. Video image from the ship near the Ashiya Canal

Weather such as rain or fog reduces visibility; however, a car navigation system can provide the same information regardless of the weather. In addition, car navigation systems can provide clear landscapes in a daylight view at night and vice versa.

On comparison of the information obtained from a car navigation system with that obtained from video images and an electronic chart, it is evident that a car navigation system is more useful as a marine navigational safety device because it is capable of displaying clear three-dimensional landscapes and facilitates ship navigation in the same manner it does for road navigation. After an assessment of the ad-vantages and disadvantages of such a system, we believe that it would be effective to use a car navigation system as a supplement to traditional two-dimensional electronic charts.

At present, a car navigation system can be purchased for several ten-thousand yen, which is significantly less expensive than a generic navigational device. Therefore, when utilized on small vessels such as pleasure boats, it would be both cost-effective and a satisfactory aid to navigation.

4 APPLICATION OF CAR NAVIGATION SYSTEM FUNCTION TO A MARINE NAVIGATION SYSTEM

We investigated the car navigation system's functions in detail to determine its effectiveness as a marine navigation system. Our objective is to integrate the advantages of a car navigation system with the GIS navigation system that we are currently developing in an attempt to upgrade the whole system. The GIS is capable of an overlapped display with various navigational information that enables the provision of detailed and effective navigational information. When the advantages of a car navigation system are combined with the GIS, it becomes capable of providing more effective information.

The following outlines the results of our investigation of the primary functions of car navigation systems that could be applied to marine traffic (Sanyo Electric Co. 2011).

1 Minimize the difference between the actual and current locations detected on the map because of the errors in GPS satellite frequency caused by autonomous navigation (estimation of the current location by judging the direction of travel using the self-contained gyroscope and speed sensor) and map matching.
   As with cars, small marine vessels such as pleasure boats have the ability to switch between high and low speed. Therefore, having a function to minimize navigational errors can be an effective navigational aid. The function would be particularly appropriate while sailing in narrow waters such as a harbor.
2 Provide route guidance by displaying a potential route when a destination is specified.
   Currently, no vessel navigation systems are capable of route finding. A function that considers weather and various ocean phenomena and determines the best and alternative routes would be a useful future endeavor.
3 Determine destination from the data stored in the software upon inputting a telephone number or an address.
   This function would be useful for commercial vessels entering a port. In addition, when a pleasure boat needs to import at small fishing ports, this function would be highly useful when searching for port information.
4 Indicate names of buildings on the map display when the cursor hovers over each facility (object). Furthermore, upon clicking "Setting" or "Tenant information," tenant information is indicated.

These functions have already been implemented in our navigation assistance device. For example, upon clicking on the marine traffic aid icons, such as primary lighthouses and lighted buoys, detailed information on each aid, such as lighting, light characteristic, color, shape, and photos, can be provided. A function that provides additional navigational information would increase the utility of the system.

5 Saving locations and routes to a SD card by accessing a specific website (NAVI) from a computer; the saved data can be displayed on a navigation system.
As vessels often take the same route, saving routes would be useful for a marine navigation system as a supplementary log book function.

6 Routes to destinations are searchable. In relation to route guidance, the ability to search and correct routes during navigation would enable more efficient navigation by minimizing travel time and fuel consumption.

7 Provide voice guidance on names of intersections, entrances/exits of highways, names of roads, junctions, etc.
Of course, there are no intersections in marine traffic. However, if voice guidance could indicate the estimated distance or time to a veering point, a port of entry, an entrance, or a crossing route, it would be helpful for ship operators to anticipate navigational changes. We are considering installing a voice guidance function in our navigation assistance device.

8 Display an enlarged three-dimensional view and provide route guidance when approaching primary junctions or exits.
To assist entering or exiting a port by a specific route, a three-dimensional geographical display would be an effective navigational simulation. Therefore, it would be a useful function to include in a navigation assistance device.

9 Simultaneous criteria-based searches (recommended, toll highways, distance, road width) for different routes to a specific destination.
It would be useful if various options for routes were searchable using relevant criteria, such as navigation time, fuel consumption, minimum waiting time, safety level.

10 Display an announcement (sign board) indicating a direction to a national highway or general road during route guidance.
The display of general warnings or guideposts for navigation while clearing the Akashi Channel or other narrow aqueducts would be an effective function for a navigation assistance device.

11 Display a virtual three-dimensional intersection during route guidance.
If a three-dimensional landscape image is displayed while approaching a veering point, a narrow aqueduct, or a destination port, it would assist navigation and increase security.

12 Display realistic three-dimensional images of buildings in metropolitan areas.
Three-dimensional displays of buildings and marine traffic aids that can be navigational targets would help effective verification of these objects.

13 Provide information on traffic congestion in text, diagrams, or map displays.
The provision/display of marine traffic congestion information received from an organization such as Marine Traffic Information Service Center would assist safe navigation.

14 Indicate one-way traffic restrictions.
During navigation of a particular route that has regulatory traffic restrictions such as the Kurushima Channel displaying the details of the restrictions could alert ship operators in a timely manner.

15 Display a top-down view of the surrounding area.
The ability to view the vessel's location from a three-dimensional top-down perspective would facilitate broader understanding of the surrounding waters. Our navigation assistance device has already adopted this function.

16 Provide map color options.
By choosing a particular color, the geography can be expressed more realistically. Our navigation assistance device has already adopted this function for sea bottom geography and water depth displays.

17 Copy and replay music, picture, and video files using self-contained flash memory.
Viewing recorded navigational information and simulations, including voice, image, and video data, could be used as an effective training tool.

18 Capable of receiving terrestrial digital television broadcasting through a self-contained tuner.
The ability to receive up-to-date weather forecasts and other pertinent information via television signals would be very helpful for navigation.

From the above investigation of car navigation system functions, we consider that a car navigation system could be converted to a marine navigation system. Our proposed marine navigation system would be capable of providing highly useful information if combined with the navigational information that is generally indicated on a marine chart. Furthermore, if all the car navigation system functions are added to the marine navigation assistance device that we have been developing, it would be more effective.

5 CONCLUSION

To improve navigational security and safety, we investigated whether car navigation systems could be applied to marine traffic. The applicability was evaluated through several comparisons of our navigation assistance device using GIS with paper charts, electronic charts, and video images.

As a result, we reached the following preliminary conclusions.

1 A car navigation system is capable of providing three-dimensional geographical information, which is not possible with a two-dimensional chart. A marine navigational system that includes a three-dimensional chart would be useful.
2 A car navigation system has a self-contained GPS. If integrated in a marine navigational system, determining ship speed and position as well as plotting and displaying routes would be possible.
3 With a car navigation system, it is possible to obtain surrounding geographical information. If geographical information such as the location of bridges and buildings was available, we would be able to provide more detailed information than is offered by a generic chart.
4 A car navigation system is not capable of providing necessary navigational information such as water depth, the location of lighthouses, and other navigational aids. If such navigational information could be added to a car navigation system, it could become an effective marine navigation system
5 We found that it would be possible to provide highly effective navigational information by integrating the functions from a car navigation system into the navigation assistance device that we have been developing.

In the future, we will conduct more studies of the effective provision of information to improve navigational security and safety.

This study was conducted as a part of the basic re-search program (B): "Prevention of marine disaster by numerical navigation system using marine ITS" (Project number: 22310100) and was supported by a Grant-in-Aid for Scientific Research.

REFERENCE

Imazu, H. et al. 2003. Study of the Integration and Presentation of Navigational Information. *The Journal of Japan Institute of Navigation*, 109: 133-140.

Japan Coast Guard, 2007. Condition of generation and rescue of marine disaster and accident involving a human life. pp.16-19.

Kawasaki A. 2006. Illustration, Step up for practice of GIS. Kokin Shoin, p.174.

Sadohara S. 2005. Illustration, Study on Near Example of ArcGIS, Kokin Shoin, p.176.

Sanyo Electric Co. 2011. Manual of SSD Portable-Navigation. No. NV-SD760FT p.A-14 - A-38.

Shiotani, S. et al. 2011. ON STUDY OF SUP-PORT OF NAVIGATIONAL INFORMATION FOR PREVENTION OF ACCIDENT ON MARINE TRAFFIC IN COASTAL SEA AREA -INFORMATION OF WATER DEPTH-. *Journal of Civil Engineering D3(Infrastructure Planning)* 28: I_1039-I_1047.

Shiotani, S. et al. 2011. STUDY ON NAVIGA-TIONAL SIMULATION FOR SAFTY OF SHIP SAILING IN COASTAL SEA AREA. *Journal of Civil Engineering D2 (Ocean Development)*. 67(72): I_838-I_843.

Wakabayashi, N. et al. 2002. Acquisition and Transfer System for Ship Navigation using IP Net-work –Implementation on T.S. Fukae maru-. *The Journal of Japan Institute of Navigation* 106: 29-37.

Remote Spatial Database Access in the Navigation System for the Blind

K. Drypczewski, Ł. Kamiński, Ł. Markiewicz, B. Wiśniewski & A. Stepnowski
Gdansk University of Technology, Gdansk, Poland

ABSTRACT: The article presents the problem of a database access in the navigation systems. The authors were among the main creators of the prototype navigation system for the blind - "Voice Maps". In the implemented prototype only exemplary, limited spatial data were used, therefore they could be stored in the mobile device's memory without any difficulties. Currently the aforementioned system is being prepared for commercialization - the resulting increase of spatial data scale and complexity required a modification of the way the data are stored and accessed. Consequently, the decision was made to maintain a central spatial database. After that modification, the mobile application fetches necessary batch of spatial data remotely from the central server. The authors present the advantages and disadvantages of this new approach.

1 INTRODUCTION

The main goal of the "Voice Maps" project was to design and implement a fully operational navigation system supporting movement of the blind in the urban areas. The development process was divided into two main parts: the creation of the prototype (which has been already finished) and the product commercialization phase (ongoing).

The major mobile navigation systems' providers use different approaches to the vector data (Shekhar et al. 1999) storage mechanism. Some of them prefer remote database access (Google Maps), while the others (Nokia Maps) tend to keep all the necessary map data in a local mobile device's memory. However, all of the applications supporting navigation of the blind, which have been investigated by the authors, use the second approach, where processed geographical data are stored locally on the navigational device, i.e. mobile device, smartphone or tablet (e.g. Sendero GPS, Trekker, LoadStone).

In the first phase of the "Voice Maps", which consisted of research and preliminary software development, the system prototype was successfully implemented. It consisted of a main mobile application and exemplary spatial data, that were gathered by the authors and cooperating project-team members, using both existing data sources (e.g. OpenStreetMap) and in-situ data acquisition methods (Kaminski et al. 2011).

The application Graphical User Interface (GUI) had to be designed for the blind users. It consists of a touchscreen-based main menu, where options are chosen by moving the finger on the screen, which is divided into 12 equal square fragments (Fig.1).

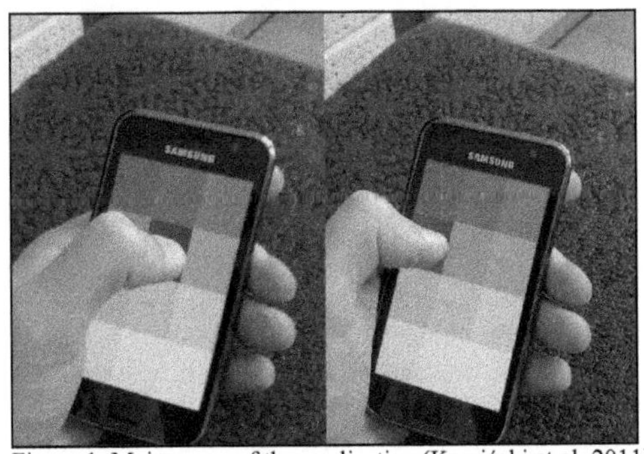

Figure 1. Main menu of the application (Kamiński et al. 2011).

Available options are read to the user using the speech synthesis. All the input data, e.g. addresses or Point of Interest (POI) names needed by the program, is entered using dedicated and customized software keyboards (e.g. gesture or voice recognition keyboard) built-in into the application (Fig.2).

At first the design was to create various versions of the program for different mobile operating systems, which would be linked by shared Java libraries. In later stages of the development process a decision was made to focus on Google Android operating system and its compatible devices. The main reason was the fact that other planned platforms - mainly Symbian and BlackBerry (supported by the application implemented in JavaME technology) (Fig.3), did not offer sufficient functionality and have lost most of its market share.

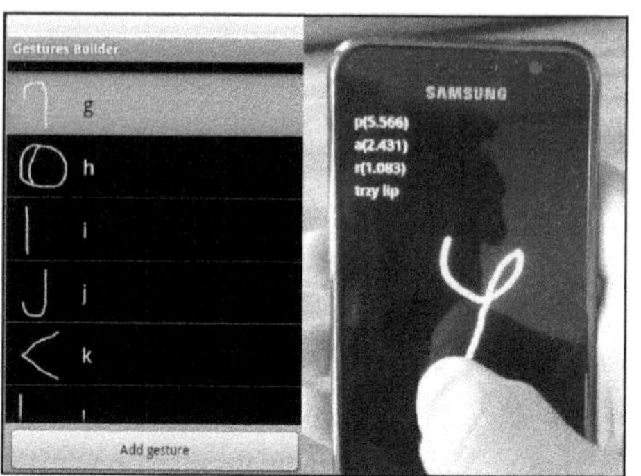

Figure 2. Gesture keyboard for the blind (Kamiński et al. 2011).

Figure 3. "Voice Maps" application on Android, Symbian and JavaME (Kamiński 2012).

The Android operating system definitely dominates current smartphone market (Gartner 2012) and offers key features for navigational systems for the blind.

During the present project phase (preparation for the commercialization) it was necessary to asses the current version of the system, especially the mobile application, in terms of using the system on the large scale - not only in the limited testing environment. It quickly became apparent, that the data locally stored on the device were not the optimal option. The spatial data scale and complexity changed significantly and the system had to be redesigned in order to modify the data storage and access.

After the modification, the mobile application downloads through the internet necessary parts of spatial data on-the-fly (as it is needed) remotely from the central server. This approach is new in the navigational systems for the blind and its advantages and disadvantages, along with the data transition, are being analyzed in this paper.

2 THE PREVIOUS SOLUTION – DATA STORED LOCALLY

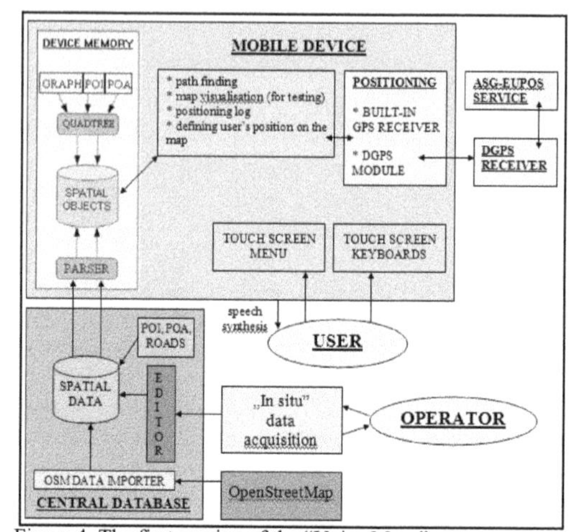

Figure 4. The first version of the "Voice Maps" system.

Figure 4 shows the simplified diagram of the "Voice Maps" system prototype (after research phase). In the bottom the system modules responsible for the spatial data acquisition, storage and access are visualized. In the first version of the system all the spatial data were stored in the central database. After spatial data conversion into "Voice Maps" native format (for example organization into quadtree structure was required) the data were then copied (as a whole) into the mobile device. The optimal pathfinding from the current user's position to the chosen objective (i.e. address or POI) was done locally and completely on the mobile device, without the need to connect through the internet to the central database (Stepnowski et al. 2011).

Mobile Global Positioning System (GPS) receiver was used for positioning. Its precision is sufficient in most cases (Zandbergen & Barbeau 2011). There is an option to extend it with Differential Global Positioning System (DGPS) module in order to further increase its accuracy.

In the course of the system prototype operational analysis authors have identified problems with the system's design and recognized the modules needed to be redesigned and changes, what required their re-implementation. The solution of the spatial data storage and access performed well in the research (prototype) phase of the system development, when only limited and not complex spatial data were available. With the acquisition of consecutive data from other parts of Gdansk and other cities it became apparent, that the first solution (linked with the spatial data) is increasingly inconvenient in the terms of "Voice Maps" navigation system.

The updates of the spatial data in the previous version had to be done manually on each of the devices (copying all the data from the central database) and, even with the new version of the notification mechanism implemented in the application, user action was required during each update. Additionally, storage of the considerable amount of data on the mobile device, in terms of the software and hardware limitations (which are becoming less and less significant with each generation of smartphones), imposed the implementation of the complex algorithms, that were optimizing the spatial data access time, i.e. quadtree structure. The optimal pathfinding between two points was also done on the mobile device, which had limited computational power compared to the standalone computer.

In order to solve all the aforementioned problems the authors decided to change the model of storage and access to the spatial data used by the main mobile application. All the required changes are described in the next chapters of this paper.

3 THE NEW SOLUTION - REMOTE DATA ACCESS

During the preparation phase it was decided that the best solution is to move all the gathered data into the central database, from which only necessary pieces of information will be downloaded via wireless connection into mobile application. All the main advantages and disadvantages of that decision are presented in the next two paragraphs of this article. The mobile application has become a client, used mainly to send requests to the database, provide the user interface, positioning, navigation and to store user's configuration preferences. The server has become more extensive as three new modules were added to its architecture (Fig.5).

Those modules are:

- Data sharing module, which allows mobile application to download selected data from database,
- POIFetcher, that allows to search for requested Point of Interest,
- Routino, a specialized module which finds a path between two given points.

The new solution is to provide the mobile application with fragments of data that describe area in the closest surroundings of their current location. Provided data dynamically change when users move to a different location, so application always contains only information that is needed to guide user toward his destination.

All the data are not stored on the device in this solution, so application cannot provide full search of the path whenever the requested destination point is outside the temporary graph – bacause of this fact part of the new solution was to transfer path-searching algorithms onto the server. The mechanism of path-searching is provided by the aforementioned, specialized module (Routino).

To keep up most accurate point of interest data, POIs also were moved to the server side central database. The new solution is to keep POIs data on the server and provide users with categories and search mechanisms.

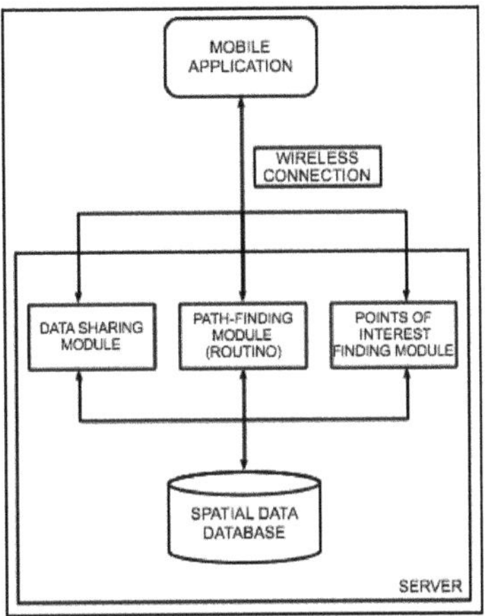

Figure 5. Changes in the system architecture forced by the adoption of remote database access.

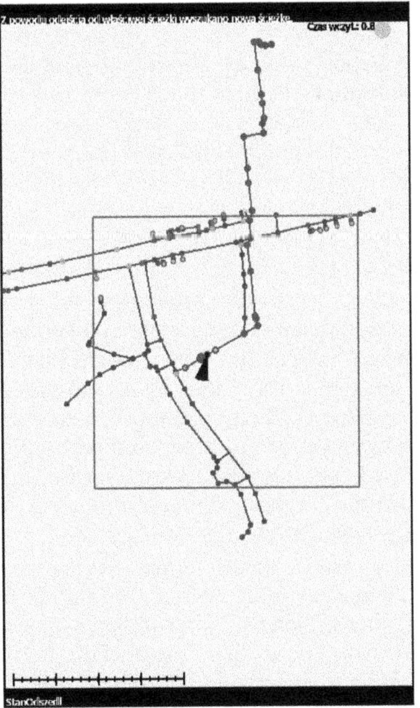

Figure 6. Graph of nearest surrounding synchronized with found path.

Application is supplied with current data, users can select interesting category (for example "Shops") and server will find path to the nearest POI from that category and provide the shortest path.

Because the graph used in application is temporary and given route is static, there is a need to synchronize these two sources of data every time a graph of the nearest area changes. This functionality is performed on the application side. The effect of this synchronization is described on Fig.6.

4 THE ADVANTAGES OF THE NEW SOLUTION

Applications purposed for the blind should be maximally easy to use. All the activities taken by its users have to be repetitious and also should take minimum effort to be learnt and remembered (Farrel 1956). Number of actions performed by the blind to achieve their goal must be minimal. The simplicity of the program interface is especially important in unusual situations. One of them is the process of updating the application - program should allow blind to get the newest version of the software in the easy-to-approach way. Since the most important data, the graph representation of geographical features, is kept on the server side, no action from the user is required to receive updated data. Moreover, some enhancements and improvements can be added without changing the client side software.

Navigation system for the blind is different from typical car navigation systems. Providing direction for sightless pedestrian requires much more precise and accurate data. Whereas the target users of "Voice Maps" have the limited ability to orientate in the field and to identify the obstacles in the road - given information must be as exact as possible. High density graph representation of metropolitan areas' geographical features may contain a large amount of vertices and edges. Real-time processing of this dataset requires significant computational power and considerable memory capacity. Since "Voice Maps" system is targeted for the Android-based smartphones the issue of their capabilities becomes important. Moving most of the geospatial data from client to server side not only reduces the size of the software but also lowers the system requirements. It is especially important for the older devices.

What is also worth mentioning, the spatial database and elements of the server side software can be used in other geographic information systems (GIS). High-density data and advanced graph algorithms can be easily adapted to become elements of different (car, pedestrian) navigation applications.

It is important to state, that the new solution makes application development less difficult. Since the server log is available, thorough and wide testing of software is easier to perform. Moreover, it enables the possibility to add several enhancements to the system. Amongst them is functionality to remotely monitor the behaviour of the blind. It would be much more difficult to achieve with the previous system architecture.

5 THE DRAWBACKS OF THE REMOTE DATA ACCESS

Remote spatial database access in "Voice Maps" requires stable wireless internet connection. Since the navigation system for the blind must be very reliable, problems with network (lost connection, transfer speed and limits) may become very important. However, nowadays, coverage of the cellular network in the metropolitan areas is usually high and should be sufficient. When user receives information about geographical features located in a fixed distance from his position, additional data are sent only if he leaves the aforementioned area. This kind of approach not only limits the needed transfer to minimum but also, if the connection is temporarily lost, ensures data availability. It is also worth mentioning that this new solution (module for spatial database access) is not the only software component that needs stable internet connection. Since the voice recognition software is incorporated in "Voice Maps", wireless internet connection is necessary.

In the previous solution, the most time-consuming operations, e.g. finding a shortest path between user location and requested vertex in a graph, were performed on the client side. In the new solution all of those operations must be executed on the server, which requires a sophisticated server infrastructure. Moreover, considering the characteristics of blind users, the quality of that service must be very high, only the reliable navigational system can ensure customer satisfaction. To sum up, the new solution enforces keeping considerably expensive data center.

Communication with remote spatial database created non-trivial problems in software development. Amongst them are the issue of synchronization with database processes (response time, network delay) and client-side threads. On the other hand, new version of "Voice Maps" does not require advanced methods for optimizing the size of the files containing information about the geographical features (e.g. quadtree).

6 SUMMARY

Early test results of the new solution are very promising. The application now focuses on a user and navigation interaction, instead of performing difficult and time-consuming calculations. There is still room for improvement in the near future, followed by more advanced tests. Most of functionality trials will be performed by end users. In addition, a great number of reliability tests is still to be performed, especially tests of the server.

The main advantage of the new solution is an increase of time efficiency for both the users and developers of the system. Updates are no longer an issue, as they are now performed on the server - user always gets the most current data. The application is lighter and needs less computational power. Main calculations are performed on the server side and downloading data became the most time-consuming task. However, remote data access causes the main disadvantage, i.e. a need for the server and a reliable network connection. It is an architectural problem that needs to be solved in the last development phases, while the second one is not very significant in era of fast and omnipresent Internet.

A significant advantage of the presented solution's strengths over its drawbacks, as well as encouraging tests' results, confirm our belief that remote database access is the best solution in the navigation systems for the blind, therefore it should be further developed and improved as part of the "Voice Maps" project.

ACKNOWLEDGEMENTS

This research work is a part of the "Voice Maps" project, which was supported by Narodowe Centrum Badan i Rozwoju (National Center of Research and Development) within the "IniTech" program (project code: ZPB /37/679 01/IT2/10).

We are highly grateful to Mr. Rafał Charłampowicz and Dr. Ryszard Kowalik for participating in our system's tests, and supporting the team with useful suggestions.

REFERENCES

Farrel, G. 1956. *The Story of Blindness*. Harvard Unversity Press.

Gartner. 2012. *Worldwide Sales of Mobile Phones Declined 3 Percent in Third Quarter of 2012; Smartphone Sales Increased 47 Percent*, source: http://www.gartner.com/it/page.jsp?id=2237315

Google Maps Developer Documentation, source: https://developers.google.com/maps/documentation/

Kamiński, Ł., Chybicki, A., Bruniecki, K. & Stepnowski, A. 2011. Smartphone application supporting independent movement of the blind. *The Future with GIS. Hrvatski Informaticki Zbor - GIS Forum 2011, Croatia.*

Kamiński, Ł. 2012. *Mobilny system informacji przestrzennej wspomagający nawigację osób niewidomych w terenie miejskim* (in Polish). PhD thesis, Gdańsk, Poland.

LoadStone, source: http://www.loadstone-gps.com

Nokia Maps User Guide, source: http://nds1.nokia.com/files/support/global/phones/guides/Maps_3_0_UG_SR1_1_No_Touch_EN.pdf

Routino, source: http://www.routino.org/

Sendero GPS, source: http://www.senderogroup.com

Shekhar S., Chawla S., Ravada S., Fetterer A., Liu X. & Lu C. 1999. Spatial Databases - Accomplishments and Research Needs. *IEEE Transactions on Knowledge and Data Engineering*, Vol. 11(1).

Stepnowski, A., Kamiński, Ł., Demkowicz, J. 2011. Voice Maps – the system for navigation of blind in urban area. *Proceedings of the 10th WSEAS International Conference on Applied Computer and Applied Computational Science, Venice, Italy.*

Trekker, source: http://www.humanware.com/en-usa/products

Zandbergen P. & Barbeau S. 2011. Positional Accuracy of Assisted GPS Data from High-Sensitivity GPS-enabled Mobile Phones. *Journal of Navigation*, Vol. 64(03): 381-399.

Integration of Inertial Sensors and GPS System Data for the Personal Navigation in Urban Area

K. Bikonis & J. Demkowicz
Gdansk University of Technology, Gdansk, Poland

ABSTRACT: GPS and Inertial Navigation Systems (INS) have complementary properties and they are therefore well suited for integration. The integrated solution offers better long-term accuracy than a stand-alone INS and batter integrity, availability and continuity or a stand-alone GPS receiver, making it suitable for demanding applications. The complementary features of INS and GPS are the main reasons why integrated GPS/INS systems are becoming increasingly popular. GPS/INS systems offer position, velocity, acceleration, attitude and angular velocity measurements with high accuracy, output rate and reliability in one package.
In the paper integration algorithm of inertial sensors (accelerometers and gyroscopes) and GPS system data for the urban area personal navigation is presented. For data integration algorithm Extended Kalman Filter (EKF) is proposed.

1 INTRODUCTION

The GPS system is widely used in navigation. The GPS receiver can offer long-term stable absolute positioning information with output rate at around 1 to 10 Hz. However, the system performance depends largely on the signal environments. In an INS system, the angular rate and specific force measurements from the Inertial Measurement Unit (IMU) are processed to yield the position, velocity and attitude solution. Such systems can navigate autonomously and provide measurements at a higher data rate (e.g., 100 Hz). However, the system has to be initialized and calibrated carefully before application. Moreover, the sensor errors are growing unboundedly over time. Due to the complimentary characteristics of GPS and INS, they are often integrated to obtain a complete and continuous navigation solution [1-6].

The inertial sensors used in IMU are made in MEMS (Micro Electro-Mechanical Systems) technology. MEMS technology enables miniaturization, mass production and cost reduction of many sensors. In particular, MEMS inertial sensors that include an acceleration sensor and an angular velocity sensor (gyroscope, or simply "gyro") are the most popular devices. Almost all MEMS acceleration sensors have a seismic mass and support spring made of silicon. The structure of MEMS gyros is somewhat similar to that of acceleration sensors – a mass supported by a spring is continuously vibrated in the device, and the Coriolis force generated by the applied angular velocity affects the movement of the mass (vibrating gyroscope). The mass in a MEMS device is very small, and therefore, the inertial forces acting on the mass, especially the Coriolis force, are also extremely small. Thus, the design of the circuit that measures the movement in mass due to the force is important in addition to the design of the mechanical structure. Recently MEMS inertial sensors have been built with an integrated circuit, with sensor structure on a single device chip [7].

A typical structure of a MEMS acceleration sensor is shown in figure 1 [7], where a silicon mass is supported by silicon springs and the displacement of the mass due to acceleration is measured by capacitance change between the mass and fixed electrodes. Since the mass is very small and the displacement is also small, the resolution of the device is generally limited to around 0.1 mg $Hz^{-1/2}$.

The basic structure of MEMS gyroscopes is similar to acceleration sensors, i.e., a mass is supported by springs. The main difference in operation is that the angular velocity is obtained by measuring the Coriolis force on the vibrating mass. Thus, the movement of the mass should have at least two degrees of freedom. The device is shown in figure 2 [7].

Inertial sensors have numerous applications. INS is a self-contained system that integrates three acceleration and three angular velocity components with respect to time and transforms them into the navigation frame to deliver position, velocity, and attitude components. The three orthogonal linear accelerations are continuously measured through three-axis accelerometers while three gyroscopes monitor the three orthogonal angular rates in an inertial frame of reference. In general, IMU, which incorporates three-axis accelerometers and three-axis gyroscopes, can be used as positioning and attitude monitoring devices. However, INS cannot operate appropriately as a stand-alone navigation system.

The presence of residual bias errors in both the accelerometers and the gyroscopes, which can only be modeled as stochastic processes, may deteriorate the long-term positioning accuracy. Hence, the INS/GPS data integration is the desirable solution to provide navigation system that has better performance in comparison with either a GPS or an INS stand-alone system.

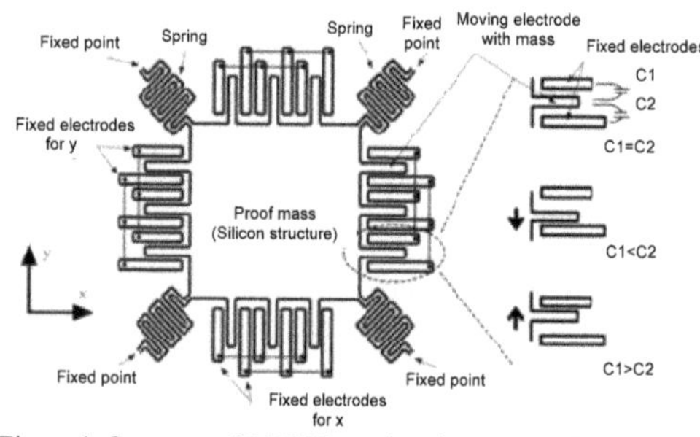

Figure 1. Structure of MEMS acceleration sensor (2-axis).

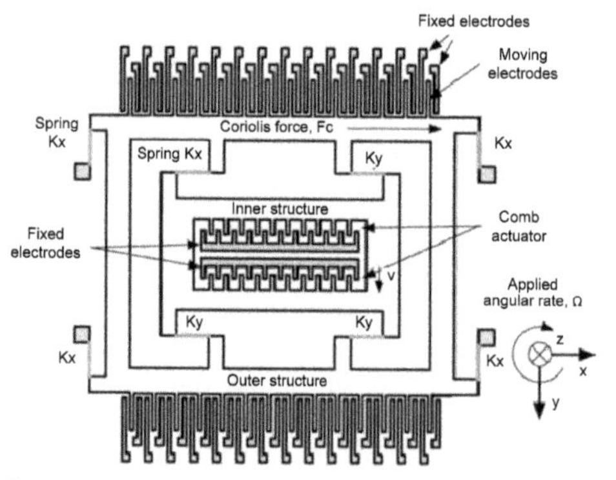

Figure 2. Conceptual structure of an MEMS gyroscope.

2 IMU DESCRIPTION

We use a commercially available IMU, model MTi-G from Xsens Technologies. Figure 3 shows this sensor. Its size is 58x58x22 mm (WxLxH), and it weights 50 grams.

The IMU has three orthogonally-oriented accelerometers, three gyroscopes, three magnetometers and GPS reciver. The accelerometers and gyroscopes are MEMS solid state with capacitative readout, providing linear acceleration and rate of turn, respectively. Magnetometers use a thin-film magnetoresistive principle to measure the earth magnetic field. The performance of each individual MEMS sensor within the MTi IMU are summarized in table 1 and GPS receiver in table 2.

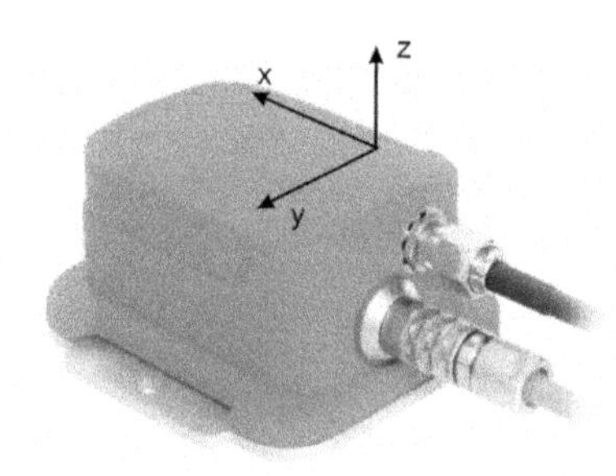

Figure 3. MTi-G Xsens IMU with annotated sensor cartesian coordinates.

The MTi-G sensor has a built-in algorithm that provides the absolute heading and attitude of the unit, which is expressed as the rotation matrix R_{GS}. It can be used to directly transform the readings from the sensor (S) to the global (G) Cartesian coordinates frames. The typical absolute orientation errors are summarized in table 3. Performance is quite good whenever the earth magnetic field is not disturbed, for example by metallic objects, power lines, personal computers, or any device containing electro-magnetic motors.

Table 1. Performance of individual sensors in Xsens IMU.

	A	G	M
Axes	3	3	3
Full Scale FS	±50 m/s^2	±300 °/s	± 750 mGuass
Linearity	0.1% of FS	0.2% of FS	0.2% of FS
Bias stability	0.02 m/s2	1 °/s	0.1 mGuass
Bandwidth	30 Hz	40 Hz	10 Hz
Max update rate	512	512	512

A – accelerometers, G – gyroscopes, M – magnetometers

Table 2. GPS receiver parameters.

Receiver type	50 channels L1 frequency, C/A code
GPS update rate	4 Hz
Start-up time cold start	29 s
Tracking sensitivity	-160 dBm
Timing Accuracy	50ns RMS

Table 3. Performance of attitude and heading as provided Xsens fusion algorithm in matrix R_{GS}.

Static accuracy (roll/pitch)	< 0.5°
Static accuracy (heading)*	< 1°
Dynamic accuracy	2° RMS
Angular resolution	0.005 °

*in homogeneous magnetic environment

3 ALGORITHM FOR INS AND GPS DATA INTEGRATION

The conventional IMU navigation algorithm is to integrate the gyroscopes and accelerometers data. However, the position values obtained by this method are reliable for only a short period of time. This is due to the accelerometer's inherent drift error as well as the gyro rate drift error, which means that when double integration of the acceleration measurements, the drift error is also accumulated over time and increases dramatically with time. So the estimated position will be far away from the actual position.

Other IMU navigation method can based on step detection algorithm and use orientation information directly from IMU. There are several step detection algorithms that have been proposed by researches in the literature [1]. Figure 4 shows IMU and GPS antenna localization on the body.

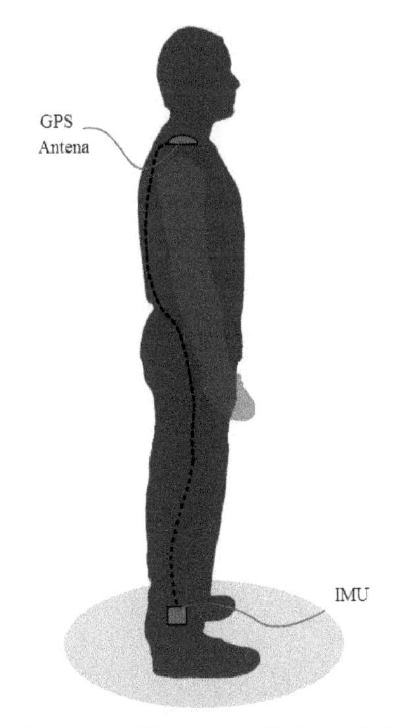

Figure 4. IMU and GPS antenna localization on the body.

The INS/GPS data integration algorithm consists in Extended Kalman Filter (EKF) usage [8, 9]. EKF uses Taylor series where the idea of a linear approximation to describe a function in the neighborhood of some point by a linear function is applied. The algorithm works in a two-step prediction/correction process. In the prediction step, the Kalman filter produces estimates of the current state variables. Because of the recursive nature of the algorithm, it can be run in real time. The present input measurements and the previously calculated state is used; no additional past information is required [10]. The very idea is presented in the figure 5 where $\hat{x}_k^-$, $\hat{x}_k$ are á priori and á posteriori system state, P_k^-, P_k are á priori and á posteriori covariance matrix, H is measurement matrix, K_k is Kalman gain, R, Q are process and state variance of the system, z_k is measurement matrix, A is process model.

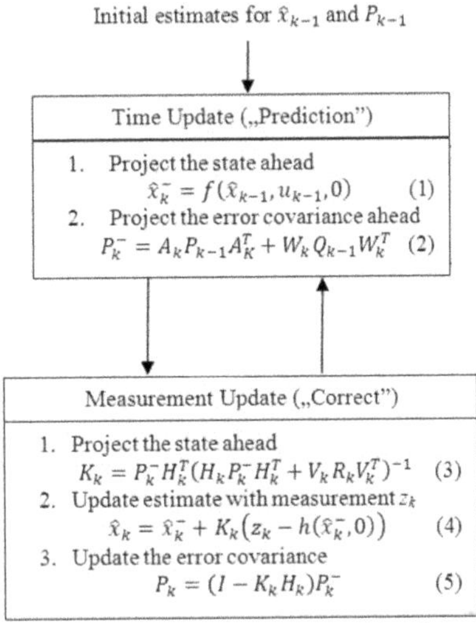

Figure 5. EKF sensor data integration algorithm diagram [5].

Prediction can be describe as follows (1) and (2) where $f(x,u,w)$ is nonlinear function, which uses a previous filter state as well as control impact and the process noise. A_k (or A^J) is a Jacobian, f with respect to x, W_k (or W^J) is a Jacobian, f with respect to w function. Where A and W are follows:

$$A^J_{[i,j]} = \frac{\delta f_{[i]}}{\delta x_{[j]}} (\hat{x}_{k-1}, u_{k-1}, 0), \qquad (6)$$

$$W^J_{[i,j]} = \frac{\delta f_{[i]}}{\delta x_{[j]}} (\hat{x}_{k-1}, u_{k-1}, 0). \qquad (7)$$

And correction is applied as follows (3), (4) and (5) where H_k (or H^J) is Jacobian, derivative of the function h with respect to x, V_k (or V^J) is Jacobian, function h derivative of with respect to v

and $h(x,v)$ is nonlinear function of the state and measurement relation. Jacobians H and V can be expressed as follows:

$$H^{J}_{[i,j]} = \frac{\delta h_{[i]}}{\delta x_{[j]}}(\tilde{x}_k, 0) \tag{8}$$

$$V^{J}_{[i,j]} = \frac{\delta h_{[i]}}{\delta x_{[j]}}(x_k, 0) \tag{9}$$

where:

$$x_k = f(x_{k-1}, u_{k-1}, 0). \tag{10}$$

All Jacobians ought to be recalculated in every iteration step where model of GPS/IMU is generated.

The state estimate vector consists of 20 elements

$$\hat{x} = [x\ y\ z\ v_x\ v_y\ v_z\ q\ b_x\ b_y\ b_z\ \varphi_x\ \varphi_y\ \varphi_z\ \varphi_{bx}\ \varphi_{by}\ \varphi_{bz}\ g]^T \tag{11}$$

with initial values

$$\hat{x} = [0\ 0\ 0, 0\ 0\ 0, 1\ 0\ 0\ 0, 0.001\ 0.001\ 0.001, 0\ 0\ 0, 0.001\ 0.001\ 0.001, g]^T \tag{12}$$

where $x\ y\ z$ is location [m], $v_x\ v_y\ v_z$ is velocity [m/s], q is quaternion vector, $b_x\ b_y\ b_z$ is bias [G], $\varphi_x\ \varphi_y\ \varphi_z$ is rotation phase [rad], $\varphi_{bx}\ \varphi_{by}\ \varphi_{bz}$ is rotation biases, g is gravitation.

The observation covariance matrix R is a diagonal matrix consisting of GPS position error (5 m) and velocity error (0.5 m/s) in direction $x\ y\ z$.

The covariance noise diagonal matrix

$$Q = [0.5\ 0.5\ 0.5, 0.5\ 0.5\ 0.5, 0.00001\ 0.00001\ 0.00001\ 0.00001, 0.01\ 0.01\ 0.01, 0.001\ 0.001\ 0.001, 0.001\ 0.001\ 0.001, 0.01]^T \tag{13}$$

4 RESULTS

The purpose of this study is to track the position of a pedestrian walking outside. One foot of the pedestrian is mounted with an IMU, which is used to measure the acceleration and angular rate of the walking foot. The GPS module is attached to a straight pole with the GPS antenna on the top of it so that the GPS position signal can be obtained more easily. The pedestrian localization is achieved by integrating the inertial and GPS information.

The IMU/GPS based pedestrian localization algorithm is firstly implemented when a pedestrian is walking along a 15m×15m square. The result of the IMU/GPS algorithm using EKF is shown in figure 6.

Another test was conducted in an environment shown in figure 7 and the results of the IMU/GPS algorithm using EKF is shown in figure 8.

It is obvious that EKF corrected trajectory is more accurate than the trajectory calculated by IMU and GPS separately. Result for EKF is nearly GPS trajectory when GPS errors are between 1 to 3 meters.

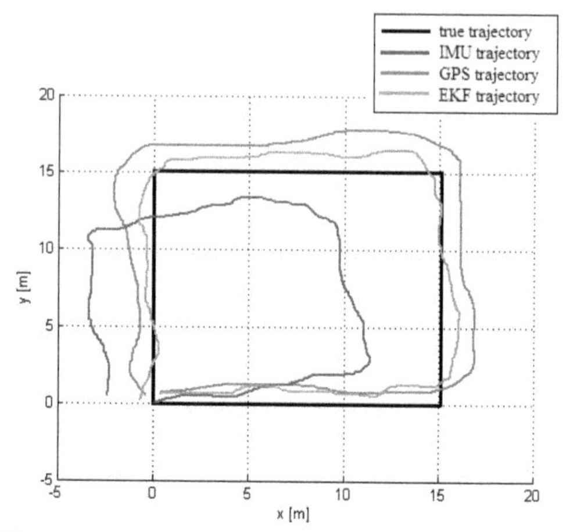

Figure 6. The pedestrian trajectory represented by different methodologies for 15mx15m square.

Figure 7. The environment test area.

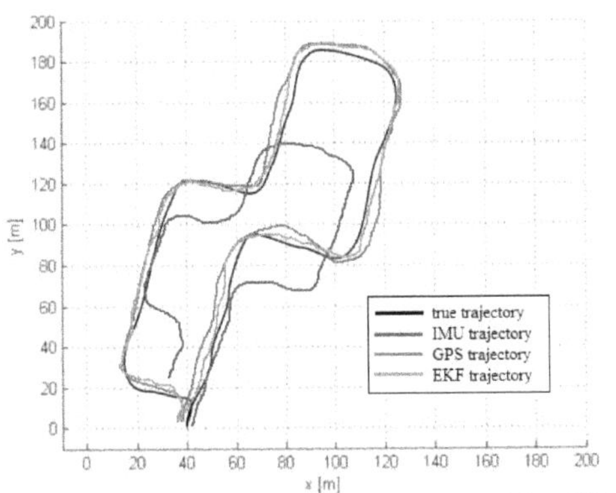

Figure 8. The pedestrian trajectory represented by different methodologies for environment.

5 CONCLUSIONS

In this paper, we proposed a EKF to estimate the location of pedestrian using IMU/GPS information. Obtained results show that d EKF algorithm for data integration from IMU and GPS is accurate and robust. However, the situation of long-term GPS outage is not considered in this paper. Our future work will focus on the improvement of the localization accuracy in long-term operations.

REFERENCES

[1] Feliz R., Zalama E., Garcia-Bermejo J. G. 2009. Pedestrian cracking using inertial sensors. *Journal of Physical Agents* Vol. 3, No. 1.

[2] Abdelkrim N. Nabil A. 2010. Robust INS/GPS sensor fusion for UAV localization using SDRE nonlinear filtering. *IEEE Sensors Jurnal* Vol. 10, No. 4.

[3] Vik B., Fossen T. I. 2001. A Nonlinear Observer for GPS and INS Integration. *Proceedings of the 40th IEEE Conference on Decision and Control.* Orlando, Florida, USA.

[4] Grejner-Brzezinska D. A., Toth Ch. K., Sun H., Wang X., Rizos Ch. 2011. A Robust Solution to High-Accuracy Geolocation: Quadruple Integration of GPS, IMU, Pseudolite, and Terrestrial Laser Scanning. *IEEE Transactions on Instrumentation and Measurement* Vol. 60,No. 11.

[5] Waegli A., Guerrier S., Skaloud J. 2008. Redundat MEMS-IMU integrated with GPS for performance assessment in sports. *Position, Location and Navigation Symposium IEEE/ION.* Monterey, California, USA.

[6] Wang M. 2006. Data Fusion of MEMS/GPS Integrated System for Autonomous Land Vehicle. *Proceedingsof the 2006 IEEE International Conference on Infrmation Acqusition.* Weihai, Shandong, China.

[7] Maenaka K. 2008. MEMS inertial sensors and their applications. *5th international conference on networked sensing systems.* Kanazawa, Japan.

[8] Ling Ch., Housheng H. 2012. IMU/GPS Based Pedestrian Localization. *2012 4th Computer Science and Electronic Engineering Conference (CEEC).* University of Essex, UK.

[9] Leutenegger S., Siegwart R. Y. 2012. A Low-Cost and Fail-Safe Inertial Navigation System for Airplanes. *2012 IEEE Conference on Robotics and Automation.* Saint Paul, Minnesota, USA.

[10] Kedzierski J. 2007. Filtr Kalmana – zastosowania w prostych układach sensorycznych. *Koło naukowe robo tyków KoNaR.* Wroclaw University of Technology, Poland.

Chapter 7

Air Navigation

Accuracy of GPS Receivers in Naval Aviation

W.Z. Kaleta
44th Naval Air Base, Siemirowice, Poland

ABSTRACT: The paper presents researches of GPS Navigation Systems Accuracy, which are sophisticated navigational devices mounted on Polish MPA (Maritime Patrol Aircraft) An-28B1R and Mi-14PL helicopter used by Polish Navy for maritime missions. Accuracy is an error between measured values of some navigational data and their real values. That is why this error has a main meaning for the safety of maritime navigation and air navigation, the same for the safety of voyage and flight. It's value should be as small as possible for the success of the mission. The main reason of this article is to evaluate the accuracy of the GPS Navigation Systems: Bendix / King GPS KLN 90B, GARMIN GPS 155XL through the en-route flight phase as a part of reconnaissance and search and rescue missions.

1 INTRODUCTION

Many times using words navigation system, GPS etc. "we are thinking" about geographical position and its accuracy, which in navigational system is a position error in chosen dimension and probability of user position [Januszewski, J. 2010]. Aviation navigation is the most complex kind of navigation because of third dimension existence [Specht, C. 2007]. That is why we should remember, in aviation two kinds of navigation system's faults are important the horizontal position error and the vertical position error while in maritime navigation "we are focused" on horizontal position error only. Highly accurate navigation is the most important factor during a voyage and flight phase. At the very beginning of GPS work an oceanic en-route flight or air corridor flight were secured by the system mentioned above [Specht, C. 2007]. The knowledge of the GPS Navigation System accuracy became one of the elements which have a major meaning for the safety of practical navigation. Today's flight navigation is based in most aircrafts on Global Positioning System (GPS), which uses satellites signals to evaluate geographical coordinates, speed of flight etc. The GPS use is simply and accurate, that is why it's became widely used and more popular each day. One of the GPS device is Bendix / King GPS Navigation System KLN 90B which is a part of An-28 B1R navigational equipment and the another is GARMIN GPS 155XL a part of navy helicopter Mi-14PL. This article presents researches of KLN 90B and GARMIN GPS 155XL accuracy during a flight plan route using Polish Navy MPA aircraft An-28B1R and Polish Navy helicopter Mi-14PL. Also the influence of the accuracy for the safety of navigation during flight phase, initial approach, intermediate approach, non-precision approach, departure, category I precision approach the same success of maritime reconnaissance and search and rescue missions.

2 GPS AIRCRAFT NAVIGATION SYSTEM ICAO REQUIREMENTS

2.1 *Space and control segment accuracy*

- position errors do not include atmospheric and receiver errors shall not exceed limitations shown in Table 1,
- time transfer errors shall not exceed 40 nanoseconds 95 % of the time,
- range domain errors shall not exceed the following limits:
  - any satellite not larger than 30 metres or 4.42 times the broadcast user range accuracy not exceed 150 metres,
- range rate error of any satellite 0.02 metres per second,
- range acceleration error of any satellite 0.007 metres per second,

- root-mean-square range error over all satellites 6 metres [International Civil Aviation Organization.1996].

Table 1. ICAO GPS aircraft system requirements for signal in space horizontal and vertical errors limitations.

Position error type	Global average 95% of the time	Worst site 95% of the time
Horizontal position error	13 m	36 m
Vertical position error	22 m	77 m

2.2 *GPS receivers requirements for different flight phases*

The combination of GPS elements and a fault-free GPS user receiver shall meet the signal-in-space requirements defined in Table 2.

Table 2. ICAO GPS aircraft system signal-in-space performance requirements for different flight phases

Typical operation	Accuracy 95%	
	Horizontal	Vertical
En-route	3.7 km	N/A
Initial approach	0.74 km	N/A
Intermediate approach	0.74 km	N/A
Non-precision approach	0.74 km	N/A
Departure	0.74 km	N/A
Category I precision approach	16 m	6.0 m to 4.0 m

The concept of a fault-free user receiver is applied only as a means of defining the performance of combinations of different GPS elements. The fault-free receiver is assumed to be a receiver with nominal accuracy and time-to-alert performance. Such as receiver is assumed to have no failures that affect the integrity, availability and continuity performance [International Civil Aviation Organization.1996].

3 CHARACTERIZATION OF GPS NAVIGATION SYSTEMS

3.1 *Bendix / King GPS Navigation System KLN 90B*

The KLN 90B is an extremely sophisticated navigational device, which can provide high accurate navigation during en-route flight over most parts of the world. This system consists of four major parts such as:

- panel mounted KLN 90B GPS sensor – navigation computer – which contains the GPS sensor, the navigation computer, a CRT display, and all controls required to operate the unit. It also houses the database cartridge which plugs directly into the back of the unit,
- data base cartridge – which is an electronic memory containing a vast amount of information on airports, navigational aids, intersections, special use airspace, and other items of value to the pilot. The database provides two primary functions: to make pilot interface with the GPS sensor much easier cause of an automatically looks up and display the latitude and longitude associated with the identifier and to serve as a very convenient means to store and easily access a vast amount of aeronautical information. The database is designed to be easily updated by the user by using a laptop computer and AlliedSignal furnished 3.5 inch diskettes and may also be updated by removing the obsolete cartridge and replacing it with a current one,
- an antennas – two can be used (KA91 and KA92) which are "patch" antennas always designed to be mounted on the top of the aircraft,
- an altitude input – which is required to obtain full navigation and operational capabilities, using it an altitude may be provided to the GPS receiver from an encoding altimeter, blind encoder, or one of the air data computers mentioned above. Altitude is used as an aid in position determination when not enough satellites are in view and also used in several altitude related features such as three dimensional special use airspace alerting, height above airport, and altitude alerting. If it's needed some of extra system components may be added or interfaced to the KLN 90B for increasing its capabilities and features [Allied Signal Inc. 1997].
  The extra components may be:
- external course deviation indicator (CDI),
- fuel management system,
- air data system providing with true air speed data which is used for wind determination,
- ARTEX ELS-10 an emergency locator transmitter,
- autopilot [Allied Signal Inc. 1997].

3.2 *Garmin Navigation System GPS 155XL*

The GARMIN GPS 155XL is a powerful navigational tool that provides pilots with accurate navigational data, including non-precision approaches, standard instrumental departures (SIDs) and standard instrumental arrivals (STARs) also can provide high accurate navigation during en-route flight over most parts of the world. GARMIN 155XL system has four major parts architecture such as:

- panel mounted sensor / navigation computer - which contains the GPS sensor, the navigation computer, a CRT display, and all controls required to operate the unit. It also houses the database cartridge which plugs directly into the back of the unit,

- data base cartridge – which is an advanced electronic memory containing high value information such as:
  - airports data necessary for en-route navigation, approach and early phase of landing,
  - navigational aids such as VOR beacons used in radio navigation,
  - special used airspace guarantee the safety of flight,
- an antennas mounted outside the aircraft which are the "patch" kind,
- an altitude input which is necessary to obtain full navigational information of the aircraft from an encoding altimeter, blind encoder or any air data computers mentioned before [Garmin corporation. 1999].

4 CHARACTERIZATION OF AN AIRPORT AIR TRAFFIC CONTROL SURVEILLANCE RADAR

Airport Surveillance Radar (Avia W) is an integrated primary and secondary radar system which is deployed at the airport for an air traffic control. It interfaces with both legacy and digital automation systems and provides six-level national weather service calibrated weather capability that will result in significant improvement in situational awareness for both controllers and pilots. The primary surveillance radar uses a continually rotating antenna mounted on a tower to transmit electromagnetic waves that reflect, or backscatter, from the surface of aircraft. The radar system measures the time required for a radar echo to return and the direction of the signal. From this, the system can measure the distance of the aircraft from the radar antenna and the azimuth, or direction, and calculate the geographical coordinates. The primary radar also provides data of the rainfall intensity.

The secondary radar uses a second radar antenna attached to the top of the primary radar antenna to transmit and receive area aircraft data for barometric altitude, identification code, and emergency conditions. Most of today's aircrafts have transponders that automatically respond to a signal from the secondary radar by reporting an identification code and altitude [http://www.faa.gov…].

ASR Avia W technical parameters:
- probability of calculating the object's real position up to 100 kilometres – 100%,
- range – up to 100 kilometres,
- height of search – up to 10 kilometres,
- length of wave – 23 centimetres,
- probability of detection – 80%,
- minimum surface of the object to detect – 2 square meters,
- antenna's speed of rotary motion – 10 to 15 rotations per second,
- maximum wind speed enables station work – 50 meters per second [Technical specification, OTU1 - AVIA W].

5 KLN 90B NAVIGATION ACCURACY DURING EN-ROUTE FLIGHT

The experiment was conducted in Siemirowice Naval Air Base October 2009 and September 2011. Its main objective was to measure faults between geographical coordinates evaluated by GPS Navigational System KLN 90B and the real position of the aircraft monitored by Airport Surveillance Radar (ASR) – Avia W.

5.1 *Task*

Flight plan route consisted of eleven turning points: Leba – Objazda – Trzebielino – Kramazyny – Wzdzydze – Stara Kiszewa – Przywidz – Gdansk – Gdynia Oksywie – Bialogora – Leba. After departure from Cewice military airfield the aircraft headed North to Leba were the experiment was started. During flight via flight plan route in every 15 degrees counter clockwise measured by personnel in Avia after their call the position of the aircraft was marked on the map and geographical coordinates from GPS Navigational System KLN 90B were written by co-pilot. After " 360 degrees " flight via flight plan route above the Leba the task (the experiment) was over and the aircraft headed South to Cewice military airfield.

6 GARMIN GPS 155XL NAVIGATION ACCURACY DURING EN-ROUTE FLIGHT

The research was took in Darlowo Naval Air Base October 2010 and August 2012. The main goals of this experiment were to examine work of Navigation System GARMIN 155XL and calculate its accuracy of evaluating geographical coordinates in comparison with real geographical position fixed by Airport Surveillance Radar (ASR) – Avia W.

6.1 *Task*

Flight plan route consisted of six turning points: Maritime Point 01 (54°50'00"N 016°20'00"E) – Maritime Point 02 (54°44'00"N 016°48'00"E) – Tursko – Karlino – Maritime Point 03 (54°40'00"N 015°47'00"E) – Maritime Point 04 (54°50'00"N 016°20'00"E). After departure from Darlowo military airfield the helicopter headed North to the first turning point (Maritime Point 01). From this position research was began, Mi-14PL flight clockwise to the next five turning points which left. In every 15 degrees co-pilot wrote geographical coordinates from GARMIN 155XL CRT display, the same did Avia operator from his radar display. Appropriate time of position marker was controlled by use of radio communication. After reaching last one turning point (Maritime Point 01) the experiment was over and the aircraft headed straight to Darlowo navy airfield.

Table 3. Experiment results of Navigation System KLN-90B accuracy in calculating geographical coordinates.

Number	2009			2011		
	Geographical coordinates		Faults	Geographical coordinates		Faults
	Real	Measured	m	Real	Measured	m
1	54°47'54"N 017°49'00"E	54°48'16"N 017°47'16"E	3262	54°49'12"N 017°56'14"E	54°49'10"N 017°55'47"E	832
2	54°46'06"N 017°38'12"E	54°46'19"N 017°37'03"E	2156	54°49'33"N 017°48'32"E	54°49'04"N 017°47'42"E	1758
3	54°43'54"N 017°27'00"E	54°43'54"N 017°27'21"E	646	54°48'48"N 017°37'43"E	54°48'54"N 017°37'14"E	909
4	54°48'01"N 017°24'00"E	54°48'51"N 017°23'27"E	1648	54°45'48"N 017°26'13"E	54°45'33"N 017°25'25"E	1541
5	54°38'06"N 017°08'00"E	54°37'49"N 017°07'58"E	506	54°41'52"N 017°18'24"E	54°41'12"N 017°18'54"E	1499
6	54°32'06"N 017°02'54"E	54°31'46"N 017°03'40"E	1533	54°37'17"N 017°10'11"E	54°36'18"N 017°10'11"E	1744
7	54°25'36"N 017°04'00"E	54°25'08"N 017°03'56"E	837	54°31'49"N 017°05'01"E	54°31'48"N 017°04'44"E	523
8	54°18'00"N 017°05'00"E	54°18'09"N 017°04'27"E	1049	54°24'35"N 017°03'15"E	54°24'24"N 017°03'02"E	515
9	54°12'00"N 017°05'00"E	54°11'56"N 017°05'55"E	1695	54°18'54"N 017°04'33"E	54°18'42"N 017°04'18"E	582
10	54°04'54"N 017°12'54"E	54°05'05"N 017°12'39"E	564	54°12'41"N 017°09'26"E	54°12'33"N 017°09'28"E	244
11	54°02'00"N 017°25'00"E	54°02'23"N 017°24'06"E	1794	54°07'44"N 017°17'56"E	54°07'36"N 017°18'07"E	413
12	53°58'42"N 017°46'00"E	53°58'48"N 017°46'10"E	355	54°04'38"N 017°28'21"E	54°04'36"N 017°28'22"E	67
13	53°57'48"N 017°58'30"E	53°57'16"N 017°58'31"E	942	54°03'09"N 017°34'39"E	54°03'01"N 017°34'18"E	687
14	53°59'36"N 018°10'18"E	53°59'23"N 018°10'15"E	393	54°00'58"N 017°42'38"E	54°00'48"N 017°42'57"E	654
15	54°05'48"N 018°14'53"E	54°05'24"N 018°14'02"E	1719	54°00'55"N 017°49'56"E	54°00'57"N 017°50'20"E	740
16	54°11'30"N 018°18'56"E	54°10'04"N 018°18'41"E	2567	54°03'39"N 017°57'20"E	54°03'41"N 017°57'37"E	526
17	54°12'49"N 018°21'00"E	54°12'39"N 018°21'00"E	294	54°07'16"N 018°10'17"E	54°07'16"N 018°10'32"E	461
18	54°17'56"N 018°30'00"E	54°17'29"N 018°30'59"E	1979	54°10'55"N 018°18'15"E	54°10'48"N 018°18'07"E	321
19	54°24'18"N 018°35'00"E	54°24'08"N 018°35'13"E	496	54°13'32"N 018°25'37"E	54°13'35"N 018°25'26"E	349
20	54°31'12"N 018°32'30"E	54°31'16"N 018°33'06"E	1113	54°16'28"N 018°27'43"E	54°16'27"N 018°27'34"E	278
21	54°37'13"N 018°25'07"E	54°37'14"N 018°25'14"E	217	54°21'51"N 018°28'14"E	54°21'55"N 018°28'02"E	387
22	54°39'52"N 018°19'12"E	54°40'09"N 018°18'28"E	1442	54°25'14"N 018°29'19"E	54°25'21"N 018°29'09"E	370
23	54°43'56"N 018°10'00"E	54°44'44"N 018°09'22"E	1832	54°28'15"N 018°25'15"E	54°28'18"N 018°25'17"E	107
24	54°47'07"N 017°46'42"E	54°48'31"N 017°47'15"E	2674	54°33'53"N 018°24'40"E	54°34'06"N 018°24'25"E	599

Table 4. Experiment results of Navigation System GARMIN 155XL accuracy in calculating geographical coordinates.

Number	2008			2012		
	Geographical coordinates		Faults	Geographical coordinates		Faults
	Real	Measured	m	Real	Measured	m
1	54°48'42"N 016°34'54"E	54°48'24"N 016°35'00"E	565	54°48'47"N 016°34'55"E	54°48'30"N 016°35'00"E	527
2	54°50'18"N 016°21'49"E	54°50'13"N 016°22'06"E	543	54°50'18"N 016°21'55"E	54°50'10"N 016°22'05"E	388
3	54°49'07"N 016°10'48"E	54°49'18"N 016°11'00"E	493	54°49'09"N 016°10'45"E	54°49'20"N 016°11'00"E	565
4	54°46'43"N 015°59'37"E	54°47'00"N 016°00'06"E	1025	54°46'44"N 015°59'40"E	54°47'00"N 016°00'00"E	777
5	54°42'30"N 015°50'08"E	54°42'42"N 015°50'43"E	1134	54°42'33"N 015°50'04"E	54°42'45"N 015°50'10"E	402
6	54°37'25"N 015°43'07"E	54°37'36"N 015°43'38"E	1008	54°37'26"N 015°43'14"E	54°37'33"N 015°43'21"E	300
7	54°31'19"N 015°39'06"E	54°29'47"N 015°39'13"E	2748	54°31'13"N 015°39'02"E	54°29'58"N 015°39'11"E	251
8	54°24'43"N 015°37'19"E	54°24'42"N 015°37'30"E	339	54°24'44"N 015°37'21"E	54°24'55"N 015°37'36"E	566
9	54°18'13"N 015°39'00"E	54°18'08"N 015°39'06"E	237	54°18'12"N 015°39'00"E	54°18'05"N 015°39'10"E	371
10	54°12'12"N 015°43'25"E	54°12'19"N 015°43'12"E	451	54°12'09"N 015°43'32"E	54°12'20"N 015°43'10"E	751
11	54°06'53"N 015°50'30"E	54°07'07"N 015°50'18"E	556	54°06'53"N 015°50'36"E	54°07'00"N 015°50'20"E	534
12	54°03'00"N 015°59'19"E	54°03'24"N 015°59'13"E	737	54°03'02"N 015°59'28"E	54°03'18"N 015°59'05"E	852
13	54°00'13"N 016°09'18"E	54°00'13"N 016°09'06"E	369	54°00'18"N 016°09'16"E	54°00'11"N 016°09'00"E	534
14	53°59'30"N 016°20'18"E	53°59'18"N 016°20'07"E	491	53°59'28"N 016°20'19"E	53°59'28"N 016°20'00"E	584
15	53°59'48"N 016°32'00"E	53°59'58"N 016°32'14"E	523	53°59'57"N 016°32'05"E	53°59'58"N 016°32'11"E	187
16	54°02'37"N 016°42'05"E	54°02'25"N 016°41'55"E	470	54°02'38"N 016°42'09"E	54°02'28"N 016°41'56"E	497
17	54°06'43"N 016°51'13"E	54°06'30"N 016°50'00"E	2277	54°06'41"N 016°51'12"E	54°06'30"N 016°50'00"E	2237
18	54°11'48"N 016°59'00"E	54°11'43"N 016°58'53"E	261	54°11'47"N 016°59'01"E	54°11'36"N 016°58'54"E	390
19	54°18'07"N 017°03'24"E	54°17'54"N 017°03'09"E	600	54°18'00"N 017°03'22"E	54°17'52"N 017°03'11"E	413
20	54°24'38"N 017°04'47"E	54°24'14"N 017°04'45"E	712	54°24'34"N 017°04'57"E	54°24'11"N 017°04'48"E	734
21	54°30'54"N 017°03'37"E	54°30'30"N 017°03'25"E	800	54°30'55"N 017°03'41"E	54°30'36"N 017°03'31"E	640
22	54°37'19"N 016°59'18"E	54°36'36"N 016°59'25"E	1290	54°37'08"N 016°59'19"E	54°36'58"N 016°59'26"E	366
23	54°42'25"N 016°52'49"E	54°42'25"N 016°53'09"E	615	54°42'21"N 016°52'45"E	54°42'19"N 016°53'00"E	465
24	54°46'30"N 016°43'25"E	54°46'19"N 016°43'30"E	360	54°46'37"N 016°43'22"E	54°46'28"N 016°43'36"E	506

7 ANALYSIS OF THE EXPERIMENT RESULTS

During each flight 24 records of the aircraft position were marked by co-pilot and Avia radar operator. For each record, error of evaluated geographical coordinates position was calculated (Table 3 and 4). For KLN 90B Navigational System highest value of position error is 3262 metres while the highest value of position error of GARMIN 155XL Navigation System is 2748 metres, both errors meet the terms of ICAO GPS receiver requirements for en-route navigation. However all 24 records were taken into consideration and the following average accuracies were calculated:

- position accuracy of KLN 90B – 733 metres,
- position accuracy of GARMIN 155XL – 594 metres.

Because of average accuracy was calculated in this experiment, standard deviation was estimated:

- for KLN 90B – 135 metres,
- for GARMIN 155XL – 114 metres.

Comparing these two standard deviations, geographical positions estimating layout in receivers KLN 90B and GARMIN 155XL is almost similar because the difference is only 21 metres. That's why these two receivers are very much alike.

8 CONCLUSIONS

After the experimental flight results were analyzed and the accuracy with standard deviation were calculated. Comparing the experiment results with the ICAO requirements for GPS receivers (Table 2) following conclusions were drawn from.

The value of the average accuracy of the GPS Navigation System KLN 90B is 733 metres and the average accuracy of the GPS Navigation System GARMIN 155XL is 594 metres, both systems average accuracy is enough to enable and maintain high level of the en-route navigation which is a part of tasks provided for naval aviation such as maritime reconnaissance and search and rescue missions. Accuracy of the GPS receiver below one kilometre enables handle the aircraft through the flight plan route simultaneously maintains geographical orientation which is necessary to complete missions mentioned above. As well the two kinds of GPS receivers accuracy enables maintaining such flight phases as:

- initial approach,
- intermediate approach,
- non-precision approach,
- departure (Table 2).

However category I precision approach is impossible to secure and maintain (Table 2).

REFERENCES

Allied Signal Inc. 1997. *KLN 90B Pilot's guide.*

Garmin corporation. 1999. *Pilot's guide and references.*

International Civil Aviation Organization.1996. *Aeronautical Telecommunications, annex 10 to the convention on international civil aviation.*

Januszewski, J. 2010. *Satellite systems GPS, Galileo and the others*. Warsaw: PWN (in Polish).

Specht, C. 2007. *System GPS*. Pelpin: Bernardinum (in Polish).

Technical specification, OTU1 - AVIA W *Airport Surveillance Radar – AVIA W* (in Polish).

http://www.faa.gov/air_traffic/technology/asr-11

Comparative Analysis of the Two Polish Hyperbolic Systems AEGIR and JEMIOLUSZKA

S. Ambroziak, R. Katulski, J. Sadowski, J. Stefański & W. Siwicki
Gdansk University of Technology, Poland

ABSTRACT: Global Navigation Satellite System (GNSS) is seen by terrorists or hostile countries as a high value target. Volpe Center report contains the following statement: "During the course of its development for military use and more recent extension to many civilian uses, vulnerabilities of Global Navigation Satellite Systems (GNSS) – in the United States the Global Positioning System (GPS) – have become apparent. The vulnerabilities arise from natural, intentional, and unintentional sources. Increasing civilian and military reliance on GNSS brings with it a vital need to identify the critical vulnerabilities to civilian users, and to develop a plan to mitigate these vulnerabilities [1]." GNSS can also be targeted by more common criminals - computer hackers and virus writers. Therefore, there is a need for maintenance and continued development of independent radionavigation and radiolocation systems. This article will compare two Polish hyperbolic systems in terms of radio parameters, functionality and usability.

1 INTRODUCTION

In the Department of Radiocommunication Systems and Networks at Gdansk University of Technology, in cooperation with the OBR Marine Technology Centre in Gdynia and with the support of the Hydrographic Office of Polish Navy a ground-based radiolocation system, which was named AEGIR [2,3] has been developed, built and tested in real environment.

There is also another Polish radiolocation system called JEMIOLUSZKA [4] developed over 10 years ago by OBR Marine Technology Centre in Gdynia.

This article will compare these two systems in terms of radio parameters, functionality and usability.

2 TWO POLISH HYBERBOLIC SYSTEMS

2.1 *JEMIOŁUSZKA*

For the purposes of determining the current ship position the system was built in the form of three (B, C and D) ground stations and the number of receivers placed on ships. A set of three stations cooperating with each other creates a so-called chain. In the chain there is a priority base station, which is usually marked with the letter D. It controls the operation of the other ground stations and currently operating radiolocation receivers. The D station controls other stations by broadcasting special signals that synchronize the generators in all devices (ground stations and receivers). For the purpose of radiolocation each station: B, C and D transmits radio signals at two carrier frequencies, in specific moments in time, and with constant phase shift (relative to the priority station). Radio signal implemented in JEMIOLUSZKA uses FDMA / TDMA (Frequency-Division Multiple Access / Time Division Multiple Access).

Radiolocation receiver in the estimation of its position measures the phase difference between the signals transmitted from ground stations. Constant phase difference measured by the receiver corresponds to the so-called line item generated by a pair of main and sub-station. On the basis of at least two phase differences the receiver estimates the position in the local coordinate system, which are then converted to the WGS-84 (World Geodetic System 1984). For the purpose of the initial calibration of the JEMIOLUSZKA receiver a DGPS (Differential Global Positioning System) module is installed.

Each ground station of JEMIOLUSZKA system consists of two containers: equipment and social ones. In the equipment container there is a transmission equipment and optional receiving components. The operating personnel is placed in the social container. Ground station is equipped with a combustion generator, which provides a stand-alone work without the need of electricity supply from the outside.

JEMIOLUSZKA system is characterized by the selected parameters of the radio link:

- the carrier frequency of the system: in the range 1600 - 2 000 kHz,
- transmission channel bandwidth - up to 2,5 kHz,
- The maximum power of the signal transmitted by the station coastline - 50 W.

2.2 *AEGIR*

The AEGIR system [5] has been built as a demonstrator of technology. The system consists of a locator and three reference stations.

The locator has been made in the technology of Software Defined Radio [6]. It consists of: an antenna, a broadband receiver, an analog to digital converter (in the form of data acquisition card) and a digital signal processor (in form of PC). This approach allows to shape flexibly functionality of the locator.

Ground stations, have the ability to "listen to" neighboring stations. It is assumed that the system should consists only of such stations. The main element of the station is a radio signal generator, whose task is to broadcast modulated signal with data that are generated by industrial computer. The task of the receiver is to listen to a nearby station and to determine difference in synchronization between reference signal and signals from the neighboring stations. To enable listening to neighboring stations, ground station has been equipped with a coupler and a SPDT switch, which periodically changes transmitting antenna into a receiving one.

All devices are based on a universal radiocommunication equipment. The entire system functionality is provided by software installed on computers.

AEGIR system is characterized by the selected parameters of the radio link:

- the carrier frequency of the system431.5MHz
- transmission channel bandwidth - up to 10 MHz,
- The maximum power of the signal transmitted by the station coastline - 30 W.

3 SYSTEMS COMPARISON

Tab. 1 shows a comparison of selected usable and functional parameters of JEMIOLUSZKA and AEGIR systems.

Table 1. Summary of selected usable and functional parameters of JEMIOLUSZKA and AEGIR systems.

Parameter name	AEGIR	JEMIOLUSZKA
Used technology	System is fully digital. Built as a technological demonstrator based on universal radio communication devices. The software determines the functionality and developed digital signal processing algorithms. Easy to implement additional features without having to change the hardware. In the final version - develop dedicated hardware layer and the formation of functionality through a software layer.	System is analog. All functional blocks in different units in the system are dedicated and developed in analogue technology. The only digital module measures the phase difference and the conversion of the results (mainly a user interface). Devices require frequent and periodic inspection of the individual parameters of electronic circuits, especially analog phase loops, which are responsible for the quality of synchronization in the system.
Work organization	Fully asynchronous system. There are no master (main) or slave (sub) stations. Failure or damage to one reference station does not affect the work of the entire system. The receiver need to receive radio signals from at least three reference stations.	Chain workflow system, one master station and two or three sub-stations. Damage or failure of the main station results in shutting down the whole chain.
Method of access to the radio channel	CDMA (Code Division Multiple Access) access to the radio channel with a relatively wide band - the target version with recommended bandwidth of 10 MHz. Such a signal is resistant to intentional interference and can receive radio signals below the thermal noise. The use of pseudo-random sequences in the transmitted spread spectrum signals naturally protects from unauthorized access.	FDMA / TDMA (Frequency-Division Multiple Access/ Time-Division Multiple Access) access to the radio channel with narrowband transmission of 2.5 kHz. Such a signal is easy to disrupt by a harmonic signal with a frequency equal to the center frequency of the transmission channel.
Carrier frequency	The choice of the carrier frequency for the implementation of the system was conditioned by adequate frequency resources of the Office of Electronic Communications. Built technological demonstrator is currently working on a carrier frequency equal to 431.5 MHz (UHF - Ultra High Frequency). Maximum range (for this frequency) is the optical horizon.	Carrier frequency of the system is in the range from 1600 kHz to 2000 kHz (band MF - Medium Frequency). The selected frequency range allows a far-reaching signal transmission beyond the optical horizon. The multipath propagation for this frequency range has little effect on the degradation of the received signal quality.

Bandwidth	1 MHz (target bandwidth: 10 MHz). The accuracy of location estimation in the method of TDOA (Time Differential Of Arrival), depends on the sample rate received signals at the receiver. The increase in sampling frequency, results in improvement of accuracy of TDOA method. Thus, systems using spread spectrum are characterized by potentially higher accuracy of locating objects in comparison with narrowband systems, because they simply require higher sampling rates (due to the wide bandwidth).	2.5 kHz - radiolocation broadcast signals have pulse character. Due to the width of the channel, the system is economical on spectrum, in comparison to the AEGIR system.
Data transmission system	Transmitted information is encrypted - it can be used in the localization process only by authorized users. You can also upload additional data to the system, such as the status of reference stations, the occurrence of emergency situations, etc.	The localization process is carried out only basing on received pulses transmitted by the coastal stations - these impulses do not carry information about the system. Knowing the basics of the system, unauthorized users can fully benefit from it. Basing on series of measurements first you can specify the position of the reference station, then you can estimate the position of the object.
The applied method of estimating the position	In the process of estimating the position of the locator the TDOA method is used. Basing on the appointed distance differences (determined on maximum correlation function), direct and unambiguous position estimation of the locator is designated (using for example the Chan algorithm [7]).	System JEMIOLUSZKA is also based on TDOA method, however, unlike the AEGIR system, to determine the distance difference between the locator and the pairs of stations phase relations between received signals are used. During the process of estimating of the locator position, several solutions are obtained which requires selecting a right one, belonging to the relevant line items. In the system line items are repeated every 150 m.
Initialization of the system	System initialization requires entering into the memory of each of the reference station geographic coordinates and turn all electronic modules on. After about 10 minutes the system is ready for operation. At that time, a reference stations determine differences between the reference signal and its neighboring stations and place the data with geographic coordinates in an encrypted message, which is then transmitted to a locator. Based on data collected from at least three reference stations and measurements made by the receiver, the position is estimated automatically.	System initialization also requires entering geographic coordinates to the memory of each of the reference station and then turning all electronic modules on. Next, the start coordinates of the locator are entered - typed directly from the keyboard or inserted automatically using the built-in DGPS signal receiver.
System mobility	The tests carried out in real environment proved that starting a single reference station with installation of an antenna on the lighthouse took one person no more than 0.5 hour.	The starting time of the costal station depends on setting out the 20m high antenna of the radiolocation system and its proper coordination with the transceiver.
Used coordinate system	WGS-84, PUWG 2000	WGS-84
The maximum power of the transmitted signal	The power of 30 W provides radio coverage of the A1 zone, on condition that an antenna is at right height above sea level.	The power of 50 W provides coverage up to 150 km from the coastline.

4 CONCLUSION

During the process of designing the AEGIR system, the following technical assumptions, which according to the authors were important to match the criteria of special applications, has been made. Modern radiolocation systems for Navy vessels should be completely independent of other radiolocation/navigation systems such as GPS, GLONASS or in the future GALILEO. A new method for asynchronous operation of such a system has been developed. It gives up chain organization of reference stations. An important issue in the radiolocation system is its ability to obtain the most accurate information on position of the localized object. It is well known that the distance between ground stations is depends on the shape of the coastline and this geometry affects estimation. In such conditions selected radio link parameters should ensure appropriate resolution measurements of reference signals, which directly affects the accuracy of objects location. It was decided, therefore, to use a CDMA channel access. This method, on the ground of low density of the spectral signals of the radio channel, is preferred for usage in special applications, due to the possibility of receiving signals that are below the thermal noise level. It is also immune to narrowband interferences (intentional or accidental). Considering bandwidth of occupied radio channel (target bandwidth of 10 MHz) carrier frequency was placed in the UHF band. The choice of carrier frequency depends on:

the technical feasibilities of the transceivers, frequency availability and propagation conditions affecting the potential range of the system. The last, but not least important issue, is to ensure personnel-free and reliable operation of the various components of the system. It is all about minimizing interactions of operator and locator with reference station devices. The role of the operator should be reduced only to supervision and control of the system accuracy.

REFERENCES

[1] Vulnerability Assessment of the U.S. Transportation Infrastructure that Relies on GPS, John A. Volpe National Transportations Systems Center, USA, 2001,

[2] R. Katulski, J. Stefański, W. Siwicki, J. Sadowski, S.J. Ambroziak: „Asynchronous system and method for estimating position of persons and/or objects", European patent 11460023,

[3] R. Katulski, J. Stefański, W. Siwicki, J. Sadowski, and S.J. Ambroziak: „Asynchronous system and method for estimating position of persons and/or objects" (in Polish), Patent application no P393181- 08.12.2010,

[4] Siemieniec W., Ławniczak J., Zając R., Radionawigacyjny system JEMIOŁUSZKA, OBR Centrum Techniki Morskiej, 1997,

[5] Ambroziak S., Katulski R., Sadowski J., Siwicki W. and Stefański J.: „Ground-based, Hyperbolic Radiolocation System with Spread Spectrum Signal – AEGIR", Navigational Systems and Simulators, Marine Navagation and Safety of Sea Transportation / ed. Adam Weintrit - London, UK : CRC Press/Balkema, 2011,

[6] Katulski R., Marczak. A., and Stefański J.; *„Software Radio Technology"* (in polish), Telecommunication review and telecommunication news No. 10/2004, pp. 402-406.

[7] Chan Y. T., Ho K. C., A Simple and Efficient Estimator for Hyperbolic Location, IEEE Transactions on Signal Processing, vol. 42, no. 8, str. 1905-1915, 1994.

The Analysis of Implementation Needs for Automatic Dependent Surveillance in Air Traffic in Poland

M. Siergiejczyk & K. Krzykowska
Warsaw University of Technology, Warsaw, Poland

ABSTRACT: Currently, the most popular surveillance systems in Poland are radar systems. This does not mean, however, that are the most effective ones. More and more talking there is about the implementation of new projects in the field of radar technology in Poland. Focus will also be on the aspect of costs related to the implementation of innovative solutions - they are lower than the radar. This therefore means that there is always a system that operates at the highest level (radar) is the best choice (the most effective, least expensive). Extremely promising for surveillance are systems using technology of satellite systems to their operation. So it is with ADS – B. Geographical conditions and imperfection radar systems in Poland cause frequent loss of information about aircraft flying at low altitudes. This phenomenon was a prerequisite to reflect on another, better form of surveillance in the region.

1 INTRODUCTION

The Automatic Dependent Surveillance ADS - B is a low cost effective monitoring system that provides periodic transmission of aircraft parameters (identification, location) via data link transmission mode. Information from ADS - B is broadcast, regardless of which user will receive it (another aircraft, air traffic controller), and without waiting for an answer of the user. However, it is required that information is available in areas of air traffic control surveillance. Each user, both in space and in the ground station can choose how to use the system: receiving, processing or displaying information.

2 ARCHITECTURE OF ADS SYSTEM

ADS - B system is automatic. Automatic, in this case, means that it acts itself and does not require the flight crew or air traffic control services to share information about the position of the aircraft. The ADS - B system is dependent in sense of relying on the source and method of transmitting information of the position of the aircraft (in this case it relies on the global navigation satellite system GNSS).

Currently, ADS is mainly a consulting acting system in the aircraft and not operational. Future use of ADS - B tend to support search and rescue services and monitor the aircraft by fleet operators.

ADS consists of two basic components:

1 part of avionics in airgraft (display in the cockpit);
2 ground station (ground - based transceiver GBT).

Automatic Dependent Surveillance System is considered a key element of air traffic management systems in the future, for example - the European SESAR and American NextGen. However, the wide-range implementation plans are for the years 2020 to 2030.

In contrast to currently used techniques of surveillance, the ADS - B installed in the aircraft - independently determine its location and other parameters and transmits them to ground stations and other users. Updating them is done once per second.

Information provided by air traffic controllers include: aircraft identification, altitude, on which the aircraft speed, the planned path approach, the pressure.

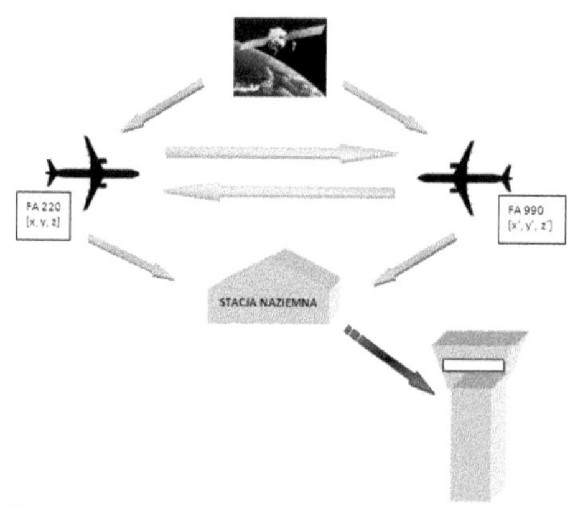

Figure 1. Architecture of ADS system

Determining the position of the vehicle by means of satellite navigation systems depends on the accepted principles of operation of the system and the type of the parameter being measured. Most important principle of operation is to determine the location of the user on the basis of measured values of the position in relation to the satellites. How to determine the position of the satellite depends, among others on the type of orbit. In case of a geostationary satellite, the position coordinates are known and are almost constant relative to the ground segment. In case of elliptical orbits - the coordinates of the satellite depend on the time and are determined by various methods, such as GPS and GLONASS - the coordinates are calculated based on the receiver's knowledge of the elements of the orbits of all the satellites used. Both GPS, GLONASS satellites as well as emitted signals on two frequencies modulate phase information derived from satellite digital memory.

Form of a message sent by a vehicle ADS is a package of information contained in a 56 - bit data field 112 - bit extended squitter transmitted at frequency of 1090 MHz. This package contains a set of defined parameters of the vehicle. Message ADS - B has two functions: ADS - B Out (associated with the transmission of information) and ADS - B In (associated with receiving information).

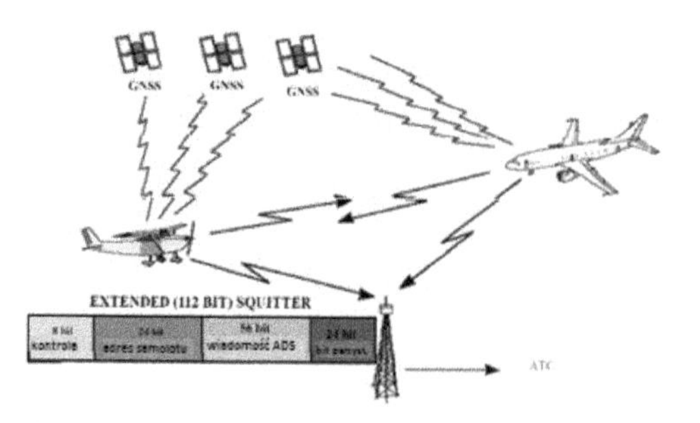

Figure 2. Format of message from ADS system

It is worth focusing on the use of ADS - B in the radar environment. Integrating data from ADS - B and radars can provide the following improvements in surveillance:

- ability to use data from ADS - B if the radar data do not impose themselves and garbling phenomenon occurs;
- ability to use data from ADS - B for ground traffic management, especially when there were problems with the transponder;
- provide surveillance services in areas not covered by radar coverage;
- the possibility of reducing the number of surveillance radars needed for a specific area and fill the gaps in the observation by the ADS - B;
- providing dual, independent of each other surveillance methods of the area;
- improve the location accuracy of aircraft.

In case of system support, ADS - B surveillance in areas not covered by it must meet a number of additional features such as the ability to integrate with other automated systems, display warnings including conflict detection for panels of air traffic controllers and pilots and provide satisfactory accuracy.

The ADS – B system is very popular in the United States, for what it's worth special attention, that among other things there two isolated data link designed for ADS - B: 1090 MHz extended squitter (1090 ES) and relay Universal Access (*Universal Access Transceiver UAT*). ES 1090 is a link dedicated to aviation and military communications, while UAT is designed for general aviation. For air traffic controller in the sense of the type of link of displayed data - it does not matter, because the data are displayed in the same way. In addition, these links also support services outside of ADS - B - TIS - B (*Traffic Information Service - Broadcast*) and FIS - B (*Flight Information Services - Broadcast*).

TIS - B is the service, which is to broadcast information about air traffic to the ground stations. The data source is radar surveillance systems. The advantage of this application is to increase the pilot's situational awareness. FIS - B, in turn, is engaged in broadcasting meteorological information (such as METAR, SPECI, TAF) and other flight information (NOTAM). The advantage of FIS - B is to allow the pilot to obtain information about the current weather situation in the air and at airports.

3 IMPLEMENTATIONS NEEDS FOR ADS SYSTEM

ADS - B is governed by specific standards. In case of aviation proper legal framework is extremely important. Assumptions concerning the implementation of this system can be found among others in the Executive Commission Regulation

(EU) No. 1207/2011 of 22 November 2011 laying down the requirements for the performance and interoperability of surveillance systems in the single European sky. This regulation establishes requirements for systems that aim to provide surveillance data to ensure the effectiveness of the European air traffic management network. According to this decree - Member States of the European Union are committed to the safety assessment of all existing surveillance systems and surveillance data processing . More importantly, there are included requirements for surveillance data provided by the cooperating surveillance networks such as ADS.

4 ANALISYS OF MEASUREMENTS

Radar systems used by air traffic services units should have a very high degree of reliability, availability, and integrity. There should be little likelihood of system failure or significant deterioration of its characteristics, which may result in total or partial work stoppage. It is important to provide back up for the basic radar equipment. Radar systems should have the ability to display visual warnings of the flights. These include notice of the situation of conflict, prejudice the achievement of the minimum safe altitude, or an unintentional duplication of SSR codes. Analysis of radar data quality using computer systems should be performed on the basis at least three six-hour recording situation, with an average flow of air traffic.

Nearly 50% of all air traffic in Poland is transit traffic at certain levels of flight. The most popular and preferred by carriers levels are FL300 and FL 380, according to data from EUROCONTROL. It is also worth noting that continuous growth is expected to air traffic. In Poland, in general. In 2012, it has increased by 7.3% compared to 2010. In 2017 was estimated to increase the number of operations performed during the day at 500 - 1000 compared to 2010.

Under the current rules - in FIR Warszawa is provided radar air traffic services. Controller using one of the call signs (eg. WARSAW RADAR, APPROACH WARSAW, KRAKÓW DIRECTOR) is an indication to the pilot that the area in which it is located, this service is provided. In the whole area of FIR WARSAW there is a multiradar system consisting of 8 stations. In addition to that, they all have a maximum range of 250 NM. When using secondary surveillance radar, aircraft identification can be made using one of the following procedures:

- recognition of the distinctive emblem of the aircraft on radar label;
- recognition of the label previously assigned individual radar code which has been tested attitude;
- direct identification on the label the trademark of radar-equipped aircraft in mod S;
- by transferring radar identification;
- observing the execution of a command setting specific code;
- observe the execution of the instruction set IDENT.

In case of assigning individual aircraft code, make sure that you set the remote code corresponds to the code assigned to it on the flight. Only after this check individual code used as the basis for identification. While the loss is determined by the two-way communication - you should check that the aircraft can pick up the signal transmission of air traffic control authority. This can be traced through the ADS - B ordered the pilot to perform specific maneuvers that would be displayed on the ADS. If the aircraft does not confirm the possibility of receiving a transmission - must be given the separation from other aircraft providing that the pilot will act in accordance with the applicable procedures in this situation.

Polish radar system has been designed to provide full coverage of aviation radar. Therefore, coverage is available above FL95 and at communication airports (Okęcie, Balice, Rębiechowo, Gdańsk, Poznań Ławica). Flying below in the class G, the better the aircraft is visible the closer to the radar antenna is located. It is important that the pilot on each flight had a transponder in mode C. This allows to read the air traffic services his position. Geographical conditions and imperfection radar systems in Poland cause frequent loss of information about aircraft flying at low altitudes. This phenomenon was a prerequisite to reflect on another, better form of surveillance in the region.

Examine the extent of the use of automatic surveillance of air traffic over the Polish can be done by performing measurements of the number of aircraft equipped with ADS – B. In order to verify the assumptions - such measurements were carried out in selected areas, such as in Olsztyn (53.779 ° N 20.489 ° E) and Warsaw (52.259 ° N 21.020 ° E). Receiver used for the measurement was AirNav RadarBox. It is a decoder ADS - B operating in real time. Taking advantage of the area's long range (up to 250 km) embraced a small antenna included in the package. The disadvantage is the extremely high sensitivity of the obstacles. The unit is equipped with the appropriate software, which allows to track aircraft within range of the receiver. It should detect all devices equipped with transponders, working in mode S. In general, the AirNav RadarBox receiver consists of hardware that decodes the data ADS - B and sends them to a computer where they are processed by software and displayed on the 3D interface. Received data can be shared with other users of ADS - B over the network, which allows to see traffic around the world.

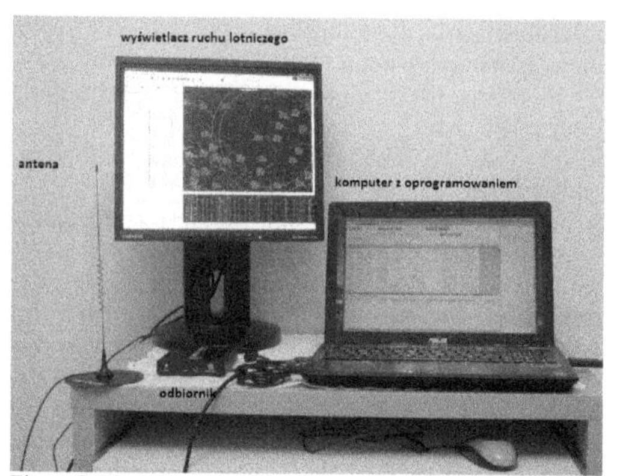

Figure 3. Measurement station (computer & software)

In the Figure 3 the measurement station is shown. It consists of a computer with an appropriate application, an antenna, receiver of signal and a display. No more components are needed to show by application a level of airplanes equipped with ADS-B. Below at the charts some measurement are shown.

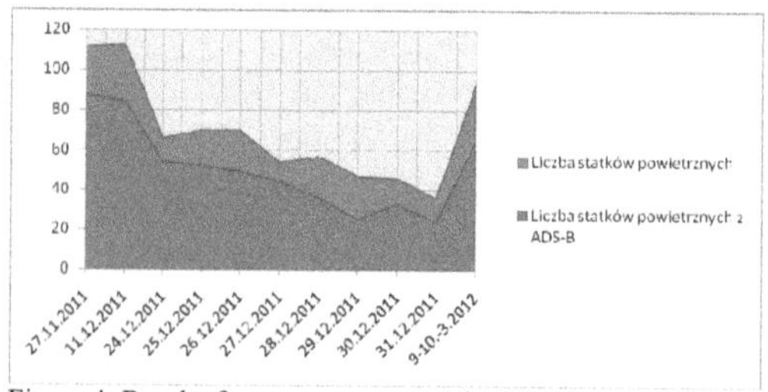

Figure 4. Results from measurements done in Olsztyn

In Figure 4 results from Olsztyn are shown. This is the region of interests in case of ADS-B. Mainly because, there is a lack of surveillance at all. Unfortunately, there some flight levels, often used, where a situation of losing a signal from an airplane currently occurred. Equipped level is around 70%.

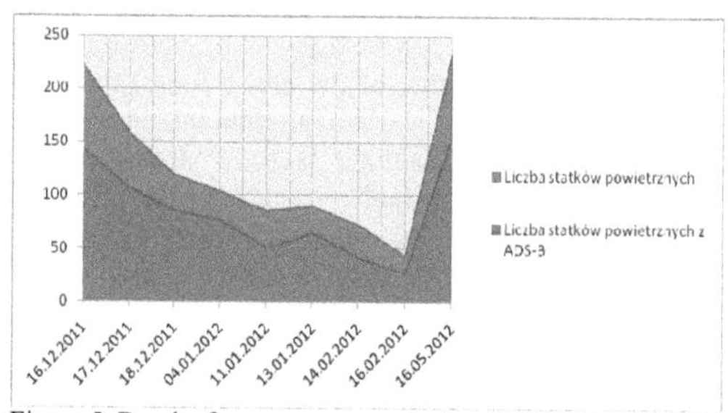

Figure 5. Results from measurements done in Warsaw

In Figure 5 results from Warsaw are shown. The main difference between Olsztyn and Warsaw is type of traffic. Apparently, there are two airports close to the capital city, which make traffic much different than it is in Olsztyn. Though, equipped level is still around satisfying 70%.

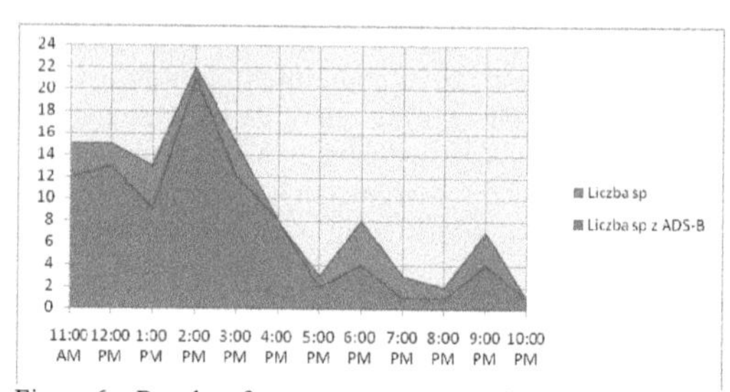

Figure 6. Results from measurements done in north-east Poland 27.11.2011

In Figure 6, there is an example of measurements done in north-east Poland during one day. Such measurement were done for each of the day mentioned in figures 4 and 5.

The measurements described above can be concluded that the system of automatic surveillance (ADS) is used in a large range of aircraft flying over the area in Polish. The level of equipment of aircraft in the system fluctuates around 70%. Therefore, it is worth considering rightness of the use of such a system as a support air traffic control services in surveillance. In line with existing trends in the world, a step towards the implementation of the surveillance technology in Poland would be to build a ground station that supports ADS – B. In addition, in the future, along with the progress in reaching the certification of ADS, the system could perhaps serve as a third layer of radar coverage in areas where it is necessary. The validity of the ADS can also confirm its strengths, such as:

- the use of satellite systems is currently the cheapest form of acquiring data on vehicle position, velocity and other parameters;
- by ADS pilots will have more situational awareness;
- the system can significantly reduce the workload of the pilot and air traffic controller.

5 CONCLUSIONS

Currently, the most popular surveillance systems in Poland are radar systems. This does not mean, however, that are the most effective ones. More and more talking there is about the implementation of new projects in the field of radar technology in Poland. As a result, it becomes commonplace using multilateration surveillance - particularly in the airports. It is worth noting that the construction of hyperbolic systems are often associated with the provision of the ADS system - B. Focus will also be on the aspect of costs related to the implementation of innovative solutions - they are lower than the radar. This therefore means that there is always a system that operates at the highest level (radar) is the best choice (the most effective, least expensive). Extremely promising for surveillance are systems using technology of satellite systems to their operation. So it is with ADS – B.

REFERENCES

Guidance for the provision of Air Traffic Services Using ADS – B in Radar Airspace (ADS – B – RAD), European Organisation For The Safety Of Air Navigation, Brussels 2008.

Air Traffic Bulletin,U.S. Department of Transportation Federal Aviation Administration, Washington 2005.

Guidance for the provision of Air Traffic Services Using ADS – B in Radar Airspace (ADS – B – RAD), European Organisation For The Safety Of Air Navigation, Brussels 2008.

Guidance for the provision of Air Traffic Services Using ADS – B in Non Radar Airspace (NRA), European Organisation For The Safety Of Air Navigation, Brussels 2008.

ROZPORZĄDZENIE WYKONAWCZE KOMISJI (UE) NR 1207/2011 z dnia 22 listopada 2011 r. ustanawiające wymogi dotyczące skuteczności działania i interoperacyjności systemów dozorowania w jednolitej europejskiej przestrzeni powietrznej.

COMMISSION REGULATION (EC) No 262/2009 of 30 March 2009 laying down requirements for the coordinated allocation and use of Mode S interrogator codes for the single European sky.

COMMISSION REGULATION (EU) No 691/2010 of 29 July 2010 laying down a performance scheme for air navigation services and network functions and amending Regulation (EC) No 2096/2005 laying down common requirements for the provision of air navigation service.

Preliminary Safety Case for Airborne Traffic Situational Awareness for Enhanced Visual Separation on Approach, European Organisation For The Safety Of Air Navigation, Brussels 2011.

Preliminary Safety Case for ADS-B Airport Surface Surveillance Application, European Organisation For The Safety Of Air Navigation, Brussels 2011.

K. Krzykowska, M. Siergiejczyk, Naziemny system detekcji pasów startowych i dróg kołowania – ASDE – a bezpieczeństwo w ruchu lotniczym, www.czasopismologistyka.pl, listopad 2012 r.

Chapter 8

Maritime Communications

Multiple Access Technique Applicable for Maritime Satellite Communications

S.D. Ilcev
Durban University of Technology (DUT), South Africa

ABSTRACT: In fixed satellite communication systems, as a rule, especially in global mobile applications for Maritime Mobile Satellite Communications (MMSC) many users are active at the same time. The problem of simultaneous communications between many single or multipoint mobile satellite users, however, can be solved by using Multiple Access (MA) technique. Since the resources of the systems such as the transmitting power and the bandwidth are limited, it is advisable to use the channels with complete charge and to create a different MA to the channel. This generates a problem of summation and separation of signals in the transmission and reception parts, respectively. Deciding this problem consists in the development of orthogonal channels of transmission in order to divide signals from various users unambiguously on the reception part. In this paper are introduced fundamentals, characteristics, advantages and disadvantages of the following MA techniques: Frequency Division Multiple Access (FDMA), Time Division Multiple Access (TDMA), Code Division Multiple Access (CDMA), Space Division Multiple Access (SDMA) and Random (Packet) Division Multiple Access (RDMA) suitable for mobile applications onboard ships.

1 INTRODUCTION

Fixed and mobile satellite communication systems are using five principal forms of MA techniques:

1 Frequency Division Multiple Access (FDMA) is a scheme where each concerned Earth Station, such as Coast Earth Station (CES) or Ship Earth Station (SES), is assigned its own different working carrier Radio Frequency (RF) inside the spacecraft transponder bandwidth.
2 Time Division Multiple Access (TDMA) is a scheme where all concerned Earth stations use the same carrier RF and bandwidth with time sharing and non-overlapping intervals.
3 Code Division Multiple Access (CDMA) is a scheme where all concerned Earth stations simultaneously share the same bandwidth and recognize the signals by various processes, such as code identification. Actually, they share the resources of both frequency and time using a set of mutually orthogonal codes, such as a Pseudorandom Noise (PN) sequence.
4 Space Division Multiple Access (SDMA) is a scheme where all concerned Earth stations can use the same RF at the same time within a separate space available for each link.
5 Random (Packet) Division Multiple Access (RDMA) is a scheme where a large number of satellite users share asynchronously the same transponder by randomly transmitting short burst or packet divisions.

Currently, these methods of multiple access are widely in use with many advantages and disadvantages, together with their combination of hybrid schemes or with other types of modulations. Hence, multiple access technique assignment strategy can be classified into three methods as follows: (1) Preassignment or fixed assignment; (2) Demand Assignment (DA) and (3) Random Access (RA); the bits that make up the code words in some predetermined fashion, such that the effect of an error burst is minimized.

In the preassignment method channel plans are previously determined for chairing the system resources, regardless of traffic fluctuations. This scheme is suitable for communication links with a large amount of traffic between receivers (Rx and transmitters (Tx). Since most SES users in MMSC do not communicate continuously, the preassignment method is wasteful of the satellite resources. In Demand Assignment Multiple Access (DAMA) satellite channels are dynamically assigned to mobile users according to the traffic requirements.

Due to high efficiency and system flexibility, DAMA schemes are suited to MSC systems.

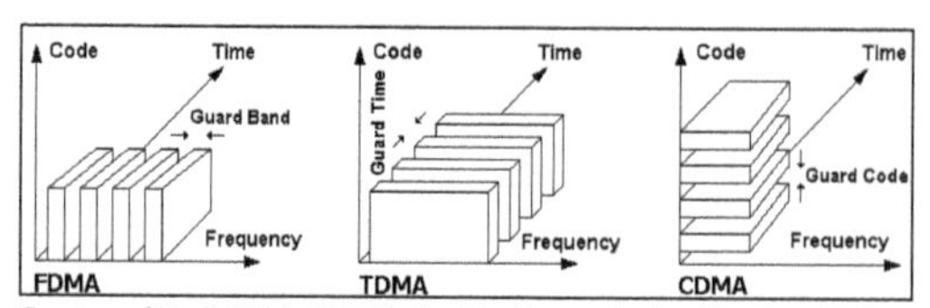

Courtesy of Book: "Global Mobile Satellite Communications" by Ilcev [01]
Figure 1. Multiple Access Techniques

In RA a large number of mobile users use the satellite resources in bursts, with long inactive intervals. So, to increase the system throughout, several mobile Aloha methods have been proposed. Therefore, the MA techniques permit more than two Earth stations to use the same satellite network for interchanging information. In such a way, several transponders in the satellite payload share the RF bands in use and each transponder will act independently of the others to filter out its own allocated RF and further process that signal for transmission. Thus, this feature allows any maritime CES located in the corresponding coverage area to receive carriers originating from several SES and vice versa and carriers transmitted by one SES can be received by any CES. This enables a transmitting Earth station to group several signals into a single, multi-destination carrier. Access to a transponder may be limited to single carrier or many carriers may exist simultaneously. The baseband information to be transmitted is impressed on the carrier by the single process of multi-channel modulation [01, 02, 03, 04].

2 FREQUENCY DIVISION MULTIPLE ACCESS (FDMA)

The most common and first employed MA scheme for satellite communication systems is FDMA concept shown in Figure 1. (FDMA), where transmitting signals occupy non-overlapping RF bands with guard bands between signals to avoid interchannel interference. The bandwidth of a repeater channel is therefore divided into many sub-bands each assigned to the carrier transmitted by an SES continuously. In such a way, the channel transmits several carriers simultaneously at a series of different RF bands. Because of interchannel interference, it is necessary to provide guard intervals between each band occupied by a carrier to allow for the imperfections of oscillators and filters. The downlink Rx selects the required carrier in accordance with the appropriate RF. When the satellite Tx is operating close to its saturation, nonlinear amplification produces intermodulation (IM) products, which may cause interference in the signals of other users. In order to reduce IM, it is necessary to operate the transponder by reducing the total input power according to input back off and that the IF amplifier provides adequate filtering.

Therefore, FDMA allocates a single satellite channel to one mobile user at once. In fact, if the transmission path deteriorates, the controller switches the system to another channel. Although technically simple to implement, FDMA is wasteful of bandwidth because the voice channel is assigned to a single conversation, whether or not somebody is speaking. Moreover, it cannot handle alternate forms of data, only voice transmissions. This system's advantages are that it is simple technique using equipment proven over decades to be reliable and it will remain very commonly in use because of its simplicity and flexibility.

It does have some disadvantages however:

1 An FDMA method is the relatively inflexible system and if there are changes in the required capacity, then the RF plan has to change and thus, involve many CES.
2 Multiple carriers cause IM in both the SES HPA and in the transponder HPA. Reducing IM requires back off of the HPA power, so it cannot be exploited at full capacity.
3 As the number of carriers increase, the IM products between carriers also increase and more HPA back off is needed to optimize the system. The throughput decreases relatively rapidly with the number of transmission carriers, therefore for 25 carriers it is about 40% less than with 1 carrier.
4 The FM system can suffer from what is known as a capture effect, where if two received signals are very close in RF but of different strengths, the stronger one tends to suppress the weaker one. For this reason the carrier power has to be controlled carefully.

Thus, with the FDMA technique, the signals from the various users are amplified by the satellite transponder in a given allocated bandwidth at the same time but at different frequencies. Depending on the multiplexing and modulation techniques employed, several transmission hybrid schemes can be considered and in general may be divided into two categories, based on the traffic demands of Earth stations on MCPC and SCPC.

1. Multiple Channels Per Carrier (MCPC) – Its main elements are multiplexer, modulator and transmitter using a satellite uplink, when CES multiplexes baseband data is received from a terrestrial network and destined for various SES terminals. Then the multiplexed data are modulated and transmitted to the allocated RF segment, when the bandwidth of the transponder is shared among several SES units, each with different traffic requirements. The transponder bandwidth is divided into several fixed segments, with several time

frequency divisions allocated to these SES units and between each band segment is a guard band, which reduces the bandwidth utilization efficiency and the loss is directly related to the number of accessing SES in the network, see Figure 1 (FDMA). Depending on the number of receiving SES units, a total number of carriers will pass through the satellite transponder. The signals received from different SES units extract the carrier containing traffic addressed to CES by using an appropriate RF filter, demodulator, baseband filter and demultiplexer. The output of the demodulator consists in multiplexed telephone channels, a baseband filter is used to filter out the desired baseband frequency and a demultiplexer retrieves individual telephone channels and feeds them into the terrestrial network for onward transmission. Each baseband filter of LES receive stations in this scheme corresponds to a specific one in the LES transmitting station.

2. Single Channel Per Carrier (SCPC) – For certain applications, such as the provision of MMSC service to remote areas or individual SES, traffic requirements are low. In reality, assigning multiple channels to each SES is wasteful of bandwidth because most channels remain unutilized for a significant part of the day. For this type of application the SCPC type of FDMA is used. In the SCPC system each carrier is modulated by only one voice or by low to medium bit rate data channel. Some old analog systems use Companded FM but most new systems are digital PSK modulated. In the SCPC scheme, each carrier transmits a single carrier. The assignment of transponder channels to each SES may be fixed Pre-Assigned Multiple Access (PAMA), about 5 to 10 channels, or variable Demand-Assigned Multiple Access (DAMA), when a pool of frequency is shared by many SES terminals. When necessary, each SES requests a channel from RF management of the Network Control Station (NCS), which may always attempt to choose the best available channel or a lower quality channel until an unoccupied channel has been found. The SCPS solution requires an Automatic Frequency Control (AFC) pilot to maintain the spectrum centering on a channel-by-channel basis. This is usually achieved by transmitting a pilot tone in the centre of the transponder bandwidth. It is transmitted by designated reference CES and all the SES units use this reference to correct their transmission frequency. A receiving station uses the pilot tone to produce a local AFC system, which is able to control the frequency of the individual carriers by controlling the frequency of the Local Oscillator (LO). This scheme is cost-effective for networks consisting in a significant number of Earth stations, each needing to be equipped with a small number of channels. Using this scheme, Inmarsat system of A, B, C, M, Fleet 33/55/77 and FleetBroadband standards can simply provide a higher usage of channels and can utilize demand-assignment equipment.

There are few hybrids of multiplexed FDMA combined with SCPS, PSK, TDM and TDMA techniques:

1. SCPC/FM/FDMA – The baseband signals from the network or users each modulate a carrier directly, in either analog or digital form according to the nature of SCPC signal in question. Each carrier accesses the satellite on its particular frequency at the same time as other carriers on the different frequencies from the same or other station terminals. Information routing is thus, performed according to the principle of one carrier per link utilizing analog transmission with FM for SES telephone channels. For calculation of channel capacity of this scheme it is necessary to ensure that the noise level does not exceed specified defined values.

2. SCPC/PSK/FDMA – Each voice or data channel is modulated onto its own RF carrier using this scheme. The only multiplexing occurs in the transponder bandwidth, where frequency division produces individual channels within the bandwidth. Various types of this multiplex scheme are used in channels of the Inmarsat standard-B SES. In this case, the satellite transponder carrier frequencies may be PAMA or DAMA. For PAMA carriers the RF is assigned to a channel unit and the PSK modem requires a fixed-frequency LO input. For DAMA, the channels may be connected according to the availability of particular carrier frequencies within the transponder RF bandwidth. For this arrangement, the SCPC channel frequency requirement is produced by a frequency synthesizer. The forward link assigned by TDM in shore-to-ship direction uses the SCPC/DA/FDMA solution for Inmarsat standard-B voice/data transmission. This standard in the return link for channel request employs Aloha O-QPSK and for low speed data/telex uses the TDMA scheme in ship-to-shore direction. The Inmarsat-Aero in forward ground-to-aircraft direction uses packet mode TDM for network broadcasting, signaling and data and the circuit mode of SCPS/DA/FDMA with distribution channel management for service communication links. Thus, the request for channel assignment, signaling and data in the return aircraft-to-ground direction the Slotted Aloha BPSK (1/2 – FES) of 600 b/s is employed and consequently, the TDMA scheme is reserved for data messages.

3. TDM/FDMA – This arrangement allows the use of TDM groups to be assembled at the satellite in FDMA, while the PSK is used as a modulation process at the Earth station. Systems such as this are compatible with FDM/FDMA carriers sharing the same transponders and the terminal requirements are simple and easily incorporated. The Inmarsat standard-B system for telex low speed data uses this

scheme in the shore-to-ship direction only and in the ship-to-shore direction uses TDMA/FDMA. The CES TDM and SES TDMA carrier frequencies are pre-allocated by Inmarsat. Each CES is allocated at least one forward CES TDM carrier frequency and a return SES TDMA frequency. So, additional allocations can be made depending on the traffic requirements. The channel unit associated with the CES TDM channel for transmission consists in a multiplexer, different encoder, frame transmission synchronizer and modulator. So at the SES, the receive path of the channel has the corresponding functions to the transmitted end. The CES TDM channels use BPSK with differential coding, which is used for phase ambiguity resolution at the receive end.

4. TDMA/FDMA – As is known, the TDMA signals could occupy the complete transponder bandwidth. In fact, a better variation of this is where the TDMA signals are transmitted as a sub-band of transponder bandwidth, the remainder of which being available for example for SCPC/FDMA signals. Thus, the use of a narrowband TDMA arrangement is well suited for a system requiring only a few channels and has the all advantages of satellite digital transmission but can suffer from intermodulation with the adjacent FDMA satellite channels. Accordingly, the practical example of this multiple schemes is the Tlx (Telex) service of the Inmarsat Standard-B system in ship-to-shore direction, which, depending on the transmission traffic, offers a flexible allocation of capacity for satellite communication and signaling slots [01, 05, 06, 07].

3 TIME DIVISION MULTIPLE ACCESS (TDMA)

The TDMA application is a digital MA technique that permits individual Earth station transmissions to be received by the satellite in separate, non-overlapping time slots, called bursts, which contain buffered information. The satellite receives these bursts sequentially, without overlapping interference and is then able to retransmit them to the SES terminal. Synchronization is necessary and is achieved using a reference station from which burst position and timing information can be used as a reference by all other stations. Each SES must determine the satellite system time and range so that the transmitted signal bursts, typically QPSK modulated, are timed to arrive at the satellite in the proper time slots. The offset QPSK modulation is used by Inmarsat-B SES. So as to ensure the timing of the bursts from multiple SES, TDMA systems use a frame structure arrangement to support Tlx in the ship-to-shore direction. Therefore, a reference burst is transmitted periodically by a reference station to indicate the start of each frame to control the transmission timing of all data bursts. A second reference burst may also follow the first in order to provide a means of redundancy. In the proper manner, to improve the imperfect timing of TDMA bursts, several synchronization methods of random access, open-loop and closed-loop have been proposed.

In Figure 1 (TDMA) a concept of TDMA is illustrated, where each SES terminal transmits a data burst with a guard time to avoid overlaps. Since only one TDMA burst occupies the full RF bandwidth of the satellite transponder at a time, input back off, which is needed to reduce IM interference in FDMA, is not necessary in TDMA. At any instant in time, the transponder receives and amplifies only a single carrier. Thus, there can be no IM, which permits the satellite amplifier to be operated in full HPA saturation and the transmitter carrier power need not be controlled. Because all SES units transmit and receive at the same frequency, tuning is simplified. This results in a significant increase in channel capacity. Another advantage over FDMA is its flexibility and time-slot assignments are easier to adjust than RF channel assignments. The transmission rate of TDMA bursts is about 4,800 b/s, while the frame length is about 1.74 seconds and the optimal guard time is approximately 40 msec, using the open-loop burst synchronization method.

There are some disadvantages because TDMA is more complex than FDMA:

1 Two reference stations are needed and complex computer procedures, for automated synchronizations between SES terminals.
2 Peak power and bandwidth of individual SES terminals need to be larger than with FDMA, owing to high burst bit rate.

Accordingly, in the TDMA scheme, the transmission signals from various mobile users are amplified at different times but at the same nominal frequency, being spread by the modulation in a given bandwidth. Depending on the multiplexing techniques employed, two transmission hybrid schemes can be introduced for use in MMSC systems.

1. TDM/TDMA – The Inmarsat analog standard-A uses the TDM/TDMA arrangement for telex transmission. Each SES has at least one TDM carrier and each of the carriers has 20 telex channels of 50 bauds and a signaling channel. Moreover, there is also a common TDM carrier continuously transmitted on the selected idle listening frequency by the NCS for out-of-band signaling. The SES remains tuned to the common TDM carrier to receive signaling messages when the ship is idle or engaged in a telephone call. When an SES is involved in a telex forward call it is tuned to the TDM/TDMA frequency pair associated with the corresponding CES to send messages in shore-to-

ship direction. Telex transmissions in the return ship-to-shore direction form a TDMA assembly at the satellite transponder. Each frame of the return TDMA telex carrier has 22 time slots, while each of these slots is paired with a slot on the TDM carrier. The allocation of a pair of time slots to complete the link is received by the SES on receipt of a request for a telex call. Otherwise, the Inmarsat-A uses for forward signaling a telex mode, while all other MSS Inmarsat standards for forward signaling and assignment channels use the TDM BPSK scheme.

The new generation Inmarsat digital standard-B (inheritor of standard-A) uses the same modulation TDM/TDMA technique but instead of Aloha BPSK (BCH) at a data rate of 4800 b/s for the return request channel used by Inmarsat-A, new standard-B is using Aloha O-QPSK (1/2 – FEC) at a data rate of 24 Kb/s. This MA technique is also useful for the Inmarsat standard-C terminal for maritime, land and aeronautical applications. In this case, the forward signaling and sending of messages in ground-to-mobile direction use a fixed assigned TDM carrier. The return signaling channel uses hybrid, slotted Aloha BPSK (1/2 FEC) with a provision for receiving some capacity and the return message channels in the mobile-to-ground direction are modulated by the TDMA system at a data rate of 600 b/s.

2. FDMA/TDMA – The Iridium mobile system employs a hybrid FDMA/TDMA access scheme, which is achieved by dividing the available 10.5 MHz bandwidth into 150 channels introduced into the FDMA components. Each channel accommodates a TDMA frame comprising eight time slots, four for transmission and four for reception. Each slot lasts about 11.25 msec, during which time data are transmitted in a 50 Kb/s burst. Each frame lasts 90 msec and a satellite is able to support 840 channels. Therefore, a user is allocated a channel occupied for a short period of time, during which transmissions occur [01, 08, 09, 10, 11].

4 CODE DIVISION MULTIPLE ACCESS (CDMA)

The modern CDMA solution is based on the use the modulation technique also known as Spread Spectrum Multiple Access (SSMA), which means that it spreads the information contained in a particular signal of interest over a much greater bandwidth than the original signal. In this MA scheme the resources of both frequency bandwidth and time are shared by all users employing orthogonal codes, shown in Figure 1 (CDMA). The CDMA is achieved by a PN (Pseudo-Noise) sequence generated by irreducible polynomials, which is the most popular CDMA method. In this way, a SSMA method using low-rate error correcting codes, including orthogonal codes with Hadamard or waveform transformation has also been proposed.

Concerning the specific encoding process, each user is actually assigned a signature sequence, with its own characteristic code, chosen from a set of codes assigned individually to the various users of the system. This code is mixed, as a supplementary modulation, with the useful information signal. On reception side, from all the signals that are received, a given mobile user is able to select and recognize, by its own code, the signal, which is intended for it, and then to extract useful information. The other received signal can be intended for other users but they can also originate from unwanted emissions, which gives CDMA a certain anti-jamming capability. For this operation, where it is necessary to identify one CDMA transmission signal among several others sharing the same band at the same time, correlation techniques are generally employed. From a commercial and military perspective this MA is still new and has significant advantages. Interference from adjacent satellite systems including jammers is better solved than with other systems. This scheme is simple to operate as it requires no synchronization of the Tx and is more suited for a military SES. Small antennas can be very useful in these applications, without the interference caused by wide antenna bandwidths.

Using multibeam satellites, frequency reuse with CDMA is very effective and allows good flexibility in the management of traffic and the orbit/spectrum resources. The Power Flux Density (PFD) of the CDMA signal received in the service area is automatically limited, with no need for any other dispersal processes. It also provides a low probability of intercept of the users and some kind of privacy, due to individual characteristic codes. The main disadvantage of CDMA by satellite is that the bandwidth required for the space segment of the spread carrier is very large, compared to that of a single unspread carrier, so the throughput is somewhat lower than with other systems. Using this scheme, the signals from various users operate simultaneously, at the same nominal RF, but are spread in the given allocated bandwidth by a special encoding process. Depending on the multiplexing techniques employed the bandwidth may extend to the entire capacity of the transponder but is often restricted to its own part, so CDMA can possibly be combined in the hybrid scheme with FDMA and/or TDMA.

The SSMA technique can be classified into two methods: Direct Sequence (DS) and Frequency Hopping (FH). A combined system of DC and FH is called a hybrid CDMA system and the processing gain can be improved without increases of chip rate. The hybrid system has been used in the military Joint Tactical Information Distribution System

(JTIDS) and OmniTRACS, which is Ku-band mobile satellite system, developed by the Qualcomm Company. In a more precise sense, the CDMA technique was developed by experts of the Qualcomm Company in 1987. At present, the CDMA system advantages are practically effective in new satellite systems, such as Globalstar, also developed by Qualcomm, which is devoted to mobile satellite handheld terminals and Skybridge, involved in fixed satellite systems. This type of MA is therefore attractive for handheld and portable satellite equipment with a wide antenna pattern. Antennas with large beam widths can otherwise create or be subject to interference with adjacent satellites. In any case, this MA technique is very attractive for commercial, military and even TT&C communications because some Russian satellites use CDMA for command and telemetry purposes.

The Synchronous-CDMA (S-CDMA) scheme proves efficiently to eliminate interference arising from other users sharing the same carrier and the same spot beam. Interference from other spot beams that overlap the coverage of the intended spot is still considerable. This process to ensure orthogonality between all links requires signaling to adjust transmission in time and frequency domains for every user independently.

4.1 *Direct Sequence (DS) CDMA*

This DS-CDMA technique is also called Pseudo-Noise (PN) modulation, where the modulated signal is multiplied by a PN code generator, which generates a pseudo-random binary sequence of length (N) at a chip rate (R_c), much larger than information bit rate (R_b). The chip rate sequence is introduced by the following relation:

$$R_c = N \cdot R_b$$

This sequence is combined with the information signal cut into small chip rates (R_c), thus, speeding the combined signal in a much larger bandwidth ($W \sim R_c$). The resulting signal has wider RF bandwidth than the original modulated signal. In such a way, the transmitting signal can be expressed in the following way:

$$s(t) = m(t)\, p(t) \cos (2\pi f_c t) = m(t)\, p(t) \cos \omega_c t$$

where m(t) = binary message to be transmitted and p(t) = spreading NP binary sequence. Consequently, at the receiver the signal is coherently demodulated by multiplying the received signal by a replica of the carrier. Neglecting thermal noise, the receiving signal at the input of the detector of Low-Pass Filter (LPF) is given by the following equation:

$$r(t) = m(t)\, p(t) \cos \omega_c t\, (2 \cos \omega_c t) = m(t)\, p(t) + m(t)\, p(t) \cos 2\omega_c t$$

The detector LLF eliminates the HF components and retains only the LW components, such as u(t) = m(t) p(t). This component is then multiplied by the local code [p(t)] in phase with the received code, where the product $p(t)^2 = 1$. At the output of the multiplier this gives:

$$x(t) = m(t)\, p(t)\, p(t) = m(t)\, p(t)^2 = m(t) \quad [V]$$

The signal is then integrated over one bit period to filter the noise. The transmitted message is recovered at the integrator output, so in fact, only the same PN code can achieve the despreading of the received signal bandwidth. In this process, the interference or jamming spectrum is spread by the PN codes, while other user's signals, spared by different PN codes, are not despread. Interference or jamming power density in the bandwidth of the received signal decreases from their original power. Otherwise, the most widely accepted measure of interference rejection is the processing gain (G_p), which is given by the ratio R_c/R_b and value of $G_p = 20 - 60$ dB. The input and output signal-to-noise ratios are related as follows:

$$(S/N)_{Output} = G_p\, (S/N)_{Input}$$

In the forward link, the CES transmits the spread spectrum signals spread with synchronized PN sequence to different MMSC users. Since orthogonal codes can be used, the mutual interference in the network is negligible and the channel capacity is close to that of FDMA. In the return link, the signals transmitted from different SES users are not synchronized and they are not orthogonal. The first case is referred to as synchronous and the second case as asynchronous SSMA. The nonorthogonality causes interference due to the transmission of other SES in the satellite network and as the number of simultaneously accessing users increases, the communication quality gradually degrades in a process called Graceful Degradation.

4.2 *Frequency Hopping (FH) CDMA*

The FH-CDMA system works similarly to the DS system, since a correlation process of de-hopping is also performed at the receiver. The difference is that here the pseudo-random sequence is used to control a frequency synthesizer, which results in the transmission of each information bit rate in the form of (N) multiple pulses at different frequencies in an extended bandwidth. The transmitted and received signals have the following forms:

$$s(t) = m(t) \cos \omega_c(t)\, t \quad \text{and}$$

$$r(t) = m(t) \cos \omega_c(t)\, t \cdot 2 \cos \omega_c(t)\, t = m(t) + m(t) \cos 2\omega_c(t)\, t$$

At Rx the carrier is multiplied by an unmodulated carrier generated under the same conditions as at Tx. The second term in Rx is eliminated by the LPF of the demodulator. The relation of processing gain for FH is:

$G_p = W/\Delta f$

where W = frequency bandwidth and Δf = bandwidth of the original modulated signal. At this point, coherent demodulation is difficult to implement in FH receivers because it is a problem to maintain phase relation between the frequency steps. Due to the relatively slow operation of the frequency synthesizer, DS schemes permit higher code rates than FH radio systems [01, 05, 06, 11, 12].

5 SPACE DIVISION MULTIPLE ACCESS (SDMA)

The significant factor in the performance of MA in a satellite communications system is interference caused by different factors and other users. In the other words, the most usual types of interference are co-channel and adjacent channel interference. The co-channel interference can be caused by transmissions from non-adjacent cells or spot beams using the same set of frequencies, where there is minimal physical separation from neighboring cells using the same frequencies, while the adjacent channel interference is caused by RF leakage on the subscriber's channel from a neighboring cell using an adjacent frequency. This can occur when the user's signal is much weaker than that of the adjacent channel user. Signal to Interference Ratio (SIR) is an important indicator of call quality; it is a measure of the ratio between the mobile phone signal (the carrier signal) and an interfering signal. A higher SIR ratio means increasing overall system capacity.

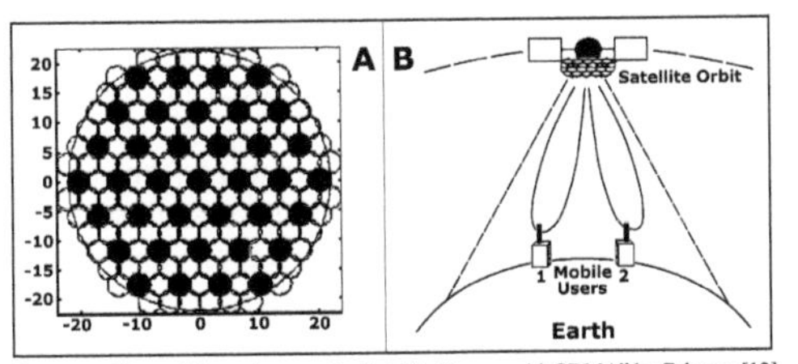

Courtesy of Paper: "Smart Antenna for Satellite Comm. with SDMA" by Zaharov [12]

Figure 2. The Beam Patterns and Adaptive Antenna Applications for SDMA

Taking into account that within the systems of MMSC, every user has their own unique spatial position, this fact may be used for the separation of channels in space and as a consequence, to increase he SIR ratio by using SDMA. This method is physically making the separation of paths available for each satellite link. Terrestrial networks can use separate cables or radio links but on a single satellite, independent transmission paths are required. Thus, this MA control radiates energy into space and transmission can be on the same frequency: such as TDMA or CDMA and on different frequencies, such as FDMA.

In using SDMA, either FDMA or TDMA are needed to allow CES to roam in the same satellite beam or for polarization to enter the repeater. Thus, the frequency reuse technique of same RF is effectively a form of SDMA scheme, which depends upon achieving adequate beam-to-beam and polarization isolation. Using this system reverse line means that interference may be a problem and the capacity of the battery is limited. On the other hand, a single satellite may achieve spatial separation by using beams with horizontal and vertical polarization or left-hand and right-hand circular polarization. This could allow two beams to cover the same Earth surface area, being separated by the polarization. Thus, the satellite could also have multiple beams using separate antennas or using a single antenna with multiple feeds. For multiple satellites, spatial separation can be achieved with orbital longitude or latitude and for intersatellite links, by using different planes. Except for frequency reuse, this system provides on-board switching techniques, which, in turn, enhance channel capacity. Additionally, the use of narrow beams from the satellite allows the Earth station to operate with smaller antennas and so produce a higher power density per unit area for a given transmitter power. Therefore, through the careful use of polarization, beams (SDMA) or orthogonal (CDMA), the same spectrum may be reused several times, with limited interference among users.

The more detailed benefits of an SDMA system include the following:

1 The number of cells required to cover a given area can be substantially reduced and interference from other systems and from users in other cells is significantly reduced.
2 The destructive effects of multipath signals, copies of the desired signal that have arrived at the antenna after bouncing from objects between the signal source and the antenna can often be mitigated.
3 Channel reuse patterns of the systems can be significantly tighter because the average interference resulting from co-channel signals in other cells is markedly reduced and separate spatial channels can be created in each cell on the same conventional channel, so intra-cell reuse of conventional channels is possible.
4 The SDMA station radiates much less total power than a conventional station, one result is a reduction in network-wide RF pollution, another is a reduction in power amplifier size and the

direction of each spatial channel is known and can be used to accurately establish the position of the signal source.

5 The SDMA technique is compatible with almost any modulation method, bandwidth, or frequency band including GSM, PHP, DECT, IS-54, IS-95 and other formats. The SDMA solution can be implemented with a broad range of array geometry and antenna types.

Another perspective of the realization of SDMA systems is the application of smart antenna arrays with different levels of intelligence consisting in the antenna array and digital processor. Since the frequency of transmission for satellite communications is high enough (mostly 6 or 14 GHz), that the dimensions of an array placed in orbit is commensurable with the dimensions of the parabolic antenna, is a necessary condition to put such systems into orbit. Thus, the SDMA scheme mostly responds to the demands of LEO and MEO constellations, when the signals of users achieve the satellite antenna under different angles (±22° for the MEO). In this instance, ground level may be split into the number of zones of service coverage determined by switched multiple beam pattern lobes in different satellite directions, or by adaptive antenna separations, which is shown in Figure 2 (A). There are two different beam-forming approaches in SDMA for satellite communications: (1) The multiple spot beam antennas are the fundamental way of applying SDMA in large satellite systems including MSS and (2) Adaptive array antennas dynamically adapt to the number of users.

5.1 *Switched Spot Beam Antenna*

Switched Multi-Beam Antennas are designed to track each subscriber of a given cell with an individual beam pattern as the target subscriber moves within the cell (spot). It is possible to use array antennas and to create a group of few overlapping beams that together result in omnidirectional coverage. This is the simplest technique comprising a basic switching function between separate directive antennas or predefined beams of an array. Beam-switching algorithms and RF signal-processing software are incorporated into smart antenna designs. For each call, software algorithms determine the beams that maintain the highest quality signal and the system continuously updates beam selection, ensuring that customers get optimal quality for the duration of their call. One might design overlapping beam patterns pointing in slightly different directions, similar to the ones shown in Figure 2 (A). Every so often, the system scans the outputs of each beam and selects the beam with the largest output power. The black cells reuse the frequencies currently assigned to the mobile terminals, so they are potential sources of interference. In fact, the use of a narrow beam reduces the number of interfering sources seen at the base station. Namely, as the mobile moves, the smart antenna system continuously monitors the signal quality to determine when a particular beam should be selected. Switched-beam antennas are normally used only for the reception of signals, since there can be ambiguity in the system's perception of the location of the received signal. In fact, these antennas give the best performance, usually in terms of received power but they also suppress interference arriving from directions away from the active antenna beam's centre, because of the higher directivity, compared to a conventional antenna, some gain is achieved. In high-interference areas, switched-beam antennas are further limited since their pattern is fixed and they lack the ability to adaptively reject interference. Such an antenna will be easier to implement in existing cell structures than the more sophisticated adaptive arrays but it gives only limited improvement.

5.2 *Adaptive Array Antenna Systems*

These systems select one beam pattern for each user out of a number of preset fixed beam patterns, depending on the location of the subscribers. Thus, these systems continually monitor their coverage areas, attempting to adapt to their changing radio environment, which consists in (often mobile) users and interferers. Thus, in the simplest scenario, that of a single user and no interferers, the system adapts to the user's motion by providing an effective antenna system pattern that follows the mobile user, always providing maximum gain in the user's direction. The principle of SDMA with adaptive antenna system application is quite different from the beam-forming approaches described in Fig.2B.

The events processed in SDMA adaptive array antenna systems are as follows:

1 A "Snapshot", or sample, is taken of the transmission signals coming from all of the antenna elements, converted into digital form and stored in memory.

2 The SDMA digital processor analyzes the sample to estimate the radio environment at this point, identifying users and interferers and their locations.

3 The processor calculates the combining strategy for the antenna signals that optimally recovers the user's signals. With this strategy, each user's signal is received with as much gain as possible and with the other users/interferers signals rejected as much as possible.

4 An analogous calculation is done to allow spatially selective transmission from the array. Each user's signal is now effectively delivered through a separate spatial channel.

5 The system now has the ability to both transmit and receive information on each of the spatial channels, making them two-way channels.

As a result, the SDMA adaptive array antenna system can create a number of two-way spatial channels on a single conventional channel, be it frequency, time, or code. Of course, each of these spatial channels enjoys the full gain and interference rejection capabilities of the antenna array. In theory, an antenna array with (n) elements can support (n) spatial channels per conventional channel.

In practice, the number is somewhat less because the received multipath signals, which can be combined to direct received signals, takes place. In addition, by using special algorithms and space diversity techniques, the radiation pattern can be adapted to receive multipath signals, which can be combined. Hence, these techniques will maximize the SIR or Signal to Interference and Noise Ratio (SINR).

5.3 *SDMA/FDMA*

This modulation arrangement uses filters and fixed links within the satellite transceiver to route an incoming uplink RF to a particular downlink transmission antenna. A basic arrangement of fixed links may be set up using a switch that is selected only occasionally. Thus, an alternative solution allows the filter to be switched using a switch matrix, which is controlled by a command link.

5.4 *SDMA/TDMA*

This solution is similar to the one previously explained in that a switch system allows a TDMA receiver to reconfigure the satellite. Under normal conditions a link between beam pairs is maintained and operated under TDMA conditions. The utilization of time slots may be arranged on an organized or contention basis. Switching is achieved by using the RF signal. Thus, on board processing is likely to be used in the future, allowing switching to take place by the utilization of baseband signals. The signal could be restored in quality and even stored to allow transmission in a new time slot in the outgoing TDMA frame. This scheme is providing up and downlinks for the later Intelsat VI spacecraft, known as SDMA/SS/TDMA.

5.5 *SDMA/CDMA*

This arrangement allows access to a common frequency band and may be used to provide the MA to the satellite, when each stream is decoded on the satellite in order to obtain the destination addresses. Thus, on-board circuitry must be capable of determining different destination addresses, which may arrive simultaneously, while also denying invalid users access to the downlink. However, on-board processors allow the CDMA bit stream to be retimed, regenerated and stored on the satellite. Because of this possibility the downlink CDMA configurations need not be the same as for uplink and the Earth link may thus, be optimized [01, 06, 08, 11, 13, 14].

6 RANDOM DIVISION MULTIPLE ACCESS (RDMA)

For data transmission, a bit stream may be sent continuously over an established channel without the need to provide addresses or unique words if the channel is not charred. In fact, where charring is implemented, data are sent in bursts, which thus, requires unique words or synchronization signals to enable time-sharing with other users, to be affected in the division of channels. Each burst may consist in one or more packets comprising data from one or more sources that have been assembled over time, processed and made ready for transmission. However, this type of multiplex scheme is also known as Packet MA. Packet access can be used in special RDMA solutions, such as Aloha, where retransmission of blocked packets may be required.

Random access can be achieved to the satellite link by contention and for that reason is called a contention access scheme. This type of access is well-suited to satellite networks containing a large number of stations, such as SES, where each station is required to transmit short randomly-generated messages with long dead times between messages. The principle of RDMA is to permit the transmission of messages almost without restriction, in the form of limited duration bursts, which occupy all the bandwidth of the transmission channel. Therefore, in other words, this is MA with time division and random transmission and an attribute for the synonym Random Division Multiple Access is quite assessable.

A user transmits a message irrespective of the fact that there may be other users equally in connection. The probability of collisions between bursts at the satellite is accepted, causing the data to be blocked from receipt by the Earth station. In case of collision, the destination Earth station receiver will be confronted with interference noise, which can compromise message identification and retransmission after a random delay period. The retransmissions can occur as many times as probably are carried out, using random time delays. Therefore, such a scheme implies that the transmitter vies for satellite resources on a per-demand basis. In this case, it will provide that no other transmitter is attempting to access the same resources during the transmission burst period, when an error-free transmission can occur. The types of random protocols are distinguished by the means provided to overcome this disadvantage. The performance of these protocols is measured in terms of the throughput and the mean transmission delay. Throughput is the ratio of the volume of traffic delivered at the destination to the maximum capacity of the transmission channels. The transmission time, i.e., delay is a random variable. Its mean value indicates the mean time between the generation of a message and its correct reception by the destination station.

6.1 *Aloha*

The most widely used contention access scheme is Aloha and its associated derivatives. This solution was developed in the late 1960s by the University of Hawaii and allows usage of small and inexpensive Earth stations (including SES) to communicate with a minimum of protocols and no network supervision. This is the simplest mode of operation, which time-shares a single RF, divided among multiple users and consists in stations randomly accessing a particular resource that is used to transmit packets. When an Aloha station has something to transmit, it immediately sends a burst of data pulses and can detect whether its transmission has been correctly received at the satellite by either monitoring the retransmission from the satellite or by receiving an acknowledgement message from the receiving party. Should a collision with another transmitting station occur, resulting in the incorrect reception of a packet at the satellite, the transmitting station waits for a random period of time, prior to retransmitting the packet. Otherwise, a remote station (SES) uses Aloha to get a hub station (CES) terminal's attention.

Namely, the SES terminal sends a brief burst requesting a frequency or time slot assignment for the main transmission. Thus, once the assignment for SES is made, there is no further need for the Aloha channel, which becomes available for other stations to use. After that, the main transmissions are then made on the assigned channels. At the end, the Aloha channel might be used again to drop the main channel assignments after the transmission is completed. The advantages of Aloha are the lack of any centralized control, giving simple, low-cost stations and the ability to transmit at any time, without having to consider other users.

In the case where the user population is homogenous, so that the packet duration and message generation rate are constant, it can be shown that the traffic carried S (packet correctly interpreted by the receiver), as a function of total traffic G (original and retransmitted message) is given by the relation:

$$S = G \exp(-2G) \quad \text{[packet/time slot]}$$

where (S = transmission throughput) and (G) are expressed as a number of packets per time slot equal to the common packet duration. The Aloha protocol cannot exceed a throughput of 18% and the mean transmission time increases very rapidly as the traffic increases due to an increasing number of collisions and packet retransmissions. The Aloha mode is relatively inefficient with a maximum throughput of only 18.4% (1/2). However, this has to be counter-weight against the gains in simple network complexity, since no-coordination or complex timing properties are required at the transmitting SES.

6.2 *Slotted Aloha*

This form of Aloha or S-Aloha, where the time domain is divided into slots equivalent to a single packet burst time; there will be no overlap, as is the case with ordinary Aloha. The transmissions from different stations are now synchronized in such a way that packets are located at the satellite in time slots defined by the network clocks and equal to the common packet duration.

Hence, there cannot be partial collisions; every collision arises from complete superposition of packets. In effect, the timescale of collision is thus, reduced to the duration of a packet, whereas with the Aloha protocol, this timescale is equal to the duration of both packets. This situation divides the probability of collision by two and the throughput becomes:

$$S = G \exp(-G) \quad \text{[packet/time slot]}$$

This protocol enables collisions between new messages and retransmission to be avoided and increases the throughput of S-Aloha in the order of 50–60% by introducing a frame structure, which permits the numbering of time slots. Each packet incorporates additional information indicating the slot number reserved for retransmission in case of collision. For the same value of utilization as basic Aloha, the time delay and probability of packet loss are both improved. The major disadvantages of S-Aloha are that more complex equipment in the Earth station is necessary, because of the timing requirement and because there are fixed time slots, customers with a small transmission requirement are wasting capacity by not using the time slot to its full availability.

6.3 *Slot Reservation Aloha*

This solution of an extension for the slotted-Aloha scheme allows time slots to be reserved for transmission by an Earth station. In general terms this mode of operation is termed a Packed Reserved Multiple Access (PRMA). Slot reservation basically takes two forms:

1 Implicit – When a station acquires a slot and successfully transmits, the slot is reserved for that station for as long as it takes the station to complete its transmission. The network controller then informs all stations on the network that the slot is available for contention once more. There is only the problem that a station with much data to transmit could block the system to other users.
2 Explicit – Every user station may send a request for the reservation of a time slot prior to transmission of data. A record of all time slot occupation and reservation requests is kept. Actually, a free time slot could be allocated on a priority basis. Some kind of control for the reservation of slots is necessary and this could be accomplished by a single or all stations being informed of slot occupancy and reservation requests [01, 06].

7 CONCLUSION

The performances and capacities of MMSC for CDMA, FDMA and TDMA/FDMA have been analyzed many years ago for an L/C-band RF network with global coverage. For the particular MMSC systems under discussion and for the particular antenna configurations, both CDMA and FDMA offer similar performance, FDMA yielding slightly higher channel capacities at the design point and CDMA being slightly better at higher EIRP levels. As the MMSC system grows and the antenna beam size decreases, CDMA appears to be a very efficient MA system, because it is not limited by L-band bandwidth constraints. However, CDMA is wasteful in feederlink bandwidth, and the choice of a multiple access system must take all parameters into consideration, such as oscillator stability, interference rejection, system complexity etc. as well as system cost before deciding on a particular multiple access system.

The communication satellites for MMSC provide multiple-beam antennas and employ frequency reuse of the allocated L-band frequency spectrum. It appears that despite the fact that FDMA and FDMA/TDMA are orthogonal systems, they nevertheless suffer from bandwidth limitations and sensitivity to interbeam interference in L-band. The CDMA scheme is better at absorbing Doppler and multipath effects, and it permits higher rate coding, but it suffers from self-jamming and from bandwidth constraints in the feederlink. In general, all three multiple access systems show similar performance. However, at the chosen design point for aggregate EIRP, number of beams, and allocated bandwidth, FDMA provides still the highest system channel capacity. Recently is developed SDMA as an advanced solution where all concerned SES terminals can share the same frequency at the same time within a separate space available for each link. On the other hand, the RDMA scheme is suitable for large number of users in MMSC, where all SES terminals share asynchronously the same transponder by randomly transmitting short burst or packet divisions. In addition is developed several mobile Aloha methods, which successfully increase the system throughout.

REFERENCES

[01] Ilcev D. S., "Global Mobile Satellite Communications for Maritime, Land and Aeronautical Applications", Book, Springer, Boston, 2005.
[02] Freeman R.L., "Radio systems design for telecommunications (1-100 GHz)", John Wiley, Chichester, 1987.
[03] Kantor L.Y., "Sputnikovaya svyaz i veschanie", Radio i svyaz, Moskva, 1988.
[04] Solovev V.I. & Others, "Svyaz na more", Sudostroenie, Leningrad, 1978.
[05] Maral G. & Other, "Satellite Communications Systems", Wiley, Chichester, 2009.
[06] Susi A. & Others, "Multiple Access in Mobile Satellite Communications", PSN, ASSI (Asosiasi Satelit Indonesia), Electro Online, 1999.
[07] Group of Authors, "Handbook - Mobile Satellite Service (MSS)", ITU, Geneva, 2002.
[08] Zhilin V.A., "Mezhdunarodnaya sputnikova sistema morskoy svyazi – Inmarsat", Sudostroenie, Leningrad, 1988
[09] Ohmory S., Wakana. H & Kawase S., "Mobile Satellite Communications", Artech House, Boston, 1998.
[10] Venskauskas K.K., "Sistemi i sredstva radiosvyazi morskoy podvizhnoy sluzhbi", Sudostroenie, Leningrad, 1986.
[11] Maini A.K. & Agrawal V., "Satellite Technology - Principles and Applications", John Wiley, Chichester, 2007.
[12] Zaharov V. & Others, "Smart Antenna Application for Satellite Communications with SDMA", Journal of Radio Electronics, Moscow, 2001.
[13] Kadish J.E., & Others, "Satellite communications fundamentals", Artech House, Boston-London, 2000.
[14] Kantor L.Y., "Sputnikovaya svyaz i veschanie", Radio i svyaz, Moskva, 1988.

Classification and Characteristics of Mobile Satellite Antennas (MSA) for Maritime Applications

S.D. Ilcev
Durban University of Technology (DUT), South Africa

ABSTRACT: In many respects the Mobile Satellite Antenna (MSA) infrastructures currently available for mobile solutions and Maritime Mobile Satellite Communications (MMSC) constitute the weakest links of the system. If the mobile antenna has a high gain, it has to track the satellite, following both mobiles and satellite orbital motions. Namely, sometimes this is difficult and expensive to synchronize and have to be introduced in the understanding manner. On the other hand, if the vehicular antenna has low gain, it does not need to perform tracking but the capacity of the communications link is limited. According to the transmission direction in this paper will be introduced transmitting and receiving or so-called transceiving MSA, as a part of all types of Mobile Earth Station (MES), only receiving MSA is a part of the special Inmarsat EGC receiver and only transmitting MSA built in Cospas-Sarsat emergency satellite beacons for maritime, land and aeronautical applications. The classifications of omnidirectional and directional tracking MSA are discussed and in addition are examined low, medium and high gain MSA. The current and forthcoming GEO and Non-GEO mobile satellite operators have conducted research on all network segments, including different types of MSA and their future development and improvements. The Engineering Test Satellite-V (ETS/V) experiments conducted in Japan for the transmission of voice, video and different data rate digital communications between ships, land vehicles and aircraft were successful. Moreover, were developed essential antenna system for maritime and aeronautical high-speed broadband and low-speed data devices by using briefcase-size transportable equipment were among the experiments.

1 TYPES OF MOBILE SATELLITE ANTENNAS (MSA)

In particular, taking in consideration the kind of mobiles, the antenna systems for MMSC can be classified into shipborne, vehicleborne, airborne, transportable, MSA for personal satellite terminals and other types.

1.1 *Shipborne MSA*

The different types of shipborne satellite antenna systems were developed for installation on board ocean-going ships and inland sailing vessels, on sea platforms and other offshore infrastructures. In general, these antennas must have strong and rugged constructions, with corresponding mechanical and electrical particulars. The Inmarsat-A Ship Earth Station (SES) is the inheritor of the first generation of Marisat and was the first Inmarsat operating standard of Mobile Satellite Communications (MSC) and especially for MMSC. In fact, this analog standard started in 1982 to use Inmarsat standard-A transceiving MSA known as Above Deck Equipment (ADU) and recently is not in use any more. In the meantime Inmarsat-C and EGC were developed with small omnidirectional antennas. In addition, Inmarsat-B digital standard started to be in service on ships at the end of 1993, using the second generation of Inmarsat satellite constellation. This standard is compatible with Inmarsat-A and uses the same antenna specification.

After employing more powerful third generation of constellation, the Inmarsat system developed Inmarsat-M, mini-M, and D+ standards with special shipborne antenna systems. Since 2003 were developed three new Inmarsat standards known as Fleet-33, Fleet-55 and Fleet-77 using special tracking MSA. In 2007 was presented new Inmarsat standard known as FleetBroadband using special broadband MSA via fourth generation of spacecraft Inmarsat-4.

All MSC antennas are transceiving antennas, except for the Inmarsat-EGC receiver, which can use a receiving antenna only. On the other hand, the Cospas-Sarsat system developed shipborne EPIRB, with small built in UHF antennas. Thus, rugged MSA with added-value system functions compatible with shipborne operation for Inmarsat, Iridium, Globalstar and other systems and compact antennas for small lifeboat operation are some of the fascinating requirements for shipborne antenna designers of in the future [01, 02].

1.2 *Other Types of MSA*

The other types of MSA can be vehicleborne for Vehicle Earth Station (VES) onboard roads and railways vehicles and airborne MSA for Aircraft Earth Station (AES). Then MSA can be used for transportable, personal, broadband (DVB-RCS), emergency beacons, receiving MSA for broadcast radio and TV, GNSS (GPS and GLONASS) and MSA for Communications, Navigation and Surveillance (CNS) equipment.

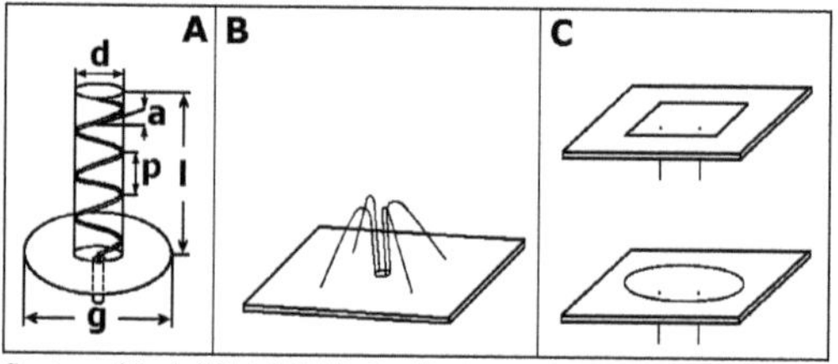

Courtesy of Book: "Global Mobile Satellite Communications" by Ilcev [02]
Figure 1. Types of Low-Gain Omnidirectional Antennas

2 LOW-GAIN OMNIDIRECTIONAL ANTENNAS

The antennas for MSC are classified into omnidirectional and directional. The gain of omnidirectional antennas is low and generally from 0 to 4 dB in the L-band, which does not require the capability of satellite tracking. There are three types of low-gain omnidirectional antennas, which are very attractive for all mobile applications owing to the small size, lightweight and circular polarization properties. These antennas are also used as elements of directional antennas for special configurations.

2.1 *Quadrifilar Helix Antenna (QHA)*

The QHA low-gain model is composed of four identical helixes wound, equally spaced, on a cylindrical surface. The helix elements are fed with signals equal in amplitude and 0, 90, 180 and 270° in relative phase. This antenna can easily generate circular polarized waves without a balloon or a 3 dB power divider, which are required to excite a balanced fed dipole and circular polarized cross-dipoles. It can also be operated on a wide frequency bandwidth of up to 200% because it is a traveling-wave-type antenna.

The components of QHA are ground plane (g), pitch (p), pitch angle (a), length (l) and diameter (d), presented in Figure 1 (A). The diameter of the ground plane is usually selected to be larger than one wavelength and the number of turns is N = l/p. However, it is well known that the parameters for (a) are about 12 to 15 and the circumference of the helix (πd) is about 0.75 to 1.25 wavelengths. Circular polarized waves with good axial ratios can be transmitted along the (z) axis direction (axial mode).

The gain of a helical antenna depends on the number of (N) turns and typical gain and half-power beam width are about 8 dBi and 50° when N = 12 ~ 12 but is usually about 3 dBi. This antenna is employed as a receiving antenna for GPS and as a transceiver antenna for L-band Inmarsat-C SES, VES and AES (low air drag Aero-C) applications covered by different kinds of radomes. All three types of antennas can also be combined with GPS receiving. It is also a component of AMSC, MSAT, MSAT-x, Iridium, Globalstar, Mobilesat ETS-V, forthcoming ICO and other MSC terminals.

In general, QHA, as a mobile antenna is the best solution and has two advantages over a conventional unifilar helical antenna. The first is an increase in bandwidth; namely, it can generate axial mode circular polarized waves in the frequency range from 0.4 to 2.0 wavelengths of the helix circumference. The second is lowered frequency for axial mode operation. The principle disadvantage is an increase in the complexity of the feed system. The area of the ground plane is usually about 3 times the diameter of the helix.

2.2 *Crossed-Drooping Dipole Antenna (CDDA)*

A dipole antenna with a half-wavelength (λ/2) is the most widely used and it is also the most popular, having been used in antenna systems such as the parabolic antenna for MSC. A half-wavelength dipole is a linear antenna whose current amplitude varies one-half of a sine wave, with a maximum at the centre. As a dipole antenna radiates linearly polarized waves, two crossed-dipole antennas have been used in order to generate circular polarized waves. The two dipoles are geometrically orthogonal and equal amplitude signals are fed to them with π/2 in-phase difference. In order to optimize the radiation pattern, a set of dipole antennas is bent toward the ground, as shown in Figure 1 (B) and for that reason it is called a drooping dipole antenna. Otherwise, the CDDA serves as a transceiver MSA onboard L-band Inmarsat-C SES and VES applications mounted inside a radome for MMSC and Land Mobile Satellite Communications (LMSC).

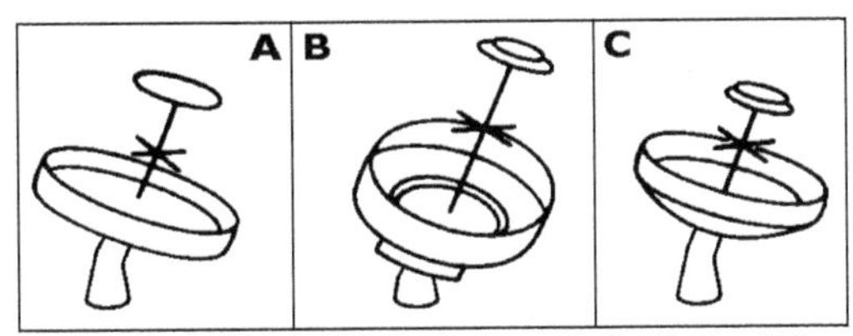

Courtesy of Book: "Global Mobile Satellite Communications" by Ilcev [05]
Figure 2. Types of Directional Medium-Gain Aperture Antennas

The CDDA is the most interesting for LMSC, where required angular coverage is narrow in elevation and is almost constant in azimuth angle. By varying the separation between the dipole elements and the ground plane, the elevation pattern can be adjusted for optimum coverage for the region of interest. The general characteristics of this antenna are: gain is 4 dBi minimums, axial ratio is 6 dB maximum and the height of the antenna is about 15 cm. This antenna has a maximum gain in the boresight direction.

2.3 *Microstrip Patch Antenna (MPA)*

A microstrip disc (patch) antenna is very low profile and has mechanical strength, so it is considered to be the best type for mobiles such as cars and especially in aircraft at the hybrid L to Ku-band, which requires low air drag. In general, a circular disk antenna element has a circular metallic disc supported by a dielectric substrate material and printed on a thin dielectric substrate with a ground plane. In order to produce a circularly polarized wave, a patch antenna is excited at two points orthogonal to each other and fed with signals equal in amplitude and 0 and 90° in relative phase. However, a higher mode patch mobile antenna can also be designed to have a similar radiation pattern to the drooping dipole. To produce conical radiation patterns (null on axis) suitable for maritime and land mobile satellite applications, the antenna is excited at higher mode orders. In Figure 1 (C) is illustrated the basic configuration for a circular patch antenna (above it is shown square patch with the same characteristics), which has two feed points to generate circular polarized waves. The resonant frequency excited by basic mode and given as:

$$f = 1{,}84c/2\pi a \sqrt{\epsilon_r}$$

where (a), (c) and (ϵ_r) are the radius of circular disc, the velocity of light in free space and the relative dielectric constant of the substrate, respectively. In LMSC, a MPA antenna with higher order excitation is considered better because it can optimize the gain in elevation angle to the satellite in the same way as a CDDA. In fact, the area of a higher mode circular MPA is about 1.7 times larger in radius on the gain is about 6 to 8 dBi. Thus, the circular patch is also suitable as a satellite navigation-receiving antenna for GPS or GLONASS receivers [01, 02, 03, 04, 05].

3 MEDIUM-GAIN DIRECTIONAL ANTENNAS

The medium-gain directional MSA are solutions with a typical gain between 12 and 15 dBi, although some antennas can have even bigger gains. These MSC antenna systems can provide voice, Fax and HSD for Inmarsat-M shipborne and vehicleborne applications and for Inmarsat airborne standards, including other systems developed by ESTEC.

3.1 *Aperture Reflector Antennas*

The aperture reflector antennas are good solutions with medium-gain characteristics used in MSC, with three basic representatives such as SBF, modified SBF and improved SBF antennas, illustrated in Figure 2 (A), (B) and (C), respectively. The main characteristics of these three antennas are shown in Table 1.

Table 1. Particulars of Aperture Types of Antennas

Characteristics	SBF Antenna	Modified SBF Antenna	Improved SBF Antenna
Effective Gain	14.5 dB	15 dB	15 dB
Half-Power Bandwidth	34°	34°	34°
Directive Gain	14.8 dB	15.5 dB	15.5 dB
First Sidelobe Level	–21 dB	–22 dB	–22.5 dB
Axial Ratio	–1.3 dB	–1.1 dB	–1.1 dB
Aperture Efficiency:			
Effective - Directive Gain	65%–75%	75%–80%	76%–85%
RF/VSWR Bandwidth, under 1,5	3%	7%	9%
Diameter of Large Reflector (DR):			
Bigger (DR 1)	40 cm (2,05λ)	40 cm (2,05λ)	40 cm (2,05λ)
Smaller (DR 2)	–	27 cm (1.38λ)	–
Diameter of Small Reflector (Dr):			
Bigger (Dr1)	9cm (0.46λ)	9.5 cm (0.48λ)	9 cm (0.46λ)
Smaller (Dr2)	–	8.5 cm (0.43λ)	8 cm (0.41λ)
Width of a Rim	4.9 (0.25 λ)	–	4.9 (0.25 λ)
Distance Between (DR) and (Dr)	9.7 cm (0.49 λ)	19.5 cm (0.99 λ)	12.9 cm (0.66 λ)
Distance Between Exciter & Dr	4.9 cm (0.25 λ)	–	5.7 cm (0.29 λ)
Distance Between (Dr1) & (Dr2)	–	–	1.8 cm (0.09 cm)
Slanting Angle of a (DR)	0^0	-	15^0

Moreover, due to the excellent radiation characteristics of SBF antennas, all three types of aperture antenna with half-power beam width of about 34° have been in their time proposed for shipborne antenna of Inmarsat-M. The SBF antennas consist in the stabilized platform with two gyroscopes for azimuth and elevation angles, diplexer, HPA and LNA, which are enclosed under the protective cupola of a radome. In order to stabilize the antenna, two gyro wheels rotate in opposite directions on a platform.

3.1.1 *Short Backfire (SBF) Plane Reflector Antenna*

The SBF plane reflector antenna that was developed experimentally by H.W. Ehrenspeck in the 1960s is well known as a highly efficient antenna of distinctly simple and compact construction. Its high directivity and low sidelobe characteristics make it a single antenna with high, even values, which is applicable to MSC, tracking and telemetry. Therefore, an SBF antenna is very attractive for gains in the order of 13 to 15 dBi peak RHCP and can be mounted primarily on small but on any size of ships. Otherwise, this type of antenna consists in two circular planar reflectors of different diameter, separated generally by about one-half wavelength, forming a shallow leaky cavity resonator with a radiation beam normal to the small reflector. Namely, the antenna is fed by a dipole at around the midpoint between two reflectors and it has almost a quarter-wavelength rim on the larger reflector. It has the problem of a narrow bandwidth of about 3% because of its leaky cavity operation. The Rx terminal G/T is –12 dBK and the EIRP of the Tx terminal is 28 dBW.

The basic configuration of the SBF antenna consists in a cross-dipole element, which is required to generate a circularly polarized wave, large and small reflectors and a circular metallic rim. The antenna has the strong directivity normal to the reflector and its performance is superior to that of other types of mobile antennas with the same diameter, however, it has the problem of narrow frequency band characteristics. This antenna has many beneficial characteristics, such as efficiency and the simplicity of construction and is also considered a favorite option for a compact and high-efficiency shipboard antenna. It is produced by many world manufacturers approved by Inmarsat or ESTEC. For instance, one of the most known manufacturers of Inmarsat-M MES with antenna is Thrane & Thrane with their partners, while the manufacturer of ESTEC SBF antenna is G&C McMichael.

3.1.2 *Modified SBF Plane Reflector Antenna*

A modified SBF antenna differs from the conventional SBF antenna in that there is either an additional step on the large reflector or a change in the shape of the large reflector from a circular to a conical plate in order to improve the gain characteristics and frequency bandwidth of the Voltage Standing Wave Ratio (VSWR). The dual reflector improves the input impedance characteristics covering the frequency range between transmitting and receiving sides.

The conventional SBF model is a resonant-type antenna, producing input impedance characteristics that are narrow in bandwidth, so wider bandwidth is required to cover the 1.6/1.5 GHz range for MES of the Inmarsat system. In effect, the improvement in the input impedance is greatly dependent on the size and the separation of the small reflectors. The VSWR can be reduced from 1.7 and 1.5 (at 1.54 and 1.64 GHz) to below 1.2 for each RF [01, 02, 03, 04].

3.1.3 *Improved SBF Conical Reflector Antenna*

The main research activities of the ETS-V program in MSC have been focused on studying the reduction of fading, using compact and high-efficiency antennas with a gain of around 15 dBi, so the electrical characteristics of a simple SBF antenna have been improved by changing its main reflector from a flat disk to a conical or a step plate and by adding a second small reflector.

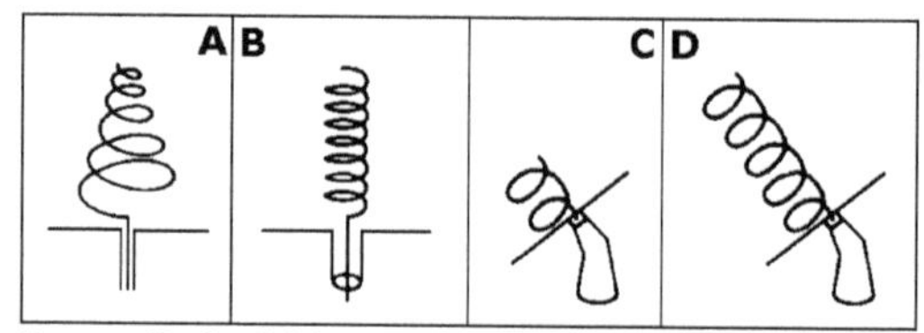

Courtesy of Book: "Global Mobile Satellite Communications" by Ilcev [05]
Figure 3. Types of Helical Wire Antennas

The gain is improved by 1 dB without changing sidelobe levels. Comparisons of electrical and other parameters of three types of SBF antennas are shown in Table 1. Stabilization of the antenna is obtained by a two-axis stabilized method and satellite pointing is carried out by a tracking program using output signals from the ship's gyroscope. It is also considered to be a suitable option for mounting aboard ships [02, 04].

4 WIRE ANTENNAS

The wire antennas are monosyllabic construction or combinations of elements, such as different shapes of wire spirals and helixes, dipoles and patches. These types of antennas have a very simple construction, with any reflector specified for medium-gain directional antennas and, with some modification, respond well to the demands of MSC applications. Here are not included Inverted V-Form Cross Dipole, Crossed-Slot, Conical Spiral and Planar Spiral Antennas, but will be introduced just Helical Wire Antennas.

4.1 *Helical Wire Antennas*

Since an axial mode helical antenna has good circular polarization characteristics over a wide frequency range, it has been put into practical use as a single wire antenna or as an array element. With respect to the structure, this antenna can be considered a compromise between the dipole and the loop antennas and the radiation mode varies with the pitch angle and the circumference of the helix. In particular, a helix with a pitch angle of 12 to 15° and a circumference of about 1 λ, has a sharp directivity towards the axial direction of the antenna.

This radiation mode is called the axial mode, which is the most important mode in helical antennas. Several studies have been carried out on the properties of the axial mode helical antenna with a finite reflector. The current induced on the helix is composed of four major waves, which are two rapidly attenuating waves and two uniform waves along the helical wire. These waves include the traveling wave and the reflected wave. Thus, in a conventional helical antenna, the uniform traveling wave will be dominant when the antenna length is fairly large, with typical versions such as a conical helix shown in Figure 3 (A) and a cylindrical helix in Figure 3 (B).

A conical helix is interesting for L-band MSS enabling HPBW in the order of 100° and circular polarization without hybrid gain of 4 to 7 dBi. Cylindrical antennas can be monofilar or multifilar, also suited for L-band MSS, while in a short-cut cylindrical helix antenna, the rapidly attenuating traveling wave will be dominant, especially in a two-turn (N = 2) helical antenna.

1. Conical Helix Antenna – This antenna can be regarded as a low-gain development of the cylindrical helix antenna and is suitable for wide-beam width applications with good efficiency. Thus, with suitable choices of cone angle and turn spacing, it is possible to achieve a beam width in the order of 100°. This type of antenna can also achieve an input VSWR of 1.5:1 or better than 5% frequency bandwidth merely by incorporating a simple quarter-wavelength transformer. The typical size for an L-band application is in the order of 15 cm in length and the ground plane is about 20 cm in diameter. The resultant gain is approximately 4 to 7 dBi, which is between low and medium-gain requirements.

2. Two-Turn Cylindrical Helix Antenna – This antenna has two-turns of wires, forming a simple helical antenna solution with reflector, illustrated in Figure 3 (C). This model has relatively high antenna gain and excellent polarization characteristics for its size. Radiation patterns characteristically are calculated with respect to (E_0) and (E_ϕ) planes. The gain of this antenna is 9 dBi and the axial ratio is about 1 dB, with reflector diameter (d) around 1λ. Such types of antenna have comparatively high performance in spite of their small size and compact construction. From the above-mentioned considerations, a highly efficient antenna for the Inmarsat-M MES can be realized by applying this antenna to elements of an array antenna.

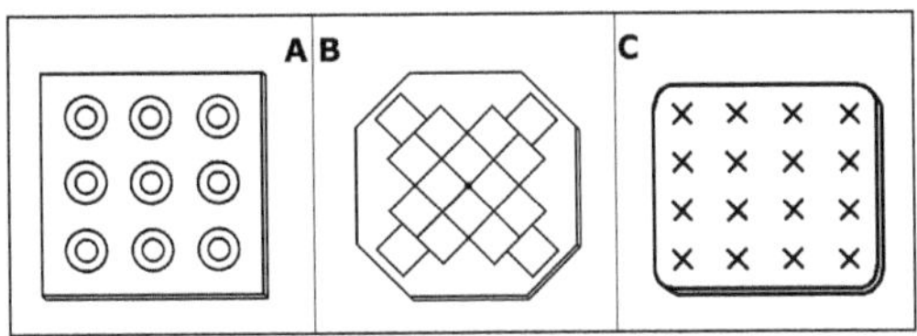

Courtesy of Book: "Global Mobile Satellite Communications" by Ilcev [02]
Figure 4. Types of Microstrip, Cross-Slot and Dipole Array Antennas

3. Five-Turn Cylindrical Helix Antenna – This antenna solution is illustrated in Figure 3 (D). The main electrical characteristics are: gain is 12.5 dBi of peak RHCP for Tx and 11.5 dBi for Rx; sidelobe level has value of about –13 dB; axial ratio is 3 dB; beam width of 3 dB has angle of –47°; terminal G/T has –16 dBK and terminal EIRP has 29 dBW. This antenna solution is designed and developed by the European research institution ESTEC. In addition, stabilization of this antenna is obtained by gravity elevation on double-gimbaled suspension. The pendulum aligns itself with the vertical when not subject to other acceleration. However because the centre of rotation of the pendulum is distant from that of the ship pitch and roll movement induces horizontal acceleration to which the pendulum is sensitive. In order to limit perturbations, the resonant frequency of the pendulum must be low with respect to the excitation frequencies in pitch and roll and the damping (friction) must be minimum. Low resonance frequency is achieved by minimizing the distance between the centre of gravity of the rotating part and its centre of rotation. This also reduces torque due to horizontal acceleration but at the same time reduces the stabilizing torque due to gravity [02, 03, 05].

5 ARRAY ANTENNAS

Several different type of antenna can be arrayed in space to make a directional pattern or one with a desired radiation pattern. This type of integrated and combined antenna is called an array antenna consisting in more than two elements, such as microstrip, cross-slot, cross-dipole, helixes or other wire elements and is suitable for MSC. Each element of an array antenna is excited by equal amplitude and phase and its radiation pattern is fixed.

5.1 *Microstrip Array Antenna (MAA)*

The type of MAA is a nine-element flat antenna, disposed in three lines spaced at 94 mm, namely about a half wavelength at 1.6/1.5 GHz and whose antenna volume is about 300 x 300 x 10 mm, see Figure 4 (A). As shown in this figure, the element arrangements of the MAA solutions are 3 x 3 rows square arrays in order to obtain similar radiation patterns in different cut planes. The MAA beam scanning is performed by controlling four-bit variable phase shifters attached to each antenna element. This type of antenna is very applicable for the MES Inmarsat-M and the Inmarsat-Aero standard.

5.2 *Cross-Slot Array Antenna (XSA)*

The XSA type of antennas are a 16-element solution with 97 mm spacing and their volume is about 560 x 560 x 20 mm, shown in Figure 4 (B). Evident is the element arrangement of the XSA, which is a modified 4 x 4 square array in order to obtain similar radiation patterns in different cut planes.

The XSA antenna beam scanning is carried into effect to control four-bit variable phase shifters associated to each antenna element. Otherwise, this type of antenna is also suitable for the MES Inmarsat-M maritime and land, including Inmarsat-Aero solutions.

5.3 *Cross-Dipole Array Antenna*

The cross-dipole array antenna is composed of 16 crossed-dipoles fed in phase with a peak gain of 17 dBi and with the feeding circuit behind the radiating aperture, shown in Figure 4. (C). The main electrical characteristics of this antenna are: gain is 15 to 17 dBi with peak RHCP transmit; axial ratio has a value of 0.7 dB; beam width of 3 dB is –34°; terminal G/T is –9.5 dBK and EIRP terminal value is 32 dBW.

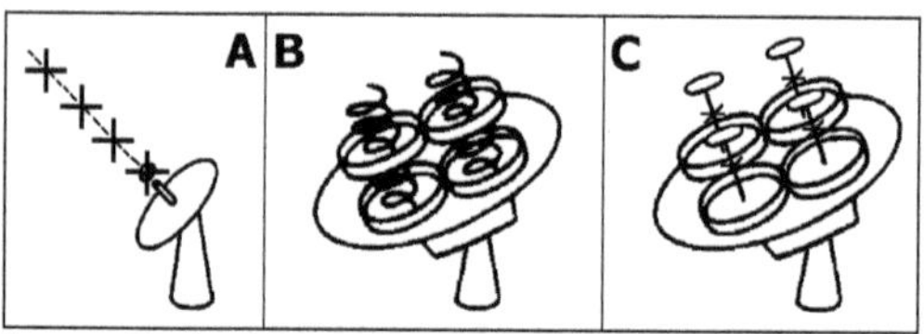

Courtesy of Book: "Global Mobile Satellite Communications" by Ilcev [05]
Figure 5. Types of Four-Element Array Antennas

Otherwise, the antenna system consists in a stabilization mechanism for tracking, flat antenna array, diplexer, HPA and LNA, which are all protected by a plastic radome. Stabilization of the antenna is obtained by a single-wheel gyroscope, when the azimuth pointing is controlled by the output from the ship's gyrocompass. Similar to the previous two models, this antenna is also suitable for maritime and aeronautical applications.

5.4 *Four-Element Array Antennas*

There have been several four-element antenna models developed, such as Yagi-Uda, Quad-Helix and four elements SBF array.

1. Yagi-Uda Crossed-Dipole Array Antenna – This array antenna has been developed for use on board ships and is protected with a radome, shown in Figure 5 (A). The feeder of this antenna is a simple formation of four in-line crossed-dipoles fixed in the middle of the reflector. This endfire array has circular polarization and the gain is between 8 and 15 dBi.

2. Quad-Helix Array Antenna – The quad-helix array antenna solution is composed of four identical two-turn helical wire antennas in the shape of a square and whose elements are oriented in the manner illustrated in Figure 5 (B). According to previous studies, the effect of mutual coupling between each element of this antenna is not negligible and this mutual coupling mainly degrades the axial ratio. The axial ratio of a single helical antenna is about 1 dB but this value is degraded to about 4.5 dB in the case of the array antenna with an array spacing of 0.7λ. However, the best properties of antenna gain and axial ratio can be obtained at a rim height of about 0.25λ.

The antenna gain is improved by 0,4 dB and the axial ratio is also improved by 3.5 dB, compared to that of the quad-helix array antenna without rims. The performance characteristics of this small antenna are essentially, gain is about 13 dB (HPBW is 38°) and aperture efficiency is about 100%. It appears that this type of helix-integrated antenna is also well suited for the shipborne Inmarsat-M standards.

3. Four-Element SBF Array – This antenna is developed on the basis of a conventional SBF antenna as an integrated array with four SBF elements, see Figure 5 (C). The antenna provides high aperture efficiency, circular polarization and almost high-performance gain between 18 and 20 dBi. Because of the high gain characteristics, this array is very suitable for maritime applications as a shipborne antenna [04, 05, 06, 07].

5.5 *Phased Array Antenna*

Phased array antennas were developed for maritime, land and aeronautical applications to provide a design for a thin antenna that can be installed on land vehicles and aircraft. Otherwise, these mobile antennas were developed initially for aircraft and are well known for their complexity and high cost. As a result, emphasis was placed on the selection of manufacturing techniques, materials and component types, in addition to meeting the RF and pointing requirements and keeping the cost down.

As mentioned previously, the radiation pattern of an array antenna is fixed, however, the radiation pattern can be scanned in space by controlling the phase of the exciting current in each element of the array. This type of antenna is called a phased array antenna, which has many advantages in terms of MSC applications such as compactness, light weight, high-speed tracking performance and potentially, low cost.

5.5.1 *Phased-Array Antenna for AMSC*

A directional medium-gain antenna is considered a key technology in AMSC. This type of antenna was developed and tested by CRL for the Japanese ETS-V program. At the same time, Inmarsat approved all aeronautical antenna standards for installations on board commercial and military aircraft. Taking account of the electrical and mechanical requirements of AMSC, a phased array with low-profile antenna elements was chosen for a directional main antenna, while a microstrip antenna was chosen as an antenna element because of its very low profile, very light weight and mechanical strength, which satisfy the requirements for airborne antenna [04, 05, 08, 09, 10].

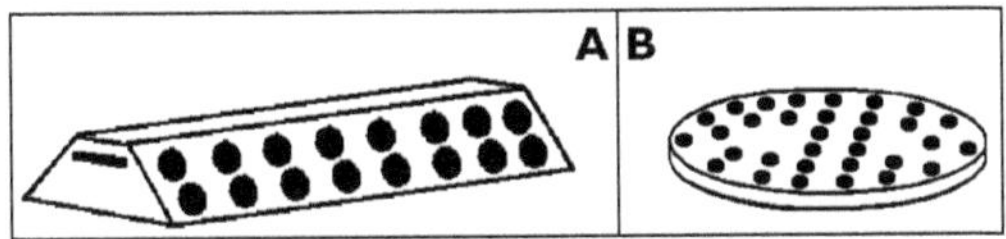

Courtesy of Book: "Global Mobile Satellite Communications" by Ilcev [02]
Figure 6. Types of Aeronautical and Maritime Array Antennas

However, one disadvantage is the very narrow frequency bandwidth, usually 2 to 3%. The antenna adopted is a two-frequency resonant element because it provides a compact array and a simple feed line configuration. On the other hand, this type of antenna has very poor axial ratio values. The problem was overcome by using the sequential-array technique, where a thin substrate with high dielectric constant is used over a wide frequency bandwidth with excellent axial ratios. The microstrip phased-array antenna is mounted on top of the fuselage and has two planes with 16 circular patch elements, 2 of which are in elevation and 8 in azimuth, see Figure 6 (A). In practice, this antenna can be used for installation onboard sea hovercraft and seacraft.

5.5.2 *Phased-Array Antenna for MMSC*

The new shipborne phased-array antenna has been developed for utilization on board big ocean-going vessels. Besides other directional types of shipborne antenna, this unit is designed to serve as ADE of the Nera F77 transceiver or other brand SES for new Inmarsat Fleet 77 service. The Nera F77 antenna is a mechanically steered circular disc with 32 low-profile radiating elements arranged in two circles with 16 and 12 elements and 4 elements are located in the middle of the disc, which is illustrated in Figure 6 (B). The ADE unit can be mounted on a mast or directly on the deck and is covered by a radome. The dimensions of antenna mast mount are 180 x 132 cm with a weight of 65 kg and deck-mounted antennas are 108 x 91 cm and 50 kg in weight. The gain of the antenna is about 15 dBi in both transmission satellite links.

6 HIGH-GAIN DIRECTIONAL APERTURE ANTENNAS

High-gain directional aperture antennas are more powerful transmission reflectors and panels used for Inmarsat maritime and transportable applications. The typical gain of these antennas is more than 20 dBi, EIRP is a maximum of 33 dBW and G/T is about –4 dBK. There are two basic types of directional parabolic antennas: dish and umbrella and the third new solution for Inmarsat transportable units is the Quad flat panel antenna designed by the South African-based company OmniPless and introduced in the next section.

6.1 *Parabolic Dish Antenna*

The first generation of parabolic dish antennas used a reflector in diameter of max 1.2 m, whereas on newer models it is likely to have reduced in size to approximately 0.7 to 0.8 m, see Figure 7 (A). Because a large proportion of the Rx signal gain and Tx EIRP is produced by the antenna, the area of the dish can only be reduced if the transmitting power from the satellite transponder is increased, when the receive preamplifier gain can be increased without an appreciable increase in noise. The parabolic reflector is most often used for high directivity for radio signals traveling in straight lines, as do light rays. They can also be focused and reflected just as light rays can, namely, a microwave source can be placed at focal point of antenna reflector.

The field leaves this antenna as a spherical wave front. As each part of the wave front reaches the reflecting surface it is phase-shifted 180°. Each part is then sent outward at an angle that results in all parts of the field traveling in parallel paths. Due to the special shape of a parabolic surface, all paths from the focus to the reflector and back into space line are the same length. When the parts of the field are reflected from the parabolic surface, they travel to the space line in the same amount of time. This antenna is a large microwave parabolic consisting in the reflector of dish shape, feeder structure, waveguide assembly, servo and drive system and protective radome.

1. Stable Platform – This is the antenna support assembly, which must remain perfectly stable when the ship is pitching and rolling in extremely bad weather conditions. Namely, it is essential that the stable platform holds the reflector in its A/E angular positions despite movement of the ship. The platform usually consists in a large solid bed mounted in such a way that four gyro compasses are able to sense movement and correct any errors detected, holding the platform level. In practice, it is a form of electronic gimbal.

2. Tracking System – The antenna is controlled in A/E angles by stepping motors, which in turn are electronically controlled in a simple feedback system. This electromechanical antenna arrangement enables the dish to maintain a lock on a satellite despite navigation course changes. In such a way, as the ship changes course, both A/E control corrections must be made automatically [02, 04, 05, 10, 11].

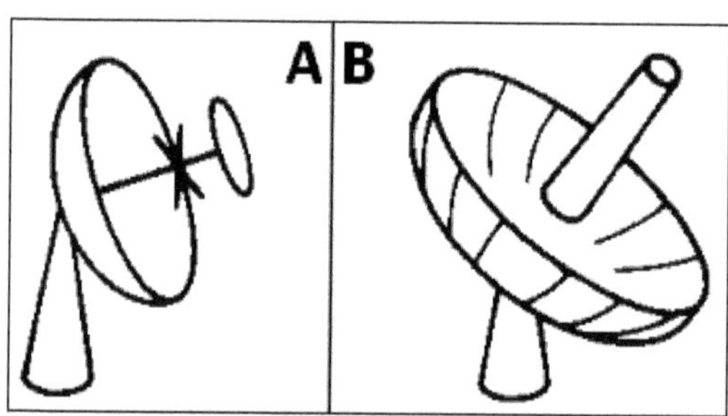

Courtesy of Book: "Global Mobile Satellite Communications" by Ilcev [05]
Figure 7. High-Gain Directional Aperture Antennas

3. Computer Control – The antenna unit processor controls all ADE functions, which include satellite tracking and electronic control.

4. RF Electronics – This segment contains the Tx HPA and the Rx RF front-end LNA stages, plus all the critical Bandpass signal filter stage.

5. Multiplexer Unit – In modern equipment it is common practice to reduce the number of cables between ADE and BDE. Hence, this is achieved by multiplexing up/down signals or commands between ADE and BDE onto the one coaxial feeder.

Therefore, this antenna is connected to BDE transceivers both incorporated in an Inmarsat-A and B SES terminal. Each manufacturer inevitably produces a different design of ADE and BDE, however they do in fact perform the same functions.

6.2 *Parabolic Umbrella Antenna*

An umbrella-type antenna is a deployable, compact and lightweight parabolic type suitable for transportable Inmarsat-A and B transceivers, illustrated in Figure 7 (B). Otherwise, this antenna has almost all the same technical characteristics as a parabolic dish antenna [02, 06, 07, 08, 09, 10, 11].

7 CONCLUSION

The modern design, configuration and characteristics of MSA are very important for implementing successful MSC and networks for all mobile applications and especially for MMSC. The good merit of antenna, physical and technical parameters are the prerequisite to provide compact and lightweight MSA for implementing in the harsh environments and very extreme operating temperatures. The convenient MSA structure can be selected among the size of the network, location and environment where it is going to be used. In fact, will be also necessary to establish a convenient compromise between these parameters constrained by the radiation characteristics of certain antenna type.

REFERENCES

[01] Aragón-Zavala A. & Saunders S., "Antennas and Propagation for Wireless Communication", John Wiley, Chichester, 2008.

[02] Ilcev D. S., "Global Mobile Satellite Communications for Maritime, Land and Aeronautical Applications", Book, Springer, Boston, 2005.

[03] Fujimoto K. & James J.R., "Mobile Antenna Systems Handbook", Artech House, Boston, 1994.

[04] Law E.P., "Shipboard Antennas", Artech, 1983.

[05] Ilcev D.S., "Mobile Communications, Navigation and Surveillance (CNS)", Manual, DUT, Durban, 2011.

[06] Gallagher B. "Never Beyond Reach", Book, Inmarsat, London, 1989.

[07] Group of Authors, "Antenna Design", Peter Peregrinus, London, 1986.

[08] Ilcev D.S., "Shipborne Satellite Antenna Mount and Tracking Systems", TransNav - International Journal on Marine Navigation and Safety of Sea Transportation, Volume 6, Number 2, Gdynia, June 2012.

[09] Group of Authors, "Inmarsat Maritime Communications Manual", Issue 4, Inmarsat, London, 2002.

[10] Maral G. & Others, "Satellite Communications Systems", John Wiley, Chichester, 1994.

[11] Ohmory S., Wakana. H & Kawase S., "Mobile Satellite Communications", Artech House, Boston, 1998.

Development of Cospas-Sarsat Satellite Distress and Safety Systems (SDSS) for Maritime and Other Mobile Applications

S.D. Ilcev
Durban University of Technology (DUT), South Africa

ABSTRACT: This paper is describing the Cospas-Sarsat system as a global Satellite Distress and Safety Systems (SDSS) integrated in network with Search and Rescue (SAR) forces and mission for sea, land and air environments, which is in use since 1982 for personal and all transportation applications. In the past time this Network is already credited with saving thousands of lives, and with present technology improvements will provide successfully these facilities in the future as well. The Cospas-Sarsat infrastructure as a part of Global Maritime Distress and Safety System (GMDSS) consists the Space and Ground segments. The Cospas-Sarsat Space Segment is represented by both near-polar Low Earth Orbit (LEO) known as LEOSAR subsystem and Geostationary Earth Orbit (GEO) known as GEOSAR subsystem. The Cospas-Sarsat system is also developing new Medium Earth Orbit (MEO subsystem known as MEOSAR. In this paper are introduced all three Space Segments of the Cospas-Sarsat subsystems and Ground Segment. In the frame of Ground Segment is examined small RF satellite beacons transmitters (Tx) known as Emergency Position Indicating Radio Beacons (EPIRB) for Maritime, Personal Locator Beacons (PLB) for Land and Personal and Emergency Locator Transmitter (ELT) for Aeronautical applications. Then are discussed Ground Segment integration structures, which includes LEOSAR, GEOSAR and MEOSAR subsystems, LUT (Local User Terminals) with separate receiving ground stations (LEOLUT. GEOLUT and MEOLUT) and an adequate regional Mission Control Centres (MCC) associated to an appropriate Rescue Coordination Centre (RCC) and local SAR infrastructures.

1 INTRODUCTION

Vessels sink or become disabled, airplanes crash in disasters, land vehicles and expeditions get lost in the wilderness, and other causal emergencies occur that jeopardize many lives and property at sea, in the vast inland regions and air space of the entire world. Continuous advances in technology and huge improved use of radio have been an historic objective of those charged with solving the SAR problems at sea, on land and in the air. Early in the history of mobile radio, only a hundred years ago, radio was installed onboard ships to improve the safety of life and property at sea. While these improvements started well and assisted saving many lives and much property, they were not very effective when there was a sudden, catastrophic loss of the supporting platform. The most effective improvements occurred at the end of the preceding century by enhancement of SOLAS (Safety of Life at Sea) Radio MF/HF/VHF and recent development of safety system as an integration of new Digital Selective Call (DSC) Radio MF/HF/VHF with satellite Cospas-Sarsat and Inmarsat systems. This integration is known as Global Maritime Distress and Safety System (GMDSS) of International Maritime Organization (IMO) and Future Air Navigation System (FANS) of International Civil Aviation Organization (ICAO) can be used.

1.1 *Introduction of SDSS*

Availability of high technology and requirements for improvement to safety mobile radio were combined in the early 1970s to enable a system to be developed to counter the alarming catastrophic losses. Need of a large number of small beacon receivers for ships and aircraft was recognized as generally small, inexpensive and very lightweight. The only monitoring platforms were overlying aircraft and hence aeronautical distress bands were used on VHF 121.5 MHz and UHF 243 MHz for signal alerting from distress beacon transmitters.

With regard to previous endeavors, the US NASA and NOAA and CNES in France had years of experience with satellites to locate and collect data from globally deployed weather platforms, high altitude balloons, floating buoys and migrating animals, etc, while Canada located ELT via Radio Amateur Satellite OSCAR.

Two separate experiments were undertaken. The first was designed to use two of the above-mentioned VHF aeronautical frequencies and the second to use the UHF 406 MHz frequency band based on the Doppler effect. Consequently, the former Soviet Union developed its own Cospas System for SAR missions, and USA, Canada and France developed Sarsat system. Operational and trial experience of the Cospas-Sarsat system by SAR organizations started on 9 September 1982 when three Canadian persons, involved in a light aircraft crash, were rescued by the Cospas system (using Cosmos 1383). As of May 1986, the system operating at 121.5 MHz had contributed to saving 576 lives worldwide, 244 in maritime, 21 on land and 311 in air. Subsequently, the system has been used in hundreds of SAR events and has been responsible for the saving of several thousand lives globally [01, 02, 03].

Courtesy of Book: "Satellite Aided SAR" by SNES [01]
Figure 1. Cospas-Sarsat Emblem

The Cospas-Sarsat system is a joint international satellite-aided SAR system established and operated by organizations in Canada, France, Russia (ex-USSR) and the United States. This new satellite system was initially developed under a Memorandum of Understanding (MoU) among Agencies of the former USSR, USA, Canada and France, signed in 1979. Following the successful completion of the demonstration and evaluation phase started in September 1982, a second MoU was signed on 5 October 1984 by the Department of National Defense (DND) of Canada, the Centre National d'Etudes Spatiales (CNES) of France, the Ministry of the Merchant Marine (MORFLOT) of the former USSR and the National Oceanic and Atmospheric Administration (NOAA) of the USA. Hence, the system was then declared operational in 1985. On 1 July 1988, the four states providing the space segment signed the International Cospas-Sarsat Program Agreement, which ensures the continuity of the System and its availability to all states on a non-discriminatory basis. In January 1992, the government of Russia assumed responsibility for the obligations of the former USSR. Otherwise, a number of other states, non-parties to the agreement, have also associated themselves with the program and participate in the operation and the management of the system, see the first Cospas-Sarsat emblem in the Figure 1.

Initially the first generation of Cospas-Sarsat system was formed by two compatible, interoperable and international LEO satellite systems:

1 The Cospas system was developed by former Soviet Union experts and in Russian is written КОСПАС (Космическая система поиска аварийных судов и самолетов), which when translated into English, means as: Space system for search of distress vessels and airplanes. In the English acronym it is not indicated that Cospas also includes aircraft.
2 The Sarsat acronym was derived from Search and Rescue Satellite-aided Tracking and the system was designed by Canada, France and the USA.

After extensive technical testing and implementation on two Cospas (Cosmos – in Russian, Космос) and one Sarsat (NOAA) spacecraft, so Cospas-Sarsat system demonstration and evaluation phase began in February 1983 until July 1985.

1.2 *Concept of Cospas-Sarsat System*

The Cospas-Sarsat is a satellite-aided SAR system initially designed to locate distress beacons transmitting from ships (EPIRB), land (PLB) and aircraft (ELT) on the frequencies of 121.5/243 MHz only and than on 406 MHz it. In the meantime, Cospas-Sarsat has ceased satellite processing of 121.5/243 MHz beacons on 1st February 2009 and replaced with 406 MHz beacons, owing to the limitations of the 121.5/243 MHz beacons and the superior capabilities of the 406 MHz systems. Recently, the Cospas-Sarsat system using LEO or Polar Earth Orbit (PEO) satellites is referred as the new Cospas-Sarsat LEOSAR system. The PEO satellites used in the LEOSAR system can provide a global but not continuous coverage for detection of distress beacons using a Doppler shift location technique. However, the non-continuous coverage introduces delays in the alerting process since the user in distress must "wait" for a satellite to pass into the visibility of the distress beacons.

Since 1996 Cospas-Sarsat participants have been experimenting with 406 MHz payloads on satellites in GEO together with their associated ground stations to detect the transmissions of Cospas-Sarsat 406 MHz beacons. These experiments have shown the possibility of almost immediate alerts at 406 MHz, providing the identity of the beacon transmitting encoded data, such as the beacon position derived from GPS or GLONASS receivers (Rx). This development is referred as the UHF 406 MHz GEOSAR system, which alerts produced by emissions from the first generation 406 MHz beacons do not include any position information, as the Doppler effect cannot be applied to signals relayed through GEO satellites. A new type of 406 MHz beacon allows for the encoding of positional data in the transmitted 406 MHz message using onboard GPS or integrated GPS Rx in 406 MHz beacon, thus providing for quasi-real-time alerts and position information from the GEOSAR system [02, 03, 04, 05].

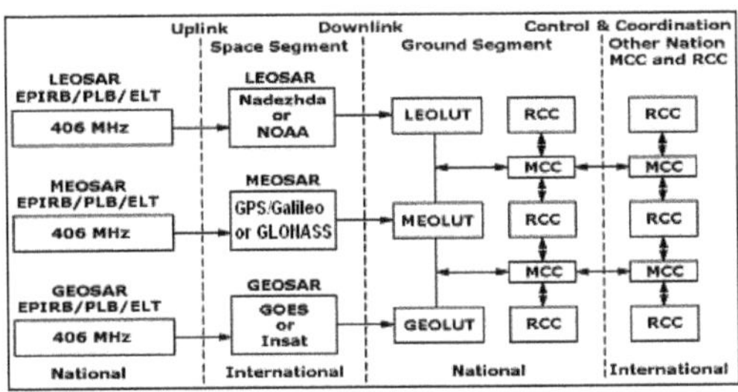

Courtesy of Book: "Global Mobile Satellite Communications" by Ilcev [03]

Figure 2. Cospas-Sarsat LEOSAR, MEOSAR and GEOSAR System Block Diagram

The latest development of Cospas-Sarsat system is that USA, Russia and European Commission/European Space Agency (EC/ESA) have announced their plans to include 406 MHz SAR repeater instruments on their respective constellations of Medium Earth Orbit (MEO) together with GNSS constellation of current US GPS and Russian GLONASS including forthcoming European Galileo. These modern MEOSAR programmes, if realized both Galileo and MEOSAR, will provide significant benefits to Cospas-Sarsat 406 MHz SAR alerting, which include: Near global MEO coverage with accurate independent location capability (no reliance on a navigation receiver); Robust distress beacon to satellite communication links and high level of satellite redundancy and availability; and Resilience to distress beacon-to-satellite link obstructions (satellite motion alleviates line-of-sight beacon-to-satellite blockages). Therefore, as illustrated in the system block diagram in Figure 2, the Cospas-Sarsat system is composed of the following four basic subsystems:

1 The first subsystem is represented by small emergency satellite EPIRB, PLB and ELT transmitter beacons designed to radiate alert and distress signals in the UHF 406 MHz bands for LEOSAR, MEOSAR and GEOSAR spacecraft systems.
2 The second subsystem is the LEOSAR, MEOSAR and GEOSAR satellite configurations, which are capable of receiving these distress messages and retransmitting them at corresponding RF to the LEOLUT, MEOLUT or GEOLUT stations for processing. The 406 MHz LEOSAR data are also processed and stored on board the spacecraft for direct or later stored data transmission.
3 The third subsystem is LUT (LEOLUT, MEOLUT and GEOLUT) that receive the relayed distress signal from beacons via LEOSAR, MEOSAR or GEOSAR satellites. These signals are then processed within the LUT stations to provide position and location of the distress signals transmitted by LEOSAR, MEOSAR or GEOSAR satellite beacons and these data are then transferred to the MSS.
4 The fourth subsystem is the Control and Coordination system infrastructure, which is fully accomplished by the MCC base in each of the participating countries also by every communications arrangement between these various MCC offices. Thus, their duties are to exchange distress incident data, or any other relevant information may be arranged between all participating MCC, as required. The final destination MCC office is then responsible for providing the data to the appropriate national RCC for SAR action.

The Cospas-Sarsat satellite configuration ensures full mutual interoperability between LEOSAR, MEOSAR and GEOSAR operational satellite subsystems, respectively, because a set of compatible parameters at each subsystem interfaces were they agreed, documented and established. So, certain key parameters, particularly on the space LEOSAR, MEOSAR and GEOSAR hardware were verified separately before launching.

1.3 *Overall Cospas-Sarsat System Configuration*

Unless, as an alternative and compulsory, every merchant ship can be fitted with an L-band satellite EPIRB operating in sea areas A1, A2 and A3 only, according to GMDSS, the carriage of a float-free satellite EPIRB operating on the 406 MHz in the Cospas-Sarsat system is required on all SOLAS vessels. Also it is mandatory to carry a satellite ELT on every aircraft. Aircraft also can carry EPIRB in the case of landing at sea, when has to act as a ship in distress. Land vehicles or individuals in expedition groups can be equipped with satellite PLB for use out of urban localities [03, 06, 07].

Within the Cospas-Sarsat current system configuration, shown in Figure 3, it is important to note that two distinct LEOSAR spacecraft have been considered, the US NOAA and Russian Nadezhda substantially different in their implementation and GEOSAR satellite system with two types of GEO satellites, the US GOES and Indian Instant-2/3. The LEOSAR subsystem has the following two data transmission systems:

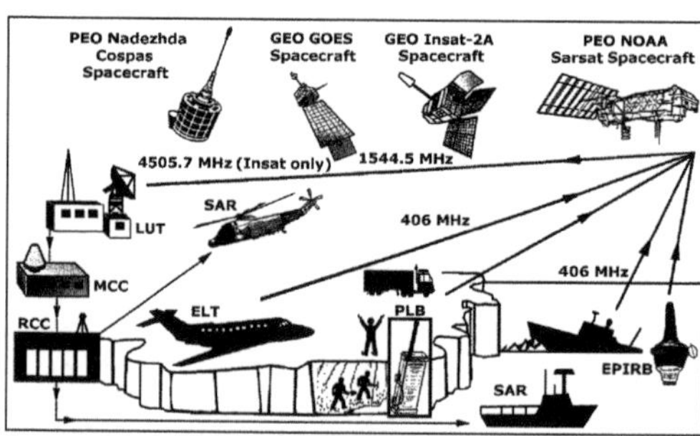

Courtesy of Book: "Global Mobile Satellite Communications" by Ilcev [03]
Figure 3. Basic Concept of Cospas-Sarsat LEOSAR and GEOSAR System

1. Repeater Data System – The repeater data solution on board the spacecraft that relays the 121.5/243/406 MHz signals directly to a visible LUT station is not in use any more.

2. Processes Data System – A 406 MHz data processor is included on both LEOSAR types of spacecraft (Cospas and Sarsat), which receives and detects all three UHF applications of satellite beacons. Thus, the Doppler shift is measured, which will be explained in the next section and the satellite beacons identification and status data are recovered. This information is time tagged, formatted as digital data and later on transferred to the repeater downlink for real-time transmission to any LUT station in view. Simultaneously, the data is stored on both types of spacecraft for later transmission to the NOAA ground stations in the case of a Sarsat and to any LUT in the case of a Cospas Nadezhda spacecraft. On the other hand, the new generation of Cospas-Sarsat GEO system does not need a process data system because GEOLUT stations are always in direct view of all GEOSAR 406 MHz beacons within the corresponding satellite coverage.

1.3.1 *Cospas-Sarsat LEOSAR Subsystem*

The Cospas-Sarsat LEOSAR VHF 121.5/243 MHz Subsystem used the original 121.5 MHz aeronautical emergency frequencies. As discussed, this frequency at 121.5 MHz served Cospas-Sarsat until 1st February 2009, and has been used EPIRB, PLB and ELT for almost 2 decades.

The Cospas-Sarsat LEOSAR UHF 406 MHz Subsystem of satellite beacons has been introduced in 1997 as a second generation shown in Figure 3, which allow for the transmission in the 406 MHz of encoded positional data acquired by the integrated GPS/GLONASS Rx. However, frequencies in the 406.0–406.1 MHz band have been exclusively reserved only for beacons operating with satellite systems. The 406 MHz beacons have been specifically designed to provide improved performance in comparison with the older VHF 121.5 MHz beacons. They are more sophisticated because of the specific requirements of the stability of the transmitted frequency and the inclusion of a digital message, which allows the transmission of encoded data, such as unique beacon identification. The 406 MHz satellite beacons are using two modes of alert transmissions:

1. 406 MHz Local Mode is just receiving 406 MHz distress signals, then the onboard SAR Processor (SARP) recovers the digital data from the beacon signal, measures the Doppler shift and time-tags the information. The result of this processing of signals is formatted as digital data and transferred to the repeater downlink for transmission to any local LUT in the view of spacecraft.

2. 406 MHz Global Mode – The 406 MHz SARP system provides near global coverage by storing data derived from the on-board processing of beacon signals, in the spacecraft memory unit. The content of the memory is continuously broadcast on the satellite downlink. Thus, each beacon can be located by the LUT, which tracks the satellite providing 406 MHz global coverage with ground segment processing redundancy.

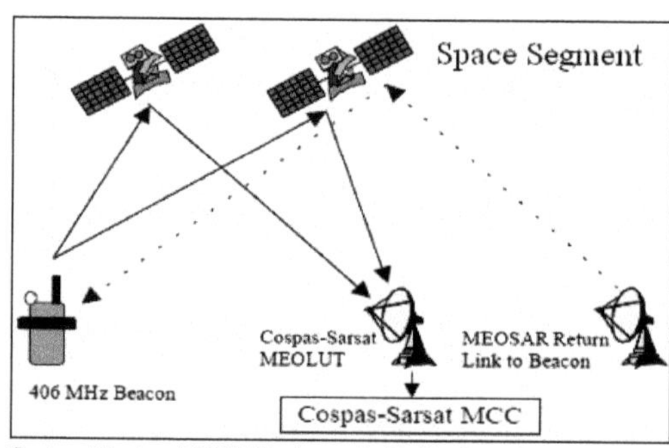

Courtesy of Brochure: "MEOSAR Implementation Plan" by Cospas-Sarsat [08]
Figure 4. MEOSAR System Concept of Operations

The 406 MHz global mode also offers an additional advantage over the local mode in respect of alerting time when the beacon is in a LUT coverage area. As the beacon message is recorded in the satellite memory at the first satellite pass in the visibility of the beacon, the waiting time is not dependent upon achieving simultaneous visibility with the LUT. The total processing time can be considerably reduced through the broadcast of the satellite beacon message to the first available LUT [03, 07, 09, 10].

1.3.2 *Cospas-Sarsat GEOSAR Subsystem*

The basic 406 MHz GEOSAR in integration with the LEOSAR system is shown by Figure 3. It consists in 406 MHz repeaters carried onboard various GEO satellites and the associated ground GEOLUT facilities. The GEOLUT stations have the capability to detect the transmissions from Cospas-Sarsat type approved 406 MHz beacons relayed by the GEO. A single GEO satellite provides GEOSAR uplink coverage of about one-third of the globe, except for the Polar Regions. Three GEO satellites equally spaced in longitude can provide continuous coverage of all areas of the globe between approximately 70° N and 70° S. As a GEOSAR satellite remains fixed relative to the Earth, there is no Doppler effect on the Rx RF and this positioning technique cannot be used to locate the distress beacon. To provide rescuers with beacon position information, such information must be either acquired by the beacon through an internal or an external GPS/GLONASS Rx and encoded in the beacon message, or derived, with possible delays, from the LEOSAR system.

1.3.3 *Cospas-Sarsat MEOSAR Subsystem*

The USA, Russia and the European Commission/European Space Agency (EC/ESA) have announced their plans to include 406 MHz SAR repeater instruments on their respective constellations of MEO together with GNSS constellation of current US GPS and Russian GLONASS including forthcoming European Galileo. These modern MEOSAR programmes, if realized both Galileo and MEOSAR, will provide significant benefits to Cospas-Sarsat 406 MHz SAR alerting, which include: Near global MEO coverage with accurate independent location capability (no reliance on a navigation receiver); Robust distress beacon to satellite communication links and high level of satellite redundancy and availability; and Resilience to distress beacon-to-satellite link obstructions (satellite motion alleviates line-of-sight beacon-to-satellite blockages).

These modern MEOSAR programmes, if realized both Galileo and MEOSAR, will provide significant benefits to Cospas-Sarsat 406 MHz SAR alerting, which include: Near global MEO coverage with accurate independent location capability (no reliance on a navigation receiver); Robust distress beacon to satellite communication links and high level of satellite redundancy and availability; and Resilience to distress beacon-to-satellite link obstructions (satellite motion alleviates line-of-sight beacon-to-satellite blockages).

Work to date has focused on ensuring that the MEOSAR system will be fully compatible with Cospas-Sarsat 406 MHz beacons, and establishing requirements for interoperability between the three MEOSAR satellite constellations. However, an implementation plan for integrating MEOSAR components into the Cospas-Sarsat System (document C/S R.012) has been developed in 2006 and can be downloaded from the Cospas-Sarsat website. An innovative MEOLUT system future development is also available to further augment existing SDSS Network, which Space and Ground segment is illustrated in Figure 4.

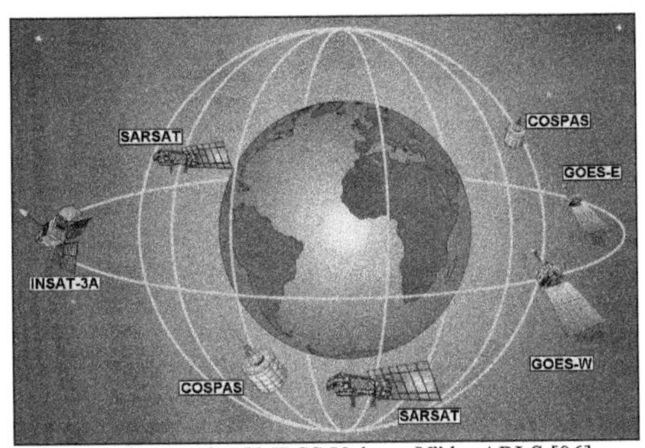

Courtesy of Manual: “GMDSS Volume V” by ARLS [06]
Figure 5. Cospas-Sarsat LEOSAR and GEOSAR Space Segment

2 COSPAS-SARSAT SPACE SEGMENT

As discussed earlier, the Cospas-Sarsat space segment has been designed and implemented to meet the requirements of the distress and safety mission as the result of a corporate effort from the various partners involved in this Program. At first, only the Cospas and Sarsat PEO satellite configuration was developed, currently known as the LEOSAR space segment consisting two Cospas and two Sarsat spacecraft, then an additional second GEOSAR constellation infrastructure of the Cospas Sarsat system was recently developed, configured by one Insat and two GOES satellites, as is illustrated in Figure 5.

In Table 1 is presented initial and current LEOSAR satellites, and in Table 2 GEOSAR satellites.

The third new projected Cospas-Sarsat space segment is MEOSAR, shown in Figure 5. The primary missions for the satellites used in the three MEOSAR constellations are the Global Positioning System (GPS), Galileo and GGLONASS global navigation satellite services. As a secondary mission, the SAR payloads will be designed within the constraints imposed by the navigation payloads. The three MEOSAR satellite constellations will utilize transparent repeater instruments to relay 406 MHz beacon signals, without onboard processing, data storage, or information specific to the Distress Alerting Satellite System (DASS)/GPS, demodulation/remodulation. The DASS, SAR/Galileo and SAR/GLONASS payloads will operate with downlinks in the 1544 – 1545 MHz band. A description of the issues that influence the selection of MEOSAR downlinks, and the frequency plan for MEOSAR downlinks are also provided. The MEOSAR satellites orbit the Earth at altitudes ranging from 19,000 to 24,000 km, which constellations are summarized at Table 3.

Table 1. Initial and Current LEOSAR Space Segment

Spacecraft	Payload	406 MHz SARP		406 MHz	121.5 MHz	243 MHz
		Global	Local	SARR	SARR	SARR
Nadezhda-1	Cospas-4	NO	NO	NO	Ceased	Ceased
Nadezhda-3	Cosaps-6	NO	NO	NO	Ceased	Ceased
Nadezhda-5	Cosaps-8	NO	NO	NO	Ceased	Ceased
Nadezhda-6	Cosaps-9	NO	NO	NO	Ceased	Ceased
Nadezhda-7	Cosaps-10	NO	NO	NO	Ceased	Ceased
NOAA-11	Sarsat-4	NO	NO	NO	Ceased	Ceased
NOAA-14	Sarsat-6	NO	NO	NO	Ceased	Ceased
NOAA-15	Sarsat-7	O	O	O	Ceased	Ceased
NOAA-16	Sarsat-8	O	O	O	Ceased	Ceased
NOOA-17	Sarsat-9	O	O	O	Ceased	Ceased
NOOA-18	Sarsat-10	O	O	O	Ceased	Ceased
METOP-A	Sarsat-11	O	O	O	Ceased	Ceased
NOOA-19	Sarsat-12	O	O	O	Ceased	Ceased

Legend: O = Operational; NO = Not Operational; SARP= SAR Processor; SARR = SAR Repeater

Table 2. The Initial GEOSAR Operational and Planned Spacecraft

Spacecraft	Launch Date	Position	Status	Rx RF MHz	Tx RF MHz	Country/Firm
GOES-E	04/1994	75° W	O	406	1544.5	USA
GOES-W	04/1997	135° W	O	406	1544.5	USA
Insat-2B	1993	93.5° E	O	406	4505.7	India
Insat-3A	Projected	93.5° E	P	406	4505.7	India
Insat-3D	Projected	83° E	P	406	4507	India
Luch-M E	Projected	95° E	P	406	11381.05	Russia
Luch-M W	Projected	16° W	P	406	11381.05	Russia
MSG	Projected	0°	P	406	1544.5	Eutelsat

Legend: O = Operational and P = Planned

Table 3. MEOSAR Satellite Constellations

MEOSAR Satellite Constellations	DASS/GPS	SAR/Galileo	SAR/GLONASS
Number of Satellites: Total Operational In-orbit Spare With MEOSAR Payload	 27 24 3 All GPS Block III Satelites	 30 27 3 TBD	 24 24 TBD(3) All GLONASS-K Satellites
Altitude (km)	20,182	23,222	19,140
Period (min)	718	845	676
Orbital Planes: Number oof Planes No of Satellites per Plane(1) Plane Inclination (Degrees)	 6 4 55°	 3 9(2) 56°	 3 8 64.8°

Notes: (1) Not including spare satellites; (2) Plus one spare in each plane; and (3) TBD - To Be Determined.

3 COSPAS-SARSAT GROUND SEGMENT

The Cospas-Sarsat ground segment contains three subsystems: Cospas-Sarsat distress and emergency beacons, such as maritime EPIRB, land or personal PLB and aeronautical ELT, then LUT ground stations and MCC formations associated to RCC and SAR infrastructures and missions.

3.1 *Cospas-Sarsat Emergency Satellite Beacons*

The Cospas-Sarsat satellite beacons are designed for use for maritime (EPIRB), personal or land (PLB) and aeronautical (ELT) applications. The first analog type of satellite beacons using 121.5 MHz band is not in use any more, and the current second type transmits a digital identification code on 406 MHz band. A list of all manufacturers of satellite beacons is presented in the Cospas Sarsat System Data brochure.

3.1.1 *Emergency Position Indicating Radio Beacons (EPIRB) for Maritime Applications*

The EPIRB maritime beacons are designated for use on board ships, hovercraft, oil/gas sea platforms and even aircraft. The 406 MHz EPIRB beacons are divided into two categories:

a) Category I EPIRB beacons are activated either manually or automatically. The automatic activation is triggered when the EPIRB is released from its bracket. Hence, these units are housed in a special bracket equipped with a hydrostatic release. This mechanism releases the EPIRB at a water depth of 1 – 3 m.

The buoyant EPIRB then floats to the surface and begins transmitting the distress signals. This type of beacon has to be mounted outside the vessel's cabin, on the deck where it will be able to "float free" of the sinking vessel. Both categories of EPIRB beacons can be detected by both LEOSAR and GEOSAR systems.

b) Category II EPIRB beacons are manual activation only units and in this sense, they should be stored in the most accessible location on board, where it can be quickly accessed in an emergency situation. At all events, before eventual use and testing of EPIRB beacons it will be necessary to follow the manufacturer's recommendations and guidelines for general beacon testing and inspection procedures. In addition, it will be also very important to register EPIRB beacons, which will help rescue forces to find it faster in an emergency and allow it to make an important contribution to the safety of others by not needlessly occupying SAR resources that may be needed in an actual emergency.

All EPIRB transmitters designs are MED/FCC approved and specially based on the IMO/SOLAS/GMDSS Regulations as float free to operate with the Cospas-Sarsat system serving LEOSAR, GEOSAR and new MEOSAR subsystem [03, 06, 07, 08, 11, 12].

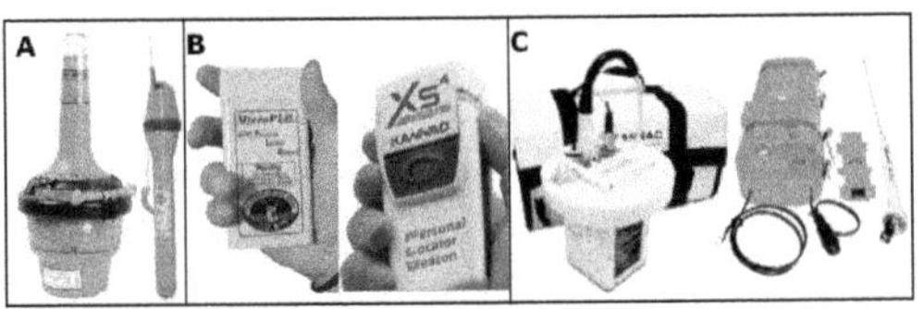

Courtesy of Manual: "Mobile CNS" by Ilcev [07]
Figure 6. Cospas-Sarsat EPIRB, PLB and ELT Satellite Beacons

There are many manufacturers of EPIRB units but the most eminent is Jotron with main products 121.5/406 MHz Tron 40 GPS MKII, illustrated in Figure 3 (A Left) and manually activated Tron 45SX, shown in Figure 3 (A Right). Tron 40 GPS MKII is GMDSS compliant 121.5/406 MHz EPIRB transmitter (Tx) integrated with 12 channel GPS receiver (Rx), obtains latitude and longitude co-ordinates updated every 20 minutes, is MED approved and has exchange/upgrades available.

The Tron 40 GPS is designated further to enhance the life-saving capabilities of conventional beacons such as similar Tron 40S without GPS Rx, which can be effectively used with the LEOSAR system only. The integrated 12 channel GPS in Tron 40 EPIRB receives GPS positional information and provides vital GEOSAR operations in a shorter time. Whenever a distress message transmitted by Tron 40 GPS is detected by LEOSAR, the delayed alert remains the same as for Non-GPS EPIRB, such as Tron 40S, within about 90 minutes but the position accuracy is improved from a radius of 5 km to 100 m. Upon detection of a distress message sent by Tron 40 GPS, the alert is immediate, within a maximum of 5 minutes. Both Tron 40S and 40 GPS are designed to be installed on board ships in the special FB-4/FBH 4 bracket, which is a float free and automatic release bracket, while a manual MB-4 or float free and heated brackets may also be used. Both beacons provide six states self-test with strobe light confirmation, including service or exchange beacons are available and are serviceable and programmable onboard ships. The position accuracy of digital VHF EPIRB units is improved considerably from a radius of 5 km to an amazing 100 m.

3.1.2 *Personal Locator Beacons (PLB) for Land Applications*

The PLB beacons are portable units that operate in much the same way as EPIRB units, they are activated manually and can be used onboard ships as well. These beacons are designed to be carried by an individual walking or moving in a vehicle, ship and aircraft using LEOSAR or GEOSAR 406 MHz subsystems.

A state-of-the-art miniature PLB is being developed through Small Business Innovative Research (SBIR) contracts funded by NASA. This lightweight and easily carried unit is made by Microwave Monolithics, Inc., in Simi Valley, California and meets the complete Cospas-Sarsat beacon specifications and certification, shown in Figure 6 (B Left).

The Kannad 400 MHz XS-ER GPS beacons have contributed for saving of onboard crews around the world, shown in Figure 6 (B Right). It has three-stage operation: lift flip cover and pull anti-tamper cover to deploy the antenna, push and hold ON button to activate the PLB. Once activated it works for at least 24 hours. It is perfect to carry in addition onboard ships and aircraft solutions as well.

3.1.3 *Emergency Locator Transmitters (ELT) for Aeronautical Applications*

The ELT terminals have been standard aircraft equipment for many decades and pre-date satellite tracking. Early versions of ELT beacons started to operate at a ceased frequency of VHF 121.5 MHz and later VHF 243 MHz. When much ELT electronic equipment initiated operating at the same frequency, there was a large increase in the level of false alerts. As a result, a new ELT technology emerged, using digital signals operating at a frequency band of UHF 406 MHz, which improved this defect with quicker tracing and determination. As discussed, due to the obvious advantages of 406 MHz beacons and the significant disadvantages of the older 121.5 MHz beacons, the International Cospas-Sarsat Program together with ICAO have made a decision to phase out the 121.5 MHz satellite alerting on 1st February 2009. In any event, all pilots are highly encouraged both by Cospas-Sarsat and their participants to consider making the switch to 406 MHz ELT. In this section will be introduced the following four ELT units:

The Kannad 406 AS ELT is an aeronautical survival ELT beacon intended to be removed from the aircraft and used to assist SAR teams in locating survivors of a crash at sea and on the ground, shown in Figure 6 (C Left). 2. The ACK Technologies E-04 ELT is designed for installation in aircraft used for general and commercial aviation, illustrated in Figure 6 (B Right) [03, 07, 11, 12, 13, 14].

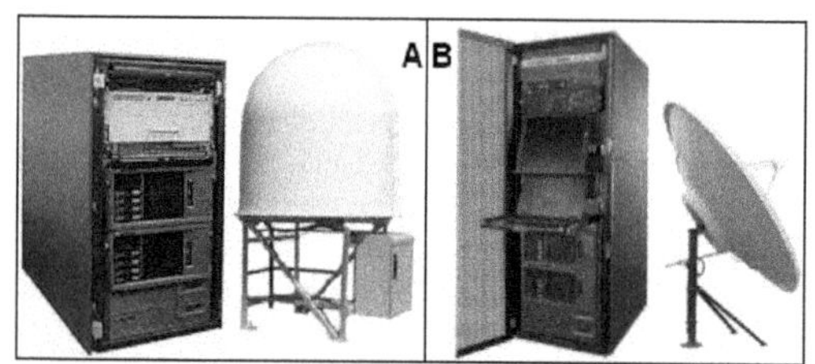

Courtesy of Brochures: "LEOLUT and GEOLUT Products" by EMS [13]
Figure 7. LEOLUT and GEOLUT Receivers with Antenna Systems

3.2 *Local User Terminals (LUT)*

The Cospas-Sarsat LUT stations are state-of-the-art systems for acquisition of LEOSAR, GEOSAR and new MEOSAR distress beacon data. Using a state of the art software-defined receiver, the LUT station s offer unprecedented flexibility in signal detection and processing. Capable of receiving all beacon types and next generation beacon signals, the LUT stations offer a path to the next generation of LEOLUT, GEPLUT and new MEOLUT technology and operations. Ensuring the safety of vessels, aircraft, and persons equipped with Cospas-Sarsat 406MHz beacons is the highest priority for the national SAR organization, because accuracy and reliability of LUT stations makes this job easier, reliable and the rescue operations more effective.

In Figure 7 (A) is shown LEOLUT Receiver and Antenna, which function is to provide Satellite downlink reception, Signal processing to extract beacon data, Location processing Communications with MCC and Self-test and performance evaluation

In Figure 7 (B) is shown GEOLUT Receiver and Antenna with function to provide Satellite tracking, Signal processing to extract beacon data, Communications with MCC and Self-test and performance evaluation.

The MEOLUT has similar equipment and the same functions as GEOLUT with only difference that station is providing Pass scheduling and Location processing of MEO satellites.

3.3 *Mission Control Centres (MCC) and Rescue Coordination Centres (RCC)*

The Cospas-Sarsat MCC sites have been set up in most of those countries operating at least one LUT, shown in Figure 8. An MCC terminals are located all over the globe and they serve as the hub of information sent by the Cospas-Sarsat system Their main MCC functions are to: collect, store and sort the data from all LUT stations and other MCC; provide data exchange within the entire Cospas-Sarsat system and distribute alert and location data to associated RCC or SPOC and other MCC hubs.

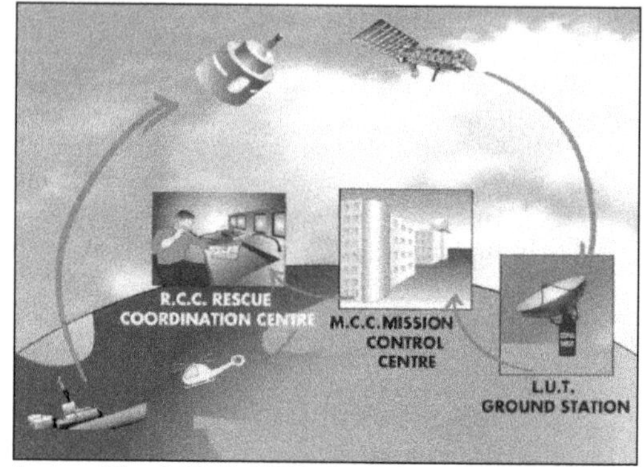

Courtesy of Manual: "Mobile CNS" by Ilcev [07]
Figure 8. Space and Ground Networks for SDSS

All Cospas-Sarsat MCC ground terminals are interconnected through nodal MCC hubs that handle data distribution in a particular region of the world. Currently, there are six (6) data distribution regions served by the United States, France, Russia, Australia, Japan and Spain. The system utilizes several communication modes to ensure the reliable distribution of alert data and system information. However, the main functions of MCC are Data processing from LUT stations, Data dissemination per Cospas-Sarsat T MCC-LUT-MCC-RCC Communications, Data archiving and Ephemeris data for LUT stations. The MCC stations is also providing Remote Control and Monitoring of LUT (LEOLUT, GEOLUT and MEOLUT) and RCC.

The SARMaster solution of RCC provides comprehensive management of Cospas-Sarsat alerts, shown in Figure 8. Alerts messages are displayed on the Geographic Information System (GIS), while associated attribute information such as frequency, time, Search Region of Responsibility (SRR), is shown in the textual view. Both composite and elemental solutions can be viewed and managed in either the map or textual view. Audible alerts and warning messages are user configurable ensuring that new data is acknowledged upon receipt. Controllers can associate the Cospas-Sarsat alerts to specific incidents, and the details being displayed and stored within that incident. SARMaster reads messages as per Cospas-Sarsat standards as well as region specific variations. The SARMaster unit supports a variety of data feeds and is configurable to meet the specific business requirements of the RCC. The RCC Workstation can communicate with MCC units or other national or international sites using many different communication protocols, such as Dial up network support; Ethernet Connection, 10-Base-T connection; Internet (including FTP, E-MAIL); Telex; AFTN and X.25. Precision SAR Manager includes support for all aspects of IMO and IAMSAR designated SAR mission planning and debriefing issues including as follows: 1. Database management system for beacon alert data, beacon registrations, SAR incidents, weather, imagery, SAR assets, etc; 2. Water temperature and current data for the SAR region; 3. Wind data for the SAR region 4. Topography for the SAR region; 5. Survivability statistics; 6. Automatic report generation; and 7. Incident Management [03, 07, 15, 16].

4 CONCLUSION

The use of satellites in LEO does not permit continuous coverage. This results in possible delays in the reception of the alert. The waiting time for detection by the LEOSAR system is greater in equatorial regions than at higher latitudes. The GEO satellites provide continuous coverage, hence they have an immediate alerting capability, but access to the GEO satellite can be masked due to obstructions, particularly on land, at high latitudes. The GEOSAR satellites do not provide coverage of the Polar Regions. The LEOSAR satellites will eventually come into the visibility of any beacon on the surface of the Earth, whatever the terrain and the obstructions, which may mask the distress transmission. In terms of coverage, the specific characteristics of LEOSAR and GEOSAR systems are clearly complementary. The rapid alerting capability of the GEOSAR system can be used by SAR forces, even when no position information is provided in the beacon message.

Such information can be used effectively to resolve a false alarm without expending SAR resources, or to initiate SAR operation on the basis of information obtained through the beacon registration data. The 406 MHz beacon signals from LEOSAR, MEOSAR and GEOSAR systems can also be combined to produce Doppler locations, or to improve location accuracy. Other aspects to the complementarity of the GEOSAR and the LEOSAR system are highlighted in the report of the UHF 406 MHz GEOSAR Demonstration and Evaluation (D&E) performed by Cospas-Sarsat Participants from 1996 to 1998.

REFERENCES

[01] Group of Authors, "Utilisation des satellites pour les recherches et le sauvetage", Cepadues, SNES, Toulouse, 1984.

[02] Group of Authors, "COSPAS-SARSAT System Documentation", COSPAS-SARSAT, London, 2001.

[03] Ilcev D. S., "Global Mobile Satellite Communications for Maritime, Land and Aeronautical Applications", Book, Springer, Boston, 2005.

[04] Solovev V.I. & Others, "Svyaz na more", Sudostroenie, Leningrad, 1978.

[05] Group of Authors, "Cospas-Sarsat Phase-Out of 121.5/243 MHz Alerting Services", Cospas-Sarsat, London, 2009.

[06] Group of Authors, "GMDSS" – Volume 5, Admiralty list of Radio Signals, Taunton, 1999.

[07] Ilcev D.S., "Mobile Communications, Navigation and Surveillance (CNS)", Manual, DUT, Durban, 2011.

[08] Group of Authors, "Cospas-Sarsat MEOSAR Implementation Plan" by Cospas-Sarsat, London, 2012.

[09] Group of Authors, "GMDSS Handbook", IMO, London, 1995.

[10] Group of Authors, "Handbook - Mobile Satellite Service (MSS)", ITU, Geneva, 2002.

[11] Website of Cospas-Sarsat: www.cospas-sarsat.int

[12] Ilcev D.S., "Maritime Communication, Navigation and Surveillance (CNS)", TransNav - International Journal on Marine Navigation and Safety of Sea Transportation, Volume 5 Number 1, Gdynia, March 2011.

[13] Zhilin V.A., "Mezhdunarodnaya sputnikova sistema morskoy svyazi – Inmarsat", Sudostroenie, Leningrad, 1988

[14] Group of Authors, Website of the: "LEOLUT and GEOLUT Products" by EMS, 2004.

[15] Venskauskas K.K., "Sistemi i sredstva radiosvyazi morskoy podvizhnoy sluzhbi", Sudostroenie, Leningrad, 1986.

[16] Tetley L. & other, "Understanding GMDSS", Edward Arnold, London, 1994.

The Propagation Characteristic of DGPS Correction Data Signal at Inland Sea – Propagation Characteristic on LF/MF Band Radio Wave

S. Okuda & M. Toba
Marine Technical College

Y. Arai
Ex-Professor of Marine Technical College

ABSTRACT: User at the Inland Sea requires high position accuracy which is 5 m (2drms) or less. Therefore the position accuracy of standalone GPS is insufficiency. Consequently it needs to use DGPS for navigator sailing the Inland Sea. We executed numerical simulation of the propagation characteristic on the extended line of bridge pier at opposite side from DGPS station, and already confirmed that bit error in DGPS correction data signal occurs, and that correction data could not form by bit error. Furthermore, we carried out numerical simulation of the propagation characteristic of DGPS correction data signal received at sailing through center of the bridge, and solved receiving condition of DGPS correction data signal before and after passing through the bridge. In this paper, we executed to inspect mutually results of electric field intensity simulation for oversea and overland propagation on some sea area of the Inland Sea and measuring results of electric field intensity for DGPS correction data signal, and evaluated the possibility of abnormal propagation comprehensively.

1 INTRODUCTION

Navigator who sails narrow channel at the Inland Sea requires high accuracy of fixed position which is 5 m (2drms) or less. In FRP 2008 (DoD, DoH.S. & DoT 2008), the requirement of position accuracy is 2-5 m for the inland waterway phase. Therefore the position accuracy of standalone GPS is insufficiency. Consequently it needs to use DGPS (Differential GPS) for navigator sailing the Inland Sea.

Decreased reliability of fixed position using GPS means that the reliability of GPS signal information decreases and there is a possibility of abnormal propagation of DGPS correction data signal, and such case occurs in the Inland Sea. Decreased reliability of GPS signal information means that there is decreased reliability of transmitting signal including satellite condition and some changes of GPS receiving condition around user including GPS receiver. Decreased reliability of transmitting signal should be compensated by RAIM (Radio Autonomous Integrity Monitoring) which is able to confirm it automatically. However, if abnormal propagation of DGPS correction data signal occurs, it is possibility to be affected by it. In this paper, we research as part of investigating received condition in the Inland Sea, but sailing circumstance in Europe where there are many river ports, large ships sailing on river around the Great Lakes have common receiving condition same as the Inland Sea. It considers that changes of GPS receiving condition are caused by incident of multi-path wave by not only sea reflection but also large offshore structures or other ships. In this paper, the main subject is to analyze the propagation characteristic of DGPS correction data signal, so to analyze changes of GPS receiving condition is future task.

Abnormal propagation of DGPS correction data signal is caused by single or multiple actions which are increase of propagation loss by overland propagation and/or some effects by offshore structure such as a big bridge. In previous papers at ITM 2011 and ITM 2012 (Okuda et al. 2011 & 2012), we executed numerical simulation of the propagation characteristic on the extended line of bridge pier at opposite side from DGPS station, and confirmed that bit error in DGPS correction data signal occurs. We also confirmed that there is a possibility that correction data could not form by bit error. In this paper, we carried out numerical simulation of the propagation characteristic of DGPS correction data signal received at sailing through center of the bridge that was future task in previous paper, and solved receiving condition of DGPS correction data signal before and after passing through the bridge. This simulation calculated a variation of signal strength by composition of superior reflection wave and surface wave. When the trail ship approaches from opposite side of using DGPS station, around just under the bridge reflection and/or scattering wave of bridge girder become to be superior, and signal strength increases because it is combined with surface wave. After passing through the bridge, signal strength increases and decreases according to combined phase because reflection wave from the bridge is combined with surface wave. Furthermore, we investigated validity of numerical simulation by checking the result of electric field measurement and also investigated effects of oversea and/or overland propagation by measuring electric field intensity every adequate distance at the Inland Sea on November 2010 and July 2012.

2 DGPS IN JAPAN

DGPS detects pseudorange error between GPS satellite and the reference station whose position is known, converts the error into correction data, and broadcasts the correction data to user on board around the reference station. Each user on board receives the correction data using MF beacon receiver. Position accuracy is improved by fixed calculation using the correction data. At present nominal position accuracy by standalone GPS is 9 m (2drms) in FRP 2008, on the other hand JCG (Japan Coast Guard) announces that position accuracy by DGPS is 1 m (2drms) or less. When using differential system, accuracy decreases depending on distance from the reference station. Then it is appropriate that position accuracy by DGPS is 1-5 m (2drms) depending on distance. DGPS has not only a function of improvement of position accuracy but also a function of integrity monitor. A function of integrity monitor informs some changes of GPS satellite healthiness or decreasing accuracy of pseudorange measurement to user quickly and break off use of the satellite data. This system is exactly the same as DGPS operated by USCG (United Status Coast Guard).

At present in Japan there are 27 DGPS stations which is the reference station, and the coverage is all coastal area except a few isolated islands. Table 1 shows DGPS specification operated by JCG. JCG calls user's attention concerning DGPS coverage (JCG DGPS center).

1 Exception of some area at Inland Sea about 200 km coverage.
2 Existence of difficult case to use by effect of terrain etc.

However JCG make no mention about an area or a phenomenon concretely.

Next, we describe about the present condition for integrity monitor. Each header of DGPS message includes operating condition of DGPS station. If position accuracy of the DGPS station becomes to be 1.5 m or more, DGPS system demands to change DGPS fix into standalone GPS fix. In Type 9, DGPS system demands to break off to use on fix calculation when correction value is 8 m or more.

Table 1. DGPS Specification in Japan

transmission rate	200 bps
transmitting power	75 W
coverage	200 km from DGPS station
transmission format	ITU-R M.823-1(RTCM SC-104)
message type	Type 3,7,9,16

Before now, it was reported about an effect of interference between nighttime ionospheric scatter propagation wave and surface wave of MF beacon wave for DGPS (Yagitani et al. 2004). They discussed some effects about distance from DGPS station and other station transmitting the same frequency, but it is different from our subject.

In this study, one of triggers is the phenomenon that transmission of differential correction value was interrupted because of propagation trouble on specific area. At present DGPS in Japan broadcasts Type 3, 7, 9, 16 based on RTCM-SC104 format (Kalafus et al. 1986). Update rate for all satellites in view is 4-5 seconds so that number of satellites in view varies. When bit error of transmission data occurs by propagation trouble and the correction data is not completed, it is shown that differential correction data is not update in the case of one time data lost.

3 ANALYSIS FOR PROPAGATION CHARACTERISTIC

3.1 *In Case of Bridge Pier*

In previous paper, we analyzed the propagation characteristic of DGPS correction data signal nearby a big bridge in order to investigate abnormal propagation of MF beacon signal for DGPS. Structure around bridge pier of big bridges at the Inland Sea is regarded as metal screen (Araki 1977) shown in Fig. 1, transmittivity around there is a little over 1 %. Electric field intensity on propagation path from DGPS station to big bridge is obtained in case of oversea and overland independently because of complex terrain (Nishitani 1980). In addition, there is diffraction loss that bridge pier is regarded as knife edge (Shinji 1992) shown in Fig. 2, and the diffraction loss is 10dB or more depending on distance from bridge pier.

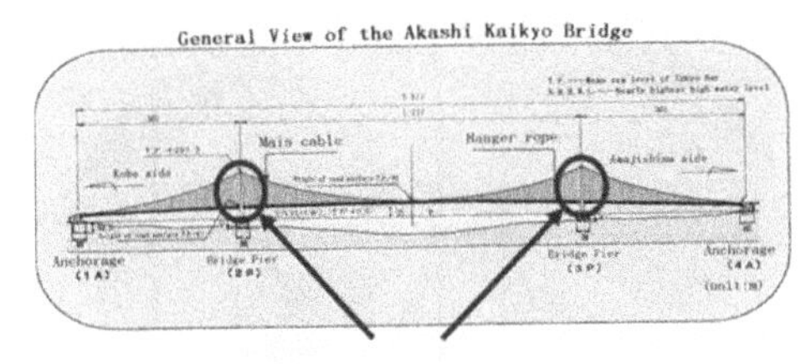

Metal Screen

Figure 1. Structure of metal screen

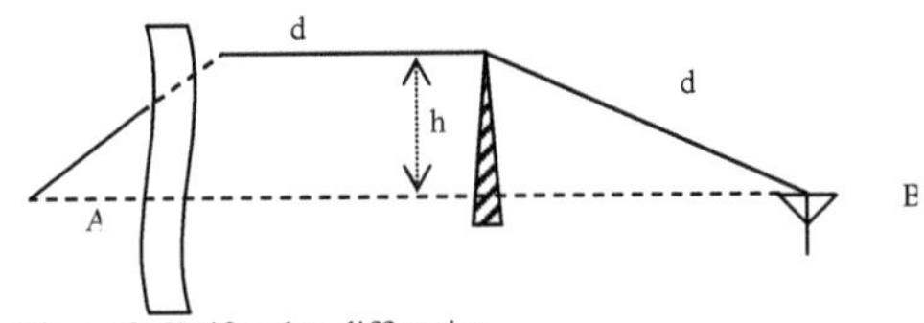

Figure 2. Knife edge diffraction

Table 2 Total electric field intensity

bridge	DGPS station	distance (km)	electric field intensity(*)	diffraction loss(*)	total intensity(*)
Kanmonkyo	Wakamiya	118.2	56	10.2	45.8
Kanmonkyo	Seto	130.8	55	10.3	44.7
Akashi-Kaikyo	Oohama	195.6	50	12.3	37.7
Akashi-Kaikyo	Muroto-Misaki	170.4	52	12.1	39.9

(*): dBμV/m

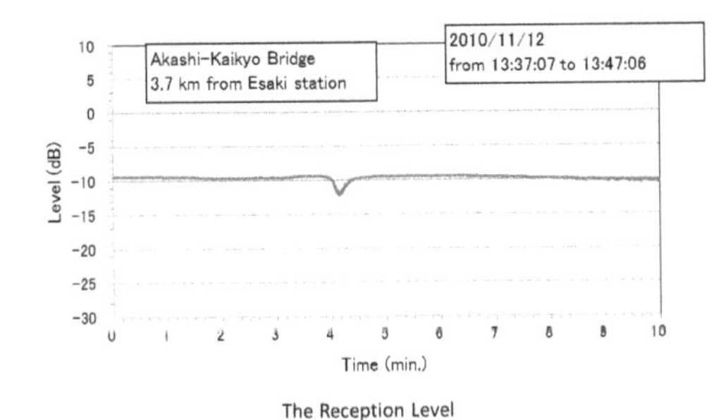

Figure 3. Electric field intensity when passing through the bridge (level down case)

Modulation system of MF beacon wave for DGPS is MSK (Minimum Shift Keying). Electric field intensity is needed 40 dBμV/m not to occur twice bit error per 1 word that bit error correcting cannot work (Saito 1996). Table 2 shows calculation results of electric field intensity nearby bridge pier. In the case of Akashi-Kaikyo Bridge, if beacon receiver uses DGPS station in Table 2, electric field intensity is 40 dBμV/m or less and always there is a possibility that DGPS correction data is missed one time.

In previous paper, there is a possibility that DGPS correction data signal causes bit error, and it is confirmed that there is a possibility that adequate position accuracy cannot be obtained (Okuda et al. 2011). Fig. 3 shows measuring data of electric field intensity measurement nearby Akashi-Kaikyo Bridge on November 2010. It was measured approximate 3dB decrease data at passing through the bridge. Because measured signal from Esaki station which is very near from Akashi-Kaikyo Bridge has large intensity, 3dB reduction does not occur some problem. However, when Esaki station signal is not utilized by some reasons which is missing to transmit etc., in case of using Muroto-Misaki station signal user's ship sails on the extended line of bridge pier, so that it may be occurred some problem to fix position. These relationship and directions of transmission from DGPS stations are shown in Fig. 4.

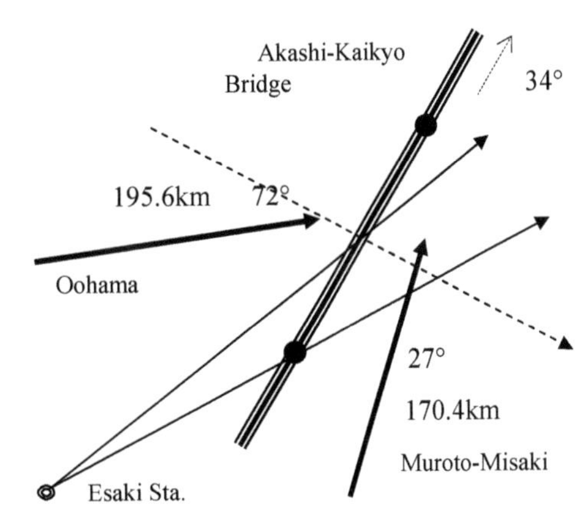

Figure 4. Geometric relation with DGPS station and Akashi-Kaikyo Bridge

3.2 *In Case of Center of Bridge*

Now, we analyze the propagation characteristic nearby a center of the bridge that was unsolved in previous paper (Okuda et al. 2011). In order to simplify this problem, it is assumed that in addition surface wave the superior reflection wave which reflects at structure of the bridge exists. Fig. 5 shows outline of direction of Minami Bisan-Seto Bridge and Oohama DGPS station used on this simulation, and trial ship sailed eastward. Fig. 6 (a) and (b) show plane view and side view respectively indicated propagation path of reflection wave. Fig. 6 (c) shows an aspect of reflection around under the bridge. Antenna height of trial ship used this simulation is 10 m and bridge girder of Minami Bisan-Seto Bridge is 65 m above sea level. Fig. 7 show one of numerical simulation results for relative value of composite signal strength (corresponding to electric field intensity) with surface wave (0dB) and reflection wave on a condition shown in Fig. 6. It was calculated that reflection coefficient was 0.5.

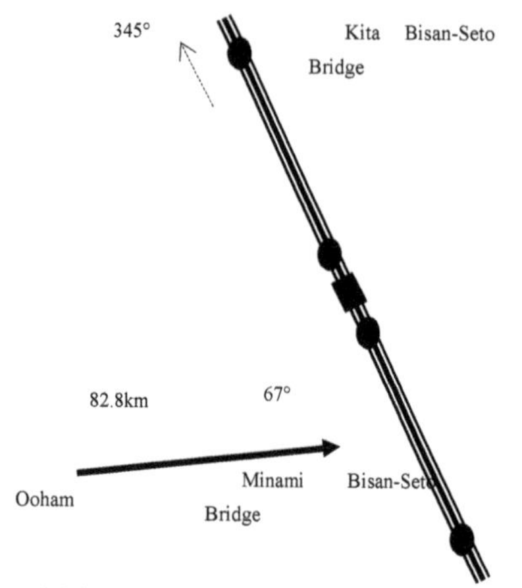

Figure 5. Bisan-Seto Bridge and direction of transmission

Relative value of signal strength to exceed 0 dB means that electric field intensity becomes to be large (to increase signal gain). Less than 0 dB means that electric field intensity becomes to be small (to decrease signal gain). If it is assumed that all of reflection wave work as noise, maximum signal to noise ratio is 6 dB. In this case, it is satisfied with advice by ITU which requires that signal to noise ratio is 7 dB or more inside DGPS coverage.

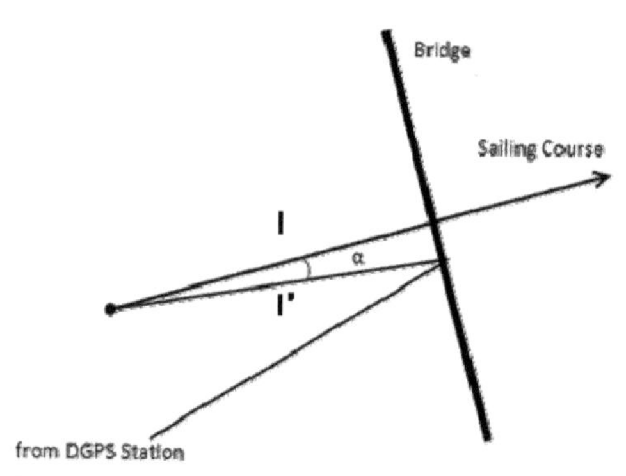

Figure 6a. Propagation path of reflection wave (plane view)

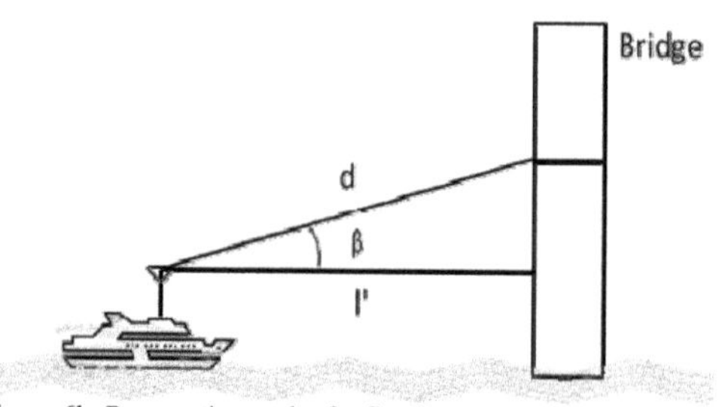

Figure 6b. Propagation path of reflection wave (side view)

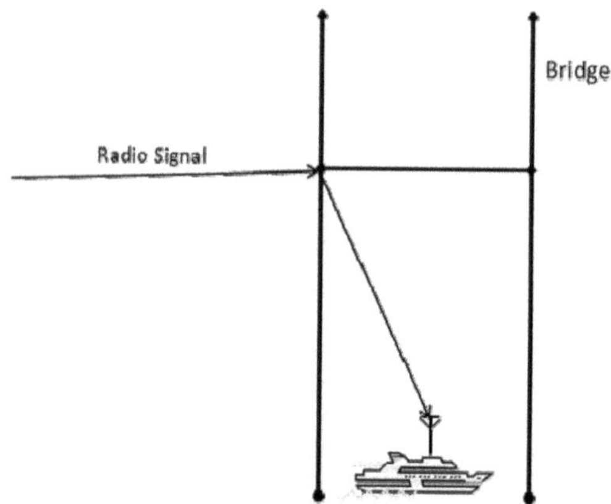

Figure 6c. Propagation path of reflection wave under the bridge

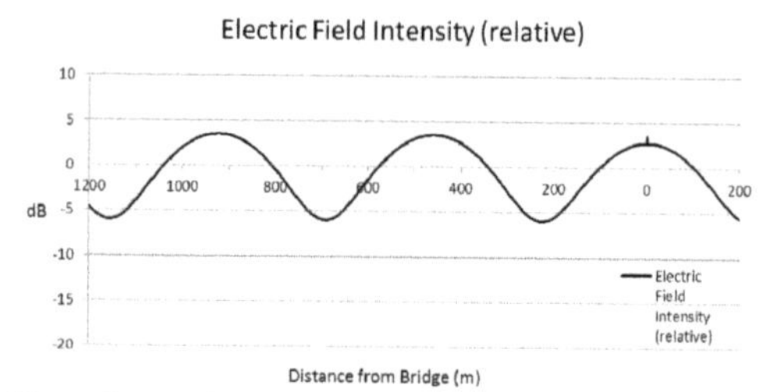

Figure 7. Result of calculation for electric field intensity (relative value) near the bridge

In fact, reflection point is not one place. There is not only mirror reflection, and effects of reflection wave decrease depending on increase of distance from the bridge. If it is assumed that there are these phenomena, it is considered that composite signal strength has some variation shown in Fig. 8.

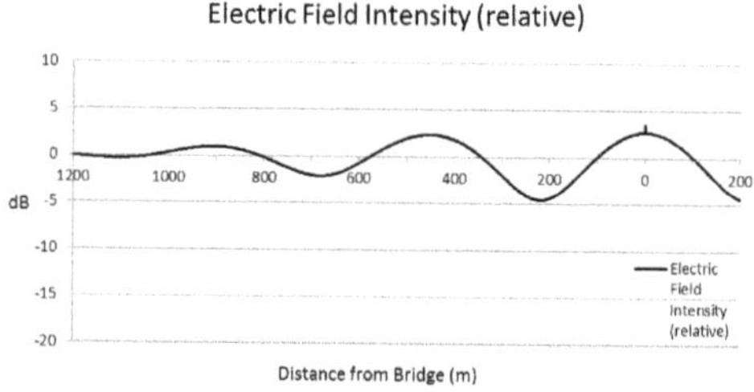

Figure 8. Corrected result of calculation for electric field intensity (relative value) near the bridge

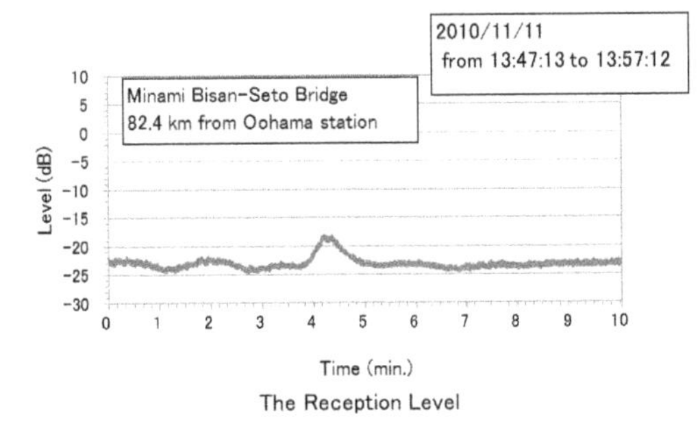

Figure 9a. Electric field intensity at passing through the bridge (Minami Bisan-Seto Bridge, using Oohama station signal)

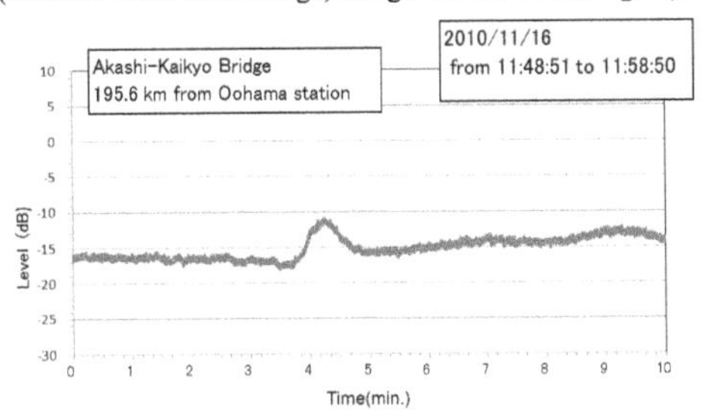

Figure 9b. Electric field intensity at passing through the bridge (Akashi-Kaikyo Bridge, using Oohama station signal)

Fig. 9 (a) shows measuring data of electric field intensity measurement whose situation is the same as this simulation. Ship to collect data sailed eastward on South Bisan-Seto Traffic Route. It was observed that signal strength increased and decreased at the front of the bridge, and signal strength increase 5 dB above minimum value around under the bridge. This simulation calculated using single reflection wave, so that measured data indicated the same tendency. To summarize, in case of sailing passing through a center of the bridge, it is observed some variation of electric field intensity around the bridge. However, when magnitude of the variation is less than 7 dB against received surface wave directly, there is no possibility that bit error occurs.

As more examples, Fig. 9 (b) shows that ship sailed westward on Akashi-Kaikyo Traffic Route, distance from Oohama DGPS station became to be close after passing through the Akashi-Kaikyo Bridge. However, compared with Fig 9 (a), periodic variation of electric field intensity was not remarkable. It might be the boundary distance as DGPS MF radio beacon wave.

3.3 *Situation of Oversea Propagation*

We investigated electric field intensity per distance to research an effect of relationship with oversea and overland propagation. We measured electric field intensity every adequate distance during sailing on traffic route or waterway. So, its propagation was almost oversea. Fig. 10 shows that it was appeared a relationship with distance to intensity although difference at 4-90 km places from DGPS station was smaller than reference (Nishitani 1980). In future study, we will collect electric field intensity included overland propagation, and try to characterize an effect of overland propagation.

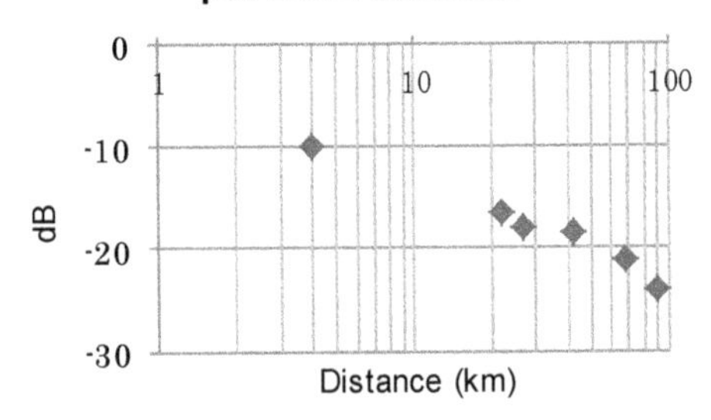

Figure 10. Electric field intensity per each distance

3.4 *Discuss for Other Bridge*

First, we discuss about nearby Kanmonkyo Bridge. In the vicinity of Kanmonkyo Bridge which locates midway between Wakamiya station and Seto station. DGPS user utilizes either of these stations. Wakamiya station is 12 km closer than Seto station from Kanmon-Kyo Bridge. Fig. 11 shows that in case of using Wakamiya station on westward sailing and in case of using Seto station on eastward sailing there is a possibility that a problem occurs on the extended line of bridge pier. Electric field intensity of DGPS correction data signal is over 40 dBμV/m required by DGPS system according to estimation in previous paper, but there is a possibility to occur bit error when overlapping each other factor.

Next, we discuss about nearby Kurushima-Kaikyo Bridge. In case of using Oohama station which is very close from Kurushima-Kaikyo Bridge, there is no problem even if diffraction loss by bridge pier exists. Consequently, when it cannot be used Oohama station, there is a possibility that same problem occurs in case of using Seto station or Esaki station.

It is considered that above-mentioned some effects of the big bridge are same as DGPS operated by USCG which is the same system in Japan. Existence of big bridge, complex terrain etc., details of DGPS operated by USCG is not clear, but it is considered that the Golden Gate Bridge and the Bay Bridge in San Francisco Bay may be existed the same problem.

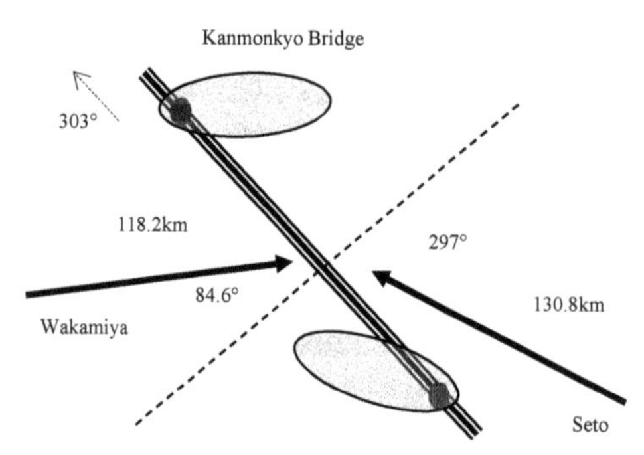

Figure 11. Geometric relation with DGPS station and Kanmonkyo Bridge

3.5 *Effect of Overland Propagation and Multi-path Signal*

Experimental navigation using our training ship Kaigi-Maru in July 2012 could be collected some electric field intensity data including overland propagation. 12-13 % of propagation path are on land, which are about 6 n.m. against 50 n.m. which are all of propagation path. In the case of 300 KHz band, if all of propagation path is land, it is estimated that electric field intensity of 50 n.m. path decreases about 10 dB (Nishitani 1980). Because decreasing electric field intensity regarding obtained data by experimental navigation is approximately 1 dB, normal sailing on traffic route is no problem.

It was compared with data for mooring at Fukae pier of Kobe port and sailing on Kobe airport off, so it was observed that electric field intensity decreased about 2 dB and data fluctuation is slightly large on mooring at Fukae shown in Fig. 12(a), (b). It is considered that 2 dB decreasing of electric field intensity are to be 5 n.m. distance longer than Kobe Airport off from Esaki station which was used station of DGPS and ratio of overland propagation are 59 %, that includes 80 % urban area. But data fluctuation at Fukae is a little less than 3 dB because there is no large construction near Fukae pier. Nearby Kanmon-Kyo, electric field intensity of Diffraction loss generated area 44-6 dBμV/m shown in section 3.1 and Table 2. Consequently, if there are some effects of multi-path signal, then electric field intensity becomes to be less than 40 dBμV/m, so that bit error occurs and data of DGPS will have a trouble to receive by occurring bit error. There is a possibility to be large effect of multipath by gantry crane at container yard except an effect of big bridge. In the case of inside port, there are some problem when pier docking and undocking. However, investigating minutely of effect of multi-path by calculation for intensity ratio of direct and reflection signal is physically impossible. In the case of existing multi-path, it is considered that receiving signal level is more than 1 Hz frequency, so we need to review methods for data collection and analyzing, and then they are future tasks.

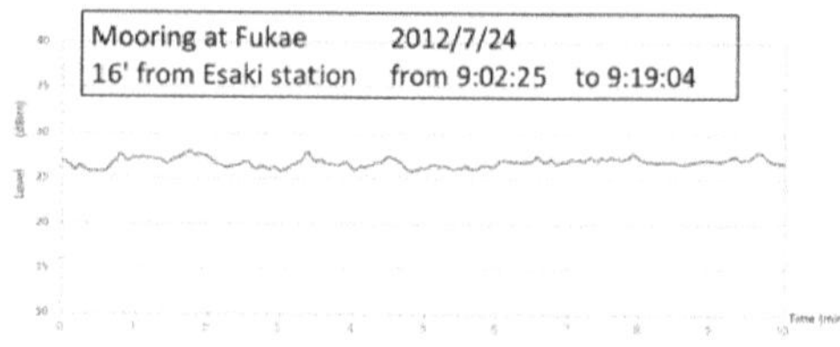

Figure 12a. Sample of Including Much Overland Propagation

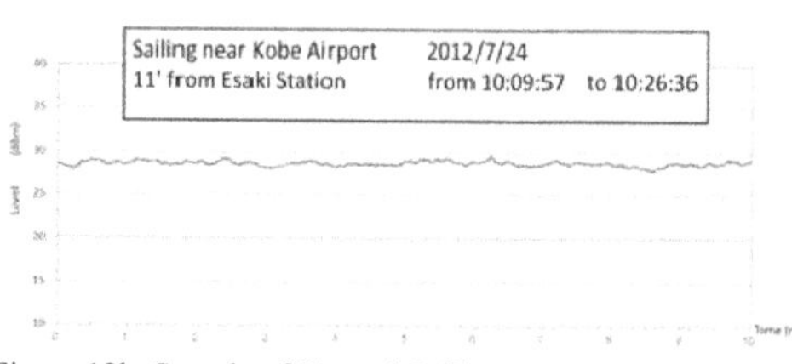

Figure 12b. Sample of Normal Sailing

4 CONCLUSION

We leaded to the propagation characteristic around a center of the bridge to apply simplified reflection wave. The result almost corresponds to measuring data of electric field intensity. To utilize this simulation results and some effects of diffraction loss by bridge pier becomes to be able to examine receiving condition of DGPS station. When DGPS user cannot utilize the DGPS station which is close and has sufficient signal strength, or when to obtain position is insufficiency from the beginning, it is confirmed that to enter on the extended line of bridge pier is that bit error may occur and there is a possibility that it cannot fix accurate position. Furthermore we measured electric field intensity at every each distance from DGPS station to research an effect of overland propagation.

Finally, as stated above, there is some possibility that DGPS correction signal have been occurred bit error caused by decreasing electric field intensity and level fluctuation by multi-path, etc. nearby large construction such as big bridge including overland propagation, so there are some case that normal DGPS position accuracy cannot obtain. Hereafter, high accuracy position fix on pier docking and undocking is essential to execute safety and efficiently navigation for navigator using DGPS. Consequently, we have to clarify these phenomenon and consider countermeasures. In this paper, it is suggested that there are not only effect of big bridge but also effect of large construction near pier, and we show necessity to investigate minutely an effect of multi-path.

REFERENCES

Araki, T. 1977, Denji Bougai to Bousi Taisaku, Tokyo Denki University Press, 138-139

Department of Defense, Department of Homeland Security and Department of Transportation: 2008 Federal Radionavigation Plan, 2008

Japan Coast Guard DGPS center, http://www.kaiho.mlit.go.jp/syoukai/soshiki/toudai/dgps/menu.htm

Kalafus, R. M., Dierendonck, A. J. VAN & Pealer, N. A. 1986, Special Committee 104 Recommendations for Differential GPS Service, The Institute of Navigation, Global Positioning System Volume III, 101-116

Nishitani, Y. 1980, Hakuyou Denshi Kougaku Gairon, Seizann-Do, 174-175

Okuda, S. & Arai, Y. 2011. The Position Accuracy of DGPS Affected by the Propagation Characteristic on MF Beacon Wave, 2011 International Technical Meeting, 718-724 January 24-26

Okuda, S., Toba, M. & Arai, Y. 2012. The Propagation Characteristic of DGPS Correction Data Signal in Japan – Propagation Characteristic near Big Bridge -, 2012 International Technical Meeting, 1383-1389, January 29-31

Shinji, M. 1992, Musen Tsuushin no Denpa Denpan, The Institute of Electronics, Information and Communication Engineers, 25-49

Saito, Y. 1996, Digital Modulation Techniques for Wireless Communications, The Institute of Electronics, Information and Communication Engineers, 70-107

Yagitani, S., Nagano, I., Tabata, T., Yamagata, K., Iwasaki, T. & Isamu Matsumoto, I. 2004, Nighttime Ionospheric Propagation of MF Radio Waves Used for DGPS – Comparison between Observation and Theory -, IEICE Transactions on Communications, Vol. J87-B, No. 12, 2029-2037

Communication Automation in Maritime Transport

Z. Pietrzykowski, P. Banaś, A. Wójcik & T. Szewczuk
Maritime University of Szczecin, Faculty of Navigation, Szczecin, Poland

ABSTRACT: To ensure safe shipping, apart from access to information, navigators have to exchange it to determine or agree on the interpretation of a current or predicted situation and to know the intentions of other transport process participants. Thanks to the standardization of information format, automatic information exchange gets increasingly wider. Another step ahead is automatic interpretation of information and automation of negotiation processes - intelligent communication. Rapid development of IT and ICT technologies creates such opportunities. This article presents the results of research on a system of automatic communication and co-operation in maritime transport.

1 INTRODUCTION

Safe navigation and vessel traffic management require access to relevant information and its proper use. More and more information systems are installed on ships and in land-based centres. These systems process data of various content and form, essential for transport process participants: navigators, shipowners, marine agents, shipchandlers, recipients of goods and services, port operators, vessel traffic services and others. The variety of physical characteristics, sources, types and scope of navigational information hampers its acquisition, collection, management, processing and presentation to decision makers. The development of information technologies makes it possible to standardize the form of navigational information and interchange processes, which may significantly contribute to the enhancement of shipping safety and effectiveness. To this end such concepts as e-maritime and e-navigation are developed along with implementation of international projects, including those executed within the EU (e.g. MARNIS [5], MonaLisa [6]).

The assurance of safe shipping, *eo ipso* the safety of personnel, cargo, ships and environment depends, apart from information access and possibility of its automatic interpretation, on communication aimed at determining or accurately specifying the interpretation and assessment of a current and predicted situation as well as intentions of other transport process participants. In maritime shipping the principles of communication between navigators steering their respective ships and navigators and land-based canters are set forth in relevant regulations. Although the regulations impose certain obligations on traffic participants, they do not eliminate possibilities of dangerous situations, resulting from failure to start communication or from errors in communications. Examples of the latter are: improper choice of means of communication, wrong information, misunderstanding or misinterpretation of interchanged information. One way to solve this problem can be the development of principles of automatic communication and co-operation, based on standards of navigational information. This means the need to extend navigational ontology with a sub-ontology for communication processes, and application of a formal language to write it down.

2 STANDARDIZATION OF INFORMATION FORM

Many systems and equipment for marine navigation, supporting decision processes on board, have been designed and developed for years. Their variety calls for standardization of navigational information format. One such example is S-57 standard for hydrographic data, used in electronic navigational charts.

Authors of the S-100 standard (version 4.0 of S-57) [3] aimed at creating possibilities for an interchange, through this standard, of more source hydrographic data and related products. In practice it means the handling of various data: matrix, raster, technical 3-D, time-varying (x, y, z, time) and new applications, for instance high-density bathymetry, bottom classifications, marine GIS. The S-100 standard may also be useful in services, based on Internet technologies, offering searching, viewing, analysis and transmission of any type of hydrographic data.

The S-100 standard offers wide possibilities of its implementation in any structures, with the use of data formats selected by the user.

The standard has the following characteristics:

- flexibility in introducing changes; future specifications of products will be based on one main data model, that will be expanded depending on the needs of various user groups;
- archive located on the IHO website will include dictionaries of features and attributes (without obligatory relations between them) and product specifications, which will enable their flexible development;
- separate folders for each user group; one of them - folder S-57 will comprise new features and attributes and additional specifications of products that may be made, specification of user data exchange standard.

The sub-standard S-101 [3] is a new ENC specification based on the S-100 concept, i.e. flexibility and arbitrariness of using source data based on catalogues and rich geometric models, information types and their attributes. Such form of data distribution enables introduction of new functionalities in ENC charts, such as dynamic presentation of tidal streams or very accurate bathymetric data. Utilizing such an amount of data may contribute to making better decisions and avoiding errors in port approach channels or when a ship proceeds in heavy traffic waters. The specification of products based on bathymetric data is found in S-102 sub-standard [8] that can be developed independently or in combination with S-101, e.g. in ECDIS.

The automatic information exchange process calls for defining the ontology of navigational information, messages sent and formats of recording. This is important in the process of transmitting an intention, question or request (demand). So far, crisp terms have been used in ship-to-ship or ship-to-shore communication, but under the S-100 standard non-crisp terms will be possible.

Data exchange automation and broadly understood co-operation also require the specification and standardization of data format and scope, and procedures of automatic translation. It seems necessary to develop a sub-standard that would define these parameters of data exchange between vessels (mobile objects) and between land-based centres and vessel operators, automatically, semi-automatically or manually.

3 NAVIGATIONAL INFORMATION ONTOLOGY. A SUB-ONTOLOGY FOR COMMUNICATION

Ontology is a description of a structure and hierarchy of notions, symbols and objects of the world or its part. The term navigational information ontology is understood as a meta-language describing the structure and form of information used in navigation, taking into account information types and scopes. An example classification, definition of set structures and their interrelations are presented in [4]. The mentioned meta-language should also be compatible with already adopted standards referring to selected areas of navigational information.

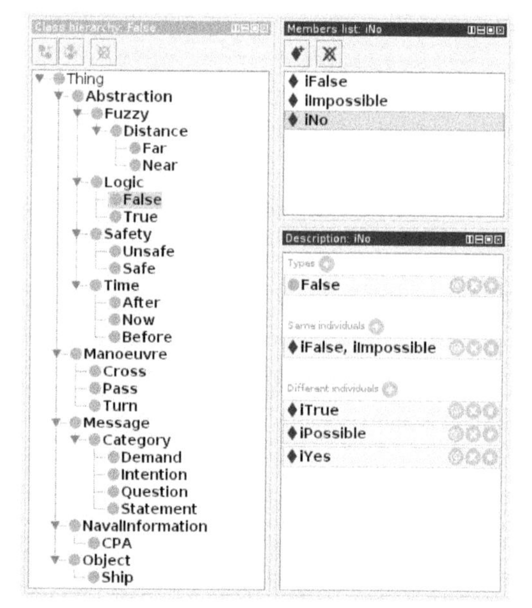

Figure 1. A window of Protége program: fragment of navigational communication ontology.

The above also refers to the manner of formulating messages, and exchanging them in ship-to-ship and ship-to-shore communications. One relevant solution is the use of appropriately constructed ontology of navigational information and a sub-ontology for communication (dialogue), where an emphasis would be put on terms connected with information exchange and negotiation processes. Research on these issues is done at the Maritime University of Szczecin [7]. The ontology was created, edited and expanded with the use of

Protége software [1], extended by an authored plug for automatic generation of ontology in the XML-Schema format. The ontology itself was developed in compliance with standards used in communication at sea [2], with assistance from navigator officers [3] and is systematically broadened. The figure 1 presents a window of the Protége program, including a fragment of navigational communication sub-ontology.

The construction of ontology requires that both terms and relations between them as well as their attributes should be defined. These are generally simple variables, storing one concrete value represented by one of the simple and enumeration type of data. Examples are such terms as true course, speed or bearing expressed by numerical values, and logical values represented by enumeration type of data (e.g. TRUE-FALSE or TRUE-FALSE-UNKNOWN).

Among others, the following are incorporated in the developed sub-ontology for communication (fragment):

- Types of messages:
  - statement (S)
  - question (Q)
  - intention (I)
  - demand (request) (D)
- Types of navigational information:
  - CPA (C)
  - ship's course (G)
  - port side, to port (L)
  - starboard side, to starboard (R)
  - forward, the bow, ahead of (F)
  - aft, the stern, astern of (K)
  - ...
- Other types of information:
  - confirmation (acknowledgement) (H)
  - ...
- Types of manoeuvres:
  - passing (P), related with L and R
  - course alteration (T), related with L and R
  - crossing course (E)
  - ...
- Abstract terms:
  - near, about (N)
  - possible (M)
  - impossible (W)
  - safe/ly (O)
  - dangerous/ly (U)
  - true (V)
  - false (X)
  - ...
- Objects:
  - Ship, related with L, R, F and S
  - ...

Writing down navigational information according to the adopted ontology necessitates coding it in a specific formal language, such as **XML** (eXtensible Markup Language). This enables automatic generation, validation and interpretation of XML messages in data communications systems. For that purpose the XML Schema can be used, as it can describe the content and structure of XML documents in XML.

4 AN EXAMPLE OF COMMUNICATION

The developed ontology of navigational information and sub-ontology for communication allow to formally write down simple messages as well as a dialog between two ships (navigators) or between a ship and a land-based centre, in which questions, intentions and other linguistic functions can be taken into consideration. To exemplify this, let us take a look at communication taking place in a very common situation: an encounter of two ships at sea. We present an automatic exchange of information that could successfully complement, and in the future partly or fully replace verbal communication.

Two motor-powered vessels (A, B as objects of type *ship*) are on crossing courses. In this situation, according to COLREGs, one ship has the right of way (ALPHA) and the other is a give-way vessel (BRAVO) (Fig. 2).

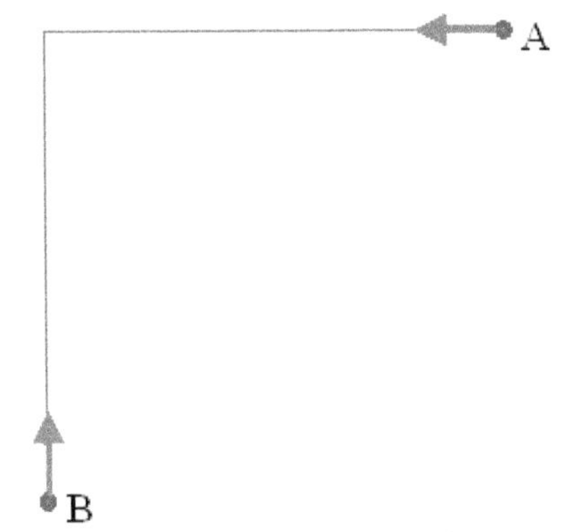

Figure 2. An example navigational situation.

When the ships are at approx. 7.5 Nm from each other, they establish contact:

1 ALPHA: CPA is 0 Nm
2 ALPHA: Pass astern of me.
3 BRAVO: OK., I will alter course to starboard.
4 ALPHA: OK.

When the distance decreases to 4.5 Nm, the following messages are exchanged:

1 ALPHA: CPA is 0.7 Nm.
2 ALPHA: Alter course to starboard to keep 1.0 Nm astern of me.
3 BRAVO: OK., I will alter course 15 degrees to starboard.
4 ALPHA: OK.

Using the ontology of navigational communication, we can present the above messages mathematically (formal notation), that enables generating and interpreting of messages and their further processing in computer systems:

1 ALPHA: CPA is 0 Nm
2 WARNING (CPA(Alpha, Bravo) is 0) → **W(C(A, B) is 0)**
3 ALPHA: Pass astern of me.
4 DEMAND (CROSS (Bravo, Alpha.Stern) is TRUE) → **F(E(B, A.K) is V)**
5 BRAVO: OK., I will alter course to starboard.
6 INTENTION(TURN(Bravo, Starboard) is TRUE) → **I(T(B, R) is V)**
7 ALPHA: OK.
8 STATEMENT (CONFIRMATION (Alpha) is TRUE) → **S(H(A) is V)**
9 ALPHA: CPA is 0.7 Nm.
10 WARNING (CPA (Alpha, Bravo) is 0.7) → **W(C(A, B) is 0.7)**
11 ALPHA: Alter course to starboard to keep 1.0 Nm astern of me.
12 DEMAND ((TURN(Bravo, Starboard) is TRUE) and (DISTANCE(Alpha, Bravo) is 1.0)) → **F ((T(B, R) is V) and (D(A, B) is 1.0))**
13 BRAVO: OK., I will alter course alter course 15 degrees to starboard.
14 STATEMENT((Bravo) is TRUE) and (COURSE(Bravo) is COURSE(Bravo)+15)) → **S((H(B) is V) and (G(B) is (G(B)+15))**
15 ALPHA: OK.
16 STATEMENT(CONFIRMATION (Alpha) is TRUE) → **S(H(A) is V)**

This form of representing certain procedures and intentions will enable using them in decision support systems. In addition, if it is presented in a readable manner (graphical or digital), it could be a valuable supplement to verbal communication, allowing to avoid ambiguity in expressing intentions and to formally acknowledge intentions.

5 NON-CRISP TERMS

Interpersonal communication in a natural language makes use of expressions with terms whose attributes assume crisp or non-crisp values. Non-crisp values, or precisely, the values of their attributes, may come directly from a natural language, e.g. "near", "far", "safely", "dangerously", "about", "safe distance", "dangerous distance". In the process of navigation, i.e. safe ship's proceeding in water area from point A to point B, such information attributes may be generated by shipboard equipment and determine ship's status as a state of a moving object in relation to other mobile or stationary objects. In the decision making process the navigator-operator accepts deviations within the assumed safety limits. Occurrence of non-crisp values appearing in a specific communication between human operators is a significant difficulty for formal description of such communication. For navigational communication ontology to offer its convenient use in a real navigational environment, it has to describe both crisp and non-crisp (fuzzy) terms.

Examples of non-crisp terms can be found in the criteria for safety assessment, namely CPA (*Closest Point of Approach*). This criterion is commonly used in ship encounter situations. Taking into account uncertainties (inaccuracies) in assessing safety is possible when we use, e.g. fuzzy logic, that allows to describe the safety level with linguistic values such as used by humans. This consists in assigning a degree of membership $\mu(x) \in \langle 0, 1 \rangle$ to crisp values, e.g. measured distance x. It means that, apart from membership (1) or no membership (0) – as in the classical set theory – membership maybe partial. In case of CPA it means that, for a value CPA_L preset by the navigator, an interval of tolerance is assumed to exist $\langle CPA_{Lmin}, CPA_{Lmax} \rangle$ such that $(CPA_{Lmin} \leq CPA_L \leq CPA_{Lmax})$, and any value of CPA is assigned a degree of membership to the fuzzy set CPA_{LF}, described by a membership function $\mu(x)$ of this set (Fig. 3) [9].

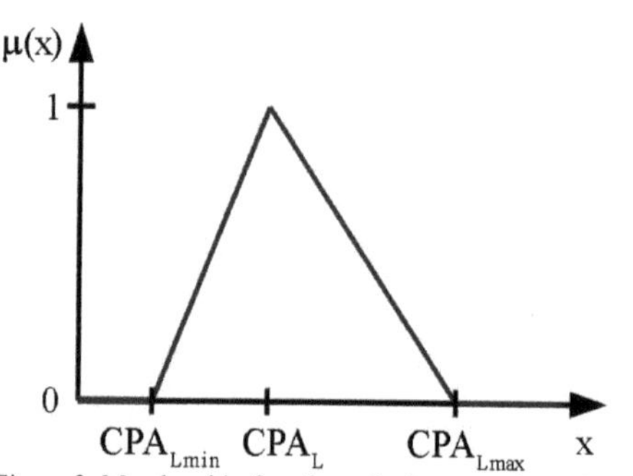

Figure 3. Membership function of a fuzzy set *fuzzy* CPA_{LF}.

The criteria of ship domain and ship fuzzy domain can be similarly considered [8].

The use of fuzzy logic, methods and tools of fuzzy sets in particular, enables a formal description of non-crisp terms, their inclusion in messages and their interpretation and processing in computer systems, e.g. in inference processes. This description requires, among others, the defining of rules for mathematical notation of a message sent. The basic notation of message (p) is as follows:

$$p \rightarrow X \text{ is } R \qquad (1)$$

where X is a variable, ***is*** – sentence-forming functor, R – relation constraining the variable X.

Dealing with a more extensive message, we can additionally define a wider range of variable origin as a function and write the message notation in this form:

$$p \rightarrow A(X) \text{ is } R \qquad (2)$$

where A is a group in the sub-ontology to which X belongs.

In a natural language there are various types of sentences, e.g. affirmative (statement) or interrogative (question). For precise interpretation of message content we can additionally adopt a function identifying sentence form, where the whole message notation is an argument of that function.

6 SUMMARY

Safe navigation often depends on information access and capabilities of automatic exchange and interpretation of relevant information. These requirements are facilitated, among others, by the standardization of information content and form. Fast advancements in IT and ICT technologies broaden possibilities of automating communication processes so far executed verbally. This, however, necessitates construction of navigational information ontology, including a sub-ontology for communication. The authors present in this article the results of research on a system of automatic intership communication and co-operation. Based on the developed sub-ontology for communication, an example communication established between two ships in an encounter situation is presented. The authors propose to extend information ontology by including non-crisp terms, typical of verbal communication.

REFERENCES

[1] Banaś P., Using the Protégé environment for building ontology for automated communication system at sea, Zeszyty Naukowe nr 30, Maritime University of Szczecin, 2012, pp. 12-17

[2] IMO, Maritime University in Szczecin, Standardowe Zwroty Porozumiewania się na Morzu, Standard Marine Communication Phrases, translated by E. Plucińska, J. Kłosiński, Maritime University of Szczecin, 1997

[3] International Hydrographic Organization, http://www.iho.int

[4] Kopacz Z., Morgaś W., Urbański J., Information of Maritime Navigation; Its Kinds, Components and Use, European Journal of Navigation, vol. 2, no. 3, Aug 2004, pp. 53-60

[5] Maritime Navigation and Information Services MarNIS, Final Report 2009, http://www.marnis.org

[6] Motorways&Electronic Navigation by Intelligence at Sea MONALISA, http://www.monalisaproject.eu

[7] Pietrzykowski Z., Chomski J., Magaj J., Niemczyk G. 2006, Exchange and Interpretation of Messages in Ships Communication and Cooperation System, Advanced in Transport Systems Telematics, Ed. J. Mikulski, Publisher Jacek Skalmierski Computer Studio, Katowice 2006, pp. 313-320

[8] Pietrzykowski Z., Uriasz J., The ship domain - a criterion of navigational safety assessment in an open sea area, The Journal of Navigation (2009), 62, The Royal Institute of Navigation, Cambridge, pp. 93-108

[9] Pietrzykowski Z., Magaj J., Wołejsza P., Chomski J., Fuzzy logic in the navigational decision support process onboard a sea-going vessel, Lecture Notes in Computer Science, Volume 6113 / 2010: Artificial Intelligence and Soft Computing - ICAISC 2010, Eds. L Rutkowski, R.Scherer, R. Tadeusiewicz, L.A. Zadeh, Springer-Verlag Heidelberg, Part I, pp. 185-193

[10] Pietrzykowski Z., Hołowiński G., Magaj J., Chomski J., Automation of Message Interchange Process in Maritime Transport, Monograph International Recent Issues about ECDIS, e-Navigation and Safety at Sea, Advances in Marine Navigation and Safety of Sea Transportation, Ed. A. Weintrit, CRC Press/Balkema, 2011, pp. 119-124

Audio Watermarking in the Maritime VHF Radiotelephony

A.V. Shishkin & V.M. Koshevoy
Odessa National Maritime Academy, Ukraine

ABSTRACT: Audio watermarking (AW) technology is proposed for application in the maritime VHF radiotelephony. AW refers to inaudible transmission of auxiliary information on the background of the voice signal without any additional radio channel resources. Robust AW algorithms are designed using the modern methods of spread spectrum and OFDM. Experimental results for the practical VHF radio channel are presented. Relating on processing complexity the designed system enables to transmit data on a rate up to 260 bit/sec in the standard VHF radio channel on an under noise level. Designed system, applied for automatic ship's identification, provides the full compatibility with the existing radio installation, and doesn't require replacement of standard VHF transceivers and operational procedures. Besides automatic identification the system may be used in the special applications, for example, by the threat of terrorist attack; generally contributes to navigation safety and information security.

1 INTRODUCTION

Audio watermarking (AW) corresponds to digital information imperceptibly embedded into the audio signal. AW for maritime VHF (Very High Frequency) communication is inspired first of all by the ability of implementation an automatic identification of radiotelephone transmissions in the channels of maritime (156...174) MHz mobile radio communication service. Applied to VHF radiotelephony, a watermarking system could overcome existing limitations, and ultimately increase safety and efficiency of maritime communication. The same application of AW may be implemented in the aeronautical (118...136) MHz mobile service. In the mentioned services analogue broadcasting channels with frequency/phase and amplitude modulation correspondingly are utilized.

For the meanwhile the identification of the sea vessels is realized by means of verbal calling of ship's call sign or numerical identification. However on account of different reasons such verbal identification may be absent, transmitted with delay, or understood with errors. This problem is illustrated in Fig. 1. Motor vessel "Arcona" transmits a certain message to all stations. But one of the receiving vessels missed the name and call sign of the transmitting ship, and another ship interpreted the name of transmitting ship as "Gargona" instead "Arcona".

It is obvious that false, incorrectly interpreted or delayed verbal identification negatively affects maritime navigation. Automatic identification could avoid misidentification and call sign confusion.

It is well known an Automatic Transmitter Identification System (ATIS) [1] for marine VHF radio that is used and mandated on inland waterways in Europe for identifying the transmitting vessel. In ATIS the identity of the vessel is sent digitally immediately after the ship's radio operator has finished talking and releases push-to-talk (PTT) button. Identification is performed by appending a short digital message in Digital Selective Calling (DSC) format. The main drawback of ATIS is post-report transmission of identification data.

In COMSAR proposal [2] the necessity of automatic identification is grounded and quite reasonably noted that the identification should be done immediately after pressing the PTT button on the contrary of ATIS releasing PTT. Another shortcoming of ATIS is principle limitation for Medium/High frequency (MF/HF) applications. This limitation results from the small bitrate (100 bit/sec) in comparison to VHF DSC rate (1200 bit/sec).

In restricted navigation environment the immediate and clear automatic identification is extremely necessary. Automatic identification would exclude the human factor and increase an efficiency of VHF radiocommunication and maritime safety in the whole.

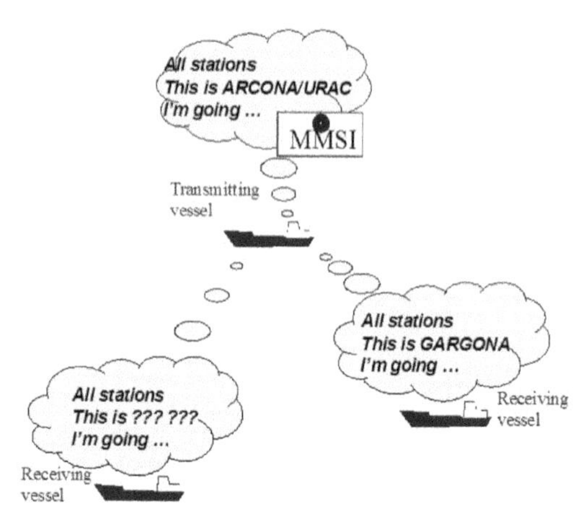

Figure 1. Audio watermarking provides automatic identification of VHF maritime radiotelephone transmissions

Only verbal identification doesn't protect against illegal radio transmission. Illegal transmissions are especially harmful on the VHF distress channel 16. Of course, unauthorized transmissions are performed anonymously. Reliable automatic identification of such transmissions could avoid the violation of radiotelephone regulation.

Another advantage of automatic identification follows from the ability of digital information inputting to another ships' navigational and information systems, for example Electronic Chart Display and Information System (ECDIS). ECDIS makes visualization of neighboring vessels in the range of VHF radio (i. e. approximately 30 nautical miles). However the transmitting vessel by no means is marked in an electronic map. Automatic identification would implement the vessel marking at electronic chart display.

One more application of AW is a covered information transmission in the special applications (for example, facing the threat of terrorist aggression).

It is essential that AW based identification doesn't require altering an existing radio installation and operational procedures. To introduce AW-identification function only new telephone receiver with the embedded processor at the transmitter side and processor with mini-display switched to common audio output at the receiver side have to be mounted. Automatic identification starts right away PTT pressing and runs continuously during all transmitting period independently from voice signal occurrence. No additional time and frequency channel recourses are required.

AW-identification supports the full compatibility with the existing transceivers and makes possibility of step-wise implementation.

2 INTERFERENCES AGAINST AW IN THE VHF RADIO CHANNEL

In the present work the AW process is considered in the frame of common communication problem. Model of communication channel with additive watermarking is presented in Fig. 2. Watermark signal w is formed on the base of embedded data m. Generally encoder may be realized in the form of a) blind (or noninformed) or b) informed encoder depending on ignoring or using the information on the carrier signal (or host signal) x. The variant b) is reflected with dotted line. The watermarked signal s is additively formed from signals x and w:

$$s = x + w \tag{1}$$

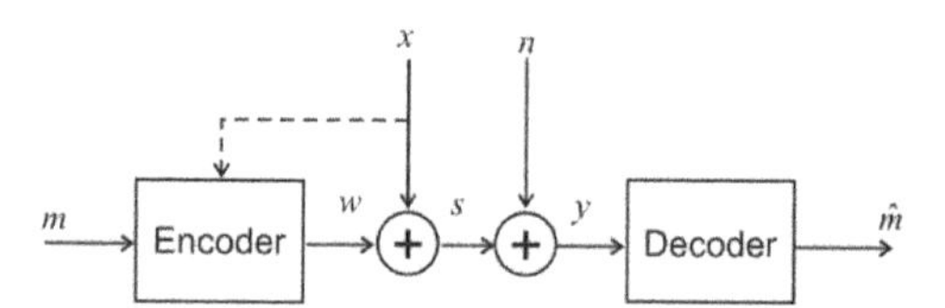

Figure 2. Model of watermarked communication channel

Power σ_w^2 of watermarked signal w is limited by the acceptable level of introduced distortions of carrier signal because watermark presence should be not audible (or quite tolerable) on the background of carrier signal x.

In the channel according the Fig. 2 two interferences act against watermark w: the first interference is itself the carrier signal with the power σ_x^2, and the second component – a noise n with the power σ_n^2.

Watermarking channel is characterized by its capacity – the maximum achievable code rate. n Assuming the both interferences are white Gaussian noises the capacity C (bit/sample) of watermarked channel with noniformed encoder, when host signal is not available to encoder, is defined by Shannon's formula:

$$C = \frac{1}{2}\left(1 + \frac{\sigma_w^2}{\sigma_x^2 + \sigma_n^2}\right) \tag{2}$$

Practically $\sigma_x^2 \gg \sigma_n^2$, and capacity is limited mainly by the host itself.

At the same time using information about the carrier signal x, it is possible to increase C for informed encoder. An idea of informed encoder goes back to Kuznetsov & Tsybakov [3] and is known as writing in memory with defective cells. Later in [4] it was shown that assuming the host is known at the transmitter, the capacity of watermarked channel is defined by the formula:

$$C = \frac{1}{2}\left(1 + \frac{\sigma_w^2}{\sigma_n^2}\right) \quad (3)$$

The formula (3) shows that appropriately considered carrier signal doesn't influence on watermark transmission and the capacity is determined only by the second noise, which is unknown at the encoder. Capacity for such channel is increasing greatly, that's why an informed encoding (i.e. "writing on dirty paper") is attractive method for watermarking on account of its potential capacity.

But besides above considered interference x there are another interferences and limitations in practical implementation of AW. These factors are:
1 intersymbol interference (ISI);
2 flat amplitude fading;
3 external additive noise;
4 nonlinear distortions (clipping);
5 resampling and desynchronization.

Hofbauer et al. [5] proposed to take into account Doppler effect that is actual for aeronautical applications.

In the radio channels ISI is usually caused by multipath propagation. The transmitting medium in VHF radio communication is the atmosphere, in which radio signal is transferred by means of electromagnetic waves. The received electromagnetic signal is usually a superposition of a line-of-sight path signal and multiple waves coming from different directions. This phenomenon is known as multipath propagation. The received signal is spread in time and the channel is said to be time dispersive.

Another physical cause of ISI is nonuniformity of frequency response of a channel. Analog low-frequency circuits of the transceiver are composed from reactive elements. Frequency dependent elements cause nonuniformity of frequency response within audio signal spectrum. When frequency response is explicitly nonuniform within signal spectrum output signal is highly differs from input one. Distortions caused by bandlimited low-frequency channel also represent ISI.

From the signal processing point of view the two physically different causes (presence of reactive elements in audio circuits and multipath radio wave propagation) lead to the same final result in the form of ISI.

Coming back to multipath propagation, one can analyze a variant when the different path lengths are very similar compared to the wavelengths of the signal components. Then the phase variations between components will be small and they will all undergo very similar amounts of cancellation or reinforcement. This case is usually termed flat fading. In watermarking flat fading is simulated by amplitude scaling attacks.

Additive noise is imposed onto the signal during transmission. The noise results from thermal noise in electronic circuits, from atmospheric noise or from other radio stations. Quantization noise from analog-to-digital converter may be attributed to additive noise. Commonly recognized model of an additive noise is additive white Gaussian noise, denoted in Fig. 2 by n.

Nonlinear distortions appear in amplitude limitations caused, for example, by the overload in audio circuits. Overload arises from redundant power of transmitting station. The simplest model of nonlinear distortions is clipping. AW in any case should be resistant against such distortions.

At the transmitter and receiver sampling processes are not synchronized. It means that sampling instants are mutually shifted. Consider sampling frequencies are equal at the transmitter and receiver. At the receiver beginning of watermark is unknown. For watermark restoration beginning of watermark should be first detected and then all decision points are counted from the starting point. Analog radiotelephone channel by all means leads to resampling and loss of the watermark beginning.

3 OFDM BASED AW EMBEDDING

AW are multiplexed directly into audio signal and therefore subjected to influence of common for audio signal transformations and interferences. Standard transformations are: amplification, modulation, filtering. In the standard band limited audio channel 300 – 3000 Hz ISI impacts destructively on watermarks.

ISI is caused mainly by nonuniformity of channel frequency response and nonlinearity of phase response in low frequency transceiver electronic circuits. The most effective measure against ISI is orthogonal frequency division multiplexing (OFDM) technology that is commonly used in numerous communication systems.

OFDM approach for AW was proposed in [6]. The main idea lies in embedding every watermarking bit completely into a certain narrowband component of host signal. AW information jointly with host signal form OFDM symbol.

Construction of OFDM symbol for AW is illustrated in Fig. 3. For simplification the figure is drawn for the next numerical values: Fast Fourier Transform (FFT) dimension $N = 8$; length of spreading sequence $L = 5$; number of watermarked subchannels $B = 2$.

Vector $\boldsymbol{x}$ of host audio samples in the time domain, a) is buffered by columns into $N \times L$ matrix, b); FFT along each column is performed, c);

B watermarking bits are embedded in frequency coefficients X_i independently along rows, d). In Fig. 3 samples in the time and frequency domains are shown by squares and circles accordingly; virgin samples are shown white and distorted ones due to watermarking are grey. As it seen watermark distortions are uniformly distributed along watermarked sequence $\boldsymbol{s}$.

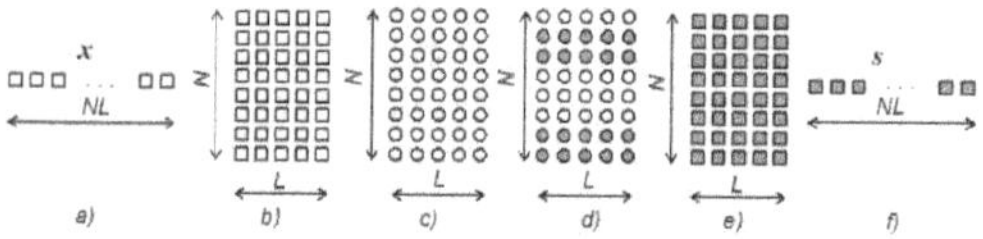

Figure 3. Forming of OFDM watermarking symbol

Embedding and detection algorithms for each subchannel are mathematically stated by formulas (4) – (10). Watermarked signal (or stegosignal) is given by:

$$\boldsymbol{s} = \boldsymbol{x} + \boldsymbol{w}(x, m), \tag{4}$$

where $\boldsymbol{s}$, $\boldsymbol{x}$, $\boldsymbol{w}$ – L-coordinates vectors of stegosignal, host signal and watermark correspondingly; $m = \pm 1$ – embedded bit; $x = (\boldsymbol{x}, \boldsymbol{u})$ – inner product of host signal and pseudorandom sequence (PRS) $\boldsymbol{u}$: $u_i = \{\pm 1\}$. For PRS m-sequence of appropriate length L may be used.

Embedding algorithm for each subchannel is based on so called improved spread spectrum (ISS) technique [7]. This algorithm is characterized by the high immunity against constant amplitude scaling.

Coordinates of vector $\boldsymbol{w}$ in (4) is determined according formula (5):

$$w_i = \tilde{w}\left(|x_i| / \|\boldsymbol{x}\|_p\right)^p u_i, i = \overline{1, L}, \tag{5}$$

where $\|\boldsymbol{x}\|_p = \left(\sum_{i=1}^{L} |x_i|^p\right)^{1/p}$ – p-norm of vector $\boldsymbol{x}$, and value w is defined as follows:

$$\tilde{w} = \begin{cases} 0, & m\tilde{x} \geq \rho, \quad (6a) \\ m\rho - \tilde{x}, & m\tilde{x} < \rho. \quad (6b) \end{cases}$$

In the formula (6) ρ – is a certain threshold that predetermines AW resistance against interferences.

Sense of expression (6) is cleared by the following. If $m\tilde{x} \geq \rho$, than it is not necessary to introduce any distortions. Otherwise, (6 b) it is necessary to introduce some distortions that would correct the inner product $\tilde{s} = (\boldsymbol{s}, \boldsymbol{u})$ to threshold $\pm\rho$ depending on the embedded bit:

$$\tilde{s} = \begin{cases} \rho, & m = 1, \\ -\rho, & m = -1. \end{cases} \tag{7}$$

Therefore, the introduced distortions for informed encoder are strongly depends on current host state.

Informed encoder provides: a) controlled introduction of distortions, setting them on the minimal required level (formula (6)), and b) guaranteed AW resistance according to the chosen threshold ρ (formula (7)).

Selection of optimum value p in (5) should be fixed to minimize audibility of introduced distortions. Generally $p > 0$. In the case $p = 1$ the total correction $\tilde{w}$ (6 b) is allocated proportionally to amplitudes of vector $\boldsymbol{x}$. In the case $p \to 0$ $\tilde{w}$ is allocated uniformly along all samples of vector $\boldsymbol{x}$. Just this case is given in the paper [7] for ISS algorithm. For uniform spreading according original ISS algorithm [7] vector $\boldsymbol{w}$ results from (5) when $p \to 0$ and contains the equal amplitude coordinates:

$$w_i = \frac{1}{L}\tilde{w}u_i, i = \overline{1, L} \tag{8}$$

It should be particularly noted that formulae (5) and (8) suffers from getting negative values of amplitudes according to (4) if $x_i < |w_i|$ and $w_i < 0$. In this case incorrect watermark detection may be occurred. An additional checking routine whether $s_i \geq 0$ should be done to eliminate this drawback. This procedure is omitted in the present paper.

Received vector is given in the form:

$$\boldsymbol{y} = \boldsymbol{x} + \boldsymbol{w} + \boldsymbol{n} \tag{9}$$

where $\boldsymbol{n}$ – additive noise vector.

In the decoder inner product (or correlation) $\tilde{y} = (\boldsymbol{y}, \boldsymbol{u})$ is calculated and detected bit is estimated from the equation:

$$\hat{m} = sign(\tilde{y}) \tag{10}$$

Embedding and detection processing for simultaneous multichannel processing according to OFDM principle is well realized in the matrix form that doesn't need using any long duration cycles.

Watermarking signal $\boldsymbol{w}$ in OFDM presentation occupies the whole frequency audio band and simulates an ordinary additive noise. By appropriate choosing of parameters in the embedding algorithm it is possible to eliminate audibility of $\boldsymbol{w}$ on the background of host signal and external noise.

4 AW DETECTION AND SYNCHRONIZATION

Estimation of watermarking bits (10) needs proper synchronization the correlation process. Synchronization and marker fields consume a certain time and decrease an effective information bitrate. Additional resources are spent also for check sum.

To settle all these tasks simultaneously we propose AW symbol detection on the whole using hash function. Hash function transforms an input data of variable size to a fixed-size string, which is called hash value. Commonly hash function is employed for checking integrity of input data. We use hash function for the purpose of OFDM watermarking symbol detection. Integrity of the detected symbol is ensured automatically.

To implement this idea let us compose watermarking data containing information bits M itself and its hash function $H(M)$: $[M, H(M)]$. On the receiving part decoder is permanently processing blocks of samples until hash function will satisfy.

Suppose current block $[s_i,\dots,s_{i+NL-1}]$, containing NL samples, gives estimations of information bits $\hat{M}$ and received hash function $\hat{H}(M)$. Decoder calculates hash function over the received information bits $H(\hat{M})$. If $H(\hat{M})=\hat{H}(M)$ then the decision of watermark symbol detection is accepted and the detected bits are believed $\hat{M}$. Otherwise next incoming block $[s_{i+1},\dots,s_{i+NL}]$ enters for processing.

Let us estimate a sufficient length of hash-function. "Good" hash function maps every input combination into unique output combination. Then lengths sequences of information bits M and hash function $H(M)$ should coincide. Suppose $length(M)=length(H)=l$. False detection will take place if and only if some random l-length input combination will give l-length hash function that coincides with the subsequent bits. The probability of such event is $p_{er}=2^{-2l}$. Taking $l=8$ one can obtain $p_{er}=2^{-16}\approx 1.5*10^{-5}$ for randomly occurred sequence. Of course, total false detection will increase proportionally the searching time t_s for subsequent AW symbol. To reduce t_s encoder is blocked after successive detection at the time slightly below the symbol duration (NL samples in our notation).

Practically cyclic redundancy check (CRC) code of length $l=8$ CRC-8 with generator polynomial x^8+x^2+x+1 may be used.

The proposed detection method in our application exceeds standard communication format of fields: synchronization – marker – data – CRC in the useful watermarked information per host sample. This method provides decoding OFDM watermarking symbol in the whole and thus saves from the necessity of synchronization and marker fields at all. Second advantage is the absence of synchronization error on correlation receiver accuracy.

These superiorities are achieved however at the expense of considerable processing loading in the receiver. But detection algorithm is based mainly on FFT matrix operations and may be easily on-line realized for audio frequency.

5 EXPERIMENTAL AND SIMULATION RESULTS

Computer simulation in MatLab environment was held using for host signal the real voice frame of length 27500 samples. Parameters of simulation are presented in Table 1.

Table 1. Parameters of simulation and experiment

Parameter, notation	Numerical value
Sampling frequency, F_s	8.0 kHz
OFDM sub-channel width	125 Hz
Number of sub-channels, B	15
WSR	- 16/-14/-12 dB
SNR	12 dB
AW raw rate, R	60/125/268 bit/sec
m-sequence length, L	31/15/7
FFT dimension, N	64
Delay in the transmitter, D_{Tx}	56/120/248 msec
Identification time	0.25/0.5/1 sec

In the Table 1 the following notations are taken:

$WSR=20lg(\sigma_w/\sigma_x)$ - Watermark-to-Signal Ratio, dB;

$SNR=20lg(\sigma_s/\sigma_n)$ - Signal-to-Noise Ratio, dB;

$D_{Tx}=NL/F_s$ - Delay time in the transmitter, sec;

$R=BF_s/NL$ - AW raw rate, bit/sec.

Identification time is estimated taking into consideration the total number of bits, needed for representation of 9-symbol decimal MMSI (Maritime Mobile Service Identity). This value is accepted by 30 bits. CRC length is taken $l=8$.

Simulation have being carried taking into consideration all interferences and limitations, discussed in Section 2.

Experimental testing was carried out in the VHF channel 17 of maritime mobile service (156.85 MHz); emission class F3E/G3E (frequency/phase modulation, analog telephony) on the base of hardware installation: maritime VHF transceivers RT-2048 Sailor, USB ADC/DAC (analog-to-digital converter/digital-to-analog converter) module E14-140 L-CARD.

In the trials all watermarking symbols along the voice frame were detected correctly and no false detected symbols were registered.

6 CONCLUSION AND FUTURE RESEARCHERS

The proper VHF voice communication is an important part in general maritime safety. Significance of particularly channel 16 was emphasized in the paper [8]. Automatic identification, in turn, would provide an effective and clearly understood communication. The considered AW based automatic identification solves a number problem towards VHF radiotelephone improvement, elimination a human factor and finally enhancement of maritime safety and security. As opposed to inland waters ATIS, AW identification is grounded not on the appending a certain digital sequence to the radiotelephone transmission, but realizes just its identification from the very beginning of the transmission, and permanently runs during the whole transmission.

AW identification doesn't require an additional frequency and time resources, alteration standard transceivers and radio communication procedures and appears only in additional noise, that can be set to minimal level (equal or below of the channel noise).

Implementation of auto identification function needs only phone receiver replacement by a new one with built-in processor at the transmitter side and switching a decoder to standard audio output at the receiver part. It easily provides compatibility of standard equipment and the equipment with identification function. Moreover, AW identification works as well in digital channels if the last will start operating in maritime radiotelephony. This ability is ensured by the AW resistance against ADC/DAC and voice compression procedures.

Automatic radiotelephone identification makes possible a further integration of communication (VHF) and navigation (AIS/ECDIS) equipment.

We see the next directions of our researches. Firstly, optimization of parameters for AW process towards decreasing watermarks audibility and increasing its robustness. Secondly, investigation of applying error correction codes for enhancement of AW detection in the receiver. And the third direction includes testing of designed system in the MF/HF bands of maritime communication.

REFERENCE

[1] ETSI EN 300698-1. Radio telephone transmitters and receivers for the maritime mobile service operating in the VHF bands used on inland waterways; Part 1: Technical characteristics and methods of measurement. 50 p.

[2] Automatic Transmission of the Identification of the Radiotelephone Station. COMSAR 16/7, 15 December 2011.

[3] Kuznetsov, A. & Tsybakov, B. Coding in a Memory with Random Parameters. Probl. Peredachi Inf, Vol. 10, 1974, No. 2, pp. 52-60.

[4] M.H.M.Costa, "Writing on dirty paper," IEEE Transactions on Information Theory, vol. IT-29, pp. 439-441, May 1983.

[5] Hofbauer, K. et al. Speech watermarking for analog flat-fading bandpass channels, IEEE Transactions on Audio, Speech, and Language Processing, 2009, Vol.17, No. 8, pp. 1624 – 1637.

[6] A.V.Shishkin. "OFDM-based audio watermarking for electronic radiotelephone identification", EWDTS'2010. St. Petersburg, Russia, 2010. pp. 190-194.

[7] Malvar, H.S., Florencio, D.A. "Improved Spread Spectrum: A New Modulation Technique for Robust Watermarking". IEEE Transactions on Signal Processing, vol. 51, no 4, April 2003, pp. 898 – 905.

[8] Brzoska S. Advantages of Preservation of Obligatory Voice Communication on the VHF Radio Channel 16, TransNav - International Journal on Marine Navigation and Safety of Sea Transportation, Vol. 4, No. 2, pp. 137 – 141, 2010.

Enhancement of VHF Radiotelephony in the Frame of Integrated VHF/DSC – ECDIS/AIS System

V.M. Koshevoy & A.V. Shishkin
Odessa National Maritime Academy, Ukraine

ABSTRACT: An existent communication suffers from inoperability and complication in initialization of clearly addressed two-way radiotelephony. A new approach to VHF/DSC communication, named COMEC, which realizes an automatic initialization of addressed VHF communication is advanced. The COMEC system is designed on the base of existing GMDSS (VHF/DSC) and navigational (ECDIS/AIS) equipment and strictly supports all operational VHF DSC-based procedures. Development of COMEC in the direction of automatic identification of radiotelephone transmissions is proposed. Probable solutions in interface selection and application advantageous for VTS and SAR-coordination are given.

1 INTRODUCTION

Very high frequency (VHF) voice communication was, is and will be in the future the important component for maritime safety and security. The present paper is focused on enhancement of VHF radio-communication on the base of the existing mandatory installations: VHF digital selective calling/radiotelephony (DSC/RT) radio and Electronic Chart Display and Information System/Automatic Identification System (ECDIS/AIS).

Concept of integration of navigation (ECDIS/AIS) and communication (VHF DSC/RT) systems was described in [1] and considered at COMSAR 14 [2].

In general system and data integration becomes more and more significant in contemporary maritime technologies. Integration between systems and information on board, including different ships' systems is one of the fundamental principles of e-Navigation programme. Some radiocommunication aspects in respect of e-Navigation have been described in [3].

The important step in ship's instruments integration is ECDIS, which is used as *interface for communication needs*. ECDIS is grounded on integration of numerous navigational tools. One of the main components of this system is AIS due to making live navigational picture on ECDIS screen. The "view from the top" on current navigation situation provides more adequate and prompt actions by an officer of the watch (OOW).

The present paper contains some results about practical implementation of VHF DSC – ECDIS/AIS integrating, further development of this integration using additional function of radio transmission automatic identification (AI), review of standard interfaces and analysis of obtained enhancements in Vessel Traffic Services (VTS) and search and rescue (SAR) operations.

2 IMPLEMENTATION OF VHF DSC – ECDIS/AIS INTERFACE

Maritime terrestrial communication in Global Maritime Distress and Safety System (GMDSS) includes two phases: 1) call (and acknowledge) in the DSC system, and 2) actually voice exchange on the working channel, which is set in the first phase or by default. In the other communication system with automatic switching of channels the first phase is realized beyond operator activity. The user doesn't make any actions for channel setting and not feeling this process at all. In the maritime radiotelephony all this burden falls on a navigator. After GMDSS coming in force DSC channel 70 became a calling (and watching) channel in VHF band instead of voice channel 16, but this innovation didn't solve the main problem of work channel settings.

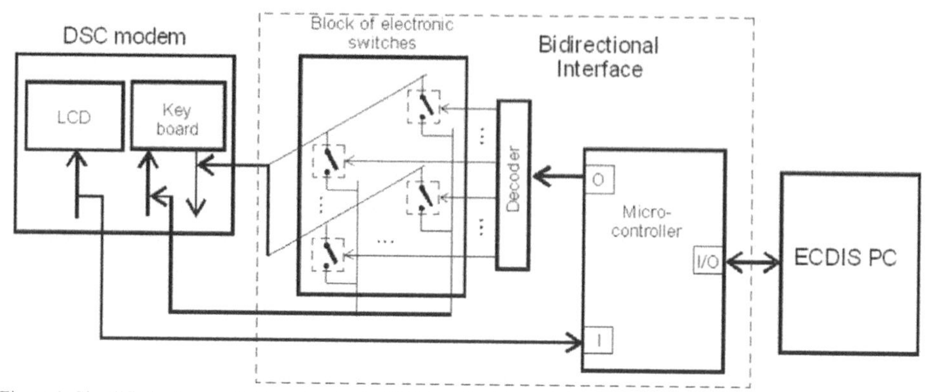

Figure 1. Simplified scheme of bidirectional ECDIS/AIS and VHF DSC interfacing

To realize ECDIS/AIS and VHF DSC integration a certain hard and soft wares were designed. In particularly, special bidirectional interface was designed (see Fig. 1) [4].

For experiments RM 2042 VHF DSC modem (Sailor) was used. In this modem the standard connectors for information input/output in DSC format are not foreseen. That's why the designing of special bidirectional interface is explained. This interface doesn't need any invasions into hard and soft wares of the DSC modem. The main idea of the interface is based on electronic emulation of key board and signal decoding from liquid crystal display (LCD) matrix.

Also an appropriate software component (plugin) was developed to customize ECDIS functionality for remote DSC controlling.

The designed integrated ECDIS/AIS - VHF DSC complex has demonstrated the possibility of practical implementation of the suggested in [2] proposal. It is essential that the integration doesn't reduce any operational functions of the composing parts: ECDIS/AIS and VHF DSC systems, while eliminating the manual routine of forming DSC call.

The involved tools and achieved new qualities in making addressed communication using ECDIS as interface for the communication means allow to name this type of communication as COMEC (COMmunication on the base of Electronic Chart). The proposed method was protected by national patent of Ukraine [5] and patented in Germany [6].

Simulated navigation environment is presented in Fig. 2. In this figure is given print screen for m/v Minerva Maya ECDIS on the first step, when she is calling to m/v Seascout. For Minerva Maya the vessel Seascout presents a potential risk. Therefore the vessel Minerva Maya must call just vessel Seascout to coordinate maneuver. Routine for Minerva Maya navigator consists in clicking Seascout AIS-mark, selection/default working channel in the menu and finally approving the DSC transmission. DSC format forming is produced automatically and any actions to DSC number identification at that are not necessary. All actions are executed quickly using standard computer methods which are not connected to DSC unit key boards of different manufactures.

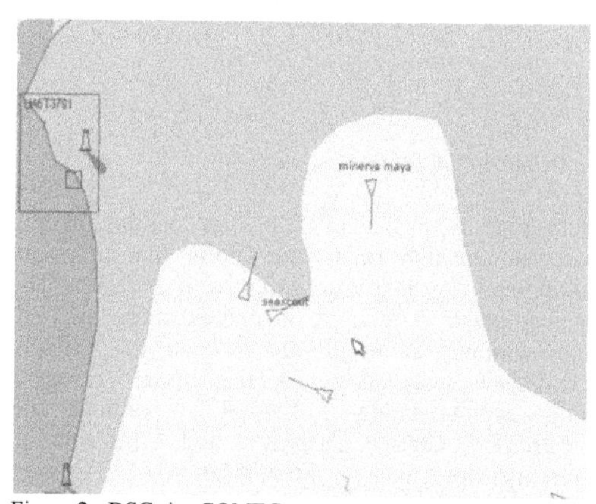
Figure 2. DSC in COMEC system: m/v Minerva Maya is making individual call to m/v Seascout

In the Fig. 3 ECDIS print screen at m/v Seascout is shown. AIS-mark of vessel Minerva Maya is blinking, pointing that it is the calling vessel. Seascout OOW is able to simultaneous estimation of navigation situation and acknowledge to calling vessel. After that vessel Minerva Maya begins an addressed communication without wasting time for ship's identification.

An additional automatic identification of radiotelephone transmissions in the frame of COMEC gives us one more aspect of the discussed integration. AI of radiotelephone transmissions on ECDIS is a developing of COMEC. AI on the base of "audio watermarks" is analysed in [6]. AI have to provide identification data of currently transmitting vessel in the form of Maritime Mobile Service Identity (MMSI) or/and ship's name/call sign. These data have to be passed from VHF equipment with AI function to ECDIS. Then the detected identification data should be compared with AIS target information and the transmitting vessel might be singled out among other vessels by special mark (letter T, for example).

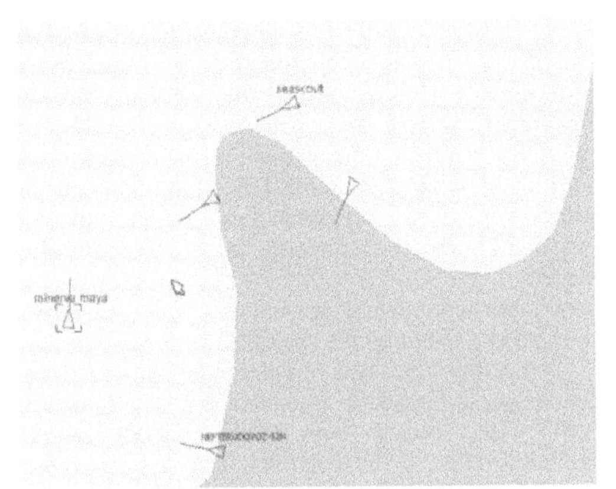

Figure 3. Marking of the calling vessel at m/v Seascout

3 SELECTING STANDARD INTERFACE

The above presented interface (Fig. 1) aimed to demonstrate only conceptual ability of the discussed integration. The wide practical implementation of the designed method is seen through the using of appropriate standard interface for DCS – ECDIS connection. Ship network challenges are presented in [8]. There are at least four possible interfaces that may be taken into consideration: Ethernet, NMEA 0183, NMEA 2000 and Light-Weight Ethernet (LWE).

3.1 *Ethernet*

Ethernet is a widely-used cable-based technology for transmitting very large amounts of electronic data between units of equipment within a local area network (LAN).

Ethernet is capable of transmitting at rates of 10 Mbit per second and more (versus 0.25 Mbit/sec for NMEA 2000). It can play a valuable role with marine electronics that process high volumes of data, for example radar, electronic charts and weather overlay information. However there is no marine standard for Ethernet and equipment from different manufacturers may not be able to communicate with each other. Ethernet rate is excessively high for our task.

3.2 *NMEA 0183*

NMEA 0183 (National Marine Electronics Association) ant its international version standard IEC 61162-1 is the most important standard for ship instruments interconnection. NMEA 0183 is based on transmission over RS-232 (or RS-422) serial lines with one talker and up to 10 listeners. Data is transmitted as 7 bit ASCII text sentences at up to 82 characters including various formatting information.

Data transmission is produced by means of so called sentences from one talker to one or more listeners at a time in broadcasting regime, and therefore it cannot be used to create networks. Data transmission is 4800 bit/sec (or 38.4 Kbit/sec for high speed specification).

NMEA 0183 supports four DSC relevant sentences: DSC, DSE, DSR and DSI.

- DSC, DSE sentences might be output from a VHF DSC radio upon reception of a distress message from another ship and position enhancement respectively.
- DSR – is DSC transponder response sentence and is used to receive data from a DSC modem.
- DSI – is DSC transponder initialize. This sentence is used to provide data to a DSC modem for use in making call.

NMEA 0183 is now the prevailing marine standard and it is perfectly adequate for point-to-point interconnection ECDIS – VHF DSC. Its rate is quite sufficient for real time data exchanging. But coming from the necessary of two-way data exchanging ECDIS/AIS – VHF DSC/RT interconnection needs two interfaces NMEA 0183 that is seen not attractive.

3.3 *NMEA 2000*

NMEA 2000 is a protocol to create a network of electronic devices. It is based on the Controller Area Network (CAN) standard and with a rate of 250 Kbit/sec it is mainly intended for real-time instrument and controller integration. It is aimed to bus LAN creation. Various instruments that meet this standard are connected to one central cable. NMEA 2000 is a bi-directional, multi-transmitter, multi-receiver serial data network.

NMEA 2000 supports multiple-talker transmissions from several data sources. Data messages are transmitted as a series of data frames, each with robust error checking, confirmed frame delivery, and with deterministic transmission times. The header for a message specifies the transmitting device, the device to which the message was sent (which may be all devices), and the message priority. This standard was also adopted as IEC 61162-3.

Currently it is widely used for leisure crafts and small ships.

Unfortunately, the particularities of NMEA 2000 standard are closed and not publicly available.

3.4 *Light Weight Ethernet*

Lightweight Ethernet (LWE) [8] or IEC 61162-450 is the latest solution for maritime implementation. It is primarily intended for high speed communication between shipboard navigation and radio communication equipment. LVE standard is based on carrying NMEA sentences over Ethernet. It has relatively low level of protocol complexity, and has being designed keeping in mind a simple migration path from the use of existing interface standards to the new specification.

4 APPLICATION IN SAR OPERATIONS AND VTS

For coordinating SAR operations it is imperative to have information on the position and identification of other ships in the vicinity. In time, clearly understood and addressed communication plays a great role in the general success of SAR operation. COMEC function allows marking out with special mark in the electronic chart distressed vessel just after receiving distress call. Herewith the distressed vessel couldn't be equipped with COMEC. It is sufficient only SAR appropriate equipment.

Vessel Traffic Services among other co-operative systems include VHF radiotelephony and AIS as invariable components. VTS provide navigational safety, efficient traffic and protection of the marine environment. VTS center always has comprehensive traffic information. It implies that any information about all monitored vessel and their intensions must be readily available for efficient and safety operation.

Considered integration ensures effective monitoring and traffic control by the VTS personnel. In this connection it should be taken into account that 1) VTS operator is usually situated in front of ECDIS, 2) he monitors simultaneously numerous targets and theirs intentions, and 3) there are no mandatory requirements for VTS installations like for ship's equipment. All these factors act in favor of COMEC application. ECDIS/AIS and VHF/DSC discussed integration has to facilitate vessel identification and monitoring, traffic management and collision avoidance.

Proposed AI also could be useful in SAR and VTS applications.

5 CONCLUSION

Existing VHF maritime telephony with preceding DSC procedure suffers from manual routine of DSC forming. It doesn't provide an addressed, prompt communication. Proposed innovative COMEC mode is based on integration of navigation ECDIS/AIS and communication VHF DSC/RT systems and allows overcoming the existing drawbacks and gives new possibilities, which are especially needed in complicated navigation for urgent communication and search and rescue operations.

It is essential that COMEC introduces no changes to the separate operation of composing systems and existing radiocommunication operational procedures. If necessary all components of integrated system may be used separately in regularly regimes. The present manual method of forming/viewing calls using DSC controller will be preserved as a supplementary means to the automatic method of forming calls in the COMEC system.

Distress call should be available only from DSC modem key board by the specified method.

Some maritime equipment manufacturers, Icom for example, produce intended for small vessels VHF DSC radios with NMEA 0183 output, including DSC, DSC sentences. The last innovation is VHF DSC radio with integrated AIS receiver. More over the transceiver is equipped a cable which connects to personal computer or chart plotter. This gives an opportunity to display AIS equipped vessels over cartography and gains other information such as MMSI, name/call sign, useful to know in case the DSC needed to arrange voice communication. Therefore, a concept of VHF radio and navigation tools integration is fully supported by the radio equipment manufacturers for small fleet.

The complete implementation of the discussed integration requires corresponding amendments in the ECDIS Performance Standards. Two additional operations for ECDIS are needed to be implemented:

1. Forming the DSC automatically directly from ECDIS. The parameters, for example, the working channel number, can be set by default (or can be chosen manually if necessary using standard computer actions). The entering of the MMSI is not required at all because AIS data already have it and MMSI can be sent to the DSC controller automatically and
2. Displaying the calling vessel by blinking AIS mark on called vessel's ECDIS (and red blinking mark in the case of distress call).

Automatic identification data also may be used in COMEC system.

REFERENCE:

[1] Miyusov M.V., Koshevoy V.M., Shishkin A.V.: Increasing Maritime Safety: Integration of the Digital Selective Calling VHF Marine Radiocommunication System and ECDIS, TransNav - International Journal on Marine Navigation and Safety of Sea Transportation, Vol. 5, No. 2, pp. 159 – 161, 2011.

[2] Proposal for simplification of VHF DSC radiocommunication and increasing DSC efficiency. COMSAR 14/7, 27 October 2009.
[3] Korcz K. Some Radiocommunication Aspects of e-Navigation, TransNav - International Journal on Marine Navigation and Safety of Sea Transportation, Vol. 3, No. 1, pp. 93 – 97, 2009.
[4] Patent of Ukraine № 94276 from 26.04.2011.
[5] Patent of Ukraine № 78762 from 25.04.2007.
[6] Patent No. : DE 11 2007 003 617 T5. Integriertes System eines digitalen Selektivrufs und einer elektronischen nautischen Seekartographie. Date of Patent : 03.12.2007.
[7] O.V.Shishkin, V.M.Koshevyy "Audio Watermarking for Automatic Identification of Radiotelephone Transmissions in VHF Maritime Communication", Watermarking - Volume 2, pp. 209 – 227. InTech, 2012. ISBN 978-953-51-0619-7 http://www.intechopen.com
[8] Rodseth O.J., Christensen M.J., Lee K. Design challenges and decisions for a new ship data network, ISIS 2011, Hamburg, 15th to 16th September 2011.

Modernization of the GMDSS

K. Korcz
Gdynia Maritime University, Gdynia, Poland

ABSTRACT: The beginning and current status of the Global Maritime Distress and Safety System (GMDSS) have been described. On the base of an analysis of the general concept, the main functions and the international regulations, some aspects of modernization of the GMDSS have been presented. The future of the GMDSS has been discussed as well.

1 INTRODUCTION

Amendments to the 1974 SOLAS Convention concerning radiocommunications for the Global Maritime Distress and Safety System (GMDSS) entered into force on 1 February 1992. Since 1 February 1999 the GMDSS has become fully implemented for all SOLAS ships.

The Maritime Safety Committee (MSC) at its 81st session decided to include, in the work programmes of the Safety of Navigation (NAV) and Radiocommunications and Search and Rescue (COMSAR) Sub-Committees, a high priority item on "Development of an e-navigation strategy".

It needs to be noted that the development of e-navigation is an ongoing process and that the "Development of an e-navigation strategy implementation plan", as a next step of the project, was included in the work programmes of the COMSAR, NAV and Standards of Training and Watchkeeping (STW) Sub-Committees by MSC 85.

Without a doubt, one of the fundamental elements of e-navigation will be a data communication network based on the maritime radiocommunication infrastructure. It follows the question on the GMDSS legitimacy.

Taking into account the above, at MSC 86, the Committee agreed to include in the COMSAR Sub-Committee work programme, a sub-item on "Scoping exercise to establish the need for a review of the elements and procedures of the GMDSS" under the work programme item on "Global Maritime Distress and Safety Systems (GMDSS)".

In the aftermath of this work MSC 90 agreed to include in the 2012-2013 biennial agenda of the COMSAR, NAV and STW Sub-Committees a high priority item on "Review and modernization of the Global Maritime Distress and Safety System (GMDSS)", with a target completion year of 2017, assigning the COMSAR Sub-Committee as the coordinating organ.

2 BEGINNING OF THE GMDSS

The original concept of the GMDSS is that search and rescue authorities ashore, as well as shipping in the immediate vicinity of the ship in distress, will be rapidly alerted to a distress incident so they can assist in a coordinated search and rescue operation with the minimum delay. The system also provides for urgency and safety communications and the promulgation of maritime safety information (MSI).

2.1 *Functional requirements*

The GMDSS lays down nine principal communications functions which all ships, while at sea, need to be able to perform (SOLAS, 2009):

1 transmitting ship-to-shore distress alerts by at least two separate and independent means, each using a different radiocommunication service;
2 receiving shore-to-ship distress alerts;
3 transmitting and receiving ship-to-ship distress alerts;
4 transmitting and receiving search and rescue co-ordinating communication;
5 transmitting and receiving on-scene communication;
6 transmitting and receiving signals for locating;

7 transmitting and receiving maritime safety information;
8 transmitting and receiving general radiocommunication from shorebased radio systems or networks;
9 transmitting and receiving bridge-to-bridge communication.

2.2 *GMDSS Sea areas*

Radiocommunication services incorporated in the GMDSS system have individual limitations with respect to the geographical coverage and services provided. The range of communication equipment carried on board the ship is determined not by the size of the ship but by the area in which it operates. Four sea areas for communications within the GMDSS have been specified by the IMO. These areas are designated as follows (SOLAS, 2009):

- Sea area A1 – an area within the radiotelephone coverage of at least one VHF coast station in which continuous DSC alerting is available.
- Sea area A2 – an area, excluding sea area A1, within the radiotelephone coverage of at least one MF coast station in which continuous DSC alerting is available.
- Sea area A3 – an area, excluding sea areas A1 and A2, within the coverage of an Inmarsat geostationary satellite in which continuous alerting is available.
- Sea area A4 – an area outside sea areas A1, A2 and A3 (the polar regions north and south of 70° latitude, outside the Inmarsat satellite coverage area).

2.3 *Equipment carriage requirements*

The type of radio equipment required to be carried by a ship is determined by the sea areas through which a ship travels on its voyage - see Sec. 2.2 (SOLAS, 2009).

2.4 *GMDSS equipment and systems*

At the beginning of the GMDSS, the following equipment and systems were provided for (Fig. 1):

- DSC - Digital Selective Calling;
- Inmarsat Satellite System;
- EPIRB - Emergency Position Indicating RadioBeacon;
- GPS - Global Positioning System (for support);
- SARTs - Search And Rescue Transponders;
- NAVTEX System;
- NBDP - Narrow Band Direct Printing;
- RTF - Radiotelephony;
- DMC - Distress Message Control (interface).

Other elements of GMDSS to be showed in Fig. 1 stand for as follows:

- CES - Inmarsat *Coast Earth Station*;
- SES - Inmarsat *Ship Earth Station*;
- LUT - COSPAS/SARSAT *Local User Terminal*;
- RCC - *Rescue Coordination Centre*.

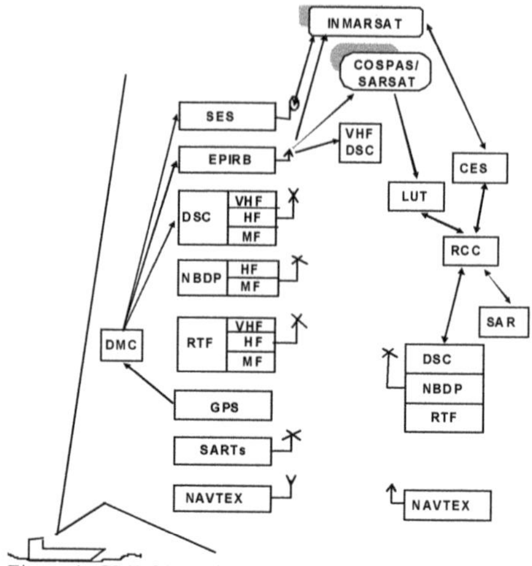

Figure 1. GMDSS equipment and systems (Korcz, 2007)

The original concept of the GMDSS includes Inmarsat A service only and three type of the EPIRBs - Inmarsat E, Cospas/Sarsat and VHF DSC (Fig. 1).

3 CURRENT STATUS OF THE GMDSS

Since implementation of the GMDSS some changes both of regulatory and technical nature have occurred.

3.1 *Technical changes of the GMDSS*

One of the most important GMDSS changes has concerned the Inmarsat. In 1999, Inmarsat became the first intergovernmental organization to transform into a private company and, in 2005, was floated on the London Stock Exchange. It caused that at present Inmarsat is recognised as a leader in mobile satellite communication field.

Inmarsat B, launched in 1993, was first maritime fully digital service.

Inmarsat C, introduced in 1994, is one of the most flexible mobile satellite message communication systems.

Inmarsat had over 170,000 registered GMDSS-capable mobile terminals at the end of October 2012, of which more than 145,000 were Inmarsat C.

Inmarsat Fleet service provides both ocean-going and coastal vessels with comprehensive voice, fax and data communications. Fleet family includes:

- Fleet 77 (introduced in 2002);
- Fleet 55 (introduced in 2003);
- Fleet 33 (introduced in 2003).

Fleet 77 fully supports the GMDSS and includes advanced features such as emergency call prioritization, as stipulated by IMO Resolution A.1001 (25). Fleet F77 also helps meet the requirements of the International Ship and Port Facility Security (ISPS) code, which enables the cost-effective transfer of electronic notices of arrival, crew lists, certificates and records.

Because Fleet 77 is IP compatible, it supports an extensive range of commercially available off-the-shelf software, as well as specialized maritime and business applications. Fleet 77 also ensures cost-effective communications by offering the choice of Mobile ISDN or MPDS channels at speeds of up to 128kbps.

FleetBroadband is Inmarsat's next generation of maritime services delivered via the Inmarsat-4 satellites. It is commercially available since the second half of 2007. The service is designed to provide the way forward for cost-effective, high-speed data and voice communications. Users have the choice of two products (FB250 and FB500). Inmarsat FB500 terminals with Maritime Safety Voice Service are compatible with GMDSS and can be used for GMDSS communications but are not currently recognized as meeting the SOLAS requirements.

It should be also noted that Inmarsat E service (L-band EPIRB) ceased to be supporting GMDSS in 2006 and Inmarsat A service – in 2007.

Instead of the Inmarsat E service, the new Cospas-Sarsat Geostationary Search and Rescue System (GEOSAR) has been introduced by Cospas-Sarsat as completion of the Low-altitude Earth Orbit System (LEOSAR). These two Cospas-Sarsat systems (GEOSAR and LEOSAR) create the complementary system assisting search and rescue operations (SAR operations).

At the same time, the Cospas-Sarsat ceased satellite processing of 121.5/243 MHz beacons on 1 February 2009.

It is also worth to note that on 1.01.2010 AIS-Search and Rescue Transmitter (AIS-SART) was introduced. So, shipboard GMDSS installations include one or more Search and Rescue Locating Devices (SARLD). These devices may be either an AIS-SART or a SART (Search and Rescue Transponder).

And at the end, as the result of the hard work of International Maritime Organization (IMO) and other bodies, two new systems have been introduced:

- Ship Security Alert System – SSAS (in 2004),
- Long-Range Identification and Tracking of ships – LRIT (in 2009).

Although the SSAS and LRIT systems are not a part of the GMDSS, in the direct way they use its communication means.

Besides the above-mentioned main changes the following should be noted as well:

- the cessation of aural 2 182 kHz listening watch;
- the transition to lower-power HF transmitters aboard ships;
- the requirement for passenger ships to have aviation VHF;
- the non-introduction of VHF DSC EPIRBs despite their inclusion in SOLAS;
- the introduction of the SafetyNET service together with Inmarsat-C and mini-C terminals;
- Inmarsat Global Limited (Inmarsat) has informed of its intention to close the Inmarsat-B Service from 31 December 2014;
- the introduction of a range of non-SOLAS Inmarsat services;
- the closure of much of the worlds "public correspondence" radiotelephony, Morse, radio-telex and VHF coastal radio networks;
- the advent of 406 MHz digital distress beacons, with or without GNSS;
- the advent of 406 MHz distress beacons with voyage data recorders;
- a reduction in the number of HF radio time-signal services;
- the advent of AIS;
- the advent of 406 MHz direction finders aboard certain SAR aircraft;
- the advent of airborne AIS transceivers aboard certain SAR aircraft;
- the advent of satellite-AIS as a commercial product;
- the development of the Galileo GNSS;
- the civil availability of Glonass GNSS;
- the development of Man Overboard devices using DSC and AIS;
- proposals for EPIRBs equipped with AIS;
- the major decline in the use of radio-telex at sea;
- the apparent non-use of narrow-band direct-printing in distress;
- the introduction of HF e-mail data (non-SOLAS) services for routine communications;
- the increase in availability of terrestrial and satellite communications networks;
- the increase in the bandwidth supported by many communications platforms, both terrestrial and satellite;
- the further development of the internet, and associated communications software, including "social media", and mobile phone "apps";
- the increased use of e-mail by crews, and for ship's business;
- the widespread take-up of data-intensive technology for ship's business (i.e. e-commerce, digital photography for cargo and ship-related business, maintenance, medical applications, stability calculations, cargo calculations, engineering analysis, numerous reporting

requirements for customs, immigration, quarantine, etc.);
- increasing broadband wireless access along coastlines;
- the further use of non-GMDSS earth stations aboard vessels;
- the massive increase in the availability of complex navigational bridge equipment in the fishing industry and in the recreational sector, and other non-SOLAS commercial sectors;
- increased satellite bandwidth available in C-band and Ku-band via geostationary satellites;
- greater access to radiocommunications equipment by crews outside the GMDSS installation (i.e. cellular phones, and broadband wireless);
- the massive deployment of cellular telephony and broadband wireless along populated coastlines;
- the advent of much cheaper and smaller radio-electronics equipment;
- increased reliability of radio-electronics;
- radio-navigation equipment/computing equipment product choice increasing and more equipment is being combined, particularly in the non-SOLAS sector;
- the creation of the International Mobile Satellite Organization (IMSO);
- the commencement of the non-GMDSS "505" Inmarsat service;
- the new Arctic NAVAREAs/METAREAs are transitioned to Full Operational Capability (FOC) on 1 June 2011;
- the advent of new satellite providers seeking participation in the GMDSS;
- the increased availability of affordable non-SOLAS satellite providers;
- the advent of software-defined radio (SDR);
- the advent of Voyage Data Recorders (VDR);

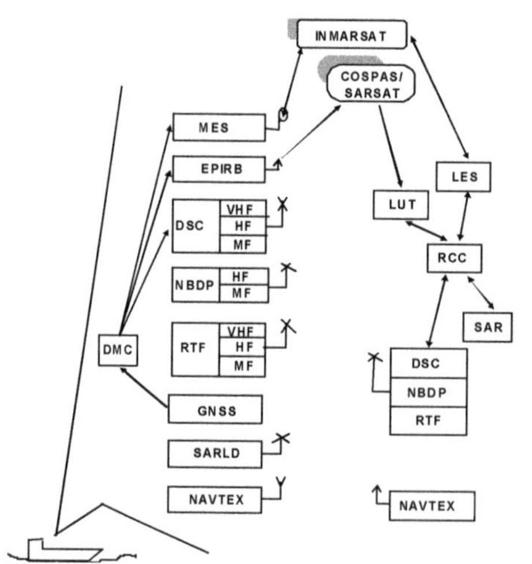

Figure 2. Current GMDSS equipment and systems

- the use of wireless local area networks aboard ships;
- the advent of short-range wireless device interconnectivity (PCs, etc.);
- the use of marine pilots carrying their own PC with a range of marine software products;
- the advent of AIS Application-Specific Messages;
- AIS-equipment aids to navigation, including virtual and synthetic AtoN;
- the advent of Integrated Bridge Systems.

Up to date equipment and systems used in the GMDSS to be showed in Fig. 2.

New elements of GMDSS to be showed in Fig. 2, in comparison with Fig. 1, stand for as follows:
- Mobile Earth Station (MES);
- Land Earth Station (LES);
- Global Navigation Satellite System (GNSS) - for support;
- Search and Rescue Locating Device (SARLD).

3.2 *Regulatory modification of GMDSS*

At a regulatory level a modification of the GMDSS is coordinated by two international organizations: International Maritime Organization (IMO) and International Telecommunication Union (ITU).

IMO modifications are mainly concerning the amendments to Chapter IV of the International Convention for the Safety of Life At Sea (SOLAS) and the proper IMO resolutions.

From the Radiocommunication point of view, the most important modification was adoption by IMO of Resolution A.1001(25) dated 29.11.2007 on Criteria for the Provision of Mobile Satellite Communication Systems in the GMDSS and revision of Chapter IV of IMO SOLAS Convention extends the International Mobile Satellite Organization (IMSO) oversight to GMDSS Services provided by any satellite operator which fits criteria [Korcz, 2011].

ITU modifications are mainly concerning the amendments to Radio Regulations. These amendments were adopted by six World Radiocommunication Conferences (WRCs), which are held every three to four years. These World Radiocommunication Conferences were held in 1995, 1997, 2000, 2003, 2007 and 2012.

The last two World Radiocommunication Conferences (WRC-07 and WRC-12), in the field of maritime radiocommunication, dealt with the following main issues:

<u>WRC-07</u>
- taking into account Resolutions **729 (WRC-97)**, **351 (WRC-03)** and **544 (WRC-03)**, to review the allocations to all services in the HF bands between 4 MHz and 10 MHz, excluding those allocations to services in the frequency range 7000-7200 kHz and those bands whose allotment plans

are in Appendices **25**, **26** and **27** and whose channelling arrangements are in Appendix **17**,with regard to the impact of new modulation techniques, adaptive control techniques and the spectrum requirements for HF broadcasting (Agenda Item 1.13);
- to review the operational procedures and requirements of the Global Maritime Distress and Safety System (GMDSS) and other related provisions of the Radio Regulations, taking into account Resolutions **331 (Rev.WRC-03)** and **342 (Rev.WRC-2000)** and the continued transition to the GMDSS, the experience since its introduction, and the needs of all classes of ships (Agenda Item 1.14);
- to consider the regulatory and operational provisions for Maritime Mobile Service Identities (MMSIs) for equipment other than shipborne mobile equipment, taking into account Resolutions **344 (Rev.WRC-03)** and **353 (WRC-03)** (Agenda Item 1.16).

WRC-12
- to revise frequencies and channelling arrangements of Appendix **17** to the Radio Regulations, in accordance with Resolution **351 (Rev.WRC-07)**, in order to implement new digital technologies for the maritime mobile service; and
- to examine the frequency allocation requirements with regard to operation of safety systems for ships and ports and associated regulatory provisions, in accordance with Resolution **357 (WRC-07)**.

From the point of view of the GMDSS modernization, among many important outputs of the WRC-12, the two documents are very important: Resolution **359 (WRC-12)** "Consideration of regulatory provisions for modernization of the Global Maritime Distress and Safety System and studies related to e-navigation" and Resolution **360 (WRC-12)** "Consideration of regulatory provisions and spectrum allocations for enhanced Automatic Identification System technology applications and for enhanced maritime Radiocommunication".

4 ROLE OF THE GMDSS IN E-NAVITATION

In the Author's opinion, the future of the GMDSS is closely connected both with the development of the e-navigation project and the radiocommunication in this process.

For realizing the full potential of e-navigation, the following three fundamental elements should be present (Korcz, 2009):
1 Electronic Navigation Chart (ENC) coverage of all navigational areas;
2 a robust electronic position-fixing system (with redundancy); and
3 an agreed infrastructure of communications to link ships and shore.

It is envisaged that a data communication network will be one of the most important parts of the e-navigation strategy plan.

In order to realize efficient and effective process of data communication for e-navigation system, the existing radio communication equipment on board (GMDSS), as well as new radio communication systems should be recognized.

Apart from the satellite systems, the GMDSS MF, HF and VHF equipment and systems (Fig. 2) can be also used as a way of data communication for the e-navigation system, provided that this equipment and systems will be technically improved by means of:
- digitization of the analogue communication MF, HF and VHF channels;
- application of high-speed channel to GMDSS;
- utilization of SDR (Software Defined Radio) technology;
- adaptation of IP (Internet Protocol) technology to GMDSS;
- integration of user interface of GMDSS equipment; and
- any other proper technology for GMDSS improvement.

This technical improvement of maritime radiocommunication equipment may mean potential replacement of the conventional equipment by *virtual* one.

With respect to the radiocommunication aspects required for e-navigation (modernization process), the following should be taken into account as well :
- autonomous acquisition and mode switching;
- common messaging format;
- sufficiently robustness;
- adequate security (e.g. encryption);
- sufficient bandwidth (data capacity);
- growth potential;
- automated report generation;
- global coverage (could be achieved with more than one technology).

In this approach to development of maritime radiocommunication it is essential that the integrity of the GMDSS must not be jeopardized.

5 NEED OF THE GMDSS REVIEWING

Taking into account the above mentioned issues it should be noted that the Sub-Committee on radiocommunications and search and rescue (COMSAR) since its 14 session in 2010 (COMSAR 14) has started work on issue "Scoping Exercise to establish the need for a review of the elements and procedures of the GMDSS". This project allowed to start the wide discussion on what the real condition and needs of the marine radiocommunication are, in particular with reference to the current discussion on the e-navigation.

Without a doubt, the review of the GMDSS, with particular reference to the Human Element, should include [COMSAR, 2012]:

- review of the existing nine functional requirements, including:
  - the possible need for inclusion of security-related communications in the GMDSS; and
  - the consideration of the possible need to develop a clearer definition of "General Communications", which is still confusing and if this category should be included within the requirements of the GMDSS;
- the need for the current order of priorities in use for radiocommunications;
- the future need for the four different areas of carriage requirements (Sea Areas A1 to A4), and Port State Control procedures if sea areas are changed;
- the future need to allow for differences for certain categories of ships, including non-SOLAS ships;
- whether distress communications should be separated from other types of communications and in consequence whether the arrangements in chapters in SOLAS could be revised;
- the issue of training and performance of crews on board ships, considering the certification and renewal of qualifications and also noting the possible reduction of technical knowledge and skills by operators;
- equipment carriage requirements for duplication, maintenance, equipment interfacing, back-up support systems and power supplies;
- the possible inclusion of Automatic Identification System (AIS) functions;
- the possible inclusion of Long-range identification and tracking of ships (LRIT) functions;
- the possible inclusion of Ship Security and Alerting System (SSAS) functions;
- the role of Narrow Band Direct Printing (NBDP);
- the role of MF/HF Digital Selective Calling (DSC) and the complexity of some of the signaling functions;
- problems which might arise due to a lack of HF stations in future;
- the usage of satellite equipment as an alternative in Sea Areas A2 currently based around MF/HF DSC;
- voice communications as an integral part of the GMDSS, benefiting search and rescue operations;
- possible new requirements for lifeboats and liferafts, for instance to provide long-range communications;
- possible alignment between chapters III, IV, V and XI-2 of SOLAS, in particular with regard to type approval, secondary equipment and maintenance arrangements and their regulatory status (i.e. mandatory or discretionary);
- the need to indicate the facilities required for capacity-building; and
- assess whether to increase the use of goal-based methodologies when reviewing the regulations and regulatory framework for GMDSS in SOLAS chapters IV and V and the STCW Convention, to provide flexibility to allow the GMDSS to adapt to new and evolving technologies without major revision of the SOLAS and STCW Conventions in future.

Of course, this reviewing process should be developed taking into account the needs of the communication users and the development of e-navigation project.

6 BEGINNING OF THE GMDSS MODERNIZATION

As a result of the work on “Scoping Exercise to establish the need for a review of the elements and procedures of the GMDSS”, the Maritime Safety Committee (MSC) at its 90 session decided to include, in the work programme of the Radiocommunications and Search and Rescue (COMSAR) Sub-Committee, a high priority item on "Review and modernization of the Global Maritime Distress and Safety System (GMDSS)".

This new work item is to review the GMDSS, and then to develop a modernization programme. The modernization programme would implement findings of the review, include more modern and efficient communications technologies in the GMDSS, and support the communications needs of the e-navigation.

At the beginning of the GMDSS modernization process a lot of issues will have to be addressed.

Analyzing the existing systems for inclusion in the modernized GMDSS, there are the opinions that mobile internet services, mobile telephone services, Broadband wireless access (BWA), e.g. Wimax/mesh networks wireless Local Area Networks and non-regulated Satellite Emergency Notification Devices (SENDs), although more and more used by the public, including non-SOLAS ships for alerting, are not the appropriate means and, therefore, should not form part of the international system.

More consideration is needed to decide which systems, relying on older or inefficient technologies, might be considered for replacement by more modern systems.

The following new equipment, systems and technologies, currently not included in GMDSS, might be included in the modernized GMDSS [COMSAR, 2013]:

- AIS, including Satellite monitoring of AIS and additional AIS channels for identification but not alerting;
- HF e-mail and data systems;
- VHF data systems;
- Application Specific Messages over AIS or VHF data systems;
- NAVDAT;
- Modern satellite communication technologies;
- Additional GMDSS satellite service providers;
- Hand-held satellite telephones in survival craft;
- Hand-held VHF with DSC and GNSS for survival craft;
- Man Overboard Devices; and
- AIS and GNSS-equipped EPIRBs.

In considering the existing order of four levels of priority (Distress calls, distress messages and distress traffic, Urgency communications, Safety communications and Other communications [Radio Regulation, 2012]), the proposal is that for voice messages the existing four priorities were still required but only two priorities were currently sufficient for controlling the radiocommunication link, for example using pre-emption [COMSAR, 2013].

In considering the nine functional requirements (see Sec. 2.1), in relation to "General communications", it should be noted that:

- in accordance with the existing definition in SOLAS regulation IV/2.1.5, the general communications means operational and public correspondence traffic, other than distress, urgency and safety messages conducted by radio;
- the (Government-owned) coast radio stations, which provided public correspondence facilities when the GMDSS was first designed, had now all largely closed down;
- facilities for public correspondence are still required; these communications have been achieved using commercial services;
- under the definition of urgency and safety communications in article 33 of the Radio Regulations the following communications were included:
  - navigational and meteorological warnings and urgent information;
  - ship-to-ship safety of navigation communications;
  - ship reporting communications;
  - support communications for search and rescue operations;
  - other urgency and safety messages; and
  - communications relating to the navigation, movements and needs of ships and weather observation messages destined for an official meteorological service.

Therefore, the term "General communications" needs to be redefined. A new definition for "General communications", could be as follows [COMSAR, 2013]: "General communications means operational traffic, other than distress conducted by radio". The new definition for "General communications" in the GMDSS would enable all the above types of communication to be achieved using the facilities of the GMDSS which were given special protection by the Radio Regulations. However, public correspondence, using commercial services, would not be given special protection.

Respecting the security-related communications, it should be noted that the requirements for maritime security were given in SOLAS chapter XI-2 and the Ship Security Alert System (SSAS) did not involve communication with other ships or with coast radio stations (communications were addressed to a designated competent authority) and therefore, the security-related communications should not be a functional requirement of the GMDSS.

In considering the issue of Maritime Safety Information (MSI), there is the view that MSI is still very important for ships and should be retained as a functional requirement for the GMDSS. In order to achieve the machine-readable solution in the short term, there might be a need for communications that were not necessarily compatible with a functional requirement of the GMDSS and therefore the new operational, performance and technical standards would be required.

Following the above, it is proposed that chapter IV (Radiocommunications) should be extended to include requirements for communications functions in addition to the GMDSS, as follows below [COMSAR, 2013]:

Every ship, while at sea, shall be capable:

1 of performing the present Global Maritime Distress and Safety System (GMDSS) functions (see Sec. 2.1); and
2 of transmitting and receiving public correspondence to and from shore-based radio systems or networks; and
3 of transmitting and receiving security-related communications, in accordance with the requirements of the International Ship and Port Facility Security Code (ISPFS Code).

In considering the GMDSS modernization the following developments that are already under way should be taken into account as well:

- proposed FleetBroadband FB500 terminal for GMDSS;
- Inmarsat MSDS (Maritime Safety and Data Service) over BGAN FleetBroadband;
- proposed Cospas-Sarsat MEOSAR network (will be probably fully operational in 2018) and retirement of the Cospas-Sarsat LEOSAR network;
- Galileo GNSS project;
- introduction of new broadcasts to ships using systems in the 495-505 kHz band; and

the following changes in the GMDSS which might be expected to occur in the near to mid-term:

- the provision of hand-held satellite telephones in survival craft;
- the provision of handheld VHF with DSC and GNSS for survival craft;
- all SOLAS EPIRBs to be fitted with GNSS;
- EPIRB-AIS;
- EPIRB-AIS to be used with/in voyage data recorders;
- MOB-AIS;
- "AIS 2.0" or "Next Generation AIS";
- development of an electronic equipment;
- using AIS in a new "ranging mode";
- additional AIS channels for channel management and data transfer;
- the use of Channels 75 and 76 for enhanced detection via satellite;
- additional simplex channels in the VHF Radio Regulations (RR) Appendix 18;
- harmonized digital bands in VHF RR;
- the further evolution of Maritime Safety Information broadcast systems, taking into account the ongoing work in IHO and WMO;
- an eventual replacement of MF NAVTEX with a new higher data-rate system in the 495-505 kHz band;
- a new channel plan for Appendix 17 HF bands to permit new digital emissions, whilst still allowing Morse and NBDP, and protection for GMDSS MSI;
- the introduction of an e-navigation communications suite independent of the GMDSS, but with interconnections with it.

7 CONCLUSIONS

More than twenty years have passed since the time when the Global Maritime Distress and Safety System (GMDSS) became introduced. Planning for the GMDSS started more than thirty years ago, so some elements of it have been in place for many years.

There have been numerous advances in the use of maritime radiocommunication to maritime safety, security and environmental protection during this period. But now there are some obsolete GMDSS equipment and systems or the ones that have seldom or never been used in practice.

The GMDSS is a system of systems, which like a "bus" going along the road, and certain technologies get off the bus and certain new technologies get on board the bus. The idea that the GMDSS and related systems are static is rather incorrect. Reference to the section 3 proves that the GMDSS continues to be a dynamic system, whilst still supporting the essential functions of the GMDSS.

On the other hand there are a lot of the new digital and information technologies outside the GMDSS, so the review and modernization of the GMDSS is needed.

During work on review and modernization of the GMDSS it is necessary first to identify real user needs and secondly to realize that the modernization of the maritime radiocommunication should not be driven only by technical requirements. In addition, it is necessary to ensure that man-machine-interface and the human element will be taken into account including the training of the personnel.

The lessons learnt from the original development and operation of GMDSS should be taken into account in the modification of GMDSS as well.

Furthermore, the continuous and open process is needed to ensure it remains modern and fully responsive to changes in requirements and evolutions of technology and it will meet the expected e-navigation requirements. To ensure it, a mechanism for continuous evolution of the GMDSS in a systematic way should be created as well.

In this approach to development of the GMDSS it is very important that the integrity of GMDSS must not be jeopardized.

It should be noted that in this context, for the process of the GMDSS modernization, very important will be regulatory decisions taken on the World Radiocommunication Conference taking place in 2015 (WRC-15).

And finally it should be noted that a key to the success of the review and modernization process of the GMDSS is not only that the work is completed on time, but also that it has the flexibility to implement changes ahead of schedule.

REFERENCES

International Maritime Organization (IMO). 2009. International Convention for the Safety of Life At Sea (SOLAS), London

International Telecommunication Union (ITU). Radio Regulations, Geneva 2012

Sub-Committee on Radiocommunications, Search and Rescue - COMSAR 16. 2012. Report to the MSC, International Maritime Organization (IMO), London

Sub-Committee on Radiocommunications, Search and Rescue - COMSAR 17. 2013. Report to the MSC, International Maritime Organization (IMO), London

Korcz K. 2007. GMDSS as a Data Communication Network for E-Navigation. 7th International Navigational Symposium on "Marine Navigation and Safety of Sea Transportation" TRANS-NAV 2007, Gdynia

Korcz K. 2009. Some Radiocommunication aspects of e-Navigation. 8th International Navigational Symposium on "Marine Navigation and Safety of Sea Transportation" TRANS-NAV 2009, Gdynia

Korcz K. 2011. Yesterday, today and tomorrow of the GMDSS. 9th International Navigational Symposium on "Marine Navigation and Safety of Sea Transportation" TRANS-NAV 2011, Gdynia

A VHF Satellite Broadcast Channel as a Complement to the Emerging VHF Data Exchange (VDE) System

F. Zeppenfeldt
European Space Agency – ESTEC, Satellite Communications Department – Future Programs, The Netherlands

ABSTRACT: The recent initiative of creating the future VHF Data Exchange (VDE) within IALA (International Association of Lighthouse Authorities) provides an opportunity to design an integrated terrestrial and satellite communications system to exchange maritime information. This paper introduces the concept of a satellite VDE component as a complement the proposed terrestrial VDE system. A satellite VHF component could disseminate relatively large amounts of data to a large number of vessels which are out of reach of the terrestrial VDE system, while at the same time making use of a large part of the VHF infrastructure of the vessels.

1 INTRODUCTION AND BACKGROUND

The recent World Radio Conference 2012 (WRC-12) has decided upon Resolution 360 **[WRC]** which is of importance to the maritime communications community. This Resolution requests that regulatory provisions and spectrum allocations for enhanced Automatic Identification System (AIS) technology applications and for enhanced maritime radio communication are investigated.

In more detail, the resolution requests as part of Agenda Item 1.16 of the upcoming WRC-15 to *"...consider modifications to the Radio Regulations, including possible spectrum allocations, to enable new AIS terrestrial and satellite applications ..."* and *"... to consider, based on the results of ITU-R studies, additional or new applications for maritime radio communication within existing maritime mobile and mobile-satellite service allocations ..."*.

This Agenda Item allows to consider new communication systems that could address maritime information exchange needs. In addition, IMO is investigations on the future GMDSS, which might drive new communication requirements, also described in **[TRANSNAV]**.

2 VHF DATA EXCHANGE SYSTEM

One of the ideas which is brought forward by IALA (International Association of Lighthouse Authorities) is the VHF Data Exchange System, or VDES. The recent IALA Maritime Radio Communications Plan **[MRCP]** is describing the VDE System in more detail.

In summary, the VDES proposes to use a digital data link over a number of (bundled) maritime VHF channels, such that data can be exchanged easily between ships, and between ship and shore.

A number of VHF channels which are part of the maritime channels described in Appendix 30 of the Radio Regulation are proposed to be used for the VDES. Bundling some of these VHF channels, and using more modern modulation and access schemes than currently in use with e.g. AIS could yield an increased information throughput.

The VDES has been presented at recent IMO/ITU joint expert groups, IMO COMSAR 17 and was discussed in the relevant ITU-R Working Party 5B, and IALA Communications and Navigation Workgroups.

It is believed that the VDES can address some of the gaps which have been identified by the maritime community in the area of e-Navigation.

Within the overall Agenda Item of the WRC-15, and within the VDES in particular, there is room for a complementary satellite segment.

3 THE PROPOSED VDE SATELLITE COMPONENT – THE PRINCIPLE

Satellite communications is an effective means to deliver information in a **broadcast** or **multicast** mode to a large number of ships, i.e. efficiently addressing many vessels using only minimal parts of the scarce maritime radio spectrum resource. A satellite downlink channel is able to address a single message to thousands of ships simultaneously within its footprint.

The frequencies which are currently under discussion for a satellite VDE downlink channel, - as part of the VDE - will allow reception of the satellite channel with **low-cost receive-only equipment**, or can make extensive use of the **existing VHF infrastructure** on the ships and require only minor modifications.

A satellite VDE downlink channel would allow to pass information to ships which are out of reach of the terrestrial VDE shore infrastructure, therefore **extending the geographical reach of services** which are carried over the VDES.

There is a large population of **smaller-size ships** - which do not carry satellite communication equipment on board - that could benefit from such a satellite VDE downlink channel. This would be of particular benefit for **vessel populations in developing countries**, small fishing boat fleets, recreational users and small leisure craft, and life rafts. Even individuals carrying a VHF receive-only device integrated in their life vest would be able to receive the VDE.

A satellite VDE downlink channel is well suited to support applications that typically address a large number of ships simultaneously or serve applications that address very remote ships. A recent report from the EfficienSea Project **[COMSAR]** provides examples of applications which are "push-addressed" or "push-multicast" ship-to-shore, and which could possibly be supported by a satellite VDE downlink channel such as MSI, METOC and SAR plan dissemination.

4 THE PROPOSED VDE SATELLITE COMPONENT – A BASIC ARCHITECTURE

The basic architecture of the VDE satellite component is depicted in figure 1.

The satellite VDE segment consists of an uplink facility that transmits information to the satellite.

The satellite will re-transmit the information into the full footprint of the satellite downlink. The satellite will use in this downlink a number of 25 kHz VHF channels. Multiple channels could also be bundled to allow a higher information transfer data rate.

It is well possible that several maritime information providers deliver information to the vessel population, i.e. more ground stations or satellites would be used requiring also multiple VHF channels. There could be also be uplink stations dedicated to regions or to specific kinds of information.

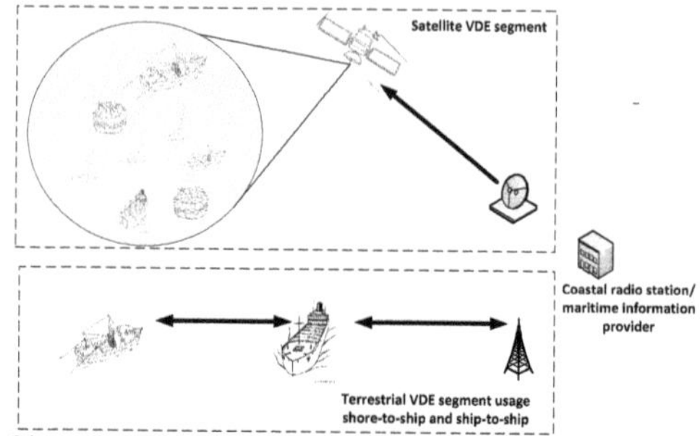

Figure 1. Basic VDE terrestrial and satellite architecture

The information will be received by ships using low-cost VHF receive-only devices, or by VDE transceivers. Within the ship, information could then be further be distributed (e.g. via NMEA or another marine electronic bus) to ECDIS plotters or other marine electronics.

5 POSSIBLE APPLICATIONS OVER A VDE SATELLITE COMPONENT

In general, a satellite VDE downlink channel could fulfil one-way information exchange requirements with characteristics as sketched above. In more detail, it is believed that possible applications could be :

1 Broadcast applications that will address the needs as coming out of the planned modernisation of the GMDSS.
2 "Push-addressed", "push-multicast", or broadcast applications that will support future e-Navigation developments, including regionally-targeted map updates.
3 Augment Maritime Safety Information (MSI) dissemination to Arctic sea area A4 or severe weather warnings, complementing the maritime safety information service and WWNWS (World-Wide Navigational Warning Service) broadcasts via HF NBDP or SafetyNET .
4 Dissemination of satellite navigation correction messages towards the Arctic regions: this will enable a satellite-based augmentation system (SBAS) such as EGNOS or WAAS at high latitude regions. It is expected that there are future applications that will benefit from the integration of GNSS and a VHF downlink.

5 Extend situational awareness by disseminating aggregated situational information to beyond-line-of-sight areas.
6 Improved situational awareness for other non-maritime platforms, such as aeronautical platforms receiving the satellite broadcast channel. This can be used for example to disseminate SAR plans.
7 Act as an alternative for acknowledgement of a distress alert, e.g. for AIS-SARTs or VHF-only equipped ships, or act as an alternative return/acknowledgement link. Intelligent combinations with SENDs (Satellite Emergency Notification Devices) or MOB (Man over Board) devices can be foreseen.
8 Allow management of AIS networks, allow for signaling to equipment such as AIS-SARTs, Aids to Navigation (AtoN) or AIS devices in general.
9 Act as an alternative NAVTEX channel, in particular for the Arctic and Antarctic NAVAREA's/METAREA's.
10 Possible integration with future use of long range AIS reception or the future Applications Specific Messaging (ASM) by satellite, as such implementing future two-way applications.
11 Allow a broadcast overlay for current VHF data networks.

The examples listed above are far from exhaustive and will need to be thoroughly discussed and reviewed within the maritime community. All examples will need to be assessed for what concerns their intended users, vessel population, geographic area, the relation with (future) GMDSS, SOLAS or non-SOLAS users, e-NAV gap analysis, possible integration in the MSI service or WWNWS (World-Wide Navigational Warning Service) , etc.. Some examples may be not be suited at all for satellite following such a review.

It is estimated that a VDE satellite downlink could deliver to many vessels simultaneously about 5-10 MByte to each vessel per satellite pass, depending on satellite downlink power, orbit and attitude.

6 SPECTRUM REQUIREMENTS

As part of the various studies that will be performed in ITU, it shall be determined where the proposed satellite VDE downlink channel can be accommodated within the maritime VHF frequency range.

Specific to the satellite usage of these VHF bands is the need to study whether a satellite downlink channels does not interfere with any other systems in its footprint, see Figure 2. Frequency allocation tables indicate that these bands are likely to be used on land by mobile or fixed services.

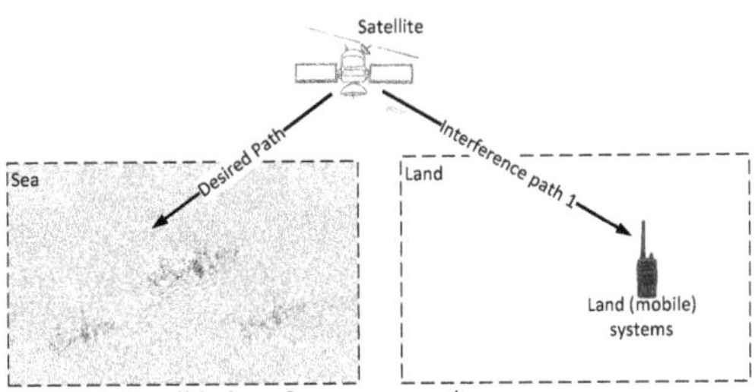

Figure 2. Possible interference scenario

If the energy of the proposed VHF downlink stays under certain power flux density criteria, it does not have to be coordinated with other systems and can offer a global service.

If coordination and specific sharing arrangements are required with land mobile systems, still a service of interest can be offered to areas like e.g. the Arctic.

In case the required ITU studies prove that a VHF downlink in maritime channels does not yield unacceptable interference levels , it is proposed that a new Space-to-Earth Mobile Satellite Service (MSS) allocation is made for these VHF channels.

In the future it should also be investigated whether the satellite VDE segment and the terrestrial VDE segment could partly share the same VHF spectrum, using signal processing technologies which are nowadays common in commercial hybrid satellite/terrestrial systems. This is very much in line with the current trends to use the scarce radio spectrum to the maximum extent possible.

It is suggested that the maritime community uses at WRC-15 this rare opportunity to request additional satellite spectrum that can be used for many safety, security, e-Navigation and future GMDSS purposes.

7 DEPLOYMENT CONSIDERATIONS

It is not likely that a satellite VDE segment will be deployed with full-fledged functionality and global coverage in "one go". However, advances in satellite and launcher technology allow to deploy an initial VDE satellite segment rather swiftly. A satellite VDE segment could then be gradually built up.

An initial satellite VDE segment could address specific geographical areas such as the Arctic sea area A4, or could serve a specific function such as METOC or MSI dissemination. In addition, a preliminary satellite VDE segment could be deployed quickly for experimental or validation purposes to gather experience with the concept, augment planned VDE terrestrial trials, and complement on-going e-Nav and other projects addressing maritime communication and navigation.

A preliminary satellite VDE segment would also allow to reach out to maritime communities in developing countries, for example with a maritime information dissemination demonstration to smaller fishing vessels in such countries.

The VDE transmit functionality on-board a future satellite is expected to be relatively small in terms of size, power and weight. Therefore, a gradual deployment of the satellite VDE segment can be well addressed using the "hosted payload" concept, which has found broad acceptance in the space industry. This concept would allow to embark VDE transmitters as a relatively independent units on "host" satellites for which there are frequent launches..

For a full deployment of a VDE satellite segment, more extensive discussions are required with regards to functionality, synergy with the terrestrial VDE segment, radio frequency compatibility, intended users, required demonstration and validation campaigns, and last but not least the business model associated to such a service.

VDE transceivers will need to be available that can tune in on the satellite frequencies. It is expected that using software defined radio (SDR) technology this can be implemented at reasonable costs. Specifically a VDE receive-only version that would receive only the satellite VDE channel is expected to be low cost. No specific antenna developments are foreseen, as the VDE equipment can use existing VHF antenna infrastructure on the vessel.

8 OPERATIONAL ISSUES

In case a future VDE system is brought to life, it is an interesting consideration of such a system will be operated. No specific business model has been elaborated for such a service. We list some considerations that could impact the architecture and operations of the VDE system, with an emphasis on the satellite component.

The use of these specific frequencies suggest that an independent body would exercise some level of coordination regarding the kind of information that is broadcast or exchanged. The way NAVTEX and SafetyNet broadcasts are coordinated comes to mind as under the WWNWS (World-Wide Navigational Warning Service).

Furthermore, the use of these frequencies does possible only allow a limited commercial usage. Nevertheless, communications service providers could offer maritime information broadcasts over such a system, under contract from e.g. a national maritime safety agency.

A VDE satellite component could possible address many smaller (and non-SOLAS) vessels in developing countries.

Multiple information providers could be co-exist, possible even multiple (small) satellites that broadcast within the framework of VDE. This will required an additional level of coordination in case e.g. interference levels would be exceeded in case of simultaneous satellite transmission.

9 FURTHER PLANNED STEPS AT THE EUROPEAN SPACE AGENCY

The Future Systems group of the ESA Satellite Communications Department will initiate some activities in this area to support the maritime community in developing further the satellite component of the VDE system.

Initial system considerations, trade-offs and operational concepts will be studied in 2013. In addition, spectrum management studies for the satellite component will initiated and their results will be submitted to the relevant regulatory bodies to secure a future Mobile Satellite Service allocation for some of these VHF channels.

ACKNOWLEDGEMENTS

Without discussing with many individuals at ESA, COMSAR, IALA, the radio regulatory and others within the maritime community the VDE satellite downlink concept would not have been conceived.

REFERENCES

[COMSAR] IMO COMSAR 16/INF/2 "Report from the EfficienSea Project"

[MRCP] IALA Maritime Radio Communications Plan http://www.iala-aism.org/iala/publications/documentspdf/doc_259_eng.pdf

[WRC] Agenda for the 2015 World Radiocommunication Conference https://www.itu.int/dms_pub/itu.../R0A0600004D0001MSWE.docx

[TRANSNAV] International Journal on Marine Navigationand Safety of Sea Transportation, "Yesterday, Today and Tomorrow of the GMDSS", December 2011, Volume 4, Number 4, K. Korcz, Gdynia Maritime University, Gdynia, Poland, http://www.transnav.eu/Article_Yesterday,_Today_and_Tomorrow_Korcz,20,315.html

Chapter 9

Methods and Algorithms

Overview of the Mathematical Theory of Evidence and its Application in Navigation

W. Filipowicz
Gdynia Maritime University, Poland

ABSTRACT: In recent past the author used to cope with problems encountered in navigation that feature imprecision and uncertainty. He stressed that methods of their solution exploit a limited number of available data and the informative context of achieved results is rather poor. He also claims that many practical problems related to navigational ship conducting and feature uncertainty can be successfully solved with Mathematical Theory of Evidence (MTE for short) and that the informative context of the obtained results is much richer when compared to those acquired by traditional methods. To adopt MTE (Dempster 1968, Shafer 1976) for discussed field of applications he has made necessary adjustments of the theory. Modifications made by the author are published in many papers, herein comes summary of his personal achievements.

1 INTRODUCTION

Mathematical Theory of Evidence, also known as Belief Theory or Dempster-Shafer Theory, exploits belief and plausibility measures and operates on belief assignments or belief structures. The theory also offers combination mechanism in order to increase the informative context of the initial evidence. The evidence is meant as a collection of facts and knowledge. In navigation, facts are results of observations such as taking bearings, distances or horizontal angles. Thus a combination scheme is expected to enable position fixing of the ship.

The theory extended for possibilistic platform creates new opportunities for modelling uncertainty. In the presented applications uncertainty is due to erroneous observations. It is widely known that errors are included in measurements. One distinguishes mainly between random and systematic errors.

Formal descriptions of problems encountered in navigation involve models that accept imprecise, erroneous and therefore uncertain values. The concept should be followed regarding quite numerous problems encountered in many different disciplines, in particular position fixing and its accuracy evaluation.

It is the navigator who has to handle a set of random points delivered by various navigational aids from which he is supposed to indicate a point as being the position of his ship. Dispersions of points are governed by two dimensional approximate distributions. The fixed position is located somewhere in the vicinity of indications at hand. It is very similar, in case of measured distances, bearings or horizontal angles. The ship's position is located within the area of crossings of appropriate isolines that intersect inside the confined area. The area of the true position is spanned over isolines' crossing points provided the available evidence features random errors and might be outside the area once systematic errors prevail. It is supposed that the navigator is able to resolve all dilemmas thanks to his knowledge, experience and also intuition. The general rule says that the smaller the area, the more accurate the observations and the better the fix. A skilful navigator is able to evaluate the acceptable dispersion of crossing points and reject some isolines if necessary.

In an analytic approach towards position fixing, the least square adjustment method is often exploited. The method seeks for the minimum sum of squared distances from each of the isolines. Weighting factors introduced during calculations diversify measurements and enable reducing the effects of the worst observations.

The traditional way of position fixing takes advantage of available measurements, their approximate random distributions and diversification of observations once the analytical approach is used. The main disadvantage of the approach is the lack of a built-in universal method of the fix a posteriori evaluation.

Expectations regarding flexibility of the upgraded models are greater. Items that should affect fixed position should also include the kind of distributions of measurements taken with a particular navigational aid and discrepancies in the parameters of such distributions. It is popular to state that the mean error of a bearing taken with radar is interval valued within the range of [±1°, ±2.5°]. The presented evaluation of the mean error appears as a fuzzy figure and as such, fuzziness should be accepted and taken into account during computations. Subjective assessment, also in form of linguistic terms, of each observation should be accepted and processed. Empirical distributions are also supposed to be included in calculations. The most important thing is the embedded ability for objective evaluation of the obtained fix along with measures indicating the probability of its locations within the surrounding area.

Meeting all the above stated expectations is impossible with traditional formal apparatus. Its ability is almost exhausted in the considered applications. Research and published works devoted to new platforms and modern environments put attention on Evidence Theory that delivers a wide range of new opportunities.

2 FIRST STEPS

The first publication (Filipowicz 2009a) has been referred to in discussions on the practicality and functionality of the Bayesian and Dempster-Shafer concepts of evidence representation and reasoning and the possibility of application of belief theory in geodetic positioning and navigational position fixing. Many authors (for example Burrus&Lesage 2003) point to numerous applications involving the first approach while examples employing other concepts are rather scarce. It is said that there are only a few meaningless practical problems solved with Theory of Evidence.

In the paper, practical nautical problems were briefly presented and the potential of the Dempster-Shafer theory exploitation was depicted. One of the presented problems was establishing imprecise distance from a navigational obstacle. The problem appears to be of data integration type, which is met within data fusion. The scheme of reasoning engaging inaccurate measurements delivered by aids of various credibility levels was presented and discussed based on fuzzy inference schemes available in the theory of evidence. The solution obtained was a set of supports for each considered fuzzy hypothesis on representation of the true distance. Support is expressed by belief and plausibility, measures exploited in the Dempster-Shafer theory.

Another problem considered in the paper was related to position fixing based on imprecise measurement data. It was assumed that available data are two dimensional random variables governed by Gaussian distributions. The assumption is often made in navigation. Hypothesis and evidence universes were defined for position fixing. Next, relations between hypothesis and evidence frames were considered as binary. Degrees of hypothesis point inclusions within measurements related sets are grades of so called location vectors. In the preliminary approach, considered vectors consisted of zero-one elements. Each vector is assigned a credibility value calculated based on confidence interval probability calculated for assumed distributions. The result of vectors associations were explored with intuitive formulae in order to obtain the fixed position. The simplified approach was further developed and published.

3 COPING WITH UNCERTAINTY

The paper (Filipowicz 2009b), published in Polish, is solid and thoroughly devoted to the fuzzy approach to position fixing. The main idea that remained behind the research and publication was introducing a more flexible approach towards position fixing. The first attempt engaging binary locations seemed inadequate since many publications devoted to nautical science emphasize that results of observations are random variable governed by various dispersions although their substitutions with Gaussian distributions are common and in many cases justified assumptions. Nevertheless, their parameters should be considered as interval valued rather than crisp ones. It is usually said that the mean error of the distance taken with medium class radar is within range of [±1%, ±1.5%] of the measured value (Jurdziński 2009). Thus binary representation of nautical knowledge is not adequate. A platform that accepts fuzziness (Yen 1990) along with multiple random distributions was introduced. This new approach was presented in the paper. At first, membership functions were discussed and expectations regarding their properties were specified in context of their nautical usage. Different functions were presented and compared from the point of view of the proposed application. Membership functions are used in order to upgrade belief assignments; that are then converted to belief structures and combined in order to make a fix.

Results of belief structure combination are a kind of encoded knowledge base that should be explored in order to seek support for various hypotheses. Hypothesis fuzzy representation and appropriate formulas deliver measures to support the proposition on representing the fix with respect to facts related to imprecise data at hand as well as to nautical knowledge. Considering position fixing, one can simplify the hypothesis representation that take the form of a singleton. Provided this type of referential, fuzzy set formulas describing belief and plausibility supports were derived and used in numerical examples included into the paper. Strong dependence of the belief support measure on allocation of hypotheses points was depicted in the publication. Therefore, plausibility support was strongly recommended as the most important factor when a fixed position is selected.

Belief structures in nautical applications contain encoded evidence related to taken measurements. The result of structures combination is a two-dimensional table that embraces enriched data enabling reasoning on the fix. From a possibilistic viewpoint this result is a belief assignment that is the distribution of possibilities regarding each hypothesis point locations within evidence related sets. Mechanisms and methods available in the theory of evidence (Denoeux 2000) can be exploited in order to derive formulas for calculating interval valued probability of representing fixed positions by each of the considered points. Interval valued limits are equal to belief and plausibility measures.

Alternatively, from a probabilistic standpoint (Liu&Hughes&McTear 1992, Klir& Parviz 1992, Filipowicz 2012a), the result of combination can be perceived as Bayesian evidence representation. It should be stressed that this standpoint is justified in a limited number of cases. In general, the final structure does not fulfill probability requirements. Nevertheless, one can use Bayesian methods to deduce a formula for calculating support probability for "being a fix" by any point out of the hypothesis universe. Not surprisingly, both approaches yield virtually the same formula. It should be noted that a possibilistic approach itself can be perceived as an extension for the probabilistic, Bayesian concept. Extension is much more flexible in respect of modelling and the ability to process uncertainty.

A preliminary version of the algorithm for selecting the fixed position based on navigational aids indications was presented and discussed in detail. Indications were considered as two dimensional random variables governed by various and approximate distribution characteristics. Inconsistency was removed using the Yager concept of normalization (Yager 1996). At the last stage of the publication, the algorithm was used for sensitivity analysis. Measures indicating selected position versus degree of uncertainty featured by initial data were compared.

4 MTE IN TERRESTRIAL NAVIGATION

The topic was considered in paper (Filipowicz 2011b). The publication, in its introductory part, contains a compilation of nautical knowledge regarding observations and their isolines, functions that are measurement projections on a chart. Application of the Mathematical Theory of Evidence in terrestrial or celestial navigation involves dealing with isolines and their gradients. Confidence intervals are established along gradient directions. The most frequently used are isolines of bearings, distances and horizontal angles and these functions were discussed in detail. For each case an example isoline, its gradient's module and direction were presented. Proposed observation evidence encoding was discussed for each considered isoline type.

A significant part of the paper was devoted to empirical type of random variables distribution. This type of distribution is encountered very often in navigation. They are usually converted to Gaussian ones although it so happens that conversions are not theoretically justified. Thus empirical distribution inclusion into evidence representation seems natural and necessary. In this case, confidence intervals are substituted by histogram bins and cumulative probabilities are replaced by relative frequencies of observations falling within the bin. Since available histograms differ, calculated frequencies are rather ranges of values than single figures. Thus belief assignments upgraded with empirical distributions are interval valued. It remains that combination scheme involving interval valued structures engages different procedures.

The relation between observation accuracy and mass of combination inconsistency was depicted in the paper. The less accurate the initial data, the greater the inconsistency mass. Disadvantages of two popular normalization schemes, known as Yager and Dempster methods (Dempster 1968, Yager 1996), were emphasized in context of the considered applications. In the Yager method, inconsistency mass increases uncertainty but the approach impairs detection of inconsistency cases. Consequently, quality of the evidence at hand is usually overestimated. In the Dempster concept, all masses assigned to non-empty sets, including those representing uncertainty, are increased by a factor that is a function of the total inconsistency mass. Final masses calculated based on initial assignment is increased during normalization with the modification factor. Unfortunately it causes confusing misbehaviour of the approach while handling low quality or contradictory evidence.

One must pay attention to the data sets presented in the paper. It is seen that Dempster normalization reduces the number of elements in the final structure. In some cases a 50% reduction in the number of result items was achieved. In view of the exponential complexity of the combination process, the approach seems to dominate over the Yager method. Despite its obvious disadvantages, the last method should not be rejected, as it features effectiveness that appears to be a serious advantage in coping with robust cases. The approach can be implemented for processing in flow association without recording complete result structures as is the case in Dempster normalization.

Specificity of the discussed field of application stipulates modified transformation of the evidence assignment. It should feature advantages of both mentioned methods. These expectations are met in the described publication. Details of the new proposal are presented and discussed in the paper that followed.

5 ABOUT ALGORITHMS

The paper (Filipowicz 2010) contains discussions on algorithms implementing the Mathematical Theory of Evidence and intended for position fixing based on terrestrial observations. Two algorithms are presented. The primary one is designated for iterative search for the fixed position while the secondary one is intended for hypothesis frame location adjustment. The idea lying behind the supplementary procedure enabled avoiding missing local maxima of calculated support measures. The concept of random shifting of the search space locations exploited in the algorithm is like that encountered in an evolutionary approach towards optimization.

The iterative search for the fixed position explores decreasing the area (see figure 1) in order to achieve required accuracy. In each loop, for a given search area, new belief assignments are created, normalized and combined. In the final stage, the search area should be small enough to guarantee a satisfying quality of the solution. Regular mesh is spanned over the search area. Thus the quality of the solution depends on the size of the mesh. It should be noted that the quality is also determined by other, widely known factors. Let us mention number and quality of observations as well as constellation of observed landmarks.

6 MORE ON EVIDENCE REPRESENTATION AND REASONING

Most problems related to maritime applications feature imprecision and uncertainty. Apart from position fixing and its accuracy evaluation, the scope also embraces collective assessment afforded in floating object detection. This can be further exploited in solving monitoring area coverage problems and planning search and rescue operations.

Measurement and indication data, along with nautical knowledge, can be encoded into belief functions. Both knowledge and data are considered as evidence that is exploited in navigation. Belief functions in nautical applications represent evidence and are subject to combination in order to increase their informative context. Evidence representations and results of their combinations could include inconsistencies wherever T-norm operations (Rutkowski 2009) are involved. Inconsistency must be removed to avoid conflicting final results. Conflict arises when belief is greater than the plausibility measure.

In the presented applications, the association of two location vectors with T-norm selects hypothesis frame points situated within a common area (Filipowicz 2011b). A null result vector means that there are no points within intersection and might indicate poor quality evidence.

It is assumed that evidence representations should be normalized at the initial and intermediate stages of processing in order to avoid contradictory results. The most popular normalization procedures feature serious disadvantages. The Yager method disables detection of inconsistency cases. In the Dempster concept, all masses assigned to non-empty sets are increased by a factor that is a function of the total inconsistency mass. It leads to the inacceptable proposition that "the higher inconsistency mass the greater probability assigned to non empty sets" or referring to position fixing, "the poorer quality data, the higher credibility attributed to the fix". Therefore the author's proposal of conversion has been submitted. The proposed conversion features the following properties:

1 Masses attributed to location vectors are not subject to unjustified changes.
2 Conflicts, which are not zero masses assigned to null sets, increase uncertainty.
3 All fuzzy sets are normal, null grades remain unchanged and subsequently conflict detection is not impaired.
4 Plausibility value as a primary factor in selecting fixed position remains intact during conversion.
5 Transformation remains basic for MTE condition, stipulating that belief measures cannot exceed plausibility value.

The condition specified in point 5 is not straightforward and needs to be proven. The proof was presented in the paper. The most important feature of the transformation is that its output contains normal fuzzy sets that proved to be enough to avoid basic conflict. Moreover, plausibility measures regarding the fix remain intact due to proposed conversion. The approach enables one to maintain the value of the plausibility measure, the primary factor determining the selection of the final solution.

7 UNDERSTANDING ACCURACY OF THE FIX

A comparison of the traditional way of position fixing and an approach based on Theory of Evidence was hinted in (Filipowicz 2012b). The main advantage of the proposed scheme of reasoning is that it utilizes the possibilistic approach. This approach is justified whenever insufficient data samples are available, as it is quite often, when dealing with estimations of measurements distributions.

In the proposed approach, knowledge included in a computational scheme is something that creates a new opportunity. A new standpoint for perceiving accuracy of the fix is possible when using reasoning mechanisms. Traditional understanding and accuracy estimating is inadequate in most cases. Appropriate formulas are intended for particular observation schemes that include, at most, three measurements. Although a basic set of data (mean errors and constellation of observed objects) are included in accuracy estimation, applying the same mean error measure for different distributions of isolines seems unjustified. The approach does not correlate quality of observations and accuracy of the obtained fix. The included figures present two cases of fixed positions and their accuracy estimations, which are the same in both cases. Intersections of isolines in one case are spread over a much larger area compared to the second case. Thus the accuracy of the first fix seems be different than in the second case. Although true, the statement seems to be somewhat contradictory and illogic. Obviously supporters of the idea can claim that as long as measurements are random variables it may happen thus. Under this assumption, accuracy estimations remain valid in both cases. Accuracy is meant as the regular area, elliptic or circular, within which the fix can be located with certain probability. The probability is the same all over the area. Unfortunately, in traditional approach, accuracy estimation does not reflect the real, a posteriori evaluated quality of the fix.

In the new approach based on MTE, accuracy estimation, along with its imprecision, are embedded into the reasoning scheme. In the proposed approach, the distribution of probabilities of the fix being located within the explored area is embedded into the methodology. Accuracy can therefore be perceived as a cohesive area within which probability of the fix location is higher than the required threshold value. It is suggested that the area should embrace points with certain percentages of plausibility value attributed to the fix. The accuracy should be perceived as a cohesive, usually irregular, area within which the plausibility of the fix varies. Its value is exactly known for each point in the area, and is supposed to be higher than certain threshold value. For example see grid of plausibility values inserted into figure 1. It should be noted that plausibility measure proved to be the primary factor in selecting the fixed position.

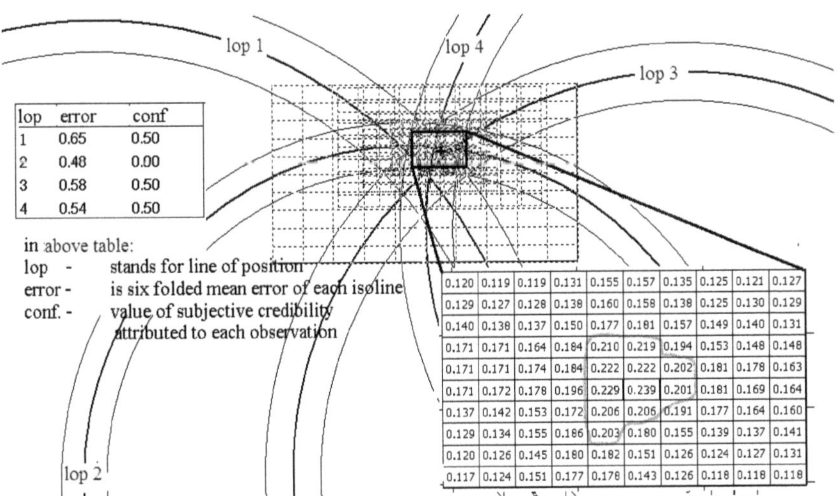

Figure 1. Result of imprecise and uncertain evidence combination generated by software implementing MTE approach to position fixing and its accuracy evaluation

8 COPING WITH SYSTEMATIC ERRORS

The title of the paper (Filipowicz 2012c) directly suggests the main problem that was solved and presented in the paper. It was an attempt to cope with the important navigational issue that is the calibration of nautical devices. Calibration refers to the indication and calculation of a permanent deflection incorporated into measurements made with a particular aid. The problem was solved with

MTE, meaning that properties of combination process involving evidence distorted with systematic error must have been examined. The presented numerical examples and concluded remarks are general; they can be useful and utilized in applications based on fuzzy evidence association.

The concept of exploiting evidence that is encoded facts and knowledge in supporting decisions in navigation is based on measurement distribution. Introduced confidence intervals define the probabilities of true isolines being located within appropriate strips established along gradient directions. Modified probabilities are incorporated into belief assignments that enable the modelling of uncertain, imprecise data. Imprecision is due to random errors but systematic deflections occurring quite often. This kind of error is to be identified and eliminated. The identification of a permanent measurement shift is an important practical and nautical issue. To reduce the effects of the systematic shift one should have observations made for landmarks situated in opposite directions or exploit differential concept. Results of the combination of evidence distorted with fixed errors feature interesting property that can be further used to get rid of their effect. The property comes out of the below presented lemma:

Lemma:

Belief and plausibility measures that are calculated based on results of the combination of two pieces of evidence related to two random variables α_1 and α_2 governed by Gaussian distributions with approximate standard deviations σ_1 and σ_2 for which appropriate isolines are separated with $d(\alpha_1, \alpha_2)$ and those obtained from association of evidence related to random variables α_3 and α_4 and governed by the same distributions with approximate standard deviations $C{\cdot}\sigma_1$ and $C{\cdot}\sigma_2$ with isolines being separated with $d(\alpha_3, \alpha_4) = C{\cdot}d(\alpha_1, \alpha_2)$ are mutually dependent on the factor that is a function on C.

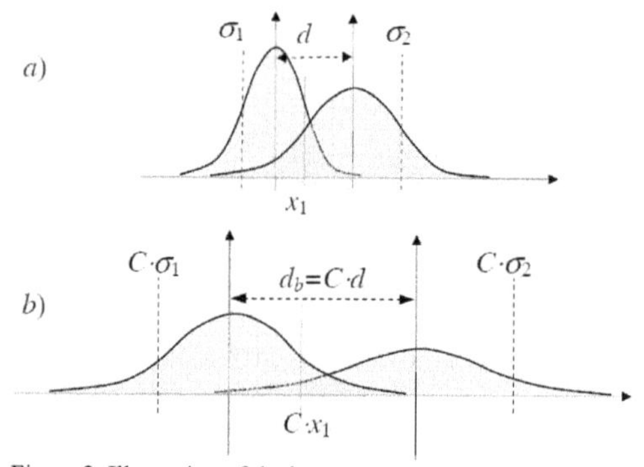

Figure 2. Illustration of the lemma

Figure 2 presents graphical interpretation of the lemma; it also depicts results of combinations marked respectively with x_1 and $C{\cdot}x_1$. To prove the proposition, results of the combination of evidence related to two random variables that projected on a plane remain separated by a certain distance were considered in the paper. Example variables referred to isolines related to the bearings taken for two objects located at nearly counter bearings. Results of observations include both systematic and random errors. Permanent error identification can be achieved with an uncertain evidence combination scheme. The presence of systematic error engaging an iterative proportional increment of isolines mean errors; iterations quit once maximum belief and plausibility measures are registered for the same hypothesis point while the mass of inconsistency remains low.

The presented benefits that can come out of the presented lemma were shown with examples devoted to compass error calculation. Two pieces of evidence, one free from systematic error and another distorted with this kind of deflection, were associated. Results of combinations were confronted in order to mine for general practical aspects. The outcome empirically proved the correctness of the presented lemma and enabled calibration of the compass. Utilization of the lemma for position fixing based upon multiple observations taken with the same tool and possibly distorted with systematic error is straightforward. At first pair or pairs of observations enabling permanent shift indication should be selected. Constant C should be extracted and further used for standard deviations of all observations adjustment. Modified evidence is to be encoded and combined afterwards.

REFERENCES

Burrus N., D. Lesage., 2003, Theory of Evidence. Technical Report no. 0307 - 07/07/03, at website www.lrde.epita.fr/ .../burrus-lesage0903_evidence

Dempster, A., P., 1968, A generalization of Bayesian inference. Journal of the Royal Statistical Society, Series B 30, pp. 205-247

Denoeux, T., 2000, Modelling Vague Beliefs using Fuzzy Valued Belief Structures. Fuzzy Sets and Systems. 116, pp. 167-199

Filipowicz Wł., 2009a, Application of the Theory of Evidence in Navigation. Academic Editorial Board EXIT, Warsaw, Knowledge Engineering and Export Systems, pp. 599-614

Filipowicz Wł., 2009b, Belief Structures and their Applications in Navigation. Methods of Applied Informatics, Polish Academy of Sciences, Gdansk, volume 3/2009, pp. 53-82

Filipowicz Wł., 2010, Fuzzy Reasoning Algorithms for Position Fixing. Measurements Automatics Control, Warsaw, volume 12, pp. 1491-1495

Filipowicz Wł., 2011a, Fuzzy Evidence in Terrestrial Navigation, CRC Press/Balkema, Leiden, Marine Navigation and Safety of Sea Transportation, pp. 65-73

Filipowicz Wł., 2011b, Evidence Representation and Reasoning in Selected Applications. Springer – Verlag, Berlin – Heidelberg, Lecture Notes in Artificial Intelligence, pp. 251-260

Filipowicz Wł., 2012a, Evidence Representations in Position Fixing, Electrical Review, 10b, Warsaw, pp. 256-260
Filipowicz Wł., 2012b, Fuzzy Evidence Reasoning and Position Fixing, IOS Press, Amsterdam, Berlin, Tokyo, Washington, Frontiers in Artificial Intelligence and Applications - Knowledge-Based Intelligent Engineering Systems, pp. 1181-1190
Filipowicz Wł., 2012c, Compass Made Good Correction with MTE, Springer – Verlag, Berlin – Heidelberg, Communications in Computer and Information Science, 329, pp. 69-78
Jurdziński, M., 2008, Principles of Marine Navigation, Gdynia Maritime University
Liu, W., Hughes, J.G., McTear, M.F., 1992, Representing Heuristic Knowledge in D-S Theory. In: UAI '92 Proceedings of the eighth conference on Uncertainty in Artificial Intelligence. Morgan Kaufmann Publishers Inc. San Francisco
Klir, G. J., Parviz B., 1992, Probability – Possibility Transformations: a Comparison. International Journal of General Systems, 21, 1
Rutkowski, L., 2009, Methods and Techniques of the Artificial Intelligence. PWN, Warsaw
Shafer, G., 1976, A Mathematical Theory of Evidence. Princeton University Press, Princeton
Yager, R., 1996 On the Normalization of Fuzzy Belief Structures. International Journal of Approximate Reasoning, 14, pp. 127 – 153
Yen, J., 1990, Generalizing the Dempster – Shafer theory to fuzzy sets. IEEE Transactions on Systems, Man and Cybernetics, 20, 3, 559—570

A New Method for Determining the Attitude of a Moving Object

S.M. Yakushin
Perm National Research Polytechnic University, Russia

ABSTRACT: In this paper a new method of the solution to the problem of the determining vehicle attitude using inertial measurements is presented. The area of study is Strapdown Inertial Navigation System (SDINS), which is composed of three gyro (angular velocity sensor) and three linear accelerometers. The attitude of the determination (angle of pitch, roll and heading) is performed under a priori uncertainty of angular and linear motion of the object and without use of any external measurement information. The solution of the problem is achieved by using some methods of analytical mechanics to determine the accelerations in the rotating reference frame. The algorithm of this method includes the numerical solution of the differential matrix equation by using successive approximation procedure and additional calculations by formulas based on measurements from inertial sensors of SDINS. Unlike the well-known methods, this method has a higher rate of the attitude determination. This paper describes a new method, algorithm, and the results of its modeling relating to the marine vessel, which has an angular and linear 3D-motion in space.

1 INTRODUCTION

Among the other things, to ensure the safety and reliability of marine navigation it is additionally necessary to have a reliable system for determining the attitude parameters of a vessel. For this purpose usually marine gyroscopic systems are used. Nowadays SDINS is mostly used. In the practice of navigation and seamanship some situations can appear because of various reasons where it is necessary to restart the SDINS, for example, in the event of technical problems in their work.

If in this case the vessel is at sea, and there is not enough accurate information from the external measuring systems about angular orientation and parameters of linear and angular motion, then the problem of restarting SDINS becomes much more difficult. In this case, it is necessary to carry out SDINS initial alignment procedure on the moving base.

Autonomous (i.e.,without additional correction) determination of the orientation of a moving object is one of the most difficult problem of the usage of the inertial navigation systems in various fields of their application. The currently known methods of solving this problem are mainly associated with the usage of the optimal filtering method, where directly or indirectly additional measurements are used, for example, from satellite navigation systems, logs, radio and Doppler systems or by use of so-called ZUPT-method.

Another way for determining the orientation of the object is a free gyro positioning system, which determines the position of a vehicle by using two free gyros, was first suggested by (Jeong & Park, 2011).

The alternative approach is the method of the virtual platform (Yakushin, 2010).

In this paper, a new method for determining the inertial orientation of the object without involvement of additional correction is suggested. This method differs from the virtual platform method because it has a higher rate determination of the attitude parameters of SDINS.

In this method to determination a connection between the inertial frame and the navigation frame a quasi-measurements of accelerometers in inertial frame are considered.

It is assumed that there are only the components of the acceleration of gravity in the axes of the navigation frame $G = [0\ 0\ g]T$ and g = const for a given location of the system.

It is obvious that in this situation a methodical error appears that will be smaller, if we identify the components of G in the inertial frame accurately. In this paper this issue is not considered, but the basis of the new method in this statement of the problem is presented.

2 NEW METHOD AND ALGORITHM

To introduce the standard coordinate systems: navigation frame, two orthogonal axes of which are related to the plane of the horizon and points North and East, and the third points local vertical upward, and the inertial frame. We will consider the rotation of the navigation frame relative inertial frame. In this case the matrix of direction cosines between the two coordinate systems can be determined by integration of the well-known matrix differential equation

$$\dot{M} = \omega M \tag{1}$$

where

$$M = \begin{bmatrix} m_{11} & m_{12} & m_{13} \\ m_{21} & m_{22} & m_{23} \\ m_{31} & m_{32} & m_{33} \end{bmatrix} \tag{2}$$

$$\omega = \begin{bmatrix} 0 & \omega_z & -\omega_y \\ -\omega_z & 0 & \omega_x \\ \omega_y & -\omega_x & 0 \end{bmatrix} \tag{3}$$

ω_x , ω_y , ω_y - components of the angular velocity of the Earth's rotation axis Ω in the navigation frame, (Y_g points North, X_g points East, Z_g points up to the local vertical). From the fact that there is a priori uncertainty velocities of the object relative to the Earth we shall assume them to be equal to zero. Then, $\omega_x = 0$ (eastern component of the angular speed of Earth's rotation), $\omega_y = \Omega\ cos\varphi$ (northern component of the angular speed of Earth's rotation), $\omega_z = \Omega\ sin\varphi$ (vertical component of the angular speed of Earth's rotation); φ - latitude of the navigation system.

For solving the problem of determining the attitude parameters in SDINS we need to determine the matrix M in the form of its elements m_{ij} $(i, j = 1, 2, 3)$ at some initial time t_0 in order to use this information in the future in operational mode of SDINS. It will be shown that this can be done by measuring the signals from accelerometers in two fixed moments of time.

Accelerations which are measured by accelerometers, respectively at time t_0 и $t = t_0 + dt$, and referred to as A_0, A_1, we can define obviously in following way:

$$A_0 = \begin{bmatrix} A_{01} \\ A_{02} \\ A_{03} \end{bmatrix} = M_0^T G \tag{4}$$

$$A_1 = \begin{bmatrix} A_{11} \\ A_{12} \\ A_{13} \end{bmatrix} = M_1^T G \frac{1}{2}, \tag{5}$$

where M_0 is the initial value (at time t_0) of the matrix M (which is unknown and to be determined), M_1 – is its subsequent value after a time interval dt, i.e., at time $t = t_0 + dt$, M_0^T is transposed matrix M_0 in the form

$$M_0^T = \begin{bmatrix} m_{11} & m_{21} & m_{31} \\ m_{12} & m_{22} & m_{32} \\ m_{13} & m_{23} & m_{33} \end{bmatrix}. \tag{6}$$

In the small interval dt for determination of M_1 the Euler's method for the numerical solution of the matrix equation (1) may be used

$$M_1 = M_0 + dt\omega M_0. \tag{7}$$

Also it is easy to see also that from the formulae (4) three elements of the matrix M_0 can be at once (at the initial time t_0) identified, namely m_{31}, m_{32}, m_{33}:

$$m_{31} = \frac{A_{01}}{g} \tag{8}$$

$$m_{32} = \frac{A_{02}}{g} \tag{9}$$

$$m_{33} = \frac{A_{03}}{g} \tag{10}$$

Then we carry out the following vector-matrix operations using equations (4, 5, 7)

$$M_1^T = M_0^T + dtM_0^T\omega^T \tag{11}$$

$$A_1 = (M_0^T + dtM_0^T\omega^T)G = A_0 + dtM_0^T\omega^T G \tag{12}$$

$$\frac{A_1 - A_0}{dt} = M_0^T\omega^T G = \begin{bmatrix} m_{11}\omega_y g - m_{21}\omega_x g \\ m_{12}\omega_y g - m_{22}\omega_x g \\ m_{13}\omega_y g - m_{23}\omega_x g \end{bmatrix}. \tag{13}$$

In light of the fact that for the elements of the direction cosine matrix M we have the following relations

$$m_{21} = m_{32}m_{13} - m_{12}m_{33} \tag{14}$$

$$m_{22} = m_{11}m_{33} - m_{31}m_{13} \tag{15}$$

$$m_{22} = m_{31}m_{12} - m_{11}m_{32}, \tag{16}$$

then, substituting them into the right side of equation (13), we obtain an algebraic system of three equations for three unknown m_{11}, m_{12}, m_{13}. The general solution for these unknown can be written as

$$m_{11} = \frac{dm11}{d} \quad (17)$$

$$m_{12} = \frac{dm12}{d} \quad (18)$$

$$m_{13} = \frac{dm13}{d}, \quad (19)$$

where

$$d = \omega_y g^3 (\omega_x^2 + \omega_y^2) \quad (20)$$

$$dm11 = g^2 (dA1(\omega_y^2 + m_{31}^2 \omega_x^2) + dA2\omega_x (m_{31} m_{32} \omega_x - m_{33} \omega_y) + dA3\omega_x (m_{32} \omega_y + m_{31} m_{33} \omega_x)) \quad (21)$$

$$dm12 = g^2 (dA1\omega_X (m_{31} m_{32} \omega_x + m_{33} \omega_y) + dA2(m_{32}^2 \omega_x^2 + \omega_y^2) + dA3\omega_x (m_{32} m_{33} \omega_x - m_{31} \omega_y)) \quad (22)$$

$$dm13 = g^2 (dA1\omega_X (m_{31} m_{33} \omega_x - m_{32} \omega_y) + dA2\omega_x (m_{32} m_{33} \omega_x + m_{31} \omega_y) + dA3(\omega_y^2 + m_{33}^2 \omega_x^2)) \quad (23)$$

$$dA1 = \frac{A_{11} - A_{01}}{dt} \quad (24)$$

$$dA2 = \frac{A_{12} - A_{02}}{dt} \quad (25)$$

$$dA3 = \frac{A_{13} - A_{03}}{dt}. \quad (26)$$

Thus, on the basis of the construction of the algorithm, using expressions (8-10), (17-19) and (14-16) we can calculate the orientation matrix M, i.e.,the angular position of the navigation frame relative to the inertial frame by use of two sampling measurements from accelerometers.

In this case, we choose that initially the inertial frame coincides with the body frame. The orientation of the body frame relative to the navigation frame can be determined in the following way.

Firstly, we obtain the numerical solution of the matrix differential equation

$$\dot{C} = \omega_b C \quad (27)$$

in relation to the orientation matrix C, which determines the angular position of the body frame relative to the inertial frame, and where ω_b is the skew-symmetric matrix whose elements are the gyro measurements of SDINS. This matrix has the same form as the matrix in (3). At the initial moment of time the matrix C is the identity matrix.

Further, calculation of the orientation matrix B as the body frame relative to the navigation frame can be made on the basis of expression

$$B = CM^T. \quad (28)$$

As a result, using the expression (28) we receive the solution for the autonomous determination the attitude of a moving object with respect to the navigation frame

Unlike the well-known methods, in this new method the reducing of the time determination of attitude is achieved , as its implementation requires only two cycles of measurements with accelerometers and angular rate sensors SDINS.

At the same time, it should be noted that the effectiveness of this method depends strongly on the accuracy characteristics of inertial sensors (accelerometers and gyroscopes), including the threshold of sensitivity, accurate accounting the position of measuring system SDINS with respect to the center of mass of the object, as well as the velocity of the object relative to the Earth. The latter circumstance due to the fact that the angular velocity ω_x and ω_y as is known depends on the vector of the linear velocity of the object relative to the Earth. We believe these velocities as unknown in the general case of autonomous attitude determination of the object.

3 SIMULATION SCHEME

To assess the performance of the new method simulated operation of his algorithm on the example of marine SDINS is fulfilled. The angular and linear movement of a ship were taken into the account while modeling in relation to the Earth, as well as errors of inertial measuring instruments. To modelling the vectors of measurements A_0, A_1 the conversion measurements of accelerometers SINS from the body frame to the inertial frame was performed.

Simulation measurements from accelerometers and angular rate sensors in the axes of the body frame was made in the following way (Yakushin, 2010).

We will consider the following equations as mathematical model of 3D-motion in space of a marine vessel with variable horizontal velocity relative to the Earth V, and also with time variables heading, ψ, pitch ϑ, roll γ and vertical velocity V_Z according to the expressions (29-35):

$$V = V_0 + A_V \sin(\omega_V t), \quad (29)$$

$$\psi = \psi_0 + A_\psi \sin(\omega_\psi t), \quad (30)$$

$$V_N = V \cos\psi, \quad (31)$$

$$V_E = V \sin\psi, \quad (32)$$

$$V_Z = A_{VZ} \sin(\omega_{VZ} t), \quad (33)$$

$$\vartheta = \vartheta_0 + A_\vartheta \sin(\omega_\vartheta t), \quad (34)$$

$$\gamma = \gamma_0 + A_\gamma \sin(\omega_\gamma t). \quad (35)$$

In accordance with the simulated motion of the object relative to the Earth will change the latitude φ and longitude λ of its location

$$\dot{\varphi} = V_N / R_2, \quad (36)$$

$$\dot{\lambda} = V_E / (R_1 \cos(\varphi)), \quad (37)$$

where R_1 and R_2 are the radiuses of curvature of the Earth reference ellipsoid, defined by well known formulas from geodesy.

According to the mathematical model of the marine vessel (29-37) we can write expressions to determine the specific force accelerations A_E, A_N, A_Z in the navigation frame *ENZ*, the axis of which are oriented to the East, North and to the local vertical up.

$$A_E = \dot{V}_E - 2V_N\Omega\sin\varphi - V_E V_N \tan\varphi / R_1 + V_Z(\Omega\cos\varphi + V_E/R_1), \quad (38)$$

$$A_N = \dot{V}_N + 2V_E\Omega\sin\varphi + V_E^2\tan\varphi/R_1 + V_Z V_N/R_2, \quad (39)$$

$$A_Z = -V_N^2/R_2 - V_E^2/R_1 - 2V_E\Omega\cos\varphi + g + \dot{V}_Z. \quad (40)$$

The measured accelerations vector in the body frame A_b with a view of an acceleration in the navigation frame A_n we will obtain by the following transformation

$$A_b = TA_n, \quad (41)$$

where T is the true direction cosine matrix between the body frame and the navigation frame. We compute the elements of this matrix on the basis the given values of the orientation angles defined by equations (30), (34), (35)

$$t_{11} = C_\gamma C_\psi - S_\gamma S_\vartheta S_\psi, \quad (42)$$

$$t_{12} = C_\gamma S_\psi + S_\gamma S_\vartheta C_\psi, \quad (43)$$

$$t_{13} = -S_\gamma C_\vartheta, \quad (44)$$

$$t_{21} = -C_\vartheta S_\psi, \quad (45)$$

$$t_{22} = C_\vartheta C_\psi, \quad (46)$$

$$t_{23} = S_\vartheta, \quad (47)$$

$$t_{31} = S_\gamma C_\psi + C_\gamma S_\vartheta S_\psi, \quad (48)$$

$$t_{32} = S_\gamma S_\psi - C_\gamma S_\vartheta C_\psi, \quad (49)$$

$$t_{33} = C_\gamma C_\vartheta, \quad (50)$$

where the notation C () = cos (), S () = sin () is used.

Now, if we add to the vector of accelerations A_b an accelerometer errors, we will obtain a simulated accelerometer signals in the body frame which we will use further in simulation program for the algorithm of the vessel attitude determination.

Next we need to do a simulation signals from the gyroscopes for a given movement of the vessel.

Let's consider as measuring signals from gyroscopes (angular rate sensors) the corresponding angular rates in axes of the body frame. In this case, taking into account change of orientation of the body frame relative to the navigation frame, described by well known Euler transformation, we will obtain following expressions for these angular rates

$$\omega_{bx} = \dot{\vartheta}C_\gamma - \dot{\psi}C_\vartheta S_\gamma + \Omega_Y(S_\psi C_\gamma + C_\psi S_\vartheta S_\gamma) - \Omega_Z C_\vartheta S_\gamma + \Omega_X(C_\psi C_\gamma - S_\psi S_\vartheta S_\gamma), \quad (51)$$

$$\omega_{by} = \dot{\gamma} + \dot{\psi}S_\vartheta + \Omega_Y C_\psi C_\vartheta + \Omega_Z S_\vartheta - \Omega_X S_\psi C_\vartheta, \quad (52)$$

$$\omega_{bz} = \dot{\psi}C_\vartheta C_\gamma + \dot{\vartheta}S_\gamma + \Omega_Y(S_\psi S_\gamma - C_\psi C_\gamma S_\vartheta) + \Omega_Z C_\vartheta C_\gamma + \Omega_X(S_\psi S_\vartheta C_\gamma + C_\psi S_\gamma), \quad (53)$$

where

$$\Omega_X = -\dot{\varphi}, \quad (54)$$

$$\Omega_Y = (\Omega + \dot{\lambda})\cos\varphi, \quad (55)$$

$$\Omega_Z = (\Omega + \dot{\lambda})\sin\varphi. \quad (56)$$

Adding to expressions from (51-53) the corresponding errors, we will obtain simulated signals of angular rate sensors in the body frame which further we will be use also as well as simulated signals of accelerometers in the simulation program of the algorithm of the vessel attitude determination.

4 SIMULATION RESULTS

One of the simulation results of the new method of autonomous inertial attitude determination for the parameters of a vessel:

$dt = 5$ s, $\varphi = 58^{\circ}$, $\lambda = 56^{\circ}$, $V_0 = 10$ m/s, $A_V = 0.1$ m/s,

$\omega_V = 1.2$ rad/s, $A_{VZ} = 1$ m/s, $\omega_{VZ} = 0.2$ rad/s,

$\psi_0 = -77^{\circ}$, $A_{\psi} = \omega_{\psi} = 0$, $\vartheta_0 = 7^{\circ}$, $A_{\vartheta} = 7^{\circ}$,

$\omega_{\vartheta} = 0.1$ rad/s, $\gamma_0 = 6^{\circ}$, $A_{\gamma} = 6^{\circ}$, $\omega_{\gamma} = 0.1$ rad/s

and taking into account the random errors of accelerometers $5 \cdot 10^{-6}$ g (3σ) and random errors of gyros 0.05°/ h (3σ) is as follows:

$e_{11} = 0.0101$, $e_{12} = -0.0723$, $e_{13} = -0.0081$;

$e_{21} = 0.0214$, $e_{22} = 0.0062$, $e_{23} = 0.0019$;

$e_{31} = 0.0021$, $e_{32} = -0.0148$, $e_{33} = -0.1607$,

where $e_{ij} = arccos(m_{ij}) - arccos(a_{ij})$, $i, j = 1 \ldots 3$ is the difference in the angular position of the calculated elements of the direction cosines of the navigation frame relative to the inertial frame and the corresponding true values. All error values e_{ij} are given in degrees. Similar in terms of errors are the elements of matrix for the body frame relative to the navigation frame. This is because the calculation of the matrix C according to (27) has no systematic errors.

5 CONCLUSIONS

This presentation of the results confirms the efficiency and sufficient accuracy of the new method of attitude determining for autonomous moving object in a low and medium intensity angular and linear motion of a vessel and the use of medium-class accuracy inertial sensor of SDINS. This can be considered as a contribution to the overall safety and reliability of marine transportation in the autonomous navigation for a wide range of vessels. As a further development of this method is supposed to optimize procedures for calculating the gravitational acceleration component by applying filtering and adaptation techniques and with the possible use of information from autonomous logs. It is also advisable to carry out a research of the opportunities to improve the accuracy of the attitude parameters by using this method in integration with an autonomous method of the virtual platform (Yakushin, 2010).

REFERENCES

Yakushin, S.M. 2010. Novel Method for Autonomous Initial Alignment of Strapdown Inertial Navigation System on Moving Base. *Proc. The European Navigation Conference on Global Navigation Satellite Systems. Braunschweig, Germany, 19-21 October 2010.*

Jeong T.-G., Park S.-C.: An Algorithmic Study on Positioning and Directional System by Free Gyros. *TransNav - International Journal on Marine Navigation and Safety of Sea Transportation, Vol. 5, No. 3, pp. 297-302, 2011*

Simulation of Zermelo Navigation on Riemannian Manifolds for dim(R×*M*)=3

P. Kopacz
Gdynia Maritime University, Faculty of Navigation, Poland
Jagiellonian University, Institute of Mathematics, Poland

ABSTRACT: We research the Zermelo navigation problem in Riemannian space for $\dim(\mathbb{R}\times M)=3$ under the force representing the action of the perturbing “wind” distribution modeled by the vector field on manifold *M*. We consider a fibered manifold $\pi:\mathbb{R}\times M\rightarrow\mathbb{R}$ representing "sea" where π is the first canonical projection. Then we perturb it by a time-dependent vector field on *M*, i.e. a projectable vector field on $\mathbb{R}\times TM$. We obtain the final form of the system of differential equations providing the solution of the problem after perturbation in general case. We present the simulation of the Zermelo navigation on Riemannian manifolds for $\dim(\mathbb{R}\times M)=3$.

1 INTRODUCTION

In 1931 E. Zermelo (Zermelo, 1931) dealt with the following geometric and control problem: *Consider a vessel sailing on the open sea in the calm waters. Let us suppose a wind comes up. How must the ship be steered in order to reach a given destination from a starting point in the shortest time?* Zermelo assumed that the open sea was $\mathbb{R}^2$ with the Euclidean metric. In an unbounded plane the wind distribution ξ has been modeled by a vector field as a function of position and time and a vessel sails with the constant velocity relative to the wind. The problem has also been studied by C. Caratheodory in (Caratheodory, 2008) and treated as the particular case of the Lagrange problem in the calculus of variations. In 2002 Z. Shen (Chern & Shen, 2005), (Bao, Robles & Shen, 2004) generalized the problem to the setting where the sea is an arbitrary Riemannian manifold (M,g). Shen finds that, when the wind is time-independent, the paths of shortest time are the geodesics of a special Finsler type. The milestone in the modern theoretical research is (Bao, Robles & Shen, 2004) where the authors consider Zermelo navigation problem on Riemannian manifolds giving the classification of Finsler metrics of constant (scalar) flag curvature in the particular case, namely the Randers metrics. The essential classification problem in general remains still open in Riemann- Finsler geometry. In (Bao, Robles & Shen, 2004) the problem was treated for the case of a "low wind perturbation". That means the Riemannian length of the wind vector is not greater than 1 everywhere on *M* and a new metric corresponding to the deviated geodesics was constructed as a Finsler metric i.e. the Riemannian length of the wind vector $g(x,-\xi_x)=\sqrt{g_{ij}(x)\xi^i\xi^j}\leq 1$. Geometrically, the problem is to find the deviation of geodesics under the action of a time-dependent vector field.

2 PRELIMINARIES AND NOTATIONS

Throughout this paper manifolds and mappings are smooth and Einstein summation convention over repeated indices is assumed. We follow the notations applied in (Palacek, Krupkova, 2012). Let a pair *(M,g)* be a Riemannian manifold where

$$g=g_{ij}dx^i\otimes dx^j$$

is a Riemannian metric with non-degenerate, symmetric and positive definite matrix (g_{ij}), and *M* is an *m*-dimensional manifold with local coordinates (x^b), $1\leq b\leq m$. We shall consider a fibered manifold $\pi:\mathbb{R}\times M\rightarrow\mathbb{R}$, where π is the first canonical projection. On $\mathbb{R}\times M$ we use coordinate charts adapted to the product structure (t, x^b), $1 \le b \le m$, where t is the global coordinate on $\mathbb{R}$. A curve $c: \mathbb{R} \to M$, defined in a neighborhood of $0 \in \mathbb{R}$, will be represented by its graph

$$\gamma:\ \mathbb{R} \to \mathbb{R} \times M,$$

$$t \mapsto (t, c(t)),$$

which is a section of the fibered manifold π. Let the wind distribution on M be represented by a time-dependent vector field on M, i.e. by a projectable vector field ξ on $\mathbb{R} \times TM$ of the form

$$\xi = \frac{\partial}{\partial t} + \xi^i (t, x^j) \frac{\partial}{\partial x^i}$$

To analyze the deformations of geodesics we consider the variational problem on $\mathbb{R} \times TM$ defined by the kinetic energy $\overline{T}$ in the form

$$\overline{T} = \frac{1}{2} g_{ij} y^i y^j \tag{1}$$

where $y^i = \dot{x}^i + \xi^i$.

3 ENERGETIC APPROACH VIA EULER-LAGRANGE EQUATIONS

The Euler-Lagrange equations of the mechanical system (1) are expressed in the form

$$\frac{\partial \overline{T}}{\partial x^k} - \frac{d}{dt}\left(\frac{\partial \overline{T}}{\partial \dot{x}^k}\right) = 0, \quad 1 \le k \le m, \tag{2}$$

where

$$\begin{aligned} \overline{T} &= \frac{1}{2} g_{ij} y^i y^j = \frac{1}{2} g_{ij} \left(\dot{x}^i + \xi^i\right)\left(\dot{x}^j + \xi^j\right) = \\ &= \frac{1}{2} g_{ij} \dot{x}^i \dot{x}^j + g_{ij} \dot{x}^i \xi^j + \frac{1}{2} g_{ij} \xi^i \xi^j. \end{aligned} \tag{3}$$

Let us denote

$$V = -g_{ij} \dot{x}^i \xi^j - \frac{1}{2} g_{ij} \xi^i \xi^j. \tag{4}$$

Hence

$$\overline{T} = T - V, \tag{5}$$

where T is the kinetic energy of the unperturbed problem and V has the meaning of the potential energy caused by the wind. Computing (2) explicitly we obtain

$$F_k - \Gamma_{kij} \dot{x}^i \dot{x}^j - g_{kj} \ddot{x}^j = 0, \tag{6}$$

where

$$F_k = \left(\frac{\partial g_{ij}}{\partial x^k} \xi^j + g_{ij} \frac{\partial \xi^j}{\partial x^k}\right) \dot{x}^i + \frac{1}{2} \frac{\partial}{\partial x^k}\left(g_{ij} \xi^i \xi^j\right) - \frac{\partial g_{kj}}{\partial x^i} \dot{x}^i \xi^j - g_{kj} \frac{\partial \xi^j}{\partial t} - g_{kj} \frac{\partial \xi^j}{\partial x^i} \dot{x}^i \tag{7}$$

and represents the force. Γ_{ijk} are the standard Christoffel symbols of (g_{ij}),

$$\Gamma_{ijk} = \frac{1}{2}\left(\frac{\partial g_{ji}}{\partial x^k} + \frac{\partial g_{ki}}{\partial x^j} - \frac{\partial g_{jk}}{\partial x^i}\right). \tag{8}$$

4 ZERMELO NAVIGATION FOR $\dim(\mathbb{R} \times M) = 3$ IN GENERAL CASE

The problem before perturbation is defined in the general form

$$\Gamma_{kij} \dot{x}^i \dot{x}^j + g_{kj} \ddot{x}^j = 0, \tag{9}$$

and the general form of the problem after perturbation yields

$$F_k - \Gamma_{kij} \dot{x}^i \dot{x}^j - g_{kj} \ddot{x}^j = 0, \tag{10}$$

where F_k, Γ_{ijk} are given by (7) and (8), respectively.

In case of $\dim(\mathbb{R} \times M) = 3$ the metric g and the vector field ξ equal, respectively

$$\begin{aligned} g &= g_{11}(x^1, x^2) dx^1 \otimes dx^1 + g_{12}(x^1, x^2) dx^1 \otimes dx^2 + \\ &+ g_{21}(x^1, x^2) dx^2 \otimes dx^1 + g_{22}(x^1, x^2) dx^2 \otimes dx^2, \end{aligned} \tag{11}$$

$$\begin{aligned} \xi &= \frac{\partial}{\partial t} + \xi^i (t, x^1, x^2) \frac{\partial}{\partial x^i} = \\ &= \frac{\partial}{\partial t} + \xi^1 (t, x^1, x^2) \frac{\partial}{\partial x^1} + \xi^2 (t, x^1, x^2) \frac{\partial}{\partial x^2} \end{aligned}. \tag{12}$$

And the force $F = (F_1, F_2)$.

From (10) we obtain the system of equations in the Zermelo problem after perturbation for $\dim(\mathbb{R} \times M) = 3$

$$\begin{cases} F_1 - \Gamma_{1ij} \dot{x}^i \dot{x}^j - g_{1j} \ddot{x}^j = 0 \\ F_2 - \Gamma_{2ij} \dot{x}^i \dot{x}^j - g_{2j} \ddot{x}^j = 0 \end{cases} \tag{13}$$

Thus

$$\begin{cases} F_1 - \Gamma_{111}(\dot{x}^1)^2 - (\Gamma_{112} + \Gamma_{121}) \dot{x}^1 \dot{x}^2 - \Gamma_{122}(\dot{x}^2)^2 - g_{11} \ddot{x}^1 - g_{12} \ddot{x}^2 = 0 \\ F_2 - \Gamma_{211}(\dot{x}^1)^2 - (\Gamma_{212} + \Gamma_{221}) \dot{x}^1 \dot{x}^2 - \Gamma_{222}(\dot{x}^2)^2 - g_{21} \ddot{x}^1 - g_{22} \ddot{x}^2 = 0 \end{cases} \tag{14}$$

Applying the particular Christoffel symbols Γ_{ijk} and the components of force F_1, F_2 we obtain two equations of the final system

$$F_1 = \begin{cases} \left(\frac{\partial g_{11}}{\partial x^1}\xi^1 + g_{11}\frac{\partial \xi^1}{\partial x^1}\right)\dot{x}^1 + \frac{1}{2}\frac{\partial}{\partial x^1}\left(g_{11}\xi^1\xi^1\right) - \frac{\partial g_{11}}{\partial x^1}\dot{x}^1\xi^1 - g_{11}\frac{\partial \xi^1}{\partial t} - g_{11}\frac{\partial \xi^1}{\partial x^1}\dot{x}^1 + \\ + \left(\frac{\partial g_{22}}{\partial x^1}\xi^2 + g_{22}\frac{\partial \xi^2}{\partial x^1}\right)\dot{x}^2 + \frac{1}{2}\frac{\partial}{\partial x^1}\left(g_{22}\xi^2\xi^2\right) - \frac{\partial g_{12}}{\partial x^2}\dot{x}^2\xi^2 - g_{12}\frac{\partial \xi^2}{\partial t} - g_{12}\frac{\partial \xi^2}{\partial x^2}\dot{x}^2 + \\ + \left(\frac{\partial g_{12}}{\partial x^1}\xi^2 + g_{12}\frac{\partial \xi^2}{\partial x^1}\right)\dot{x}^1 + \frac{1}{2}\frac{\partial}{\partial x^1}\left(g_{12}\xi^1\xi^2\right) - \frac{\partial g_{12}}{\partial x^1}\dot{x}^1\xi^2 - g_{12}\frac{\partial \xi^2}{\partial t} - g_{12}\frac{\partial \xi^2}{\partial x^1}\dot{x}^1 + \\ + \left(\frac{\partial g_{21}}{\partial x^1}\xi^1 + g_{21}\frac{\partial \xi^1}{\partial x^1}\right)\dot{x}^2 + \frac{1}{2}\frac{\partial}{\partial x^1}\left(g_{21}\xi^2\xi^1\right) - \frac{\partial g_{11}}{\partial x^2}\dot{x}^2\xi^1 - g_{11}\frac{\partial \xi^1}{\partial t} - g_{11}\frac{\partial \xi^1}{\partial x^2}\dot{x}^2 + \end{cases}$$
$$-\frac{1}{2}\frac{\partial g_{11}}{\partial x^1}\left(\dot{x}^1\right)^2 - \left(\frac{\partial g_{11}}{\partial x^2} + \frac{\partial g_{21}}{\partial x^1} - \frac{1}{2}\frac{\partial g_{12}}{\partial x^1} - \frac{1}{2}\frac{\partial g_{21}}{\partial x^1}\right)\dot{x}^1\dot{x}^2 +$$
$$-\left(\frac{\partial g_{21}}{\partial x^2} - \frac{1}{2}\frac{\partial g_{22}}{\partial x^1}\right)\left(\dot{x}^2\right)^2 - g_{11}\ddot{x}^1 - g_{12}\ddot{x}^2 = 0 \tag{15}$$

and

$$F_2 = \begin{cases} \left(\frac{\partial g_{11}}{\partial x^2}\xi^1 + g_{11}\frac{\partial \xi^1}{\partial x^2}\right)\dot{x}^1 + \frac{1}{2}\frac{\partial}{\partial x^2}\left(g_{11}\xi^1\xi^1\right) - \frac{\partial g_{21}}{\partial x^1}\dot{x}^1\xi^1 - g_{21}\frac{\partial \xi^1}{\partial t} - g_{21}\frac{\partial \xi^1}{\partial x^1}\dot{x}^1 + \\ + \left(\frac{\partial g_{22}}{\partial x^2}\xi^2 + g_{22}\frac{\partial \xi^2}{\partial x^2}\right)\dot{x}^2 + \frac{1}{2}\frac{\partial}{\partial x^2}\left(g_{22}\xi^2\xi^2\right) - \frac{\partial g_{22}}{\partial x^2}\dot{x}^2\xi^2 - g_{22}\frac{\partial \xi^2}{\partial t} - g_{22}\frac{\partial \xi^2}{\partial x^2}\dot{x}^2 + \\ + \left(\frac{\partial g_{12}}{\partial x^2}\xi^2 + g_{12}\frac{\partial \xi^2}{\partial x^2}\right)\dot{x}^1 + \frac{1}{2}\frac{\partial}{\partial x^2}\left(g_{12}\xi^1\xi^2\right) - \frac{\partial g_{22}}{\partial x^1}\dot{x}^1\xi^2 - g_{22}\frac{\partial \xi^2}{\partial t} - g_{22}\frac{\partial \xi^2}{\partial x^1}\dot{x}^1 + \\ + \left(\frac{\partial g_{21}}{\partial x^2}\xi^1 + g_{21}\frac{\partial \xi^1}{\partial x^2}\right)\dot{x}^2 + \frac{1}{2}\frac{\partial}{\partial x^2}\left(g_{21}\xi^2\xi^1\right) - \frac{\partial g_{21}}{\partial x^2}\dot{x}^2\xi^1 - g_{21}\frac{\partial \xi^1}{\partial t} - g_{21}\frac{\partial \xi^1}{\partial x^2}\dot{x}^2 + \end{cases}$$
$$-\left(\frac{\partial g_{12}}{\partial x^1} - \frac{1}{2}\frac{\partial g_{11}}{\partial x^2}\right)\left(\dot{x}^1\right)^2 - \left(\frac{\partial g_{12}}{\partial x^2} + \frac{\partial g_{22}}{\partial x^1} - \frac{1}{2}\frac{\partial g_{12}}{\partial x^2} - \frac{1}{2}\frac{\partial g_{21}}{\partial x^2}\right)\dot{x}^1\dot{x}^2 +$$
$$-\frac{1}{2}\frac{\partial g_{22}}{\partial x^2}\left(\dot{x}^2\right)^2 - g_{21}\ddot{x}^1 - g_{22}\ddot{x}^2 = 0 \tag{16}$$

Ordering in respect to the derivatives and considering the metric g with the assumptions $g_{12} = 0$ and $g_{21} = 0$, above system ((15) & (16)) in the navigation problem after perturbation yields

$$\begin{cases} g_{11}\ddot{x}^1 + \frac{1}{2}\frac{\partial g_{11}}{\partial x^1}\left(\dot{x}^1\right)^2 - \frac{1}{2}\frac{\partial g_{22}}{\partial x^1}\left(\dot{x}^2\right)^2 + \frac{\partial g_{11}}{\partial x^2}\dot{x}^1\dot{x}^2 + \\ + \left(g_{11}\frac{\partial \xi^1}{\partial x^2} + \frac{\partial g_{11}}{\partial x^2}\xi^1 - g_{22}\frac{\partial \xi^2}{\partial x^1} - \frac{\partial g_{22}}{\partial x^1}\xi^2\right)\dot{x}^2 + \\ -\frac{1}{2}\frac{\partial}{\partial x^1}\left(g_{11}\xi^1\xi^1 + g_{22}\xi^2\xi^2\right) + 2g_{11}\frac{\partial \xi^1}{\partial t} = 0 \\ \\ g_{22}\ddot{x}^2 - \frac{1}{2}\frac{\partial g_{11}}{\partial x^2}\left(\dot{x}^1\right)^2 + \frac{1}{2}\frac{\partial g_{22}}{\partial x^2}\left(\dot{x}^2\right)^2 + \frac{\partial g_{22}}{\partial x^1}\dot{x}^1\dot{x}^2 + \\ -\left(g_{11}\frac{\partial \xi^1}{\partial x^2} + \frac{\partial g_{11}}{\partial x^2}\xi^1 - g_{22}\frac{\partial \xi^2}{\partial x^1} - \frac{\partial g_{22}}{\partial x^1}\xi^2\right)\dot{x}^1 + \\ -\frac{1}{2}\frac{\partial}{\partial x^2}\left(g_{11}\xi^1\xi^1 + g_{22}\xi^2\xi^2\right) + 2g_{22}\frac{\partial \xi^2}{\partial t} = 0 \end{cases} \tag{17}$$

Formula (17) lets to present the Zermelo navigation problem and provide the solution for arbitrary metric g and perturbing vector field ξ in case of $\dim(\mathbb{R} \times M) = 3$ with respect to the general assumptions considered in the preliminaries.

5 SIMULATION OF 3-DIMENSIONAL SITUATION

We consider the simulation for the model of "sea" and "wind" consisting of the following type of metric and vector field. We put

$$(g): \begin{cases} g_{11} = g_{11}\left(x^1, x^2\right) \\ g_{22} = g_{22}\left(x^2\right) \\ g_{12} = g_{21} = 0 \end{cases} \tag{18}$$

Then we have the final system in respect to (18) presented in (19)

$$\begin{cases} g_{11}\ddot{x}^1 + \frac{1}{2}\frac{\partial g_{11}}{\partial x^1}\left(\dot{x}^1\right)^2 + \frac{\partial g_{11}}{\partial x^2}\dot{x}^1\dot{x}^2 + \\ -\left(g_{22}\frac{\partial \xi^2}{\partial x^1} - \frac{\partial g_{11}}{\partial x^2}\xi^1 - g_{11}\frac{\partial \xi^1}{\partial x^2}\right)\dot{x}^2 + \\ -\frac{1}{2}\frac{\partial}{\partial x^1}\left(g_{11}\xi^1\xi^1 + g_{22}\xi^2\xi^2\right) + 2g_{11}\frac{\partial \xi^1}{\partial t} = 0 \\ \\ g_{22}\ddot{x}^2 - \frac{1}{2}\frac{\partial g_{11}}{\partial x^2}\left(\dot{x}^1\right)^2 + \frac{1}{2}\frac{\partial g_{22}}{\partial x^2}\left(\dot{x}^2\right)^2 + \\ -\left(\frac{\partial g_{11}}{\partial x^2}\xi^1 + g_{11}\frac{\partial \xi^1}{\partial x^2} - g_{22}\frac{\partial \xi^2}{\partial x^1}\right)\dot{x}^1 + \\ -\frac{1}{2}\frac{\partial}{\partial x^2}\left(g_{11}\xi^1\xi^1 + g_{22}\xi^2\xi^2\right) + 2g_{22}\frac{\partial \xi^2}{\partial t} = 0 \end{cases} \tag{19}$$

The form of the particular final system is obtained after applying the particular metric (g_{11}, g_{22}) and perturbing vector field (ξ^1, ξ^2). In general, the vector field $\left(\frac{\partial}{\partial y^i}\right)_{1 \le i \le n}$ for wind distribution ξ is represented by $\xi = \xi^i(y)\left(\frac{\partial}{\partial y^i}\right)$. Here we choose the vector field ξ as follows

$$(\xi): \begin{cases} \xi^1 = 0 \\ \xi^2 = \xi^2\left(x^2\right) \neq 0 \end{cases} \tag{20}$$

Hence

$$\xi = \frac{\partial}{\partial t} + \xi^2\left(t, x^1, x^2\right)\frac{\partial}{\partial x^2} \tag{21}$$

Then the final system after perturbation (19) becomes

$$\begin{cases} g_{11}\ddot{x}^1 + \frac{1}{2}\frac{\partial g_{11}}{\partial x^1}\left(\dot{x}^1\right)^2 + \frac{\partial g_{11}}{\partial x^2}\dot{x}^1\dot{x}^2 = 0 \\ g_{22}\ddot{x}^2 - \frac{1}{2}\frac{\partial g_{11}}{\partial x^2}\left(\dot{x}^1\right)^2 + \frac{1}{2}\frac{\partial g_{22}}{\partial x^2}\left(\dot{x}^2\right)^2 - \frac{1}{2}\frac{\partial}{\partial x^2}\left(g_{22}\xi^2\xi^2\right) = 0 \end{cases} \tag{22}$$

As an example we consider in 3-dimensional case the particular components of the model

$$\text{"sea"}: \; g = x^1 dx^1 \otimes dx^1 + x^2 dx^2 \otimes dx^2 \tag{23}$$

$$\text{"wind"}: \; \xi = \frac{\partial}{\partial t} + x^2 \frac{\partial}{\partial x^2} \tag{24}$$

Then equation (6) takes the form

$$\begin{cases} 2x^1\ddot{x}^1 + \left(\dot{x}^1\right)^2 = 0 \\ 2x^2\ddot{x}^2 + \left(\dot{x}^2\right)^2 - 3\left(x^2\right)^2 = 0 \end{cases} \tag{25}$$

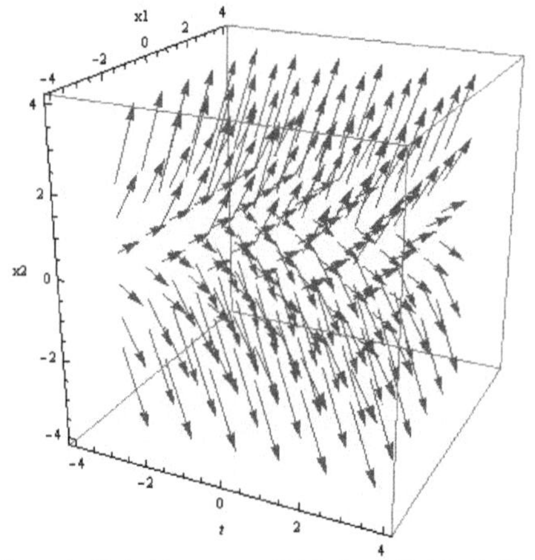

Figure 1. Perturbing vector field ξ applied in 3-dimensional simulation.

The perturbing vector field applied in the simulation is presented in Figure 1. Solving (25) we obtain the parametric form of the 3-dimensional curve $\tilde{\gamma}(t)$ representing the solution of considered Zermelo navigation problem

$$\tilde{\gamma}(t): \begin{cases} x^1(t) = C_1 e^{-t}\left(e^{3t} + e^{2C_2}\right)^{\frac{2}{3}} \\ x^2(t) = C_3 e^{-t}\left(e^{3t} + e^{2C_4}\right)^{\frac{2}{3}} \end{cases} \quad C_1, C_2, C_3, C_4 \in \mathbb{R} \tag{26}$$

while the system before the wind deformation gives the curve $\gamma(t)$

$$\gamma(t): \begin{cases} x^1(t) = C_1\left(3t - 2C_2\right)^{\frac{2}{3}} \\ x^2(t) = C_3\left(3t - 2C_4\right)^{\frac{2}{3}} \end{cases} \quad C_1, C_2, C_3, C_4 \in \mathbb{R} \tag{27}$$

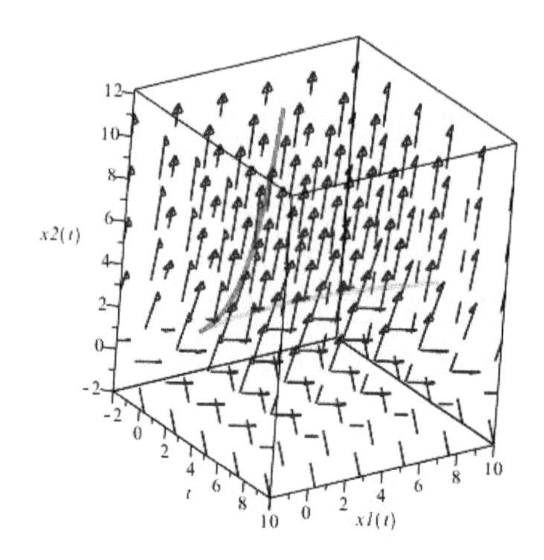

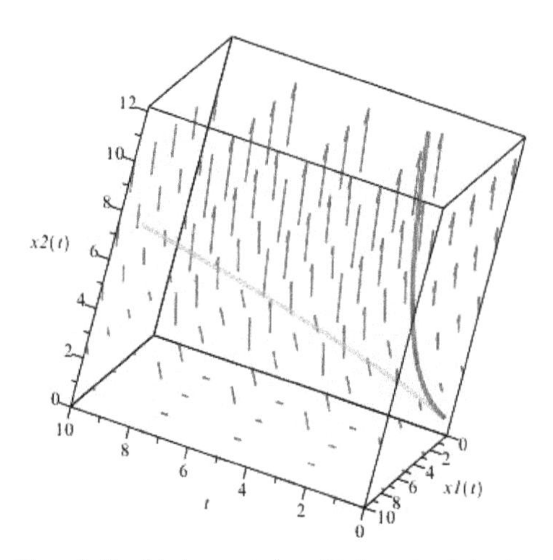

Figure 2. Graphical presentation of 3-dimensional simulation.

Figure 2 shows the whole situation for considered example. The curve after perturbation (red) is a product of curve before the wind deformation (green) and the perturbing vector field (marked by arrows).

6 CONCLUSIONS

We researched the Zermelo navigation problem on Riemannian manifolds for $\dim(\mathbb{R}\times M) = 3$. Geometrically we found the deviation of geodesics under the action of a time-dependent vector field. We considered a fibered manifold $\pi: \mathbb{R}\times M \to \mathbb{R}$ where π is the first canonical projection. We obtained the final form of the system of differential equations providing the solution of the problem after perturbation under the force representing the action of the "wind" distribution modeled by the vector field on manifold M. Considered metric g states for the geometric construction of modeled "sea" and the time-dependent vector field ξ for the "wind" distribution, i.e. a projectable vector field on $\mathbb{R} \times TM$. Perturbed final system of equations in the general case is presented in formula (17) and is followed by the 3-dimensional simulation applying particular type of metric and perturbation. The research shows that the Zermelo navigation can compose the fruitful connection between the different branches of mathematics and can also be applied in modeling the practical aspects of air and marine navigation, in particular the optimization of complex route planning in 2- and 3-dimensional modeling space.

REFERENCES

Zermelo E., *Uber das Navigationsproblem bei ruhender oder veranderlicher Windverteilung*, Zeitschrift fur Angewandte Mathematik und Mechanik, Vol. 11, No. 2, pp. 114–124, 1931.

Chern S.S., Shen Z., *Riemann-Finsler Geometry*, Nankai Tracts in Mathematics, Vol. 6, World Scientific, 2005.

Bao D., Robles C., Shen Z., *Zermelo navigation on Riemannian manifolds*, Journal of Differential Geometry, Vol. 66, no. 3, pp. 377 – 435, 2004.

Caratheodory C., *Calculus of Variations and Partial Differential Equations of the first order*, American Mathematical Society, Chelsea Publishing, 2008 (reprint).

Palacek R., Krupkova O., *On the Zermelo problem in Riemannian manifolds*, Balkan Journal of Geometry and Its Applications, Vol. 17, no. 2, pp. 77 – 81, 2012.

AUTHOR INDEX

Milton Keynes UK
Ingram Content Group UK Ltd.
UKHW051853071024
449327UK00025B/1930